Lecture Notes in Computer Science 3321

Commenced Publication in 1973
Founding and Former Series Editors:
Gerhard Goos, Juris Hartmanis, and Jan van Leeuwen

Michael J. Maher (Ed.)

Advances in Computer Science - ASIAN 2004

Higher-Level Decision Making

9th Asian Computing Science Conference
Dedicated to Jean-Louis Lassez
on the Occasion of His 5th Cycle Birthday
Chiang Mai, Thailand, December 8-10, 2004
Proceedings

 Springer

Volume Editor

Michael J. Maher
University of New South Wales
National ICT Australia, Sydney Laboratory
Locked Bag 6016, Sydney NSW 1466, Australia
E-mail: michael.maher@nicta.com.au

Library of Congress Control Number: Applied for

CR Subject Classification (1998): F.1-2, H.4, F.3, I.2, D.3, C.2.4, E.3, I.4, G.2

ISSN 0302-9743
ISBN 3-540-24087-X Springer Berlin Heidelberg New York

Springer is a part of Springer Science+Business Media

springeronline.com

© Springer-Verlag Berlin Heidelberg 2004
Printed in Germany

Typesetting: Camera-ready by author, data conversion by Scientific Publishing Services, Chennai, India
Printed on acid-free paper   SPIN: 11355922   06/3142   5 4 3 2 1 0

Preface

The 9th Asian Computing Science Conference was held in Chiang Mai in December 2004. This volume contains papers that were presented at that conference. The conference was dedicated to Jean-Louis Lassez, on the occasion of his fifth cycle (60th year) birthday. Its theme was "higher-level decision-making".

Philippe Flajolet was invited to give the opening keynote address, while Yuzuru Tanaka and Phillip Rogaway were also keynote speakers. In addition to the keynote speakers, distinguished colleagues of Jean-Louis Lassez were invited to present talks in his honour. Many of those talks are represented by papers in this volume, but I would like to thank all those speakers:

Jonathan Bernick
Alex Brodsky
Vijay Chandru
T.Y. Chen
Norman Foo
Martin Golumbic
Richard Helm
Claude Kirchner
Hélène Kirchner
Kim Marriott
Krishna Palem
Kotagiri Ramamohanarao
Vijay Saraswat
Masahiko Sato
R.K. Shyamasundar
Nicolas Spyratos
Andrew Sung
Pascal Van Hentenryck
Jean Vuillemin

Following a call for papers, 75 papers were submitted, comprising 47 from Asia, 13 from Europe, 8 from Australia, 5 from North America, and 2 from the Middle East. Submissions came from a total of 23 countries. Korea, China and Thailand were the most-represented countries. These papers underwent anonymous refereeing before the Program Committee selected 17 papers for presentation at the conference. I thank the Program Committee for doing an excellent job under severe time pressure.

I thank the General Chairs of the conference, Joxan Jaffar and Kanchana Kanchanasut, for their support. I thank Pensri Arunwatanamongkol of AIT for her outstanding work in support of the conference website throughout the reviewing process. I thank the Local Organization Committee and its chair, Sanpawat

Kantabutra, for their work on the ground in Chiang Mai. Finally, I thank the Steering Committee for inviting me to chair the Program Committee.

I also recognize the sponsoring institutions for this conference:

Asian Institute of Technology, Thailand
Chiang Mai University, Thailand
IBM Research, USA
INRIA, France
National Electronics and Computer Technology Center, Thailand
National ICT Australia
National University of Singapore
Tata Institute of Fundamental Research, India
Waseda University, Japan

Michael Maher
Program Chair
ASIAN 2004

Organization

Steering Committee

Joxan Jaffar (National University of Singapore, Singapore)
Gilles Kahn (INRIA, France)
Kanchana Kanchansut (Asian Institute of Technology, Thailand)
R.K. Shyamasundar (Tata Institute of Fundamental Research, India)
Kazunori Ueda (Waseda University, Japan)

Program Committee

Vijay Chandru (Strand Genomics and Indian Institute of Science, India)
T.Y. Chen (Swinburne University, Australia)
Philippe Codognet (Embassy of France in Tokyo and University of Paris 6, France)
Susumu Hayashi (Kobe University, Japan)
Jieh Hsiang (NTU, Taiwan)
Vitit Kantabutra (Idaho State University, USA)
Hélène Kirchner (LORIA-CNRS and INRIA, France)
Jimmy Lee (CUHK, Hong Kong, China)
Tze Yun Leong (NUS, Singapore)
Michael Maher (Loyola University Chicago, USA and NICTA, Australia)
Kim Marriott (Monash University, Australia)
Krishna Palem (Georgia Institute of Technology, USA)
Raghu Ramakrishnan (University of Wisconsin, Madison, USA)
Taisuke Sato (Tokyo Institute of Technology, Japan)
V.S. Subrahmanian (University of Maryland, College Park, USA)

Local Organization

Sanpawat Kantabutra (Chiang Mai University, Thailand)

Referees

David Austin
Steven Bird
Colin Boyd
Gerd Brewka
Christophe Cerisara
Lai-Wan Chan
Jeff Choi
Peter Chubb
Leonid Churilov
Johanne Cohen
Véronique Cortier
Christophe Doche
Alan Dorin
Spencer Fung
Claude Godart
Michael Goldwasser
Guido Governatori
Philippe de Groote
James Harland
L.C.K. Hui
Ryutaro Ichise
Yan Jin
Norman Foo
Waleed Kadous

Yoshitaka Kameya
Yukiyoshi Kameyama
Vellaisamy Kunalmani
Shonali Krishnaswam
François Lamarche
Phu Le
Dong Hoon Lee
Ho-Fung Leung
Guoliang Li
Lin Li
Jimmy Liu
Jim Lipton
Eric Martin
Ludovic Mé
Bernd Meyer
Thomas Meyer
Dale Miller
George Mohay
Yi Mu
Lee Naish
Amedeo Napoli
Nicolas Navet
Barry O'Sullivan
Maurice Pagnucco

Kuldip Paliwal
Shawn Parr
Silvio Ranise
Jochen Renz
Seiichiro Sakurai
Abdul Sattar
Jun Shen
John Shepherd
Arcot Sowmya
Peter Stuckey
Changai Sun
Willy Susilo
Antony Tang
Peter Tischer
Takehiro Tokuda
Mark Wallace
Kin-Hong Wong
Mariko Yasugi
Akihiro Yamamoto
Haruo Yokota
Jane You
Janson Zhang

Table of Contents

Keynote Papers

Contributed Papers

Invited Papers

Counting by Coin Tossings

Philippe Flajolet

Algorithms Project, INRIA-Rocquencourt, 78153 Le Chesnay, France
Philippe.Flajolet@inria.fr

Abstract. This text is an informal review of several randomized algorithms that have appeared over the past two decades and have proved instrumental in extracting efficiently quantitative characteristics of very large data sets. The algorithms are by nature probabilistic and based on hashing. They exploit properties of simple discrete probabilistic models and their design is tightly coupled with their analysis, itself often founded on methods from analytic combinatorics. Singularly efficient solutions have been found that defy information theoretic lower bounds applicable to deterministic algorithms. Characteristics like the total number of elements, cardinality (the number of distinct elements), frequency moments, as well as unbiased samples can be gathered with little loss of information and only a small probability of failure. The algorithms are applicable to traffic monitoring in networks, to data base query optimization, and to some of the basic tasks of data mining. They apply to massive data streams and in many cases require strictly minimal auxiliary storage.

1 Approximate Counting

Assume a blind man (a computer?) wants to locate a single black sheep amongst a flock of $N-1$ white sheep. The man can only ask an assistant questions with a Yes/No answer. Like for instance: *"Tell me whether the black sheep is amongst sheep ranked between 37 and 53 from the left"*. This is a variation on the theme of "Twenty Questions". Clearly, about $\log_2 N \equiv \lg N$ operations are both necessary and sufficient. (The proof relies on the fact that with ℓ bits of information, you can only distinguish between 2^ℓ possibilities, so that one must have $2^\ell \geq N$, hence $\ell \geq \lg N$.) This simple argument is of an information-theoretic nature. It implies the fact that one cannot keep a counter (the "black sheep") capable of recording counts between 1 and N with less than $\lg N$ bits.

Assume that we want to run a counter known *a priori* to be in the range $1..N$. Can one beat the information-theoretic lower bound? Yes and No! Not if we require an exact count as this would contradict the information theoretic argument. But, ... Say we relax the constraints and tolerate an uncertainty on the counts of at most 10% (say) in relative terms. The situation changes dramatically. We now just need to locate our count amongst the terms of a geometric scale, $1, A, A^2, A^3 \ldots$ (till N), where $A = 1.1$. The problem then becomes that of

M.J. Maher (Ed.): ASIAN 2004, LNCS 3321, pp. 1–12, 2004.

finding an interval in a collection of about $\log_A N$ intervals. Information theory then tells us that this cannot be effected in fewer than

$$\lg \log_A N \approx \lg \lg N + 2.86245$$

bits, but it also tells us that the "amount of information" contained in an approximate answer is of that very same order. For instance, it is conceivable that an algorithm could exist with which counts till $N = 2^{16} \equiv 65536$ can be maintained using $8 + 3 = 11$ bits instead of 16.

This is the situation which Robert Morris encountered at Bell Labs in 1977. He needed to maintain the logs of a very large number of events in small registers, since the space available at the time was too small to allow exact counts to be kept. A gain by a factor of 2 in memory against a degradation in accuracy by some 30% was perfectly acceptable. How to proceed? Information theory provides a theoretical possibility, not a solution.

Morris's solution [23], known as the APPROXIMATE COUNTING Algorithm, goes at follows. In its crudest (binary) version, you maintain a counter K that initially receives the value $K := 0$. Counter K is meant to be a logarithmic counter, in the sense that when the exact count is n, the value K of the counter at that time should be close to $\lg n$. Note that the single value that gets stored is K, which itself only requires $\lg K \approx \lg \lg n$ bits. Morris' ingenious idea consists in updating the counter K to its new value $K^\star$ according to the following procedure:

$$K^\star = K + 1 \quad \text{with probability } 2^{-K}; \qquad K^\star = K \quad \text{with probability } 1 - 2^{-K}.$$

As time goes, the counter increases, but at a smaller and smaller pace. Implementation is especially easy given a (pseudo)random number generator of sorts [21].

The notable fact here is the appeal to a *probabilistic* idea in order to increment the counter. A plausible argument for the fact that K at time n should be close to $\lg n$ is the fact that it takes 1 impulse for the counter to go from value 0 to value 1, then on average 2 more impulses to from 1 to 2, then on average 4 more impulses from 2 to 3, and so on. In other words, a value $K = \kappa$ should be reached after about

$$1 + 2 + 4 + \cdots + 2^{\kappa-1} = 2^\kappa - 1$$

steps. Thus, provided this informal reasoning is correct, one should have the numeric and probabilistic approximation $2^\kappa \approx n$. Then, the algorithm should return at each instant $n^\circ = 2^K$ as an estimate of the current value of n.

We have just exposed the binary version of the algorithm, which can at best provide an estimate within a factor of 2 since the values it returns are by design restricted to powers of 2. However, it is easy to change the *base* of the counter: it suffices to replace 2 by a smaller number q typically chosen of the form $q = 2^{1/r}$. Then, the new counter gets updated at basically r times the rate of the binary counter. Its granularity is improved, as is, we may hope, the accuracy of the result it provides.

So far, we have offered hand-waving arguments to justify the plausible effectiveness of the algorithm. It can be proved indeed for the base q algorithm that $n° := (q^K - 1)/(q - 1) + 1$ is strictly an *unbiased estimator* of the unknown quantity n [the expected value of $n°$ is n exactly], and that the chances of a *large deviation* of the counter are small. (For instance, with a fair probability, its deviation will not exceed twice the accuracy for which it was designed.) The mathematical problem is not totally obvious since probabilistic decisions pile one upon another. A complete analysis of the states of Morris' Approximate Counting Algorithm was first published by the author in 1985; see [11]. The analysis combines regular languages from theoretical computer science, pure birth processes, basic calculus, Euler's partition identities, some identities due to the German mathematician Heine in the nineteenth century (as later noted by Prodinger [24]), as well as a use of the Mellin transform [13] otherwise familiar from pure and applied mathematics.

The abstract process underlying Morris' algorithm has a neat formulation. It can be described as a stochastic progression in an exponentially hardening medium. It appears to be closely related to recent probabilistic studies by Bertoin-Biane-Yor [4] and by Guillemin-Robert-Zwart [17], the latter relative to the transaction control protocol TCP—the additive-increase multiplicative-decrease process known as AIMD. Other connections include the space efficient simulation of non-deterministic computations by probabilistic devices, as was observed by Freivalds already in the late 1970's. Interestingly enough, the pure-birth process underlying Approximate Counting also surfaces in the analysis of a transitive closure algorithm for acyclic graphs; see Andrews-Crippa-Simon [2]. Such unexpected connections between various areas of the mathematical sciences are fascinating.

2   Probabilistic and Log-Log Counting

Admittedly, nowadays, memory is not such a major concern, so that one might be tempted to regard Morris' algorithm as a somewhat old-fashioned curiosity. This is not so. Morris' ideas have had an important rôle by demonstrating that complexity barriers can be bypassed by means of probabilistic techniques, as long as a small tolerance on the quality of results is granted.

Consider now information flowing at a very high rate in a network. It is impossible to store it and we do not have time to carry much computation on the fly as we see messages or packets passing by. Can one still extract some *global information* out of the flux of data? An example, which has motivated much of the recent work of Estan, Varghese, and others is as follows. Imagine you have a large volume of data (say, packet headers to fix ideas). Is it possible to *estimate* the number of *distinct* elements? This is an important question in the context of network security [7, 9] since several attacks may be detected at router's level by the fact that they generate an unusual number of *distinct* open connections (i.e., source-destination pairs also known as "flows").

Clearly, the solution that stores everything and sorts data afterwards is far too costly in terms of both processing time and storage consumption. An elegant solution is provided by a technique combining *hashing* and *signatures*. Say we have a *multiset*[1] $\mathfrak{M}$ of elements and call *cardinality* the number of distinct elements $\mathfrak{M}$ contains. Suppose we have a "good" hash function h, which is assumed to provide long enough strings of pseudorandom bits from actual data and neglect the effect of collisions. Then the collection $h(\mathfrak{M})$ has repetitions that are of the *very same* structure as those of $\mathfrak{M}$ itself, safe for the important remark that one may now assume the elements to be random real numbers uniformly drawn on the interval $[0, 1]$.

Let us now examine the elements once hashed. Given the uniformity assumption, we expect the following patterns in values to be observed, with the corresponding frequencies as described in this table:

$.1\cdots$	$.01\cdots$	$.001\cdots$	$.0001\cdots$
$\frac{1}{2}$	$\frac{1}{4}$	$\frac{1}{8}$	$\frac{1}{16}$

Let $\rho(x)$ be the rank of the first bit 1 in the representation of x (its hashed value, rather). If hashed values are uniform and independent, the event $\rho(x) = k$ will occur with probability $\frac{1}{2^k}$. The quantity,

$$R := \max_{k \in \mathbb{N}} \left[\rho = 1, \rho = 2, \ldots, \rho = k \text{ are } all \text{ observed} \right], \tag{1}$$

is then expected to provide a good indication of the *logarithm* of the unknown cardinality of $\mathfrak{M}$ and 2^R should, roughly speaking, estimate the cardinality $n = \|\mathfrak{M}\|$. Note that the algorithm needs to maintain a bitmap table, which records on the fly the values of ρ that are observed. (A *word* of 32 bits is typically adequate.) By construction, the algorithm depends only on the underlying *set* and it is in no way affected by replications of records: whether an element occurs once or a million times still results in the same operation of setting a flag in the bitmap to 1.

The complete algorithm was published in the period 1983–1985 by Flajolet and Martin [14, 15]. The motivation at the time was query optimization in database systems. Indeed, such data tend to have a lot of replications (think of the collection of towns in which employees of an organization reside). In this context, computing an intersection of two multisets $\mathfrak{A} \cap \mathfrak{B}$ (towns of your own company's employees *and* of employees of your partner's company) benefits greatly of knowing how many *distinct* elements $\mathfrak{A}$ and $\mathfrak{B}$ may have since then the suitable nearly optimal strategy may be set up. (Options include sorting, merging, hashing, looking up elements of $\mathfrak{A}$ in $\mathfrak{B}$ after an index has been set up, and so on).

With respect to the crude binary version outlined above, the actual algorithm named PROBABILISTIC COUNTING differs in two respects:

---

[1] A multiset is like a set but with repetitions allowed.

— A general technique known as *stochastic averaging* emulates the effect of m simultaneous observations at the sole expense of a hash function calculation and a simple "switch". The purpose is to increase the accuracy from about one binary order of magnitude for a single bitmap to $O(1/\sqrt{m})$ for stochastic averaging. The idea consists in computing the basic observable (1) in each of the $m = 2^\ell$ groups determined by the first ℓ bits of hashed values, then averaging, and finally scaling the estimate by m.

— The algorithm as described so far would be *biased*. A very precise probabilistic analysis (using inclusion-exclusion, generating functions, and Mellin transforms) makes it possible to effect the proper corrections and devise an algorithm that is *asymptotically unbiased*.

Without entering into the arcanes of the analysis, the reader can at least get a feeling of what goes on by contemplating the magical constant [10–p. 437],

$$\varphi := \frac{e^\gamma}{\sqrt{2}} \prod_{m=1}^{\infty} \left(\frac{2m+1}{2m}\right)^{\varepsilon(m)} \doteq 0.7735162909,$$

where γ is Euler's constant and $\varepsilon(m) = \pm 1$ indicates the parity of the number of 1–bits in the binary representation of m. This constant, provided by a Mellin analysis, enters the design of the algorithm as it corrects a multiplicative bias inherent in the raw parameter (1). In this way, the following proves to be possible: *To estimate the cardinality of large multisets (up to over a billion distinct elements) using m words of auxiliary memory, with a relative accuracy close to*

$$\alpha = \frac{0.78}{\sqrt{m}}. \tag{2}$$

Marianne Durand and I recently realized that one could even do a bit better, namely: *To estimate the cardinality of large multisets (up to several billion distinct elements) using m short bytes (< 8 bits) of auxiliary memory, with a relative accuracy close to*

$$\alpha = \frac{1.30}{\sqrt{m}}. \tag{3}$$

Rather than maintaining a bitmap table (of one word) out of which the new observable (previously R) is computed, we choose as "observable" an integer parameter of the file:

$$S := \max_{x \in \mathfrak{M}} \rho(x).$$

This quantity is stored in binary, so that only $\lg S$ bits are needed. Since S is itself a logarithmic estimator, its storage only requires $\lg\lg N$ bits. Now, with a *full byte* (8 bits), one could estimate cardinalities till about $2^{2^8} \approx 10^{77}$, which is of the order of the number of particles in the universe. Taking each memory unit to be of 5 bits then suffices for counts till $N = 2^{2^5} \approx 4 \cdot 10^9$, i.e., four billions. When experiments are repeated, or rather "emulated" by means of stochastic averaging, precision increases (in proportion to $1/\sqrt{m}$ if m is the number of memory units). This gives rise to the LogLog Counting algorithm. There

```
ghfffghfghgghggggghghhheehfhfhhgghghghhfgffffhhhiigfhhffgfiihfhhh
igigighfgihfffghigihghigfhhgeegeghgghhhgghhfhidiigihighihehhhfgg
hfgighigffghdieghhhggghhfghhfiiheffghghihifgggffihgihfggighgiiif
fjgfgjhhjiifhjgehgghfhhfhjhiggghghihigghhihihgiighgfhlgjfgjjjmfl
```

Fig. 1. The LOGLOG Algorithm with $m = 256$ condenses the whole of Shakespeare's works to a table of 256 "small bytes" of 4 bits each. The estimate of the number of distinct words in this run is $n° = 30897$ (the true answer is $n = 28239$), which represents a relative error of $+9.4\%$

is a slight theoretical loss in accuracy as the constant 0.78 of (2) relative to Probabilistic Counting is replaced by a slightly larger value, the formulæ being

$$\alpha = \frac{1.30}{\sqrt{m}} \quad \text{and} \quad \alpha = \frac{1.05}{\sqrt{m}}, \tag{4}$$

depending on implementation choices. (The first formula repeats (3); the second one is relative to the variant known as SuperLogLog). However this effect is completely offset by the fact that LogLog's memory units (from words to small bytes) are smaller by a factor of about 4 than the words that Probabilistic Counting requires. Globally, *Probabilistic Counting is about 3 to 5 times more accurate than Probabilistic Counting, for a fixed amount of global storage.*

In summary, LOGLOG counting creates a signature from a multiset of data and then deduces a cardinality estimator. That signature only depends on the set of distinct values underlying the multiset of data input to the algorithm. It consists of m small-byte registers, resulting in an accuracy of $\approx 1/\sqrt{m}$. For instance, $m = 1024$ corresponds to a typical accuracy of 3% and its maintenance necessitates under a kilobyte of auxiliary memory. In Figure 1 (taken from [6]), we give explicitly the four line signature by which the number of distinct words in all of Shakespeare's writings is predicted to be 30,987 (the true answer is 28,239). The algorithm offers at present the best accuracy/memory trade-off known.

Yet other observables are conceivable. For instance, Giroire (2004, unpublished) has recently succeeded in analysing a class of *algorithms based on minima*, where the observable is now a collection of some of the smallest hashed values.

The algorithms discussed above involve a fascinating interplay between theory and practice. Analysis of algorithms intervenes at several crucial stages. First in order to correct the bias of raw observables suggested by probabilistic intuition and back of an envelope calculations; next to estimate variance and limit probability distributions, thereby quantifying the risk of "large deviations", which means abnormally inaccurate estimates. The LogLog algorithm for instance relies on maxima of geometric random variables, exponential generating functions, analytic depoissonization (Jacquet-Szpankowski [20, 26]), and once again Mellin transforms.

3   Sampling

Yet another algorithm originally designed for cardinality estimation in data bases and due to Wegman around 1980 turns out to regain importance in the modern context of approximate queries and sketches in very large data bases [3].

Once more, a *multiset* $\mathfrak{M}$ is given. Say we would like to extract a sample of s *distinct* value. In other words, we are sampling the domain of values, the *set*, that underlies $\mathfrak{M}$. Such a sample may be used to design adaptive data structures or to gather useful statistics on the "profile" of data, like approximate quantiles: think of the problem of estimating the median salary in a population, given aggregated heterogeneous files presenting unpredictably repeated entries.

First, as usual, there is a simple-minded algorithm that proceeds by keeping at each stage, as data flows, the exact set (without repetitions) of distinct elements encountered so far and finally extracting the desired sample of size s. This algorithm suffers from a storage complexity that is at best proportional to the cardinality of $\mathfrak{M}$ and of a time complexity that is even nonlinear if sorting or index trees (B–trees) are used. The method may be improved a bit by fixing *a priori* a sampling factor p, say $p = \frac{1}{1024} = 2^{-10}$. This is STRAIGHT SAMPLING: Elements are hashed and only elements whose hashed value starts with a sequence of 10 zeros are filtered in and kept as a distinct set in a temporary file; at the end, resample the p-sample and obtain the needed collection of size s. The algorithm has become probabilistic. It will be reasonably behaved provided the cardinality of $\mathfrak{M}$ is well in excess of $1/p$ (precisely, we need $p\|\mathfrak{M}\| \gg s$), and the storage complexity will decrease by a factor of about $1/1000$ (not bad!). However, if the nature of the data is *a priori* completely unknown and p has not been chosen in the right interval, the algorithm may fail dramatically by oversampling or undersampling. For in instance, in the case of only $n = 500$ different records and $p = 1/1000$, it is more than likely that *no* element at all is selected so that no sample of whatever size is obtained.

Wegman's remedy is known as ADAPTIVE SAMPLING. This elegant algorithm is described in the article that analyses it [12]. Assume again a sample of s distinct values is wanted. Fix a parameter b, called the bucket size, commensurate with s, e.g., $b = \frac{5}{2}s$, and prepare to manage a running list of distinct elements whose length is $\leq b$. The algorithm maintains a parameter called the sampling depth, δ whose value will get incremented as the algorithm proceeds; the sampling probability p is also a variable quantity bound to δ by the condition $p = 2^{-\delta}$. The algorithm runs then as follows:

> Start with $\delta = 0$ and accordingly $p = 1$. All the distinct elements seen are stored into the bucket until the bucket overflows ($b+1$ distinct elements have been found). At this stage, increase the sampling depth by 1, setting $\delta := \delta+1$ and $p := \frac{1}{2}p$. Then eliminate from the bucket all elements whose hashed value does not start with *one* initial zero: this has the effect of stochastically splitting the bucket and dividing its content by a factor close to 2. Resume and only consider from this point on, elements whose hashed value is of the form $0\cdots$, discarding the others. Repeat the following cycle: "increase-depth, decrease-sampling-ratio, and split-bucket" after each overflow occurs.

This algorithm is closely related to an important data structure known as the *digital tree* or "*trie*" (pronounce like"try") and its analysis benefits from techniques known for about three decades in analysis of algorithms [18, 22, 25, 26]. Let M be the (random) number of elements contained in the bucket once all elements have been scanned. It turns out that the bucket (conditioned upon the particular value of M) contains an unbiased sample. If $b/2$ suitably exceeds the desired sample size s, then, with high probability, one has $M > s$, so that a subsample of size s can be extracted from the M elements that are available. This subsample is unbiased. Et voila!

The Adaptive Sampling algorithm also serves as a cardinality estimator, and this appears to have been the primary motivation for its design. Indeed, as is intuitively plausible, the quantity $M \cdot 2^\delta = M/p$ (with p, δ given their final values) turns out to be an unbiased estimator of the cardinality of $\mathfrak{M}$. Analysis based on generating functions and Mellin transforms shows that the accuracy is now

$$\alpha = \frac{1.20}{\sqrt{b}},$$

which is slightly less favorable than previous solutions. The algorithm however has the advantage of being totally unbiased, including in its nonasymptotic regime.

At this stage, we should mention yet another algorithm for cardinality estimation. It is due to K-Y. Whang *et al.* [28] and is sometimes called HIT COUNTING. Its ideas are somewhat related to sampling. Say again we want to estimate the number of distinct elements in a file and this number is known in advance not to exceed a certain bound ν. Hash elements of $\mathfrak{M}$ and keep the skeletton of a hash table of size (say) $m := \frac{\nu}{5}$ in the following manner: a bit is set to 1 for each cell that is hit at least once during the hashing process. (This technique is akin to the famous Bloom filters.) The observable that is chosen is the proportion E of empty (not hit) cells at the end of the procedure. Let n be the unknown cardinality of $\mathfrak{M}$; a Poisson approximation shows that the mean proportion of empty cells (0-bits in the skeletton) at the end is about $e^{-n/m}$. Thus, based on the approximate equality $E \approx e^{-n/m}$, propose $n^\circ = m \log E$ as an estimator of the unknown cardinality n. The algorithm works fine in theory, it is extremely accurate for small values of n, but its memory cost, which is of $O(n)$ bits, becomes exorbitant for massive data sets. At least, Hit Counting can be recycled within the LogLog algorithm [5]: this permits to correct nonlinearities present in the case of very low cardinalities when the full asymptotic regime of LogLog counting is not yet attained.

Returning to sampling, we note that the problem of sampling with multiplicities (i.e., one samples k positions out of n) looks simpler, but sophisticated algorithms can be developed; see Vitter's reference work [27].

4   Frequency Moments

The amount of literature on the topic of estimating characteristics of very large data sets has exploded over the past few years. In addition to databases and networking, data mining has come into the picture. Hashing remains a prime randomization technique in this context and statistical observations made on hashed values can again provide very accurate information of various sorts.

An influential paper by Alon, Matias, and Szegedy [1] has proposed an important conceptual framework that includes many highly interesting quantitative characteristics of multisets, while encompassing the maintenance of approximate counters and the estimation of cardinalities. Let $\mathfrak{M}$ be a multiset and let V be the underlying set, that is, the set of distinct values that elements of $\mathfrak{M}$ assume. Let f_v be the frequency (number of occurrences) of value v in $\mathfrak{M}$. The rth *frequency moment* is defined as

$$F_r(\mathfrak{M}) := \sum_{v \in V} f_v^r.$$

Thus maintaining a counter of the number of nondistinct elements of $\mathfrak{M}$ (with multiplicity counted) is equivalent to determining F_1, for which Approximate Counting is applicable. Similarly, the problem of estimating cardinalities is expressed in this framework as that of obtaining a good approximation to F_0, a problem addressed by algorithms Probabilistic Counting, LogLog, Adaptive Sampling, and Hit Counting discussed above.

The quantity F_2, a sort of variance, when suitably normalized, provides an indication of the amount by which the empirical distribution of elements of $\mathfrak{M}$ differs from the flat (uniform) distribution. Its estimation can be approached from

$$\phi_2 := \left(\sum_{x \in \mathfrak{M}} \epsilon(x) \right)^2, \tag{5}$$

where $\epsilon(x)$ is ± 1 and can be determined by translating the first bit (say) of the hashed value $h(x)$. This estimator is however subject to great stochastic fluctuations when several experiments are performed using various bits of hashed values, so that the estimated values are much less accurate than in the cases of F_0 and F_1.

The quantities F_r for $r > 2$ are intrinsically hard to estimate in the absence of any *a priori* assumption on the frequency profile of $\mathfrak{M}$. Of great interest is F_∞, which is taken to mean

$$F_\infty := \lim_{r \to \infty} F_r^{1/r},$$

as this is the frequency of the most frequent element in the multiset $\mathfrak{M}$. The determination of this quantity is crucial in many networking applications [8], as it is part of the general problem known as *Mice and Elephants*: in a large flux of data, how to recognize the ones that occupy most of the bandwidth?

The quantities F_r when r lies in the interval $(0, 2)$ can be estimated by resorting to stable laws of probability theory. This is a brilliant idea pioneered by Piotr Indyk [19] that relates to active research in the area of dimension reduction in statistics and computational geometry. For our purposes, we may take a *stable law of index* α to be the law of a random variable X whose characteristic function (Fourier transform) is

$$\mathbb{E}\left(e^{itX}\right) = e^{-|t|^\alpha}.$$

Such laws exist for $\alpha \in (0, 2)$. Given a multiset $\mathfrak{M}$ compute the quantity (compare with (5))

$$\phi_r := \sum_{x \in \mathfrak{M}} \epsilon(x), \tag{6}$$

where each $\epsilon(x)$ now obeys a stable law of parameter r and is determined via hashing from x itself. We have also, with previous notations,

$$\phi_r := \sum_{v \in V} f_v \epsilon(v).$$

Now a fundamental property of stable laws is that if the X_j are r–stable and independent, then

$$\sum_j \lambda_j X_j \overset{\text{(distribution)}}{=} \xi Z, \qquad \text{with} \quad \xi = \left(\sum_j |\lambda_j|^r \right)^{1/r},$$

and Z being itself r–stable. There results from the last two displayed equations the possibility of constructing an estimator for F_r when $r \in (0, 2)$. It suffices to devise an estimator of the multiplier of an r–stable law: in the literature, this is usually based on medians of experiments; it has not been investigated yet whether logarithmic techniques might be competitive.

The fact that F_0 can be approached as $\lim_{\eta \to 0} F_\eta$ attaches the problem of cardinality estimation (F_0) to this circle of ideas, but a precise assessment of practical complexity issues seems to be still lacking and the corresponding algorithm does not seem to outperform our earlier solutions.

Let us mention finally that the F_r can be used for parametric statistics purposes. If the data in the multiset $\mathfrak{M}$ are known to be drawn according to a family of distributions ($\mathcal{D}_\theta$) (say a class of generalized Zipf laws), then the F_r which are computationally easily tractable (in linear time and small storage) may be used to infer a plausible value of θ best suited to a multiset $\mathfrak{M}$.

5   Conclusion

Simple probabilistic ideas combined with suitable analysis of the intervening probabilistic and analytic phenomena leads to a class of algorithms that perform counting by coin flippings. The gains attained by probabilistic design are in a number of cases quite spectacular. This is also an area where practice merges agreeably with theory (discrete models, combinatorics and probability, generating functions, complex analysis and integral transforms); see the manuscript of the forthcoming book by Flajolet & Sedgewick [16] for a systematic exposition. The resulting algorithms are not only useful but also much used in databases, network management, and data mining.

Acknowledgements. I am grateful to the organizers of *ASIAN'04* (the Ninth Asian Computing Science Conference held at Chiang Mai in December 2004) for their kind invitation. Many thanks in particular to Ms Kanchanasut, "Kanchana", for pushing me to write this text, for her patience regarding the manuscript, her numerous initiatives, and her constant support of French–Thai cooperation in this corner of science.

References

1. ALON, N., MATIAS, Y., AND SZEGEDY, M. The space complexity of approximating the frequency moments. *Journal of Computer and System Sciences 58*, 1 (1999), 137–147.
2. ANDREWS, G. E., CRIPPA, D., AND SIMON, K. q-series arising from the study of random graphs. *SIAM Journal on Discrete Mathematics 10*, 1 (1997), 41–56.
3. BABCOCK, B., BABU, S., DATAR, M., MOTWANI, R., AND WIDOM, J. Models and issues in data stream systems. In *Proceedings of Symposium on Principles of Database Systems (PODS)* (2002), pp. 1–16.
4. BERTOIN, J., BIANE, P., AND YOR, M. Poissonian exponential functionals, q-series, q-integrals, and the moment problem for log-normal distributions. Tech. Rep. PMA-705, Laboratoire de Probabilités et Modèles Aléatoires, Université Paris VI, 2002.
5. DURAND, M. *Combinatoire analytique et algorithmique des ensembles de données*. PhD thesis, École Polytechnique, France, 2004.
6. DURAND, M., AND FLAJOLET, P. LOGLOG counting of large cardinalities. In *Annual European Symposium on Algorithms (ESA03)* (2003), G. Di Battista and U. Zwick, Eds., vol. 2832 of *Lecture Notes in Computer Science*, pp. 605–617.
7. ESTAN, C., AND VARGHESE, G. New directions in traffic measurement and accounting. In *Proceedings of SIGCOMM 2002* (2002), ACM Press. (Also: UCSD technical report CS2002-0699, February, 2002; available electronically.).
8. ESTAN, C., AND VARGHESE, G. New directions in traffic measurement and accounting: Focusing on the elephants, ignoring the mice. *ACM Transactions on Computer Systems 21*, 3 (2003), 270–313.
9. ESTAN, C., VARGHESE, G., AND FISK, M. Bitmap algorithms for counting active flows on high speed links. Technical Report CS2003-0738, UCSD, Mar. 2003. Available electronically. Summary in *ACM SIGCOMM Computer Communication Review* Volume 32 , Issue 3 (July 2002), p. 10.
10. FINCH, S. *Mathematical Constants*. Cambridge University Press, New-York, 2003.
11. FLAJOLET, P. Approximate counting: A detailed analysis. *BIT 25* (1985), 113–134.
12. FLAJOLET, P. On adaptive sampling. *Computing 34* (1990), 391–400.
13. FLAJOLET, P., GOURDON, X., AND DUMAS, P. Mellin transforms and asymptotics: Harmonic sums. *Theoretical Computer Science 144*, 1–2 (June 1995), 3–58.
14. FLAJOLET, P., AND MARTIN, G. N. Probabilistic counting. In *Proceedings of the 24th Annual Symposium on Foundations of Computer Science* (1983), IEEE Computer Society Press, pp. 76–82.
15. FLAJOLET, P., AND MARTIN, G. N. Probabilistic counting algorithms for data base applications. *Journal of Computer and System Sciences 31*, 2 (Oct. 1985), 182–209.
16. FLAJOLET, P., AND SEDGEWICK, R. *Analytic Combinatorics*. 2004. Book in preparation; Individual chapters are available electronically.
17. GUILLEMIN, F., ROBERT, P., AND ZWART, B. AIMD algorithms and exponential functionals. *Annals of Applied Probability 14*, 1 (2004), 90–117.

12   P. Flajolet

18. HOFRI, M. *Analysis of Algorithms: Computational Methods and Mathematical Tools.* Oxford University Press, 1995.
19. INDYK, P. Stable distributions, pseudorandom generators, embeddings and data stream computation. In *Proceedings of the 41st Annual IEEE Symposium on Foundations of Computer Science (FOCS)* (2000), pp. 189–197.
20. JACQUET, P., AND SZPANKOWSKI, W. Analytical de-Poissonization and its applications. *Theoretical Computer Science 201*, 1-2 (1998), 1–62.
21. KNUTH, D. E. *The Art of Computer Programming*, 3rd ed., vol. 2: Seminumerical Algorithms. Addison-Wesley, 1998.
22. KNUTH, D. E. *The Art of Computer Programming*, 2nd ed., vol. 3: Sorting and Searching. Addison-Wesley, 1998.
23. MORRIS, R. Counting large numbers of events in small registers. *Communications of the ACM 21*, 10 (1977), 840–842.
24. PRODINGER, H. Approximate counting via Euler transform. *Mathematica Slovaka 44* (1994), 569–574.
25. SEDGEWICK, R., AND FLAJOLET, P. *An Introduction to the Analysis of Algorithms.* Addison-Wesley Publishing Company, 1996.
26. SZPANKOWSKI, W. *Average-Case Analysis of Algorithms on Sequences.* John Wiley, New York, 2001.
27. VITTER, J. Random sampling with a reservoir. *ACM Transactions on Mathematical Software 11*, 1 (1985).
28. WHANG, K.-Y., VANDER-ZANDEN, B., AND TAYLOR, H. A linear-time probabilistic counting algorithm for database applications. *ACM Transactions on Database Systems 15*, 2 (1990), 208–229.

On the Role Definitions in and Beyond Cryptography

Phillip Rogaway

Dept. of Computer Science, University of California, Davis, California 95616, USA
and Dept. of Computer Science, Fac. of Science, Chiang Mai University,
Chiang Mai, Thailand 50200
rogaway@cs.ucdavis.edu
www.cs.ucdavis.edu/~rogaway

Abstract. More than new algorithms, proofs, or technologies, it is the emergence of *definitions* that has changed the landscape of cryptography. We describe how definitions work in modern cryptography, giving a number of examples, and we provide observations, opinions, and suggestions about the art and science of crafting them.

1   Preamble

Over the years, I suppose I have given my share of terrible talks. In retrospect, this has most often happened because I chose a topic too technical or narrow for my audience. There are few things in life as boring as listening to a talk that you can't understand a word of. So when I was invited to give a talk here, for an audience that might—if she comes—include a single person in my field, I vowed that I would not mess up yet again by choosing a topic that my audience would find incomprehensible. This would give me the opportunity to mess up in some new and unanticipated way.

My topic today is odd and ambitious. It's about something that has been a theme of my work for some fifteen years, and yet about which I have never spoken in an abstract or general way. The theme is *the importance of definitions*. My plan is to intermix abstract musings about definitions with nice examples of them. The examples, and indeed my entire perspective, are from *modern cryptography*, my area of research. But the talk isn't meant to be a survey of cryptographic definitions; it is something more personal and philosophical.

Before we really get going, I should acknowledge that there are many uses of the word *definition*. When you say something like *let n = 10* you are making a definition, and people use the word *definition* for an informal description of an idea, too. Here I'm not interested in definitions of either flavour. For this talk, *definitions* are things that specify, in a mathematically rigorous way, some significant notion in a field. You've all seen such definitions in computer science—things like *a Turing-computable function* or *a language that is NP-complete*.

M.J. Maher (Ed.): ASIAN 2004, LNCS 3321, pp. 13–32, 2004.

2   Why Make Definitions?

Definitions in cryptography emerged rather suddenly, in 1982, with the work of Shafi Goldwasser and Silvio Micali [GM]. Before then, cryptography was all about schemes and attacks, and there was no way to have confidence in a scheme beyond that which was gained when smart people failed to find an attack. What Goldwasser and Micali did was, first of all, to *define* cryptography's classical goal, message privacy, the goal of an encryption scheme. Their definition was strong[1] and satisfying, and they proved it equivalent to very different-looking alternatives. Next they gave a *protocol* for encryption, and *proved* their protocol satisfied their definition, given a complexity-theoretic assumption. The proof took the form of a *reduction*: if the encryption protocol didn't meet the security definition, some other protocol wouldn't satisfy *its* security definition. The *definition–protocol–proof* approach has come to be called *provable security*[2].

Provable security would dramatically change the character of my field. No longer would cryptography be solely an art; in an instant, a science of cryptography was born. Literally thousands of papers would come to be written within this framework. Nowadays, roughly half of the papers in cryptography's top conferences are in the provable-security tradition. In recent years, provable-security has come to have a large impact on practice, too, delivering concrete mechanisms like HMAC [BCK] (the message-authentication method used in your web browser) as well as high-level idea that were silently absorbed into practice, such as the need for a public-key encryption scheme to be probabilistic [GM].

Provable security begins with definitions. It embraces what one might call the *definitional viewpoint*: the belief that our technical goals can be formally defined, and that by understanding the properties and realizations of these definitions we are better able to address our original goals.

Many people assume that a field's definitions are just to make something formal. They conclude that definitions are an incredible bore. But definitions aren't about formalism; they're about *ideas*. In making a definition in cryptography we are trying to capture, in precise language, some human notion or concern dealing with privacy or authenticity. When you ask a question like *What is an encryption scheme supposed to do?*, or *What does a digital signature accomplish?*, it is a definition that you should be aiming for in answer—not an algorithm, an example, a piece of code, or some descriptive English prose.

Definitions enable theorems and proofs, but they do more than that, and can be useful even in the absence of theorem and proofs. For one thing, defini-

---

[1] A definition is *strong* if it implies a lot; constructions that achieve the definition achieve other, *weaker* definitions, too. If you are proving that your protocol achieves some goal, the proof will mean more if you choose a strong definition.

[2] This paragraph greatly simplifies the actual history. Papers setting the context for [GM] include [Co, DH, BBS, Bl, Sha, SRA]. Nearly contemporaneous provable-security work, for a different problem, is [BM, Ya]. No one paper is fully responsible for the idea of doing definitions and proofs in cryptography, but [GM] played a pivotal role.

tions facilitate meaningful communication. When someone says *Here you need an encryption scheme that achieves semantic security under an adaptive chosen-ciphertext attack*, this tells you an enormous amount about what kind of object is expected. I believe that a good deal of non-productive discourse in cryptography is attributable to muddled, definitionless speech.

Definitions help you to think, and shape how you think. I have seen how, often times, one can hardly get anywhere in thinking about a problem until things have been named and properly defined. Conversely, once a definition *is* spelled out, what was obscure may become obvious. I'll give an example in Section 6.

Let me now enumerate the first set of points I have made. When the spirit moves me, I'll list random claims, viewpoints, or pieces of unsolicited advice.

▷ **1** The emergence of definitions, and the definitional viewpoint, can usher in a huge transformation in the character of that field.

▷ **2** Being formal isn't the purpose of a definition. While only something precise deserves to be called a definition, the purpose lies elsewhere.

▷ **3** Definitions are about ideas. They arise from distilling intuitively held notions. They capture the central concepts of a field. They enable theorems and proofs, allow you to engage in more productive discourse, and help you to think.

▷ **4** Definitions can be worthwhile even in the absence of theorems and proofs.

3   Pseudorandom Generators

Let's cease all this abstraction and give a first example. I want to ask: *What does it mean to generate random-looking bits?* There are several approaches to answering this question, but the one that I want to describe is due to Blum and Micali [BM] and Yao [Ya]. It focuses on this refinement of our question: *what surrogate for true randomness can you use in any probabilistic algorithm?*

I think that a good first step in answering this question is to pin down the *syntax* of the object that you are after. We're interested in the pseudorandomness of bit strings, and yet we're not going to focus on individual strings. Instead, our notion centers on an object called a *pseudorandom generator* (PRG). This is a deterministic function G that maps a string s, the *seed*, into a longer string $G(s)$. Formally, let's say that a PRG is any function $G\colon \{0,1\}^n \to \{0,1\}^N$ where $N > n \geq 1$ are constants associated to G.

We hope that $G(s)$, for a random $s \in \{0,1\}^n$, will look like a random string of length N, but this wasn't part of the definition. The definition might seem woefully inadequate for this reason. It tells us, for example, that the function $G\colon \{0,1\}^{100} \to \{0,1\}^{200}$ defined by $G(s) = 0^{200}$ is a PRG. If the intent of a PRG is that you can use $G(s)$, for a random s, in lieu of N random bits, then how is it that outputting a constant earns you the right to be called a PRG?

Nonetheless, it seems convenient, when giving definitions, to separate the syntax of the object you are defining from the measure of its worth. You shouldn't have to concern yourself with valuing an object when defining it.

▷ **5** Always separate the syntax of the object you are defining from the measure of its quality.

The quality of a PRG will be captured by a real number. Actually, the number is associated not to the PRG but to the pair consisting of a PRG, G, and a *distinguisher*, D. A distinguisher is just an algorithm, possibly a probabilistic one, equipped with a way to interact with its environment. One can think of it as an *adversary*—an agent trying to break the PRG. We equip D with an *oracle* that has a "button" that D can push. Each time D pushes the button ("makes an oracle query"), it gets a string. For a PRG $G\colon \{0,1\}^n \to \{0,1\}^N$ and a distinguisher D, consider running (G, D) in either of two different *games*:

- Game 1: every time the distinguisher D makes an oracle query, choose a random string $s \in \{0,1\}^n$ and give the distinguisher $Y = G(s)$.
- Game 0: every time the distinguisher D makes an oracle query, choose a random string $Y \in \{0,1\}^N$ and give the distinguisher Y.

The distinguisher's goal is to ascertain if it is playing game 1 or game 0. To that end, when it is ready, it outputs a bit $b \in \{0,1\}$ and halts. If it thinks it is playing game 1, it should output 1, and if it thinks it is playing game 0, it should output 0. The advantage of D in attacking G is defined as $\mathbf{Adv}_G^{\mathrm{prg}}(D) = \Pr[D^{\mathrm{Game1}} \Rightarrow 1] - \Pr[D^{\mathrm{Game0}} \Rightarrow 1]$, the probability that D outputs a 1 when it is plays game 1, minus the probability that D outputs a 1 when it plays game 0. This difference measures how well D is doing.

A large advantage, say a number like 0.9, means that the distinguisher is doing well. A small advantage, say a number like 2^{-40}, means that the distinguisher is doing poorly. A negative advantage means that the distinguisher ought to flip when it outputs 0 and 1.

Let's return to the example generator $G(s) = 0^{200}$ for each $s \in \{0,1\}^{100}$. That this is a bad PRG is captured by the fact that there is an efficient distinguisher D that gets high advantage in breaking G. All D has to do is to request from its oracle a single string Y. If $Y = 0^{200}$ then D outputs 1; otherwise, it outputs 0. The advantage of D is $1 - 2^{-200} \approx 1$, so D is doing great. The existence of this distinguisher shows that G is a poor PRG.

4   Asymptotic Versus Concrete Security

When should a PRG G be deemed *secure*? Note that it will always be *possible* to get high advantage in breaking G: consider the distinguisher D that asks for a single string Y and then returns 1 iff $Y = G(s)$ for some $s \in \{0,1\}^n$. Then $\mathbf{Adv}_G^{\mathrm{prg}}(D) \geq 1 - 2^{n-N} \geq 0.5$. But this distinguisher is extremely inefficient, even for modest n, because it needs 2^n steps to enumerate the strings of $\{0,1\}^n$. Unreasonable distinguishers will be able to get significant advantage. We would be satisfied if every *reasonable* distinguisher D earns *insignificant* advantage. How should we define *reasonable* and *insignificant*? There are two approaches, the *asymptotic* approach and the *concrete-security* approach.

The *asymptotic approach* usually equates *reasonable* with *polynomial time* and *insignificant* with *negligible*, where $\epsilon(n)$ is said to be *negligible* if for all $c > 0$ there exists a $K > 0$ such that $\epsilon(n) < n^{-c}$ for all $n \geq K$. To use this approach there needs to be a *security parameter*, n, relative to which we speak of polynomial-time or negligible advantage. For a PRG, the security parameter can be the length of the seed s. We need to go back and adjust the syntax of a PRG so that it operates on seeds of infinitely many different lengths—for example, we could redefine a PRG to be a function G: $\{0,1\}^* \rightarrow \{0,1\}^*$ where $|G(s)| = \ell(|s|)$ for some $\ell(n) > n$. Then we could say that a PRG G is *secure* if every polynomial-time distinguisher D obtains only negligible advantage in attacking G. The asymptotic approach is the traditional one, and all of the early definitions in cryptography were originally formulated in this way.

The *concrete-security approach* is what we illustrated in Section 3. It punts on the question of what is *reasonable* and *insignificant*, choosing never to define these terms. As a consequence, it can't define when an object is *secure*. The viewpoint is that a definition of advantage already *is* a measure of security, and making the "final step" of defining *reasonable* and *insignificant* is often unnecessary, and even artificial. It is ultimately the user of a scheme who will decide what is *reasonable* and *insignificant*, and not based on any formal definition. The concrete-security approach was popularized by Mihir Bellare and me.

You get different definitions for an object if you use asymptotic or concrete-security. I'd like to ask if the difference is *important*. I think the answer is both *yes* and *no*. We begin with *no*.

Not Important. The essential idea in our treatment of PRGs transcends the definitional choice of asymptotic vs. concrete security. That idea was to think about a *distinguisher* that is asked to differentiate between two kinds of things: the result of applying the PRG to a random string; or a bunch of random bits. We measure, by a real number, how well the distinguisher does this job. At this level, the asymptotic and concrete notions coincide. In addition, a small amount of experience lets one easily translate between the two notions, and the former effectively has the latter built-in. Asymptotic-security proofs embed a concrete-security one, and, again, a little experience lets you extract it. All of this argues that asymptotic vs. concrete security is not an important difference.

There were additional choices silently built into our treatment of PRGs that seem likewise tangential. Choices like allowing multiple oracle queries, and (in the asymptotic case) making adversaries "uniform" algorithms across different values of n. In general, important definitions seem to be quite robust, but maybe not in the sense that is often assumed.

▷ **6** Good definitions are robust, not in the sense that we can modify definitional choices and leave the object being defined unchanged, but in the sense that diverse elaborations leave intact a core definitional idea.

Important. I believe that the culture and character of modern cryptography has been dramatically influenced by the fact that early definitions were always asymptotic. The choice not only reflected shared sensibilities, it reinforced them.

Let's first ask *why* early definitions were always asymptotic. Provable security evolved within the theory community, in the intellectual tradition of other asymptotic notions like big-O notation and NP-completeness. An asymptotic treatment made for more convenient discourse, defining when an object is *secure*. The convenient language helped bring out broad relationships between notions. Early workers in provable-security cryptography aimed at answering questions that seemed most fundamental and aesthetic to them, and minimalist notions, particularly *one-way functions*, were seen as the best starting point for building other kinds of objects. This pushed one towards complex and inefficient constructions, and the definitional choices that would simplify their analyses.

Cryptography might have developed quite differently if concrete security had been more prominent from the start. Concrete security encourages a higher degree of precision in stating results and exploring the relationships between notions. It is a better fit for blockciphers, which rarely have any natural security parameter, and thus a better fit for goals usually achieved using blockciphers, particularly symmetric encryption and message authentication codes. Concrete security makes for a more accessible theory, with fewer alternating quantifiers and complexity-theoretic prerequisites. It encourages a more applied theory.

Practitioners were alienated by the language of asymptotic complexity, the high-level statements of results that fell under it, and the algorithmic inefficiency that seemed endemic to early work. There emerged a pronounced culture gap between cryptographic theory and practice. Theorists and practitioners ignored one other, attending disjoint conferences. Neither group regarded the other as having anything much to say.

The asymptotic approach isn't responsible for the theory/practice gap (which is, after all, endemic to many fields), but it has exacerbated it, effectively encouraging a less relevant style of theory. When Bellare and I wanted to push provable security in a more practice-cognizant direction, we saw abandoning asymptotics as a key element of our program.

Making concrete security more visible and viable has had a big impact on the type of work that now gets done. Papers are published that give tighter analyses of existing protocols; new protocols are invented so as to admit better security bounds; notions are compared by looking at the concrete security of reductions and attacks; and blockcipher-based constructions are designed and analyzed. All of these activities are fostered by the concrete-security view.

▷ **7** Definitional choices can dramatically affect the way that a theory will develop and what it is good for. They impact the types of questions that are likely to be asked, the level of precision expected in an answer, and what background is needed to understand it.

▷ **8** Definitions arise within the context of a particular scientific culture. They reflect the sensibilities of that culture, and they also re-enforce it, distancing it from concerns not their own.

▷ **9** To change the character of work within a field, change its definitions.

5   Blockciphers

Let's next look at blockciphers, one of the basic objects of cryptographic practice. Well-known blockciphers include DES and AES. I want to ask what *is* a blockcipher, and how do we measure a blockcipher's security? Our treatment is based on that of Bellare, Kilian, and Rogaway [BKR] which, in turn, builds on Goldreich, Goldwasser, and Micali [GGM] and Luby and Rackoff [LR].

A blockcipher is a function $E \colon \mathcal{K} \times \{0,1\}^n \to \{0,1\}^n$ where $\mathcal{K}$ is a finite, nonempty set (the *key space*) and $n \geq 1$ is a number (the *blocksize*) and $E(K, \cdot)$ is a permutation (on $\{0,1\}^n$) for each $K \in \mathcal{K}$. When $Y = E(K, X)$ we call X the *plaintext-block* and Y the *ciphertext-block*.

For measuring security, there are lots of adversarial goals that one might focus on. Goals like an adversary's inability to recover plaintext blocks from ciphertext blocks, or its inability to recover a key K from (X_i, Y_i)-pairs, where $Y_i = E(K, X_i)$. But experience leads in a different direction.

To define blockcipher security it is useful to distinguish the *model* (or *attack model*) from the *goal*. The model says what the adversary can do—how the system runs in the adversary's presence. The goal says what the adversary is trying to accomplish as it operates within the model. For our attack model, we consider an *adaptive, chosen-plaintext attack*: the adversary can get the ciphertext block for any plaintext block that it names, and each plaintext block that the adversary asks about may depend on the ciphertext blocks that it has received. For the goal, we'll say that the adversary is trying to distinguish the ciphertext blocks that it receives from an equal number of random, distinct strings.

Our notion of security, what is called PRP (pseudorandom permutation) security, associates a real number to a blockcipher E and a distinguisher D. In game 1, a random key $K \in \mathcal{K}$ is chosen and then, whenever the adversary asks its oracle a question $X \in \{0,1\}^n$, we return $Y = E(K, X)$. In game 0, a random permutation π is chosen among all the permutations from $\{0,1\}^n$ to $\{0,1\}^n$ and then, whenever the adversary asks its oracle a question X, we return $\pi(X)$. Distinguisher D wants to figure out if it is playing game 1 or game 0. It outputs a bit $b \in \{0,1\}$ which is its guess. We define D's advantage, $\mathbf{Adv}_E^{\mathrm{prp}}(D)$, as $\Pr[D^{\mathrm{Game1}} \Rightarrow 1] - \Pr[D^{\mathrm{Game0}} \Rightarrow 1]$, the difference in the probabilities that D outputs 1 when it plays games 1 and 0. Intuitively, a blockcipher E is "good" if no reasonable distinguisher D gets high prp-advantage.

The reader may notice that what I called the *model* has sort of vanished in the actual definition. I never formally defined the model, and it effectively got merged into the definition of the goal. This situation is common. Thinking in terms of an attack model is an important precursor to doing a definition, but the attack model might get abstracted away. For simple definitions, like blockcipher security, this is not a problem. For complex definitions, the model should be left in tact, for the definition will be easier to understand.

▷ **10** It is often useful to develop a model prior to giving a definition. Later, the definition might absorb the model. For complex goals, formalize the model and have it remain intact in the definition.

Our definition of PRP security may seem unrealistically strong. Why should the adversary be allowed to make arbitrary queries to an enciphering oracle when, in typical usages of a blockcipher, its capabilities will be much more constrained? The answer is, in large part, that it works. First, good PRP security seems to be achieved by objects like AES; the assumption doesn't over-shoot what we can efficiently create. Second, the PRP definition is strong enough to give rise to simple and provably-good constructions. Finally, natural, weaker definitions have not been shown effective for designing new constructions, nor in justifying popular constructions that predate notions of PRP security. The above experience effectively becomes the rationale for the definition. It would be wrong to focus on the fact that the definition gives the adversary unrealistic power, because the definition was never intended to directly model an adversary's real-world capabilities in attacking some particular use of the blockcipher.

▷ **11** In defining low-level primitives, simple, easily used, pessimistic definitions are better than more complex and possibly more faithful ones.

The PRP definition of blockcipher security has by now been used in numerous papers. This fact, all by itself, is the best evidence that a definition is doing its job. Definitions are best evaluated retrospectively.

If you want to make a definition of value, you need a market. A definition has to formalize a notion that people would like to have defined, and it has to do so in a way of use to that community. Of course *all* scientific work should be done with the interests of some community in mind. But for a definition, what will best serve this community has to be the focus of ones concern.

▷ **12** The primary measure of a definition's worth is how many people and papers use it, and the extent to which those people and papers say interesting things. A definition is crafted for the benefit of some community.

6   Authenticated Encryption

Authenticated encryption (AE) allows a sender to transmit a message to a receiver in such a way as to assure both its privacy and authenticity. The sender and the receiver share a secret key K. The AE goal has been known to cryptographic practice for decades, but it was only recently provided with a definition and provable-security treatment [KY, BR1, BN]. I'll follow the nonce-based treatment from [RBB]. To capture privacy, we'll formalize that ciphertexts are indistinguishable from random strings. To capture authenticity, we'll formalize that an adversary can't devise authentic messages beyond those that it has seen.

Beginning with syntax, an AE-scheme is a pair of deterministic algorithms $(\mathcal{E}, \mathcal{D})$ where $\mathcal{E}\colon \mathcal{K} \times \mathcal{N} \times \{0,1\}^* \to \{0,1\}^*$, $\mathcal{D}\colon \mathcal{K} \times \mathcal{N} \times \{0,1\}^* \to \{0,1\}^* \cup \{\bot\}$, sets $\mathcal{K}, \mathcal{N} \subseteq \{0,1\}^*$ are nonempty and finite, $|\mathcal{E}(K,N,M)| = |M| + \tau$ for some constant τ, and $\mathcal{D}(K,N,C) = M$ whenever $C = \mathcal{E}(K,N,M)$. Algorithms $\mathcal{E}$ and $\mathcal{D}$ are called the encryption algorithm and the decryption algorithm, and strings K, N, M, and C are called the key, nonce, plaintext, and ciphertext.

The nonce is supposed to be a non-repeating value, such as a counter, and the $\perp$ symbol is used to indicate that the ciphertext is inauthentic.

To quantify security we associate a real number to an AE-scheme $(\mathcal{E}, \mathcal{D})$ and an adversary A. This is done by imagining two different "contests" that A may enter, a privacy contest and the authenticity contest. The adversary may enter either contest, getting a score. Our measure of security is the real number $\mathbf{Adv}^{\text{ae}}_{(\mathcal{E}, \mathcal{D})}(A)$ which is the score that A earns from the contest that it enters.

For the privacy contest the adversary A plays one of two games. In game 1, a key $K \in \mathcal{K}$ is randomly chosen at the beginning of the game. Then, when the adversary asks an oracle query of (N, M) it gets back the string $C = \mathcal{E}(K, N, M)$. In game 0, when the adversary asks an oracle query of (N, M) it gets back a random string C of length $|M| + \tau$. Playing either game, the adversary may not repeat an N-value. When it is ready, the adversary outputs a bit $b \in \{0, 1\}$. The score that adversary gets is $\Pr[A^{\text{Game1}} \Rightarrow 1] - \Pr[A^{\text{Game0}} \Rightarrow 1]$, meaning the difference in the probabilities that A outputs 1 in games 1 and 0. Intuitively, adversary A is trying to distinguish if it is receiving actual ciphertexts for the plaintexts that it asks about (game 1) or if it is receiving an equal number of random bits (game 0).

For the authenticity contest, the adversary again has an oracle, but this time the oracle always behaves according to the above game 1: a key $K \in \mathcal{K}$ is randomly chosen and each query (N, M) is answered by $C = \mathcal{E}(K, N, M)$. As before, the adversary may not repeat an N-value as it interacts with its oracle. When the adversary is ready, it outputs a *forgery attempt* (N, C). The adversary A is said to *forge* if $\mathcal{D}(K, N, C) \neq \perp$ and A never asked an oracle query (N, M) that returned C. The adversary's score is the probability that it forges. Intuitively, adversary A is trying to produce a valid ciphertext that is different from any ciphertext that it has seen.

Authenticated encryption is an important goal of shared-key cryptography, and the prerequisites for defining it and investigating it have been available for a long time. Why would a definition for AE wait until the year 2000 to first appear? There are several answers, but I think that the most important one is that nobody noticed there was anything that needed to be defined. There was already a notion for the privacy of a shared-key encryption scheme, and there was already a notion for what it means to create a good tag, or *message authentication code*, to ensure a message's authenticity. People understood that if you wanted to achieve privacy *and* authenticity, you just did both things. It seemed to fall beneath anyone's radar that one would still need to *prove* that the composite mechanism worked; that there are several ways to do the combining and they *don't* all work; and one couldn't even speak of the composite scheme working, or not, until there was a definition for what the composite scheme was supposed to do. If you think too much in terms of mechanisms, not definitions, then you may fail to notice all of this. In general, the first step in creating a definition is to notice that there is something that needs to be defined. Provable-security cryptography began with the realization that privacy needed to be defined, which was already a crucial and non-obvious observation.

▷ **13** It is easy to overlook that something needs to be defined.

Shortly after AE was defined, a new AE-scheme was put forward by Jutla [Ju]. By better blending the parts of the algorithm used for privacy and for authenticity, Jutla was able to construct a scheme that achieves the AE-goal with nearly the efficiency of traditional, privacy-only mechanisms. Having a definition for AE was essential for figuring out if the scheme was correct.

In the absence of a definition for AE, one can easily design schemes that look sound but aren't. Indeed following the publication of Jutla's scheme and others, the U.S. National Security Agency (NSA) released, through NIST, its own scheme for efficient AE. The mechanism was called *dual counter mode*. I myself read the definition of the mode and broke it in less than a day, privately informing NIST. Donescu, Gligor, and Wagner likewise broke the scheme right away [DGW]. What is it that we knew that the folks from the super-secret NSA did not? The answer, I believe, is *definitions*. If you understood the definition for AE, it was pretty obvious that the NSA scheme wouldn't work. In the absence of understanding a definition, you wouldn't see it. Definitions, not initially intended to help people find attacks, nonetheless do just that. By specifying the rules and goal of the game, the attacker can think more clearly about strategies that conform to the rules but violate the goal. It seems to be the norm that the first person to set forth the definition for any complicated, unformalized goal will also break, without much effort, all the existing protocols for that goal. With no clear definition in mind, inventors can't do their job well.

▷ **14** Having definitions makes it easier to come up with attacks.

I would like to end this section by emphasizing that a definition like that we have given for AE does not suddenly appear; it is part of an evolving line of definitions. Here the sequence of definitional ideas flow from [GM] and [GMRi] to [BKR] and [BDJR] to [BR1, KY, BN] and [RBB]. In general, definitions in cryptography seem to be constantly evolving. They evolve for several reasons. New problems get provable-security treatments (oftentimes these problem having already been considered by practitioners). Shortcomings are discovered in prior definitions—oversights, undesirable consequences, or outright bugs. More elegant ways to do old definitions are discovered. Or the intended use of a definition changes, motivating a different way of doing things. That definitions evolve doesn't contradict point ▷ 6, which might help you to feel better when your definitional choices get antiquated.

▷ **15** Definitions emerge, change, and die more than people think.

7   Session-Key Distribution

In a distributed system, communication between parties typically takes place in *sessions*, a relatively short period of interaction between two parties, protected by an associated *session key*. A party can maintain multiple sessions, even to a particular partner. A protocol for *session-key distribution* (SKD) aims to securely distribute a pair of session keys. There are several *trust models* for SKD, a trust model saying who has what keys. Session-key distribution is addressed by the Kerberos and SSL/TLS protocols.

Definitions for SKD and the related problem of *entity authentication* begin with Bellare and Rogaway [BR2], continuing with work like [BR3, BPR, CK, Sho]. Models and definitions in this domain are more complex than those we've met so far and so, for concision, we will have to be less thorough. Our description is loosely based on [BPR], following [BR3, BR2].

Our model for SKD pessimistically assumes that all communication among parties is under the adversary's control. It can read messages produced by any party, provide its own messages to them, modify messages before they reach their destination, delay or replay them. It can start up entirely new *instances* of any party. We'll also let the adversary learn already-distributed session keys, and we'll let it *corrupt* parties, learning all that they know.

To capture all of these possibilities, we imagine providing an adversary A with an infinite collection of oracles. See Fig. 1. There's an oracle Π_i^s for each party i and each natural number s. Oracle Π_i^s represents instance s of party i. Each oracle has its own private coins and its own state, which it remembers between calls. Oracle initialization depends on the trust model.

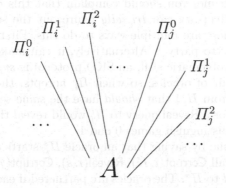

Fig. 1. The model for session key distribution envisages an adversary, A, in a sea of oracles. Oracle Π_i^s models instance s of party i. Each oracle runs the SKD protocol

The adversary's capabilities are embodied by the types of oracle queries we permit, and how we say to answer them. There are three types of queries corresponding to the adversarial capabilities we have mentioned.

(1) Send (i, s, M) — This sends message M to oracle Π_i^s. The oracle computes what the protocol says to, and sends back the response. Should the oracle *accept*, regarding itself as having now arrived at a session key, this fact is made visible to the adversary, along with a value called a *session-id* (SID), and a *partner-id* (PID). The actual *session key* that the oracle holds, sn, will not be visible to

the adversary. To initiate the protocol between an initiator i and a responder j, the adversary should send message $M = j$ to an unused instance of i.

(2) Reveal (i, s) — If oracle Π_i^s has accepted, holding some session key sn, then this query returns sn to the adversary. This query will be used to capture the idea that loss of a session key shouldn't be damaging to other sessions.

(3) Corrupt (i, K) — This is directed against a party, not an instance of a party. The adversary obtains any long-lived or secret keys associated to party i, as well as the private states of all utilized instances of i. The query may also be used to replace any keys under i's control; this is the role of the argument K.

With a model now sketched out, let's describe a measure for adversarial success. The idea is to test if the adversary can distinguish a session key embedded inside an oracle from a *random* session key drawn from the same distribution. Thus we extend the model above to include a Test query, Test(i, s), directed to some oracle Π_i^s. There are two ways to answer this query. In game 1, a query Test(i, s) returns the session key sn inside oracle Π_i^s. In game 0, it returns a random session key sn drawn from the distribution that session keys are supposed to be drawn from. Both cases require that Π_i^s has accepted. The advantage of the adversary, $\mathbf{Adv}_\Pi^{\mathrm{skd}}(A)$, is, once again, the probability that A outputs 1 in game 1 minus the probability that it outputs 1 in game 0.

If you're still with me, you should complain that this definition makes no sense, because the adversary can *trivially* figure out the session key sn in an oracle Π_i^s. In fact, there are multiple ways to do this. First, the adversary can make a Corrupt query to party i. Alternatively, it can make a Reveal query to instance Π_i^s. More problematic still, an SKD protocol is *supposed* to distribute a session key to a pair of oracles, so when Π_i^s accepts, there may well be an oracle Π_j^t (different from Π_i^s) that *should* have the same session key sn as Π_i^s. A Corrupt query to j or a Reveal query to Π_j^t would reveal the session key of Π_i^s and make it easy to distinguish games 0 and 1.

We handle this issue by saying that an oracle Π_i^s starts off *fresh*, but ceases to be so if there is a call Corrupt(i, K), Reveal(i, s), Corrupt(j, K), or Reveal(j, t), where Π_j^t is *partnered* to Π_i^s. The oracles are partnered if each has accepted and they share the same SID. In addition, we demand that only they have that SID, one oracle is an initiator and the other is a responder, and Π_i^s has a PID of j while Π_j^t has a PID of i. Now we simply demand that the adversary perform its Test query on an oracle that is *fresh*; choosing an unfresh oracle earns no credit.

One aspect of the definitional approach above is seen often enough to single out: the idea of lifting something unavoidable into a definition. That is, a common definitional motif is to formalize that which one can not avoid, and then capture the idea that there is nothing beyond that of significance.

As before, a key step in arriving at a definition for SKD was to realize that there was a significant goal to be defined. I learned this from talking to security architects when I worked at IBM. Back then, the security folks were interested in something called *Kerberos*, which I had never heard of. Nobody could adequately explain to me what problem this *Kerberos* thing aimed to do. It was frustrating.

Either the security people were nuts, obsessing on some non-existent or trivial problem, or else they weren't. I decided to assume the latter.

▷ **16** Practice that has not yet met theory is an excellent place for which to craft definitions. Practitioners won't be focusing on a pointless problem.

Soon after the appearance of [BR3], Rackoff came up with an example showing how our definition was not strong enough to guarantee security for certain applications that would use the distributed session key. We traced the problem to a simple issue: we had wrongly made the restriction that the adversary's Test query would be its final query. Removal of this restriction (as in the SKD notion described above) solved the problem.

The definition in [BR3] actually has a more basic problem. Our initial approach to defining partnering, used in [BR2], depended on the idea of a *matching conversation*: oracles Π_i^s and Π_j^t were said to be partnered if they engaged in conversations that are consistent with messages being faithfully relayed between them (except that the last message might not be delivered). This worked fine, but it focused on something that seemed syntactic and fundamentally irrelevant. So in [BR3] we tried something different, assuming an existentially guaranteed partnering function on the global transcript of Send-queries. Almost immediately, I regretted this choice. It did not carry with it the strong intuition of matching conversations, and there was no reason to think that it was the "right" way to identify a party's partner. The notion was hard to explain and hard to use.

Our exposition of $\mathbf{Adv}_{\Pi}^{\text{skd}}(A)$ used SIDs to define partnering. That idea sprang from discussions in 1995 among Bellare, Petrank, Rackoff, and me. An explicit SID seems more intuitive and elegant than an existentially-guaranteed partnering functions, and less syntactic than matching conversations, so we switched to that approach in [BPR]. A recent paper by Choo, Boyd, Hitchcock, and Maitland [CBHM] correctly criticizes how partnering was done in [BR3].

Overall, my experience, shared by others, has been that it is hard to get complex definitions exactly right. I might even venture to say that initial definitional attempts will usually have sub-optimal choices, unexpected consequences, or outright bugs. If the definition is important, the issues will eventually get identified and cleaned up.

▷ **17** Definitions, when they are first proposed, will often encompass poor decisions or errors. For a good definition, these don't greatly diminish the value of the contribution.

▷ **18** Definitional choices that don't capture strong intuition are usually wrong, and may back to haunt you.

Let me make one final point in this section. Our first paper in this space [BR2] was mostly on entity authentication (EA), not SKD, since EA was the more popular and well-known problem of the day. An EA protocol lets two parties have a conversation at the end of which each knows that he has just spoken to the other. (This is the *mutual authentication* version of the problem.) The model is nearly the same for EA and SKD. In working on EA, something that disturbed me greatly was that the problem seemed, to me, to be nearly pointless. If Alice learns that, a moment ago, Bob was present, of what value is that information? Bob may not be present anymore, and even if he is, Alice has learned nothing that will help her to have a continuing conversation. What kind of egotists are Alice and Bob that they care about saying *I am here* and nothing more?

There is some justification for working on EA. One is that I have probably overstated the pointlessness of the goal[3]. Another is that a problem becomes significant *because* people have agreed to focus on it. Still, I'm not so convinced.

▷ **19** The fact that a definitional goal is nearly pointless doesn't seem to bother people nearly as much as it should.

8   The Random-Oracle Model

The random-oracle (RO) model goes back to Fiat and Shamir [FS], but was popularized and made into an explicit design paradigm by Mihir Bellare and me [BR4]. The idea is to design definitions and protocols in an embellished model of computation in which all parties, including the adversary, are given access to a common random-oracle. This is a map $H: \{0,1\}^* \to \{0,1\}$ that associates a random bit to each and every string. One proves the protocol correct in this enriched model of computation. Then, as a heuristic final step, one instantiates the oracle H by something like a cryptographic hash function.

Experience has shown that many cryptographic problems can be solved within the RO-model by protocols that are simpler and more efficient than their best-known standard-model counterparts. For this reason, the RO-model has become popular for doing practical protocol design.

The RO-model has been the locus of much controversy. The question is what assurance, if any, should be invested in an RO-result[4]. The concern was first developed in a paper by Canetti, Goldreich, and Krawczyk [CGK], who give an RO-secure protocol whose standard-model counterpart is always insecure.

To some people, proofs in the RO-model are effectively not proofs. One well-known researcher calls them *heuristic arguments*. That isn't right. A proof in the RO-model is still a proof, it's just a proof in a model of computation that some people don't find worthwhile.

There is certainly a difference between defining a measure of worth for an object and modeling that object. But the modeling-approach isn't less scientific, and it is not at all clear that it yields something whose real-world significance is vastly inferior.

---

[3] In a smartcard setting, for example, EA may be exactly what you want.

[4] Our own paper was rather guarded on the question, but it did make it a thesis that there is value in an RO-result, and that RO-based design is better than design without definitions and proofs [BR4].

When you are working within the RO-model you are working within a specific model, and a not-so-realistic one at that. What is often not recognized is that when you are working within the standard model, you are *also* working within a specific model, and a not-so-realistic one. The standard-model *also* abstracts away key aspects of the real world—like the fact that real computation takes time, uses power, and leaks radiation. There *is* a big gap between the RO-model and reality (hash functions *aren't* like random oracles)—and there is *also* a big gap between the standard model and reality. Some recent work by Micali and Reyzin aims to close the latter gap [MR].

In cryptography, we are in the business of making models. We should always be skeptical that these models are accurate reflections of the world. Models always embed unrealistic and ultimately bogus assumptions, and yet, somehow, they often work. This is the wonder of scientific abstraction.

▷ **20** The distinction between modeling an object and defining an object is real, but its impact is often overstated. Definitions always embed a model and are always subject to model-based limitations. Modeling something is not less scientific than defining it.

The nothing-but-the-standard-model sentiment that is behind some of the RO-criticism is the same sentiment that led to the partitioning of cryptography following Dolev and Yao [DY]. That paper's sin was to model encryption, not define it, and the prevailing sentiment within my community has been to say that such a thing is not real cryptography, and doesn't belong in our venues. A separate community emerged that works on cryptographic protocols, but where the primitives of the protocols are modeled, not defined. The partitioning of cryptography was unhealthy. The field is better off when a diversity of models are encouraged, when they vie for space at the same conferences, and when the relationships among different model becomes a significant point of attention.

9   Closing Comments

No paper on definitions in contemporary cryptography can ignore the emergence of general definitions for *secure protocols* by Backus, Pfitzmann, and Waidner [BPW1] and Canetti [Ca]. These ambitious works define when one protocol is *at-least-as-secure* as another. In doing so, they provide a framework for defining arbitrary protocol goals: a protocol for a given goal is *secure* if it is at-least-as-secure as the *ideal protocol* for that goal. In this way a description of the ideal protocol for some task (e.g., secure message transmission, digital signature, or electronic voting) yields a definition of security for that task. The definitions of [BPW1, Ca] build on work that includes [Bea, GMRa, GMW].

The [BPW1, Ca] line of work represents a particularly important advance. It simultaneously defines a fundamental concept in our field and holds out the promise for more rigorously and manageably treating a variety of ambitious protocol problems. The latter hope stems, in part, from a focus on *composability*, a property built into the notion of security and intended to enable modular design and proofs. It is my view that cryptography is in a sort of "crisis of complexity" for many of the tasks that people now consider: as goals get more complex, what people call a definition and proof ceases, for me, to be rigorous and convincing. General definitional frameworks could improve the situation, and [BPW1, Ca] seem to be having considerable success. See [BPW2] as an impressive example.

Still, I am not without doubts. First, the definitions of [BPW1, Ca] are long and complex. Despite this, there are problems of precision and understandability. To be useful, definitions need to be clear and succinct. Second, the definitions that one arrives at by way of the [BPW1, Ca] framework seem less intuitive and less prescriptive than what one gets by a more direct route. There is an assumption that they are strictly stronger, but such claims are not always proven, and I doubt they're always true. Third, [BPW1, Ca]-derived definitions always involve a simulator. For some goals, simulatability does not seem to correspond to any intuitive understanding of the goal. Simulatability captures strong intuition when it is used to define zero knowledge, but when it is used to define a problem like secure key-distribution or bit commitment, it seems to play a much more technical role. Finally, I see no reason to think that all definitional considerations appropriate to a particular problem domain can be captured by the choices implicit in defining the ideal protocol. For example, is the presence or absence of receipt-freeness[5] in a voting protocol captured by what one does in the ideal voting protocol?

▷ 21 General definitional frameworks [BPW1, Ca] for secure protocols are an important direction—probably the most important definitional activity currently going on in cryptography. But it remains unclear how this will play out.

We are nearing the end of our journey, and I find that I have barely mentioned what is one of the most beautiful definitions from cryptography: the idea of *zero knowledge* (ZK), due to Goldwasser, Micali, and Rackoff [GMRa]. Zero knowledge is a testament to the power of a definition: it has created an area of its own, giving rise to derivative notions and impacting both cryptography and complexity theory. The notion of simulatability that came with ZK has spread across cryptography, becoming a central notion of the field. Briefly and informally, communications with a party P is *zero knowledge* if that which is seen from interacting with P can be created, just as effectively, *without* involving P.

It may sound flip, but I want to acknowledge that *zero knowledge* is a beautiful name. In just four syllables, it promises intrigue and paradox. How can something be *knowledge* and yet be *zero*? How can one be speaking of a mathematical notion of *knowledge* in the first place?

Names are important. They need to be succinct and suggestive. Accuracy is good, but it's not as important. (After all, you are going to provide the word or phrase with a definition; a name all by itself can't be expected to have a precise meaning.) The phrase *minimum disclosure proof* [BCC] is more accurate than *zero knowledge*, but it's nowhere near as good a term.

---

[5] A voting protocol is receipt-free if parties can't prove to someone whom they voted for [BT]. The property is desirable to help avoid coercion.

Sometimes it is hard to anticipate what will make a good name. In [BDPR] we needed a compact name for two flavours of chosen-ciphertext attack (CCA), a weak form and a stronger form. We tried everything, but nothing seemed to work. With much reluctance, we published using the terms CCA1 and CCA2. Amazingly, the terms caught on. I've even come to like them. Every time I see CCA2 in a talk or paper, I smile. Other names I am happy with are *plaintext awareness* and an *AXU-hash function* (almost-xor-universal).

The oddest thing can change a name from being terrible to OK. I used to regard *universal one-way hash function*, UOWHF, as one of the worst names ever invented for a lovely object [NY]. It is cumbersome without accurately suggesting what the object does. I warmed up to the name after hearing Victor Shoup give a talk in which he pronounced UOWHF as *woof*. Once there was a way to say UOWHF, and a fun way, the name seemed immeasurably better.

Notation should be chosen with great care. Names and notation are the currency of thought. Our minds can be imprisoned by poor notation and set free by good notation. Mathematics blossomed only after its notation matured. It is painful to read papers with ill-chosen notation but, more than that, it is hard to think deeply about things until the names and notation are right.

▷ **22** Names and notation matter. Choose them well.

Good names and good notation are just one aspect of good writing, and it is the entirety of technical writing that is important. I reject the viewpoint that the ideas of a paper and the presentation of a paper are fundamentally distinct and orthogonal. A paper is, first and foremost, pages full of marks, not some transcendent concept of what the thing's about. A paper *is* its presentation.

While the quality and impact of any scientific paper is intimately connected to the quality of its writing, I believe that this is *especially* true for papers that aim to make a definitional contribution. These are simultaneously harder to write and more important to write well. Writing definitions is hard because, in part, there will be no existing example for how best to communicate your idea. There is an extra premium on concision and elegance, and a huge penalty for ambiguity. If a proof is poorly written, people will skip over it or convince themselves that the result is right and they could re-prove it. It usually won't render the paper pointless. If a definition is poorly written and can't readily be understood, nothing that follows will make any sense. Nobody will be interested to do follow-on work. The value of the paper vanishes.

A definitional work should explain its definitions, not just give them. Definitional choices should be justified. But the exposition shouldn't intermix definitions and descriptive prose; these need to be clearly separated.

In a paper that uses an existing definition, you should fully state that definition, not just reference it. A failure to do this will typically make your paper meaningless without tracking down the other work, which the reader shouldn't have to do. What's more, the paper you reference might exist in multiple versions, and small differences between their definitions can have a huge impact.

▷ **23** Good writing, always important for a scientific paper, is even more important when the paper's main contribution is definitional.

Of the different kinds of work that I have done, it is the definition-centric/ notion-centric work that I believe to have the most value. Mechanisms come and go, are improved upon, rarely become popular, and are never really basic and compelling enough to bring satisfaction. Theorems are ignored or strengthened, forgotten, and hardly anyone reads their proofs or cares. What lasts, at least for a little while, are the notions of the field and that which is associated to making those notions precise: definitions.

I believe that definitions have been the most important thing to bring understanding to my field and make it into a science. Though I have no particular evidence for it, it is my guess that definitions can play a similar role in other areas of computer science too, where they have not yet played such a role. And that is the real reason that I have chosen this topic for today—that I might, possibly, help infect some other area of computer science with definitions and the definitional-viewpoint.

Acknowledgements

Most of what I understand about definitions I learned from my former advisor, Silvio Micali, who is a master of crafting them. If there is anything right and of value in this note, it probably stems from Silvio. My frequent co-author, Mihir Bellare, shares with me a fondness for definitions and has no doubt influenced by sensibilities about them. I received helpful comments from John Black and Chanathip Namprempre. This note was written with funding from NSF 0208842 and a gift from Cisco Systems. Most especially for this odd paper, it goes without saying that all opinions expressed here are my own.

References

[BBS]   L. Blum, M. Blum, and M. Shub. A simple secure unpredictable unpredictable pseudo-random number generator. *SIAM J. on Computing*, vol. 15, pp. 364–383, 1986.

[BCC]   G. Brassard, D. Chaum, and C. Crépeau. Minimum disclosure proofs of knowledge. *JCSS*, vol. 37, no. 2, pp. 156–189, 1988.

[BCK]   M. Bellare, R. Canetti, and H. Krawczyk. Keying hash functions for message authentication. *Crypto '96*, LNCS vol. 1109, Springer, 1996.

[BDJR]   M. Bellare, A. Desai, E. Jokipii, and P. Rogaway. A concrete security treatment of symmetric encryption: analysis of the DES modes of operation. *FOCS 1997*, 1997.

[BDPR]   M. Bellare, A. Desai, D. Pointcheval, and P. Rogaway. Relations among notions of security for public-key encryption schemes. *Crypto '98*, LNCS vol. 1462, Springer, 1998.

[Bea]   D. Beaver. Secure multiparty protocols and zero-knowledge proof systems tolerating faulty minority. *J. of Cryptology*, vol. 4, no. 2, pp. 75–122, 1991.

[Bel]   M. Bellare. Practice-oriented provable security. *Lectures on Data Security 1998*, LNCS vol. 1561, pp. 1–15, 1999.

[BGW]   M. Ben-or, S. Goldwasser, and A. Wigderson. Completeness theorems for non-cryptographic fault-tolerant distributed computation. *STOC 1988*, ACM Press, pp. 1–10, 1988.

[BKR]   M. Bellare, J. Kilian, and P. Rogaway. The security of the cipher block chaining message authentication code. *JCSS*, vol. 61, no. 3, pp. 262–399, 2000.

[Bl]   M. Blum. Coin flipping by phone. *IEEE Spring COMPCOM*, pp. 133–137, 1982.

[BM]   M. Blum and S. Micali. How to generate cryptographically strong sequences of pseudo-random bits. *SIAM J. on Computing*, vol. 13, no. 4, pp. 850–864, 1984. Earlier version in *FOCS 1982*.

[BN]   M. Bellare and C. Namprempre. Authenticated encryption: Relations among notions and analysis of the generic composition paradigm. *Asiacrypt '00*, LNCS vol. 1976, Springer, pp. 531–545, 2000.

[BPR]   M. Bellare, D. Pointcheval, and P. Rogaway. Authenticated key exchange secure against dictionary attacks. *Advances in Cryptology —Eurocrypt '00*. LNCS vol. 1807, Springer, 2000.

[BPW1]   M. Backes, B. Pfitzmann, and M. Waidner. Secure asynchronous reactive systems. Cryptology ePrint report 2004/082, 2004. Earlier version by Pfitzmann and Waidner in *IEEE Symposium on Security and Privacy*, 2001.

[BPW2]   M. Backes, B. Pfitzmann, and M. Waidner. A universally composable cryptographic library. Cryptology ePrint report 2003/015, 2003.

[BR1]   M. Bellare and P. Rogaway. Encode-then-encipher encryption: How to exploit nonces or redundancy in plaintexts for efficient encryption. *Asiacrypt '00*, LNCS vol. 1976, Springer, pp. 317–330, 2000.

[BR2]   M. Bellare and P. Rogaway. Entity authentication and key distribution. *Crypto '93*. LNCS vol. 773, pp. 232–249, Springer, 1994.

[BR3]   M. Bellare and P. Rogaway. Provably secure session key distribution: the three party case. *STOC 1995*, pp. 57–66, 1995.

[BR4]   M. Bellare and P. Rogaway. Random oracle are practical: a paradigm for designing efficient protocols. *Conference on Computer and Communications Security, CCS '93*, pp. 62–73, 1993.

[BT]   J. Benaloh and D. Tuinstra. Receipt-free secret ballot elections. *STOC 1994*, pp. 544–553, 1994.

[Ca]   R. Canetti. Universally composable security: a new paradigm for cryptographic protocols. Cryptology ePrint report 2000/67, 2001. Earlier version in *FOCS 2001*.

[CBHM]   K. Choo, C. Boyd, Y. Hitchcock, and G. Maitland. On session identifiers in provably secure protocols, the Bellare-Rogaway three-party key distribution protocol revisited. *Security in Communication Networks '04*, LNCS, Springer, 2004.

[CGK]   R. Canetti, O. Goldreich, and H. Krawczyk. The random oracle methodology, revisited. *STOC 1998*, pp. 209–218, 1998.

[CK]   R. Canetti and H. Krawczyk. Universally composable notions of key exchange and secure channels. *Eurocrypt '02*, LNCS vol. 2332, pp. 337–351, Springer, 2002.

[Co]   S. Cook. The complexity of theorem-proving procedures. *STOC 1971*, ACM Press, pp. 151–158, 1971.

[DGW]   P. Donescu, V. Gligor, and D. Wagner. A note on NSA's Dual Counter Mode of encryption. Manuscript, 2001. Available from Wagner's webpage.

[DH]   W. Diffie and M. Hellman. New directions in cryptography. *IEEE Trans. on Inf. Th.*, vol. 22, pp. 644-654, 1976.

[DY]   D. Dolev and A. Yao. On the security of public key protocols. *IEEE Trans. on Information Theory*, vol. 29, no. 12, pp. 198–208, 1983.

[FS]   A. Fiat and A. Shamir. How to prove yourself: practical solutions to identification and signature problems. *Crypto '86*, LNCS vol. 263, pp. 186–194, Springer, 1986.

[GGM]   O. Goldreich, S. Goldwasser, and S. Micali. How to construct random functions, *JACM*, vol. 33, no. 4, 210–217, 1986.

[GM]   S. Goldwasser and S. Micali. Probabilistic encryption. *JCSS*, vol. 28, pp. 270–299, 1984.

[GMRa]   S. Goldwasser, S. Micali, and C. Rackoff. The knowledge complexity of interactive proof systems. *SIAM J. on Computing*, vol. 18, no. 1, pp. 186–208, 1989.

[GMRi]   S. Goldwasser, S. Micali, and R. Rivest. A digital signature scheme secure against adaptive chosen-message attacks. *SIAM J. on Computing*, vol. 17, no. 2, pp. 281–308, 1988.

[GMW]   O. Goldreich, S. Micali, and A. Wigderson. How to play any mental game, or a completeness theorem for protocols with honest majority. *STOC 1987*, pp. 218–229, 1997.

[Go1]   O. Goldreich. The foundations of modern cryptography. Manuscript, 2000. Available from Goldreich's webpage. Earlier version in *Crypto 97*.

[Go2]   O. Goldreich. *The Foundations of Cryptography*. Cambridge University Press. Volume 1 (2001) and Volume 2 (2004).

[Ju]   C. JUTLA. Encryption modes with almost free message integrity. *Eurocrypt '01*, LNCS vol. 2045, Springer, pp. 529–544, 2001.

[KY]   J. Katz and M. Yung. Unforgeable encryption and adaptively secure modes of operation. *Fast Software Encryption, FSE 2000*. LNCS vol. 1978, Springer, pp. 284–299, 2000.

[LR]   M. Luby and C. Rackoff. How to construct pseudorandom permutations from pseudorandom functions. *SIAM J. on Computing*, vol. 17, no. 2, April 1988.

[MR]   S. Micali and L. Reyzin. Physically observable cryptography. *TCC 2004*, LNCS vol. 2951, Springer, pp. 278–296, 2004. Cryptology ePrint report 2003/120.

[NY]   M. Naor and M. Yung. Universal one-way hash functions and their cryptographic applications. *STOC 1989*, pp. 33–43, 1989.

[RBB]   P. ROGAWAY, M. BELLARE, and J. BLACK. OCB: a block-cipher mode of operation for efficient authenticated encryption. *ACM Transactions on Information and System Security* (TISSEC), vol. 6, no. 3, pp. 365–403, 2003.

[Sha]   C. Shannon. Communication theory of secrecy systems. *Bell System Technical Journal*, vol. 28, pp. 656–715, 1949.

[Sho]   V. Shoup. On formal methods for secure key exchange. Cryptology ePrint report 1999/012, 1999.

[SRA]   A. Shamir, R. Rivest, and L. Adleman. Mental poker. MIT/LCS report TM-125, 1979.

[Ya]   A. Yao. Theory and applications of trapdoor functions. *FOCS 1982*, pp. 80–91, 1982.

Meme Media for the Knowledge Federation Over the Web and Pervasive Computing Environments

Yuzuru Tanaka, Jun Fujima, and Makoto Ohigashi

Meme Media Laboratory, Hokkaido University,
N.13 W.8, Kita-ku Sapporo 060-8628, Japan
{tanaka, fujima, ohigashi}@meme.hokudai.ac.jp

Abstract. Although Web technologies enabled us to publish and to browse intellectual resources, they do not enable people to reedit and redistribute intellectual resources, including not only documents but also tools and services, published in the Web. Meme media technologies were proposed to solve this problem, and to accelerate the evolution of intellectual resources accumulated over the Web. Meme media technologies will make the Web work as a pervasive computing environment, i.e., an open system of computing resources in which users can dynamically select and interoperate some of these computing resources to perform their jobs satisfying their dynamically changing demands. Federation denotes ad hoc definition and/or execution of interoperation among computing resources that are not a priori assumed to interoperate with each other. We define knowledge federation as federation of computing resources published in the form of documents. This paper reviews and reinterprets meme media technologies from a new view point of knowledge federation over the Web and pervasive computing environments. It focuses on the following four aspects of meme media technologies: (1) media architectures for reediting and redistributing intellectual resources, (2) client-side middleware technologies for application frameworks, (3) view integration technologies for the interoperation and graphical integration of legacy applications, and (4) knowledge federation technologies for pervasive computing environments.

1 Introduction

With the growing need for interdisciplinary and international availability, distribution and exchange of intellectual resources including information, knowledge, ideas, pieces of work, and tools in reeditable and redistributable organic forms, we need new media technologies that externalize scientific, technological, and/or cultural knowledge fragments in an organic way, and promote their advanced use, international distribution, reuse, and reediting. These media may be called meme media [1] since they carry what R. Dawkins called "memes" [2].

Intellectual resources denote not only multimedia documents, but also application tools and services provided by local or remote servers. They cannot

M.J. Maher (Ed.): ASIAN 2004, LNCS 3321, pp. 33–47, 2004.

be simply classified as information contents since they also include tools and services. Media to externalize some of our knowledge as intellectual resources and to distribute them among people are generically called knowledge media [3]. Some knowledge media that provide direct manipulation operations for people to reedit and to redistribute their contents work as meme media.

Although WWW and browsers enabled us to publish and to browse intellectual resources, they do not enable people to reedit and redistribute memes published in meme media. When intellectual resources are liberated from their servers and distributed among people for their reediting and redistribution, they will become memes and be accumulated in a society to form a meme pool, which will bring a rapid evolution of intellectual resources shared by this society. This will cause an explosive increase of intellectual resources similar to the flood of consumer products in our present consumer societies, and requires new technologies for the management and retrieval of memes.

Some intellectual resources over the Web are computing resources such as computation services and database services. The Web works as an open repository of computing resources. It treats each computing resource as a service, and publishes it as a Web application. Pervasive computing denotes an open system of computing resources in which users can dynamically select and interoperate some of these computing resources to perform their jobs satisfying their dynamically changing demands. Such computing resources include not only services on the Web, but also embedded and/or mobile computing resources connected to the Internet through wireless communication. In pervasive computing, the ad hoc definition and/or execution of interoperation among computing resources is called federation. While the integration denotes interoperation among computing resources with standard interoperation interfaces, federation denotes interoperation among computing resources without a priori designed interoperation interfaces. We define knowledge federation as federation of computing resources published in the form of documents.

Federation over the Web is attracting the attention for interdisciplinary and international advanced reuse and interoperation of heterogeneous intellectual resources especially in scientific simulations [4], digital libraries [5], and research activities [6]. It may be classified into two types: federation defined by programs and federation by users. Most studies on federation focused on the former type. Their approach is based on both the proposal of a standard communication protocol with a language to use it and a repository with a matching mechanism between service providing programs and service consuming programs [7]. The origin of such an idea can be found in [8]. Federation of this type over the Web uses Web service technologies. In this paper, we focus on the latter type, i.e., federation defined by users.

Meme media technologies, when applied to the Web, can federate arbitrarily selected Web applications with each other in an ad hoc manner by extracting arbitrary input forms and output contents from these Web pages, and by defining interoperation among them only through direct manipulation. They work as knowledge federation technologies over the Web. Meme media technologies make the Web or its arbitrary subsystem work as a pervasive computing environment. They can be further applied to federation of computing resources distributed over ubiquitous computing environments including embedded and/or mobile computing resources, and make such environments work as pervasive computing environments.

This paper reviews and reinterprets meme media technologies from a new view point of knowledge federation over the Web and pervasive computing environments. It focuses on the following four aspects of meme media technologies: (1) media architectures for reediting and redistributing intellectual resources, (2) client-side middleware technologies for application frameworks, (3) view integration technologies for the interoperation and graphical integration of legacy applications, and (4) knowledge federation technologies for pervasive computing environments.

2   Meme Media as Enabling Technologies

We have been conducting research and development on meme media and meme pool architectures since 1987. We developed 2D and 3D meme media architectures 'IntelligentPad' and 'IntelligentBox' respectively in 1989 and in 1995, and have been working on their meme-pool and meme-market architectures, as well as on their applications. These are summarized in [1].

IntelligentPad is a 2D representation meme media architecture. Its architecture can be roughly summarized as follows. Instead of directly dealing with component objects, IntelligentPad wraps each object with a standard pad wrapper and treats it as a pad (Figure 1). Each pad has both a standard user interface and a standard connection interface. The user interface of every pad has a card like view on the screen and a standard set of operations like 'move', 'resize', 'copy', 'paste', and 'peel'. As a connection interface, every pad provides a list of slots, and a standard set of messages 'set', 'gimme', and 'update'. Each pad defines one of its slots as its primary slot. Most pads allow users to change their primary slot assignments. You may paste a pad on another pad to define a parent-child relationship between these two pads. The former becomes a child of the latter. When you paste a pad on another, you can select one of the slots provided by the parent pad, and connect the primary slot of the child pad to this selected slot. The selected slot is called the connection slot. Using a 'set' message, each child pad can set the value of its primary slot to the connection slot of its parent pad. Using a 'gimme' message, each child pad can read the value of the connection slot of its parent pad, and update its primary slot with this value. Whenever a pad has a state change, it sends an 'update' message to each of its child pads to notify this state change. Whenever a pad receives an 'update' message, it sends a 'gimme' message to its parent pad. By pasting pads on another pad and specifying slot connections, you may easily define both a compound document layout and interoperations among these pads. Pads can be pasted together to define various multimedia documents and application tools. Unless otherwise specified, composite pads are always decomposable and reeditable.

IntelligentBox is the 3D extension of IntelligentPad. Users can define a parent-child relationship between two boxes by embedding one of them into the local coordinate system defined by the other. A box may have any kind of 3D representations. Figure 2 shows a car composed with primitive boxes. When a user rotates its steering wheel, the steering shaft also rotates, and the rack-and-pinion converts the rotation to a linear motion. The cranks convert this linear motion to the steering of the front wheels. This composition requires no additional programming to define all these mechanisms. Suppose for example that someone published an animation of a flying eagle that is composed of a wire-frame model of an eagle that was covered by special boxes that apply deformation to a wire-frame model. One could easily replace this eagle with the wire-frame of a 3D letter 'A' to obtain a flying letter 'A'.

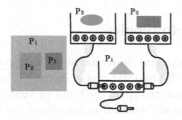   

Fig. 1. A composite pad and its slot connection structure

Fig. 2. An example box composition

A pad as well as a box uses an MVC model consisting of a display object and a model object. Its display object defines its GUI, while its model object defines its internal state and behavior. Each display object further consists of a controller object and a view object. Its view object defines its view on the display screen, while its controller object defines its reaction to user events. When a pad (a box) P_2 is pasted on (connected to) another pad (box) P_1, the IntelligentPad (IntelligentBox) constructs a linkage between their view parts. This defines a dependency from P_1 to P_2.

3   Reediting and Redistributing Intellectual Resources Over the Web

The Web is becoming a rich source of intellectual resources including texts, multimedia contents, tools, and services open for public use. Figure 3 shows, at its bottom right corner, a Web page by US Naval Observatory showing day and night over the earth. The left object in this figure is a composite pad showing the difference of arbitrarily chosen two seasons; the same time on summer solstice and winter solstice for example. This was constructed by just picking up the date input forms and the simulated result as pads from the Naval Observatory Web pages, and drag-and-dropping them on a special pad called DerivationPad.

The date input forms include the year, the month, the day, the hour and the minute input forms. When an extracted pad is pasted on a DerivationPad, it is given a unique cell name like A, B, or C on this DerivationPad. Four more pads are extracted from a time conversion Web application, and dropped on the same DerivationPad. These four pads with cell names A, B, C, and D perform a conversion from an input local time to the corresponding GST. The converted GST is obtained in the cell D. By specifying equations between the GST output cell D and the five date input form cells E, F, G, H, and I extracted from the Naval Observatory Web pages, you can make these two Web applications interoperate with each other. In Figure 3, we applied the multiplexing to the input to obtain multiple outputs.

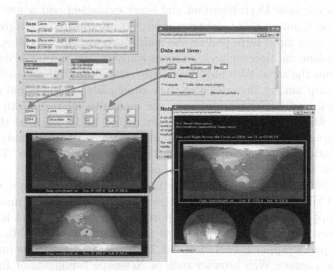

Fig. 3. Construction of a new intellectual resource, using contents extracted from Web pages as meme media objects

All these operations were performed through direct manipulation of pads. For browsing the original Web page, we use a Web browser pad, which dynamically frames different extractable document portions for different mouse locations so that its user may move the mouse cursor around to see every extractable document portion. When it frames a desired object, you can just drag the mouse to extract this object as a pad. All the pads thus extracted from Web pages in a single navigation process keep their original functional relationship even after their arrangement on the same DerivationPad. Whenever you input a new date to the extracted date input form, the corresponding extracted output pad showing a simulated result will change its display. This simulation is performed by the server corresponding to the source Web page from which all these pads are extracted.

The Naval Observatory Web application allows us to specify any time in GST. The composed pad in Figure 3 has four cells A, B. C and D extracted from a time conversion Web application. The cell A accepts a local time input. The cell B is for the selection of a local time zone. The cell C is used to specify London. The cell D shows the GST corresponding to the local time that is input to the cell A. Using the spreadsheet-like function of DerivationPad, you can just specify an equation between the cell D and the cells E, F, G, H, and I to set the year, the month, the day, the hour, and the minute of the GST time in cell D to the cells E, F, G, H, and I extracted from Naval Observatory page. Such a direct manipulation process customizes existing Web-published intellectual resources.

The multiplexer pad, when inserted between the base DerivationPad and the extracted local time input cell A, automatically inserts another multiplexer pad between the base DerivationPad and every extracted pad whose value may depend on the input to this local time input cell. If you make a copy of the extracted local time input form pad on the multiplexer pad at the cell A, then each of the dependent multiplexer pads automatically makes a copy of its child pad extracted from the same Web page. Mutually related multiplexer pads maintain the relationship among the copies of their child pads. The original copies, the second copies, and the third copies on the mutually related multiplexer pads respectively form independent tuples of extracted pads. Each of these tuples maintains input and output relationship among its constituent pads. In Figure 3, two simulation results are obtained for two different local times. This mechanism is useful for the intercomparison of more than one case.

Web documents are defined in HTML format. An HTML view denotes an arbitrary HTML document portion represented in the HTML document format. The pad wrapper to wrap an arbitrary portion of a Web document is capable of both specifying an arbitrary HTML view and rendering any HTML document. We call this pad wrapper HTMLviewPad. Its rendering function is implemented by wrapping a legacy Web browser such as Netscape Navigator or Internet Explorer. In our implementation, we wrapped Internet Explorer. The specification of an arbitrary HTML view over a given HTML document requires the capability of specifying an arbitrary substructure of the internal representation of HTML documents, namely, the DOM tree. HTMLviewPad specifies its HTML view by specifying the DOM tree node that is the root of the sub tree representing this HTML view. The DOM tree representation allows us to identify any of its nodes with its path expression such as /HTML[0]/BODY[0]/TABLE[0]/TR[1]/TD[1].

HTMLviewPad can dynamically frame different extractable document portions for different mouse locations so that its user may move the mouse cursor around to see every extractable document portion. When HTMLviewPad frames what you want to extract, you can drag the mouse to create a new copy of HTMLviewPad with this extracted document portion. The new copy of HTMLviewPad renders the extracted DOM tree on itself.

HTMLviewPad allows us to extract arbitrary Web portions during a navigation process across more than one Web page. When you start a new extraction process, you first need to initialize the current copy of HTMLviewPad through the menu selection. Then you need to specify the URI of the initial Web page from which to start a navigation process. We define the navigation log as a sequence of user's actions from the beginning to the current state of the navigation. Each user action may be an input to an input form, a mouse click on an anchor, or a menu selection. The description of an action includes the path expression specifying the node to which this action is applied, and a parameter value such as the input value or the selected menu item.

When you extract a Web content as a pad during a navigation process, HTMLviewPad creates its new copy holding both the current navigation log and the path expression specifying the extracted portion.

We use a DerivationPad to rearrange all the pads extracted during a navigation process. Whenever a new extracted pad is pasted on DerivationPad during a single navigation process, DerivationPad reads out both the navigation log and the path expression from this new pad. It replaces its stored navigation log with the new read out log. The previous log must be always a prefix of the new read out log. DerivationPad creates its new slot corresponding to the extracted HTML view identified by its path expression, and connects the extracted pad to this slot. Slot names are systematically created using the alphabet like A, B, C, and so on. These are also called cell names. DerivationPad visually labels each extracted pad with its cell name. Then DerivationPad searches the current navigation log for all the path expressions defining its slots, and, for each action with a found path expression, it replaces the parameter value of this action with a pointer to the corresponding slot. When we start a new navigation and paste the first extracted pad on DervationPad, it groups the slots defined in the preceding navigation process, and associates this group with the latest navigation log to maintain the relationship among these slots. For the new navigation process, DerivationPad uses its new storage area to create a set of sots and to maintain the relationship among them.

DerivationPad allows us to define equations among their cells. The equation $\leftarrow$ f(C1, C2, $\cdots$ Cn) defined for a cell C0 means that the value of the function f evaluated using the values of the cells C1, C2, $\cdots$ Cn should be used as the input value of the cell C0. The value of each of the right side cells is evaluated by reexecuting the stored navigation log that relates these cells.

Meme media technologies also allow us to publish such a composite pad thus obtained as a new Web page, which allows other people to access this Web page and to reuse its composite function using a legacy Web browser. Such a composite pad itself is not an HTML document. We need to convert it to an HTML document to publish it as a Web page. We use script programs to define both slots of extracted pads and DerivationPad, and slot connections among them. We represent each pad as an HTML document in an inline frame to embed its definition into another HTML definition, and use cross-frame script programming to define message passing among inline frames. Meme media technologies also allow you to extract arbitrary portions as pads from such a Web page converted from a composite pad.

4   Proxy Pads and Client-Side Application Frameworks

For each typical application, a set of generic meme media components and their typical construction structure form the development framework, called an application framework. Figure 4 shows a form interface framework for relational databases. A form interface is a visual interface that provides users with an office-form view of retrieved records. IntelligentPad works as a form-construction kit for relational databases. The base pad in this figure works as an interface to the legacy database management system. Such a pad that works as a proxy of some external system is called a proxy pad. Such external systems include industrial plants, database management systems, computer-controlled devices, and numerical computation programs running on supercomputers.

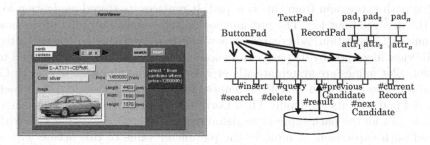

Fig. 4. A form interface framework using pads

A database proxy pad DBProxyPad performs all the details necessary to access a database. It has slots including the #query slot, the #search slot, the #insert slot, the #delete slot, the #previousCandidate slot, the #nextCandidate slot, the #result slot, and the #currentRecord slot. The whole set of pads available in IntelligentPad works as a form construction kit for the visual interface of this database. A RecordPad is a blank-sheet pad. When it receives an 'update' message, it reads out a record-type value, i.e., an association list, from its parent pad and holds this record. A RecordPad allows us to add an arbitrary number of special slots called attribute slots. It also allows us to remove a specified slot from its list of attribute slots. Each attribute slot, when requested to send back its value, reads out the stored record and gets the value of the attribute having the same name as this attribute-slot name. If the record does not have the same attribute name, this attribute slot returns the value 'nil'. When a RecordPad is pasted on the DBProxyPad with its connection to the #currentRecord slot of this proxy pad, it works as a base pad to define a form representation of each record that will be retrieved from the DBProxyPad (Figure 4).

The pad Pad_i is a display pad that shows the value of the attribute $attr_i$. Some examples of such a display pad are TextPad, ImagePad, MoviePad, and BarMeterPad. A mouse click of the ButtonPad connected to the #search slot invokes a search of the database. A click of the ButtonPad connected to the nextCandidate slot advances the record cursor to the next candidate record in the list of retrieved records stored in the #result slot.

In its typical use on a DBProxyPad, a RecordPad divides each retrieved record into a set of attribute-value pairs. Each attribute value is set to the slot with the same name as its attribute name. Depending on the value type of each attribute slot, you may connect a text viewer pad, an image viewer pad, a drawing viewer pad, or a video viewer pad to this slot. You may arbitrarily design the layout of these viewer pads on the RecordPad. A DBProxyPad with a RecordPad pasted with some viewer pads is called a form-based DB viewer, or a form interface to a database.

For the setting of a query in the #query slot of a DBProxyPad, you can simply use a text pad. An SQL query written on this text pad is sent to the #query slot of the proxy pad. When the #search slot is accessed by a set message, DBProxyPad issues the query stored in its #query slot to the corresponding database, and sets the retrieved set of records in its #result slot. It has a cursor to point one of the records in the #result slot. The pointed record is kept in the #currentRecord slot. The cursor can be moved forward or backward respectively when the #nextCandidate slot or the #previousCandidate slot is accessed by a set message.

5   View Integration of Legacy Applications

Legacy software signifies application systems that had been widely used by a community of users, or frequently by an individual user, before the introduction of a new system paradigm. Legacy software migration refers to making these legacy systems usable in a new system paradigm. Meme media systems are examples of new system paradigms. They force each manipulable object to be a meme media object. Hence, legacy software migration in meme media systems means either to wrap a legacy system with a meme media wrapper, or to provide it with its proxy meme media object. Here we will first consider legacy software migration into an IntelligentPad environment.

Any legacy software system, if it has no GUI, can be easily assimilated into an IntelligentPad environment just by developing its proxy pad. The database proxy pad in the form-interface framework assimilates a legacy database management system into an IntelligentPad environment. If a legacy system has its own GUI, its migration into an IntelligentPad environment is not an easy task. In this case, we use a pad wrapper to make the legacy system behave like a pad.

The migration of a legacy system with GUI into a visual component environment generally requires special consideration on its drawing function. The drawing by a visual component may damage, or be damaged by, the drawing by another component, which requires each visual component to have the capability of redrawing and managing damaged areas. The required capability is usually more than what is required for the GUI of the original legacy system. This difference should be programmed when we wrap this legacy system, which is usually not an easy task. An often-used solution to this problem makes the legacy draw its display output off the screen, and maps this image as a texture onto its visual component representation. Every user event on the visual component is dispatched to the off-the-screen GUI of the legacy system. This solution is sometimes called a 'shadow copy' mechanism; it was used in HP NewWave architecture [9].

We use this shadow copy mechanism to wrap a legacy system into a pad. A pad wrapper logically works as a proxy pad. A pad wrapper is a pad that makes its size same as the display area size of the original legacy system. It makes a legacy system display its GUI off the screen, and maps the display bit map image onto itself. The wrapper detects all the events including mouse events and keyboard events, and dispatches them to the same relative location on the off-the-screen display of the legacy system. This method requires that each legacy system allows us to write a program that takes its shadow copy and dispatches any detected events to it. If legacy systems are developed based on some standard development framework satisfying these requirements, they can be easily migrated into IntelligentPad environments.

Slots of such a wrapped legacy system are defined by the program of the wrapper. This program uses the API of the legacy system to access its functions. The definition of each slot requires the definition of its two methods, i.e., one invoked by a set message and the other invoked by a gimme message. For legacy systems sharing a standard API, we can develop a wizardry system that guides us to wrap any of them to a pad in a step-by-step manner without writing any code.

The wrapped legacy systems work as pads and can be pasted together with other pads. They interoperate with each other, exchanging messages through slot connections. This means that we can easily integrate such a wrapped legacy system not only with other applications implemented as composite pads, but also with other wrapped legacy systems. Such integration is called view integration since components are combined and exchange messages through their display objects. The composition using more than one wrapped Web application is an example of view integration.

An IntelligentBox environment is also a legacy system for an IntelligentPad environment, and therefore can be wrapped to work as a pad. Such an IntelligentBox environment pad allows us to export some slots of the component boxes as its slots. It can be pasted together with other pads to define interoperation among them.

In principle, we may consider view integration between a 3D meme media object and a 2D legacy system, and also between a 3D meme media object and a 3D legacy system. While the latter is not an easy task, the former is a natural extension of view integration of 2D applications. For the former, we can use the shadow copy mechanism extended to the texture mapping onto surfaces of arbitrary 3D shapes. The extended shadow copy mechanism requires much more complicated coordinate transformation for both the texture mapping and the event dispatching. The shadow copy mechanism enables us to define a 3D meme media object with a shadow copy of a 2D legacy system mapped onto its surface,

Using the API of this legacy system, we can define appropriate slots of this 3D meme media object. The obtained 3D meme media object can be combined with arbitrary 3D meme media objects. They interoperate with each other exchanging messages through slot connections. This mechanism enables us to define view integration among more than one 2D legacy system mapped onto a 3D surface and 3D meme media objects.

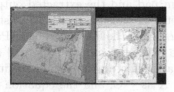  

(a) a flood simulation   (b) an avalanche simulation

Fig. 5. View integration of a 2D GIS with a 3D geography simulation object

Fig. 6. View integration of a 2D GIS with a geographic simulator meme media object

Figure 5 shows view integration of a legacy 2D GIS and a 3D simulation using 3D meme media objects. The 2D GIS shown on the right is mapped onto the corresponding 3D geography without loosing its functions to show detailed information for an arbitrary object that is directly specified by mouse click operation in the 3D environment. The 3D geography representation object has a function to retrieve the required altitude data from a legacy geographic database to determine its shape. This area is specified by the area specification on the 2D GIS. This meme media object can be combined with other 3D meme media objects such as geographic flood or avalanche simulator components. Such a geographic simulator component reads out the polygon data of the geographic shape from the 3D geography meme media object through their slot connection, and renders the simulation result using this polygon data. Figure 6 (a) and (b) respectively show a flood simulation and a avalanche simulation using the view integration of a legacy 2D GIS, a legacy geographic database, and a geographic simulator.

The view integration of a 2D application with a 3D meme media system also allows us to integrate an IntelligentPad environment with 3D meme media objects. Any intelligentPad application can be mapped onto a surface of a 3D meme media object. Users can directly manipulate pads on this surface. Since a 3D meme media system runs in a window, you may also map a 3D meme media application environment onto a surface of a 3D object in another 3D meme media application. Users can directly manipulate 3D objects in the window mapped onto the surface.

6   Wrapping Wiki to Work as a Meme Pool

Meme media technologies can be also applied to Wiki service to make it work as a worldwide repository of pads for sharing and exchanging them, i.e., as a meme pool. Wiki is a piece of server software that allows users to freely create and edit

Web page content using any Web browser [10]. Wiki supports hyperlinks and has a simple text syntax for creating new pages and crosslinks between internal pages on the fly.

In order to make a meme pool system from Wiki, you can access a Wiki page, and extract the URI input, the HTML input form, refresh button, and the output page as pads, and paste them on a DerivationPad. You need to paste a PadSaverLoaderPad as a cell of the same DerivationPad, and relate its input and output respectively to the extracted input form pad and the extracted output page. PadSaverLoaderPad makes conversion between pads on itself and a list of their save format representations in XML. Suppose that the PadSaverLoaderPad, the extracted HTML input form pad, and the extracted output page pad are assigned with cell names A, B, and C. The relationship among them is defined as follows. We define the equation for the cell A as ←C, and the equation for the cell B as ←A.

People can access any page specifying its URI, drag-and-drop arbitrary composite pads to and from the PadSaverLoaderPad on the composed pad to upload and download them to and from the corresponding Wiki server. Each page is shown by the PadSaverLoaderPad. This meme pool system based on Wiki technologies is called a Wiki piazza. Figure 7 shows a Wiki piazza. Users may manipulate and/or edit some pads on an arbitrary page of a Wiki piazza to update their states. They can complete such an update by clicking the save button on the page to upload the update to the corresponding server. Another user accessing the same page can share the updated pads by just clicking the reload button to retrieve this page again from the corresponding server. For a jump from a page to another page in a Wiki piazza system, we can use an anchor pad that can be pasted on a Wiki piazza page. This anchor pad holds a URI that can be set through its #refURI slot, and, when it is clicked, sets this URI to the #URI slot of the Wiki piazza, i.e., to the #URI slot of its base DerivationPad.

A Wiki piazza system allows people not only to publish and share pads, but also to compose new pads by combining components of those pads already published in it, and to publish them again. The collaborative reediting and redistribution of intellectual resources represented as pads in a shared publishing repository by a community or a society of people will accelerate the memetic evolution of intellectual resources in this repository, and make it work as a meme pool.

Fig. 7. A worldwide repository of pads developed by applying meme media technologies to Wiki

7   Knowledge Federation Over the Web and Pervasive Computing Environments

Pervasive computing is an open system of computing resources, and requires both a repository service and a lookup service. The repository service is used to register available computing resources. A lookup service is used to find out a desired computing resource from a repository of available computing resources, and to get its reference as a proxy object as well as a method to access it through the proxy. In addition, pervasive computing needs to allow users to define interoperation among computing resources and/or their proxy objects in an ad hoc way to perform their jobs.

We assume that all computing resources including embedded and mobile resources are all published as Web applications. Some of them are open to the public, while others are available only to closed user communities. Web applications of the latter type may use Web-based security control technologies such as the password-based access control to restrict their access. Web applications execute computing resources in their servers. When accessing Web applications, client systems execute only the Web browser code with some plug-in codes. Application codes are executed by the corresponding servers. Client systems need not execute codes of object classes unknown to them. Therefore, federation among Web applications requires no object migration across networks, and causes no class migration problem.

The meme media technologies to extract Web contents and to reedit them to compose new intellectual resources allow us to dynamically define ad hoc federation among computing resources published as Web applications. Such a composite pad composed of pads extracted from a Web application works as a proxy of this Web application. The view integration technology of meme media also allows us to interoperate such proxy pads of Web applications with legacy applications such as databases, simulators, GISs, and Microsoft Office Tools that run locally or on different machines. Meme media technologies allow us to extract arbitrary portions of any Web application page to compose its proxy pad, which allows us to define an arbitrary proxy interface of each computing resource in an ad hoc way for its federation with other resources.

For a repository service, we may use a Wiki piazza system both for resource providers to register new computing resources and for resource consumers or users to look up and to use available resources. Its pages are categorized for different usages and/or different contexts. Proxy pads of computing resources are also categorized for different usages and/or different contexts into groups. Proxy pads in the same group are published on the same Wiki piazza page. For example, some page may provide proxy pads of stock-market information-service Web applications, while some other may provide proxy pads of available services such as a printing service and a slide-projection service installed in a specific room.

Figure 8 shows a Wiki piazza page of the latter type. It provides proxy pads of services available in a meeting room. This page is automatically accessed by the following mechanism. In our laboratory building, each room is equipped with an independent access point for wireless network connection. Each access point can detect which wireless-connection PC card enters and leaves its covering area. Our laboratory installs a location reference server to manage and to provide information about the current location of each wireless-connection PC card. When you enter a room for a slide presentation carrying your own PC, you can just click a special pad to open the Wiki piazza page corresponding to this room. This special pad first accesses the location reference server to know the current room number and the URI of the Wiki piazza page corresponding to this room. Then it invokes a Wiki piazza system with this URI. The automatically accessed page in Figure 8 contains three service proxy icons. One of them is a slide-projection service proxy icon. You can make a copy of this icon, and drag it out into your own IntelligentPad environment to open the proxy pad, which is shown on the right in the window in Figure 8. This proxy pad has a file input pad, and two buttons to go to the next or to the previous slide. You can input your local PowerPoint file into this file input pad to immediately start your slide presentation using the projector in this room without any cable connections.

Fig. 8. A Wiki piazza page with proxy icons of services available in this room, and the use of slide projection service proxy by opening its icon to project a local file to a large screen

Since our system architectures and frameworks use generic meme media technologies and Web technologies, all the above mentioned functions can be applied in an ad hoc way to any computing resources in any open environments over the Web. They can be also applied to computing resources over any closed intra-networks, and furthermore to any combination of open resources over the Web and proprietary resources over a closed intra-network environment.

Lookup services in general vary from naming services to content-addressable lookup services. Here in this chapter, we have proposed a navigation-based lookup service. A content-addressable lookup service accepts a quantification condition of desired objects and returns their proxies. Such a look up service is necessary to automate federation among available resources, which is our next research goal.

8   Concluding Remarks

The current Web is a mine of intellectual resources. Their advanced use requires knowledge federation technologies that allow us to dynamically define ad hoc interoperation among arbitrarily selected some of them. Such selection requires a repository service and a lookup service for the registration and retrieval of resources. Here we assume that any intellectual resource, whether it is on the Web or embedded in a physical environment or a mobile device, will become accessible through some Web page. Based on this assumption, we have proposed the use of meme media technologies for dynamic and ad hoc knowledge federation of intellectual resources. They also allow us to federate legacy applications with such intellectual resources accessible through the Web. Wiki piazza will work as a repository and lookup service. It will also work as a meme pool to accelerate the evolution of intellectual resources and their usages. Their capability of dynamically defining federation in ad hoc situations may make them applicable to strategic planning and operation in business, administration, and academic activities.

References

1. Tanaka, Y.: Meme Media and Meme Market Architectures: Knowledge Media for Editing, Distributing, and Managing Intellectual Resources. IEEE Press & Wiley-Interscience (2003)
2. Dawkins, R.: The Selfish Gene. 2 edn. Oxford University Press, Oxford (1976)
3. Stefik, M.: The next knowledge medium. AI Mag. **7** (1986) 34–46
4. Miller, J.A., Seila, A.F., Tao, J.: Finding a substrate for federated components on the web. In: Proceedings of the 32nd conference on Winter simulation, Society for Computer Simulation International (2000) 1849–1854
5. Feng, L., Jeusfeld, M.A., Hoppenbrouwers, J.: Towards knowledge-based digital libraries. SIGMOD Rec. **30** (2001) 41–46
6. Bass, M.J., Branschofsky, M.: Dspace at mit: meeting the challenges. In: Proceedings of the first ACM/IEEE-CS joint conference on Digital libraries, ACM Press (2001) 468
7. Eugster, P.T., Felber, P.A., Guerraoui, R., Kermarrec, A.M.: The many faces of publish/subscribe. ACM Comput. Surv. **35** (2003) 114–131
8. Gelernter, D.: Mirror Worlds. Oxford University Press (1992)
9. Hewlett-Packard Cupertino, CA: HP NewWave Environment General Information Manual. (1988)
10. Cunningham, W.: Wiki design principles, Portland Pattern Repository, 27 November 2003 (2003)

Probabilistic Space Partitioning in
Constraint Logic Programming

Nicos Angelopoulos

Department of Computer Science,
University of York,
Heslington, York, YO1 5DD, UK
nicos@cs.york.ac.uk
http://www.cs.york.ac.uk/~nicos

Abstract. We present a language for integrating probabilistic reasoning and logic programming. The key idea is to use constraints based techniques such as the constraints store and finite domain variables. First we show how these techniques can be used to integrate a number of probabilistic inference algorithms with logic programming. We then proceed to detail a language which effects conditioning by probabilistically partitioning the constraint store. We elucidate the kinds of reasoning effected by the introduced language by means of two well known probabilistic problems: the three prisoners and Monty Hall. In particular we show how the syntax of the language can be used to avoid the pitfalls normally associated with the two problems. An elimination algorithm for computing the probability of a query in a given store is presented.

1   Introduction

Probabilistic reasoning is an integral part of areas such as AI and decision making. As our understanding of non-probabilistic, crisp, reasoning is improving, researchers are increasingly interested in furthering support for probabilistic aspects to a variety of formalisms. On the other hand, formalisms such as graphical models and in particularly Bayesian networks (BNs), are being extended to incorporate reasoning about other aspects of the real world. For instance dynamic BNs have been proposed to cope with time dependent change. However, BNs, by nature a propositional formalism, has had limited success in being incorporated with a high level general purpose programming language.

This paper presents a high level language which a extends a successful crisp formalism, logic programming, with probabilistic constructs. Our approach builds on work originally presented in (Angelopoulos, 2001). Although there exist a number of approaches integrating logic programming (LP) and probabilistic reasoning, they have all, but one, been based on the pure, clausal form of LP. Here we propose a framework which is different in nature and scope. Instead of attempting to integrate the two parts into a single inference engine, as most approaches do, CLP(pfd(Y)), based on ideas introduced by constraint logic programming, CLP(X) (Jaffar & Lassez, 1987), keeps reasoning in the two constituents separate, but interfaced seamlessly via the constraint store. The constraint store was also used in (Costa, Page, Qazi, & Cussens, 2002) for storing probabilistic information. However, the main thrust in Costa et.al. was on learning BNs.

M.J. Maher (Ed.): ASIAN 2004, LNCS 3321, pp. 48–62, 2004.

Their means for achieving this, is by fitting probability distributions on skolem functions. Their approach is thus much more focused on integration with learning of BNs than our work, which is about integration of inference with generic probabilistic engines.

From the other approaches, the ones that seem to lie closest to our work, can be broadly put into two categories. On one hand there are the fully compositional approaches, which attach probabilities to relations in a manner that renders them full compositional probabilistic units. Such approaches include, (Baldwin, 1987; Ng & Subrahmanian, 1992; Lukasiewicz, 1999). On the other hand generative formalisms, (Riezler, 1998; Cussens, 2000; Kameya & Sato, 2000; Poole, 1993) that sprung from areas such as probabilistic context free grammars and machine learning, which are stronger in representing distributions over objects/terms generated by the underlying program, specifically over terms generated from a single top-level query.

Fully compositional approaches tend to have problems with defining the appropriate *reference class* (H. E. Kyburg, 1983) for objects to which we seek to find a probability measure. Whereas, generative approaches have difficulty with making assertions about the probability of general statements. In the terms introduced by (Pollock, 1992), the former are capable of defining *indefinite* probabilities whereas the latter are better at defining *definite* probabilities. Compositional approaches are handicapped in that no universally acceptable procedures exist for reducing indefinite to definite probabilities, which are normally what we are interested in. Generative formalisms on the other hand, cannot be readily used for modelling indefinite probabilities which is the the kind of knowledge that is more compactly representable.

Interestingly researchers from both directions have tried to bridge the gap. For instance (Lukasiewicz, 2001) attempts to enhance a compositional approach, while (Cussens, 2000) devised an alternative semantics for a generative formalism. However, there has been less effort in reducing programming constructs to basic probability theory concepts, particularly, interpretation-free concepts, which would improve understanding over computed results.

The remainder of the paper is structured as follows. Section 2 contains basic terminology definitions. Section 3 introduces the generic framework. Section 4 presents an instance of the framework based on probability ascribing functions and conditional constraints over arbitrary predicates. Section 5 presents the three prisoners puzzle and Section 6 shows a solution for the Monty Hall problem. Section 7 briefly discusses some computational issues. The concluding remarks are in Section 8.

2   Preliminaries

We briefly clarify our use of the LP terminology. We follow (Hogger, 1990) and use Prolog conventions throughout. Predicates, or atomic formulae, are of the form $r(t, A)$. In our example r is a predicate name and t with A are first-order terms (t a constant and A a logical variable). We extend first-order terms to also include probabilistic variables. We follow the convention of denoting both logical and probabilistic variables

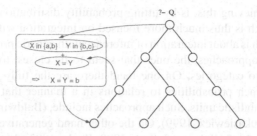

Fig. 1. Interleaving SLD-resolution and constraints inference

with names starting with a capital letter. A logic program, such as $\mathcal{P}$, is a set of definite clauses. A substitution θ, e.g. $\{A/a\}$, is a set of variables to term pairs. Applying θ to a formula will replace every occurrence of a variable in θ with its associated term. In our example $\theta r(t, A)$ is equivalent to $r(t, a)$.

Queries are of the form $?E_1, E_2, \ldots$ where each E_i is a predicate. Each query interrogates particular program and store. $?E$ succeeds iff $\mathcal{P} \cup \mathcal{S} \vdash E$ ($\vdash$ is the derives operator). Each successful derivation often provides a substitution θ, denoting that $\mathcal{P} \cup \mathcal{S} \vdash \theta E$. A Prolog engine will backtrack and find all θs. We take $\mathcal{P} \cup \mathcal{S} \vdash \neg E$ to be equivalent to $\mathcal{P} \cup \mathcal{S} \not\vdash E$. Note that in this case there is no associated θ ($\nexists \theta \cdot \mathcal{P} \cup \mathcal{S} \vdash \theta E$).

We view the store, $\mathcal{S}$, in the usual constraint logic programming fashion as a set of constraints. In this paper $\mathcal{S}$ holds CLP(fd) and probabilistic constraints. We will often abbreviate $\mathcal{P} \cup \mathcal{S} \vdash$ to $\mathcal{S} \vdash$. Furthermore we will use $P_S(X)$ to note the probability of X in the context of $\mathcal{P}$ and $\mathcal{S}$. Set definitions are given as $T = \{t \cdot B_1, B_2, \ldots, B_n\}$. Set T is defined as the collection of all instances t such that all B_i statements are true. We will also denote collections of unique objects as lists of the form $[B_1, B_2, \ldots, B_n]$. We extend the usual set operations to apply to lists. A list is similar to a set except that the order is not the natural sort order but a predefined arbitrary order in which the elements appear in the list.

3   Probabilistic Finite Domains for Probabilistic Reasoning

The key ideas of CLP(X) is that SLD resolution, the deduction mechanism of LP, is interleaved with constraints inference and that communication between the two components is facilitated by constraints posted by the LP engine and present in the constraints store. Fig. 1 shows an example of an interaction in diagrammatic form. The undirected part of graph corresponds to standard LP inference, while constraints posted, and information derived from the constraints store, are shown as directed arrows.

Constraints assist LP to overcome two of its computational drawbacks; the inflexibility of its computational model and the inability to use state information. Linear equation solving and finite domain propagation (CLP(fd)) are examples of commonly used CLP instances. In both, backtracking is replaced by forward pruning of the search space of special variables.

Based on these principles we can introduce variables which will carry probabilistic information. Instead of pruning the space, we are interested in evaluating the probabilities assigned to composite parts of the space. The pfd(Y) framework, short for CLP(pfd(Y)), allows graphical models to be built dynamically in the constraints store by logic programs. Probabilistic inference can then be seen as constraints inference with its results accessible to the logic engine, thus being able to influence the computation. The basis of pfd(Y) is quite generic and a number of different probabilistic inference engines can be used. In the remaining of this section we will examine the generic features of the framework while subsequent sections detail a particular instance. Another instance of the framework, pfd(bn) allowing integration of BN inference, was outlined in (Angelopoulos, 2003).

3.1   Probabilistic Variables

In CLP(fd), a finite domain variable V is introduced to the store with an associated set of values, $fd(X)$. These are all the allowed values for the variable. Constraints posted to the store may subsequently reduce this set. The basic ideas behind constraints between finite domain variables were introduced in (Montanari, 1974). In arbitrary store S the consistency algorithm computes $fd_S(V) \subseteq fd(V)$ the set of allowed values given S. Elaborating on these ideas, a probabilistic finite domain variable V, or probabilistic variable for short, in pfd(Y), defines a probability distribution over its finite domain. When introduced to the store not only is its domain declared, but the probabilistic behaviour of the variable is also described. The constraint store, in addition to CLP(fd) constraints, may also hold probabilistic information about such variables in the form of probabilistic constraints.

Definition 1. *For any variable, V present in a pfd(Y) store S, probabilistic inference must be able to compute $\psi_S(V) = \{(v_1, \pi_1), (v_2, \pi_2), \ldots, (v_n, \pi_n)\}$. Each v_i is an element of $fd_S(V)$ and set $\cup_{i=1}^{n}\{\pi_i\}$ should define a probability distributions, i.e. $0 \leq \pi_i \leq 1$ and $\sum_i \pi_i = 1$. We let $P_S(V = v_i)$ be the probability attached to element v_i under store S.*

Note that by the above definition $P_S(V = v_i) = \pi_i$. Also that pfd(Y) constraints are orthogonal to CLP(fd) constraints, as long as the probabilistic inference engine can recover an appropriate distribution (i.e. set of π_is). In what follows we will assume $fd_S(V)$ to be a list of distinct objects rather than a set.

3.2   Predicates as Events

In the theory of probability, an event is a subset of the space of all possible outputs of an experiment. By treating the assignment of all possible values to probabilistic variables as our space of possible outcomes, we can view predicates containing such variables as events in this space. The main intuition is that the probability assigned to an event is proportional to the space covered by combinations of its probabilistic variables which lead to successful derivations.

For example, consider predicate $lucky(X, Y)$ in the context of program $\mathcal{P}_1$

$$\mathcal{P}_1: \quad \begin{array}{l} \text{lucky(iv, hd).} \\ \text{lucky(v, hd).} \\ \text{lucky(vi, hd).} \end{array}$$

The space of all possible experiments corresponds to the Cartesian product of all possible values for X and Y. The event which corresponds to $lucky(X, Y)$ is the subset of the Cartesian product that makes the predicate true. In this example $[(iv, hd), (v, hd), (vi, hd)]$. For uniformity we use square brackets throughout since we view these sets as lists of unique elements.

Definition 2. *Let $pvars(E)$ be the vector of probabilistic variables in predicate E, $\mathcal{P}$ be the program defining E and $\mathcal{S}$ a constraint store. Let E_i index the variables in $pvars(E)$, e be a vector collecting one element from the finite domain of each variable in $pvars(E)$ and $P_{\mathcal{S}}(E_i = e_i)$ the probability attached to value e_i of variable E_i as defined above. E/e denotes predicate E with its probabilistic variables replaced by their respective elements in e. The probability of predicate E with respect to store $\mathcal{S}$ and program $\mathcal{P}$ is given by:*

$$
\begin{aligned}
P_{\mathcal{S}}(E) &= P(E \mid \mathcal{P} \cup \mathcal{S}) \\
&= \sum_{\substack{\forall e \cdot \\ \mathcal{P} \cup \mathcal{S} \vdash E/e}} P_{\mathcal{S}}(E/e) \\
&= \sum_{\substack{\forall e \cdot \\ \mathcal{P} \cup \mathcal{S} \vdash E/e}} \prod_i P_{\mathcal{S}}(E_i = e_i)
\end{aligned}
$$

For instance, reconsider $\mathcal{P}_1$ in conjunction with the store $\mathcal{S}_1$ holding variables D and C, with $\psi_{\mathcal{S}_1}(D) = [(i, 1/6), (ii, 1/6), (iii, 1/6), (iv, 1/6), (v, 1/6), (vi, 1/6)]$ and $\psi_{\mathcal{S}_1}(C) = [(hd, 1/2), (tl, 1/2)]$. The probability of a lucky combination given by the probability assigned to event $lucky(D, C)$ is $P_{\mathcal{S}_1}(lucky(D, C)) = 1/4$.

In terms of programming constructs, in pfd(Y) the probability of a predicate E can be assigned to a logical variable, L with the following syntax: L is $p(E)$. For convenience L is $p(E \mid G)$ with $p(E \mid G) = p(E \wedge G)/p(G)$, is also provided.

4 Partitioning the Store

In this section we present an instance of the pfd(Y) framework. In pfd(c) probabilistic variables are coupled with probability functions and constraints between variables mark partitions to the store. The former provides means for better integration with CLP(fd) and the latter enables arbitrary events to act as conditionals on the store.

4.1 Probability Ascribing Functions

When first introduced to the store, a variable in pfd(c) is paired with a probability ascribing function. The basic intuition is that the function will provide the unconstrained distribution at any stage of the computation. This means that it should be well defined over all possible sublists of the associated finite domain.

Definition 3. *Let Fd be a list of distinct objects representing a finite domain, ϕ_V be a probability function defined over all sublists of Fd and $Args$ a list of ground terms. Probabilistic variable V is declared with $V \sim \phi_V(Fd, Args)$.*

For a sublist T of Fd ($T \subseteq Fd$) the result of $\phi_V(T, Args)$ is a list of pairs, each pair coupling an element of T to its associated probability. Each element of T should be given a probability value and the sum of all given values should be equal to one. $Args$ parameterise aspects of ϕ_V and if it is the empty list we shorten $\phi_V(Fd, [])$ to $\phi_V(Fd)$. We will use this shorter form when ϕ_V is applied, since $Args$ only play a role in variable declarations. We will also drop the subscript from ϕ_V when the context clearly identifies a particular V. In relation to the ψ notation from Definition 1, the following holds for the empty store, $\emptyset$: $\phi_V(Fd) = \psi_\emptyset(V)$. We let $\phi_V^v(Fd)$ be the probability attached to domain element v ($v \in Fd$) when applying ϕ_V to Fd.

For example, $Heat \sim \texttt{finite\_geometric}([l, m, h], [2])$ declares a finite geometric distribution for variable $Heat$. In this case the deterioration factor is 2. The distribution in the absence of other information is $\psi_\emptyset(Heat) = [(l, 4/7)(m, 2/7), (h, 1/7)]$. A distinct feature of pfd(c) is that the two constituents of a probabilistic variable are kept separate. As a result, the variable is still capable of participating in finite domain constraints. pfd(c) is therefore orthogonal to CLP(fd) while at the same time share information. A second consequence of the separation is that probabilistic functions capture statistical behaviour of variables, in a manner which is to a large extent, independent of specific domain values. Adding CLP(fd) constraint $Heat \neq m$ to the empty store, changes the distribution of $Heat$ to $[(l, 2/3), (h, 1/3)]$. This is due to the removal of m from the list of possible values for the variable. The distribution is readjusted over the remaining elements.

4.2   Conditional Constraints

In pfd(c) the probabilistic information added to the store is the conditional constraint. Its main intuition is that of defining probability subspaces in which we know that certain events hold.

Definition 4. *Let $1 \leq i \leq m$, and $\cup_i\{D_i\}$ be a set of predicates sharing a single probabilistic variable V. Let Q be a predicate not containing V, and set $\cup_i\{\pi_i\}$ define a distribution, i.e. $0 \leq \pi_i \leq 1, \sum_i \pi_i = 1$. Conditional constraint C is declared with:*

$$D_1 : \pi_1 \oplus \ldots \oplus D_m : \pi_m \mid Q$$

We say V is the conditioned variable in C. V's distribution is altered as a result of C being added to the store. To see how the distribution changes, let q be the vector collecting one element from the finite domain of each variable in $pvars(Q)$. The conditional states that each q for which $\exists\theta$ such that $\mathcal{P} \cup \mathcal{S} \vdash \theta Q/q$ divides the current store to a number of partitions. Each partition has an associated probability, π_i, and in that subspace the values of V are constrained to be the sublist of its current finite domain that satisfy θD_i ($\mathcal{P} \cup \mathcal{S} \vdash \theta D_i$). No evidence is provided when $\nexists\theta$. Each subspace is thus weighed by $\prod_{Q_i} P_\mathcal{S}(Q_i = q_i)$ for $Q_i \in pvars(Q)$. The distribution of V according to C, $P_{\{C\}}(V = v)$, can then be computed by considering all possible subspaces and by applying ϕ_V within each subspace. The precise form is a special case of the formula given in Eq. 1.

For example, consider $A \sim \texttt{uniform}([y, n])$ and $B \sim \texttt{uniform}([y, n])$ having $\psi_\emptyset(A) = \psi_\emptyset(B) = [(y, 1/2), (n, 1/2)]$. That is, the $uniform$ function gives equal probability to all elements of a finite domain. Adding constraint C_1, $B = y : .8 \oplus B = n : .2 \mid A = y$ to the empty store, changes the probability of B to $\psi_{\{C_1\}}(B) = [(y, .65), (n, .35)]$. For $B = y$ this is derived as follows

$$
\begin{aligned}
P_{\{C_1\}}(B = y) &= \quad P_\emptyset(A = y).8\phi_B^y([y]) \\
&+ \quad P_\emptyset(A = n)\phi_B^y([y, n]) \\
&= .5 \cdot .8 \cdot 1 + .5 \cdot .5 \\
&= .65
\end{aligned}
$$

$P_{\{C_1\}}(B = n)$ is computed in a similar way by replacing .2 for .8 and $[n]$ for $[y]$. In an arbitrary store, a variable may be conditioned by more than one constraint. In this case, all different groups of constraints in which qualifying Qs are satisfied and within each such group all combinations of D_i need to be considered. It is also the case, that in an arbitrary store cyclic dependencies of variables are not allowed, i.e. a variable V conditioned by constraint with qualifying predicate Q, should not appear in a qualifying predicate of another constraint that conditions a variable in $pvars(Q)$.

Intuitively, the conditional states that when we know Q to be the case then we can divide the store, i.e. the space of all possible continuations, to probabilistic partitions. As we will see later, Q might be always true. What is interesting in this case, is that each satisfying instantiation θ can be used to partition the space accordingly.

4.3   Computing the Probability of Events

For query E and store S we use an elimination algorithm for computing the query's probability $P_S(E)$. Elimination algorithms generalise nonserial dynamic programming. Their use as a unifying framework for probabilistic reasoning in BNs was advocated in (Dechter, 1996). Given an ordering over a set of variables and a number of buckets each holding functions over subsets of the variables, elimination algorithms process and eliminate each bucket in turn. The main idea is that the effect of each bucket eliminated is summarised in functions placed in the remaining buckets. When the last bucket is considered the overall effect of the network to the asked question is computed. Dechter (1996) presents elimination algorithms for computing answers to four kinds of probabilistic questions posed against BNs.

In our algorithm, each bucket B_i, contains a variable and all the variables it depends on, while the ordering is some topological ordering of the dependency graph. The buckets are eliminated backwards. At each stage the updated distribution of the dependant variable in the bucket is calculated and it is then passed to the remaining buckets that mention this variable. The algorithm is shown in Fig. 2, its crux is the formula for computing the marginal probability of variables, which is given in Eq. 1.

Let $dep(V)$ be the set of variables which condition V, i.e. for each constraint conditioning V add $pvars(Q)$ to $dep(V)$. To construct the dependency graph for a set of variables start with adding to the empty graph the variables of this set as nodes. For each variable V added to the graph, recursive add each $X \in dep(V)$ as a node to

Algorithm: Compute probability of event
    Input: Query Q and store S.
    Output: $P_S(Q)$

Initialise:
    – Construct dependency graph G for $pvars(Q)$.
    – Find a topological ordering O of G.
    – Place $pvars(Q)$ to B_0. Place each O_i and $dep(O_i)$ in B_i.
Iterate:
    For i = n to 1
        compute $P_S(O_i)$ according to (1)
        add $P_S(O_i)$ to each remaining bucket that mentions O_i
Compute:
        updated $P_S(Q)$ based on probabilities of $pvars(Q)$ in B_0

Fig. 2. Elimination algorithm for $P_S(Q)$

the graph, if it is not already present, add the arc $V \rightarrow X$. The constructed graph is directed and acyclic since cyclic dependencies of variables are not allowed. It is thus guaranteed that a topological ordering O exists. Let O_i be the ith variable in O with $1 \leq i \leq n$.

The marginal distribution of V in store S can be found given the distributions for $dep(V)$. Intuitively, each distinct combination of values for $dep(V)$ need to be considered. For each combination the set of constraints for which the qualifying constraint is satisfied form a probabilistic subspace. The weight of this subspace is equal to the product of probabilities of the elements in the combination. The subspace is further partitioned to all possible combinations of dependent partitions in the qualified constraints. A diagrammatic example is presented later in Fig. 5.

More formally, let n be the number of constraints conditioning variable V and C_i be the ith constraint, such that $D_{i,1} : \pi_{i,1} \oplus \ldots \oplus D_{i,l_i} : \pi_{i,l_i} \mathbin{|} Q_i$ ($1 \leq i \leq n$, and l_i number of Ds in C_i). Let R be a non empty subset of $\{1, \ldots, n\}$ and T_R be a set of pairs $T_R = \{(r, t_r) \cdot r \in R, 1 \leq t_r \leq l_r\}$. Conjunctions of qualifying Qs are in $\hat{Q}^R \equiv \wedge_{r \in R} Q_r$, possible combinations of D_is from qualified conditionals are in $\hat{D}^{T_R} \equiv \wedge_{(r,t_r) \in T_R} D_{r,t_r}$ and $\neg \hat{Q} \equiv \neg \hat{Q}^{\{1 \ldots n\}}$ is the case where none of the conditionals is qualified. We assume that each element in D_{r,t_r} is independent from the others. Let $\Psi_S(v, \hat{D}^{T_R})$ be the probability value of element v when ϕ_V is applied to the subset of V's domain which satisfies $\hat{D}^{T_R}$. $\hat{q}$ is a vector of values for $pvars(Q)$ and $\hat{q}^R$ is the sub-vector of $\hat{q}$ for conjuction of predicates $\hat{Q}^R$. θ is an instantiation of the non probabilistic variables in $\hat{Q}^R$ such that $\mathcal{P} \cup \mathcal{S} \vdash \theta \hat{Q}^R / \hat{q}^R$ and for a given $\hat{q}$, R is the maximal subset of $\{1, \ldots, n\}$ for which such θ exists.

Definition 5. *The marginal distribution of variable V conditioned by n constraints in arbitrary S is computed with:*

$$P_S(V = v) = \sum_{\hat{q}\cdot\exists R} P_S(V=v \wedge \hat{Q}^R/\hat{q}^R) + \sum_{\hat{q}\cdot\nexists R} P_S(V=v \wedge \neg\hat{Q}/\hat{q})$$

$$= \sum_{\hat{q}\cdot\exists R} P_S(V=v \,|\, \hat{Q}^R/\hat{q}^R) P_S(\hat{Q}^R/\hat{q}^R) + \sum_{\hat{q}\cdot\nexists R} P_S(V=v \,|\, \neg\hat{Q}/\hat{q}) P_S(\neg\hat{Q}/\hat{q})$$

$$= \sum_{\hat{q}\cdot\exists R} (\sum_{T_R} (\prod_{\substack{r\in R \\ (r,t)\in T_R}} (\pi_{r,t})\Psi_S(v, \theta\hat{D}^{T_R})) P_S(\theta\hat{Q}^R/\hat{q}^R))$$

$$+ \sum_{\hat{q}\cdot\nexists R} P_\emptyset(V=v) P_S(\neg\hat{Q}/\hat{q}^R)) \qquad (1)$$

There are two special cases of the conditional that are of particular interest. Let $D \,\mathsf{I}_\pi\, Q$ be a shorthand for $D : \pi \oplus \neg D : (1-\pi) \,\mathsf{I}\, Q$. Note that $\neg D$ is equivalent to $\mathcal{P} \cup \mathcal{S} \not\vdash D$. Also, let $DepV \,\mathsf{I}^\ast\, \pi\, QlfV$ be a shorthand for

$$DepV \neq X : \pi \oplus DepV = X : 1 - \pi \,\mathsf{I}\, QlfV = X$$

Predicates = and $\neq$ are the standard equality and inequality predicates respectively. $DepV$ and $QlfV$ are probabilistic variables and X is a logical variable. Furthermore, in what follows, we drop π when it is equal to 1. For example, $DepV \,\mathsf{I}^\ast\, 1\, QlfV$ is shortened to $DepV \,\mathsf{I}^\ast\, QlfV$ and it is equivalent to $DepV \neq X : 1 \,\mathsf{I}\, Qlf = X$.

To recapitulate, pfd(c) instantiates the proposed framework as follows. Probabilistic information at declaration time is given by ϕ_V, which describes the probabilistic behaviour of the variable after all finite domain pruning. Conditionals add probabilistic information to the store. This information partitions the space to weighed subspaces within which different events may hold. Inference uses these partitions and the application of functions to compute updated probability distributions for the conditioned variables.

5   Three Prisoners

We use the three prisoners puzzle to illustrate that even in simple settings human intuition can lead to incorrect conclusions. This problem is a well known cautionary example, e.g. see (Mosteller, 1965; Grünwald & Halpern, 2003). Grünwald and Halpern (2003) state the problem as follows:

> Of three prisoners a, b, and c, two are to be executed, but a does not know which. Thus, a thinks that the probability that i will be executed is 2/3 for $i \in \{a, b, c\}$. He says to the jailer, "Since either b or c is certainly going to be executed, you will give me no information about my own chances if you give the name of one man, either b or c, who is going to be executed." But then, no matter what the jailer says, naive conditioning leads a to believe that his chance of execution went down from 2/3 to 1/2.

The objective in Grünwald and Halpern (2003) is to contrast and compare solutions provided by conditioning on the "naive space" to those given by conditioning in the "sophisticated space". The objective of this paper is to show how the use of syntax describing the problem in pfd(c) can lead to computations that always occur in the sophisticated space. Provided this space is built correctly, mathematically correct answers are guaranteed.

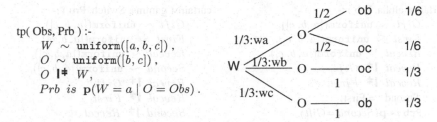

tp(Obs, Prb) :-
  $W \sim$ uniform$([a, b, c])$,
  $O \sim$ uniform$([b, c])$,
  O I≠ W,
  Prb is p$(W = a \mid O = Obs)$.

Fig. 3. Left: Clause for the three prisoners puzzle. Right: Graph for Prb is p$(W = a \mid O = Obs)$

In the three prisoners puzzle, let w_i be the win of i, i.e. prisoner i is not executed then the naive space is $W = \{w_a, w_b, w_c\}$. Let o_i be the case when i is observed, i.e. when the jailer names i as one of the prisoners to be executed, then the set of all possible observations is $O = \{o_b, o_c\}$. The sophisticated space is then a product of the two set reflecting all the possible interactions. That is $I = \{(w_a, o_c), (w_a, o_b), (w_b, o_c), (w_c, o_b)\}$. Note that these are no longer equiprobable due to luck of symmetry. Each of the two (w_a, o_i) has half the probability of the other two elements in the set.

The wrong answer is computed when one conditions in the naive space. Once w_b, or equivalently w_c, is removed because of observation o_b then $W = \{w_a, w_c\}$, which leads to the belief that $P(w_a) = 1/2$. The correct answer when o_c is observed can be computed from $\{(w_a, o_c), (w_b, o_c)\}$ and the equation $P((w_a, o_c)) = \frac{1}{2}P((w_b, o_c))$. It is then apparent that $P(w_a) = P((w_a, o_c)) = \frac{1}{3}$ while $P(w_c) = P((w_b, o_c)) = \frac{2}{3}$.

To model this puzzle in pfd(c) we need two variables and one constraint. Variable W holds the name of the prisoner avoiding execution. Since we do not have any evidence to support preferences amongst the three prisoners we attach function uniform. The variable definition is thus: $W \sim$ uniform$([a, b, c])$. Variable O holds the name of the prisoner named by the jailor. The elements of this domain are b and c, since a can never be named. Whenever the jailor can choose either of the two names, i.e. when a is to be spared, we assume that the choice is unbiased. Thus we define: $O \sim$ uniform$([b, c])$. The constraint between these two variables states, that any value for W has to be distinct from the value of O. Equivalently, when the informed jailor makes his choice between b and c, he must avoid naming the wrong prisoner, i.e. the one who is not going to be executed. Thus we have O I≠ W.

A predicate clause for the three prisoners puzzle is given in Figure 5. To ask the probability of w_a given observation b, we pose the query $tp(b, P)$. The answer will be provided by instantiating P to $1/3$. The algorithm computes Prb is p$(W = a \mid O = Obs)$ by traversing the graph in Figure 5 adding the probabilities of branches for which $W = a \wedge O = b$ is true, and those for which $O = b$ holds. Finally, $p(W = a \mid O = b) = p(W = a \wedge O = b)/p(O = b)$. Here, $\frac{1}{6}/\frac{1}{2} = \frac{1}{3}$.

Note that the order of the variables in the conditional is crucial. If instead of the correct O I≠ W one uses W I≠ O the answer will be $1/2$. The correct order is in accord with the jailor knowing the winner. Thus, observations, O are conditioned on W not the other way around.

curtains(alpha, Prb) :-
    $Gift \sim$ uniform($[a, b, c]$),
    $First \sim$ uniform($[a, b, c]$),
    $Reveal \sim$ uniform($[a, b, c]$),
    $Reveal$ I≠ $Gift$,
    $Reveal$ I≠ $First$,
    $Second$ = First,
    Prb is p($Second=Gift$).

curtains(gamma, Switch, Prb) :-
    $Gift \sim$ uniform($[a, b, c]$),
    $First \sim$ uniform($[a, b, c]$),
    $Reveal \sim$ uniform($[a, b, c]$),
    $Second \sim$ uniform($[a, b, c]$),
    $Reveal$ I≠ $Gift$,
    $Reveal$ I≠ $First$,
    $Second$ I≠ $Reveal$,
    $Second$ I≠$_{Switch}$ $First$
    Prb is p($Second=Gift$).

Fig. 4. pfd(c) clauses for strategies α and γ

6 Monty Hall

Here we examine another problem which is often used to demonstrate that reasoning with probability is hard and often leads to wrong answers. In the Monty Hall TV show, (vos Savant, 1992; Grinstead & Snell, 1997, p.137) there are three curtains. Behind these a car and two goats are concealed. A contestant chooses an initial curtain. One of the remaining curtains is then drawn revealing one of the goats. The contestant is then asked to make a final choice. Whatever is behind this final curtain is the contestant's prize. We will say that the contestant wins if the final choice is the curtain concealing the car. A more thorough analysis of the show in the context of logic and probability can be found in (Grünwald & Halpern, 2003).

What we would like to establish is whether the contestant can develop a strategy for playing the game which will increase the chances of a win. By far, most people believe that this is not possible, and that both options available to the contestant at the final choice of curtain (strategy α :stay-with-same, against strategy β switch-to-other curtain) are equiprobable. However, a closer look reveals that the latter choice is twice as likely to lead to a win. (For a justification see Grinstead and Snell (1997), for evidence of the counter-intuitiveness of the example see vos Savant (1992).) Furthermore, both choices can be seen as part of a single strategy; that of switching with probability Swt. The first, stay-with-same is equivalent to $Swt = 0$ and switch-to-other to $Swt = 1$. Lets call this unifying strategy γ. Then probability of win is $P(\gamma) = \frac{1+Swt}{3}$.

An intelligent decision making system would, ideally, be able to construct the analytical formula from a basic description of the problem. However, this is a very ambitious target. Instead, we require that the system, given a description, should perform computations that provide the same answers to those computed by the formula. In pfd(c) the program modelling the game, is used to extensionally compute the answers given by the analytical formula, from the constituents parts of the problem. The pfd(c) clauses for strategies α and γ are shown in Fig. 4. The declarative reading for the latter clause is that Prb is the probability of a win given that the player switched with probability Swt. Probabilistic variables $First$, $Gift$, $Reveal$ and $Second$ are defined as having no prior information to distinguish the three possible values in their domains. Constraint I≠ is used to declare the pair of variables that must be different (for example the revealed curtain cannot be the player's first choice $Reveal$ I≠ $Gift$). Variable $Second$

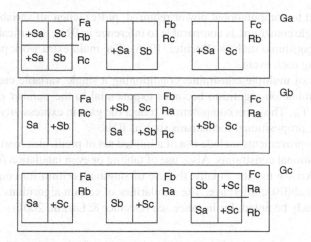

Fig. 5. Weighed subspaces for Monty Hall program. Probability of a win under strategy γ with $Swt = \frac{1}{2}$ is proportional to the area covered by squares marked with $+S_s$ labels

is defined to be different than *First* with probability *Swt* and the same as *First* with probability $1 - Swt$. $P(\gamma)$ is assigned to *Prb* with the call `Prb is p(`*Second=Gift*`)`.

Querying this program with `curtains(gamma, 1/2, Prb)` provides the answer $Prb = 1/2$. In the meta-interpreter we have constructed, when trying to reduce/prove the goal `Prb is p(`*Second=Gift*`)` all possible concrete values to *Gift*, *First* and *Reveal* are considered. Thesse variables are chosen because they do not appear on the right hand side of any conditional constraints while at the same time, variables in *Second = Gift*, the goal for which we wish to find the probability, depend on them. The distribution of *Second* can then be established for each set of values from the final conditional. Note that each such set has an associated probability. Finally the total distribution of *Second* is equal to the total of the weighed sub-distributions.

A diagrammatic form of the partitioning performed by pfd(c) is in Fig. 5. V_v stands for variable V taking value v. We have abbreviated variables from Fig. 4 to their first letter. Inner squares containing S_s values are drawn to scale. The probability of a win, under strategy γ with $Swt = \frac{1}{2}$ is equal to the area covered by $+S_s$ labelled squares, where $s \in \{a, b, c\}$ and the + prefixes assignments S_s which coincide with respective assignments to G. The current implementation of pfd(c) explores this space naively in order to find all combinations that satisfy $G = S$.

7   Discussion

With respect to inference, pfd(c) follows very closely work in probabilistic inference with graphical models such as BNs. We have not analysed the exact relationship here, since pfd(Y) can be instantiated to languages that provide discrete BN inference with any algorithm from the BN literature. pfd(bn) was outlined in (Angelopoulos, 2003).

With regard to computational power required, pfd(c) when all variables are conditioned on a single constraint, is comparable to inference in BNs. Specifically, to variable elimination algorithms, such as (Dechter, 1996). The main extra work pfd(c) needs to do is in deriving each event.

In the case of multiple constraints conditioning a single variable each summation step in the elimination algorithm becomes exponential to the number of elements in corresponding T_R. The extra computations reflect the gain in expressivity pfd(Y) offers on top of static propositional formalisms such as BNs.

Potential improvements include use of a limited set of predicates (particularly = and $\neq$) in the conditional constraints. Also, use of tabling or even tabulation for eliminating re-proving of known events. Finally, the use of single algorithm for constraint propagation and probabilistic inference. The similarities of certain algorithms for both these tasks have already be noted, for instance see (Dechter & Larkin, 2001).

8   Conclusions

In this paper we presented a generic framework for integrating logic programming with probabilistic reasoning. We have used constraints techniques to accomplish this. In particular, finite domains used as basic probabilistic objects, the constraint store for holding probabilistic information and the solver for probabilistic inference. Interaction between crisp and uncertain reasoning was furthered by treating predicates as statistical events. In this regard, the use of probability theory was free of a particular interpretation.

The main characteristic of pfd(Y) is its ability to transparently mix crisp and uncertain computation. As it has been argued by many researchers, e.g. (H. E. Kyburg, 1996), probabilistic reasoning is computationally more expensive than its crisp counterpart. By allowing crisp and uncertain reasoning within one framework we gain representational power. Whereas, by keeping them separate, probabilistic reasoning within the constraint solver and crisp reasoning in the logic engine, we contain the algorithmic behaviour of each part within distinct limits. Altogether, we anticipate that the ability to add well studied algorithms for probabilistic inference to logic programming will be of benefit to AI research and applications.

We have detailed an instance of the generic formalism and showed how it can be used by means of two well studied statistical puzzles. Of particular significance is the fact that the solutions to these puzzles are often different than those obtained by non-experts.

A prototype implementing pfd(c) has been built to provide initial feedback. This can be downloaded from http://www.cs.york.ac.uk/~nicos/sware/pfds/ or tested on-line at http://www.doc.ic.ac.uk/~nicos/cgi/pfds/ pfds.cgi. The prototype system has been used to demonstrate both the soundness of probabilistic inference from intuitive first principles and the exploitation of probabilistic information for improvements in performance within a declarative framework.

In the future, we expect to built a similar system for discrete BNs, and we would be keen to experiment with methods to incorporate finite domain propagation interaction, i.e. the ability to recover a meaningful distribution once the set of possible values for a BN variable was pruned. Finally we hope to use this framework for combining constraint propagation and probabilistic inference into a single algorithm. The two kinds of algorithms have a number of similarities and can be viewed as variable elimination algorithms, as noted by (Dechter & Larkin, 2001). Finally, we hope to refine the definition of the pfd(Y) constraint domain along the lines of existing formal definitions for constraints systems such as (Saraswat, 1992).

Acknowledgements

Thanks to James Cussens for his helpful comments, to one of the reviewers for the meticulous corrections they provided, and to another reviewer for their encouraging comments.

References

Angelopoulos, N. (2001). *Probabilistic Finite Domains*. Ph.D. thesis, Department of Computing, City University, London, UK.

Angelopoulos, N. (2003). clp(pfd(Y)): Constraints for probabilistic reasoning in logic programming. In *Ninth International Conference on Principles and Practice of Constraint Programming* Kinsale, Ireland.

Baldwin, J. F. (1987). Evidential support logic programming. *Journal of Fuzzy Sets and Systems*, *24*, 1–26.

Costa, V., Page, D., Qazi, M., & Cussens, J. (2002). CLP(*BN*) constraint logic programming for probabilistic knowledge. In *19th Annual Conference on Uncertainty in AI*, pp. 517–524 Acapulco, Mexico.

Cussens, J. (2000). Stochastic logic programs: Sampling, inference and applications.. In *16th Annual Conference on Uncertainty in AI*, pp. 115–122 San Francisco, USA.

Dechter, R. (1996). Bucket elimination: A unifying framework for probabilistic inference. In *Proceedings of Uncertainty in AI (UAI-96)* Portland, USA. Extended version in Dechter99.

Dechter, R., & Larkin, D. (2001). Hybrid processing of beliefs and constraints. In *Proceedings of Uncertainty in AI (UAI-2001)*. Extended version in Dechter2001a.

Grinstead, C. M., & Snell, J. L. (1997). *Introduction to Probability* (Second Revised Edition edition). American Mathematical Society.

Grünwald, P., & Halpern, J. (2003). Updating probabilities. *Journal of AI Research*, *19*, 243–278.

H. E. Kyburg, J. (1983). The reference class. *Philosophy of Science*, *50*, 374–397.

H. E. Kyburg, J. (1996). Uncertain inferences and uncertain conclusions. In *12th Annual Conference on Uncertainty in AI*, pp. 365–372.

Hogger, C. J. (1990). *Essentials of Logic Programming*. Oxford University Press, Oxford.

Jaffar, J., & Lassez, J.-L. (1987). Constraint logic programming. In *POPL'87: Proceedings 14th ACM Symposium on Principles of Programming Languages*, pp. 111–119 Munich. ACM.

Kameya, Y., & Sato, T. (2000). Efficient learning with tabulation for parameterized logic programs. In Sagiv, Y. (Ed.), *1st International Conference on Computational Logic (CL2000)*, pp. 269–294. Springer.

Lukasiewicz, T. (1999). Probabilistic logic programming. In *13th Biennial European Conference on Artificial Intelligence*, pp. 388–392 Brighton, UK.

Lukasiewicz, T. (2001). Probabilistic logic programming under inheritance with overriding. In *17th Conference on Uncertainty in Artificial Intelligence (UAI-2001)*, pp. 329–336 Seattle, Washington, USA.

Montanari, U. (1974). Networks of constraints: Fundamental properties and applications to picture processing. *Information Sciences*, *7*, 95–132.

Mosteller, F. (1965). *Fifty challenging problems in probability, with solutions.* Addison-Wesley.

Ng, R., & Subrahmanian, V. (1992). Probabilistic logic programming. *Information and Computation, 101,* 150–201.

Pollock, J. L. (1992). The theory of nomic probability. *Synthese, 90,* 263–300.

Poole, D. (1993). Probabilistic horn abduction and bayesian networks. *Artificial Intelligence, 64,* 81–129.

Riezler, S. (1998). *Probabilistic Constraint Logic Programming.* Ph.D. thesis, Neuphilologische Fakultät, Universität Tübingen, Tubingen, Germany.

Saraswat, V. (1992). The category of constraint systems is Cartesian-closed. In *Proceedings, Seventh Annual IEEE Symposium on Logic in Computer Science,* pp. 341–345 Santa Cruz, California. IEEE Computer Society Press.

vos Savant, M. (1992). *Ask Marilyn.* St. Martins, New York.

Chi-Square Matrix: An Approach for Building-Block Identification

Chatchawit Aporntewan and Prabhas Chongstitvatana

Chulalongkorn University, Bangkok 10330, Thailand
Chatchawit.A@student.chula.ac.th, Prabhas.C@chula.ac.th

Abstract. This paper presents a line of research in genetic algorithms (GAs), called building-block identification. The building blocks (BBs) are common structures inferred from a set of solutions. In simple GA, crossover operator plays an important role in mixing BBs. However, the crossover probably disrupts the BBs because the cut point is chosen at random. Therefore the BBs need to be identified explicitly so that the solutions are efficiently mixed. Let S be a set of binary solutions and the solution $s = b_1 \ldots b_\ell$, $b_i \in \{0, 1\}$. We construct a symmetric matrix of which the element in row i and column j, denoted by m_{ij}, is the chi-square of variables b_i and b_j. The larger the m_{ij} is, the higher the dependency is between bit i and bit j. If m_{ij} is high, bit i and bit j should be passed together to prevent BB disruption. Our approach is validated for additively decomposable functions (ADFs) and hierarchically decomposable functions (HDFs). In terms of scalability, our approach shows a polynomial relationship between the number of function evaluations required to reach the optimum and the problem size. A comparison between the chi-square matrix and the hierarchical Bayesian optimization algorithm (hBOA) shows that the matrix computation is 10 times faster and uses 10 times less memory than constructing the Bayesian network.

1 Introduction

This paper presents a line of research in genetic algorithms (GAs), called building-block identification. The GAs is a probabilistic search and optimization algorithm [2]. The GAs begin with a random population – a set of solutions. A solution (or an individual) is represented by a fixed-length binary string. A solution is assigned a fitness value that indicates the quality of solution. The high-quality solutions are more likely to be selected to perform solution recombination. The crossover operator takes two solutions. Each solution is splited into two pieces. Then, the four pieces of solutions are exchanged to reproduce two solutions. The population size is made constant by discarding some low-quality solutions. An inductive bias of the GAs is that the solution quality can be improved by composing common structures of the high-quality solutions. Simple GAs implement the inductive bias by chopping solutions into pieces. Next, the pieces of solutions are mixed. In GAs literature, the common structures of the high-quality solutions are referred to as building blocks (BBs). The crossover operator mixes

M.J. Maher (Ed.): ASIAN 2004, LNCS 3321, pp. 63–77, 2004.

and also disrupts the BBs because the cut point is chosen at random (see Figure 1). It is clear that the solution recombination should be done, while maintaining the BBs. As a result, the BBs need to be identified explicitly.

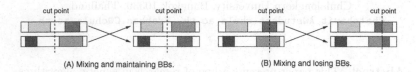

(A) Mixing and maintaining BBs.   (B) Mixing and losing BBs.

Fig. 1. The solutions are mixed by the crossover operator. The BBs are shadowed. The cut point, chosen at random, divides a solution into two pieces. Then, the pieces of solutions are exchanged. In case (A), the solutions are mixed while maintaining the BBs. In case (B), the BBs are disrupted

For some conditions [2, Chapter 7–11], the success of GAs can be explained by the schema theorem and the building-block hypothesis [2]. The schema theorem states that the number of solutions that match above average, short defining-length, and low-order schemata grows exponentially. The optimal solution is hypothesized to be composed of the above average schemata. However, in simple GAs only short defining-length and low-order schemata are permitted to the exponential growth. The other schemata are more disrupted due to the crossover. The trap function is an adversary function for studying BBs and linkage problems in GAs [3]. The general k-bit trap functions are defined as:

$$F_k(b_1 \ldots b_k) = \begin{cases} f_{\text{high}} & ; \text{ if } u = k \\ f_{\text{low}} - u \frac{f_{\text{low}}}{k-1} & ; \text{ otherwise,} \end{cases} \quad (1)$$

where $b_i \in \{0,1\}$, $u = \sum_{i=1}^{k} b_i$, and $f_{\text{high}} > f_{\text{low}}$. Usually, f_{high} is set at k and f_{low} is set at $k-1$. The additively decomposable functions (ADFs), denoted by $F_{m \times k}$, are defined as:

$$F_{m \times k}(K_1 \ldots K_m) = \sum_{i=1}^{m} F_k(K_i), \quad K_i \in \{0,1\}^k. \quad (2)$$

The m and k are varied to produce a number of test functions. The ADFs fool gradient-based optimizers to favor zeroes, but the optimal solution is composed of all ones. The trap function is a fundamental unit for designing test functions that resist hill-climbing algorithms. The test functions can be effectively solved by composing BBs. Several discussions of the test functions can be found in [5, 11].

To illustrate the difficulty, the 10×5-trap functions ($F_{10 \times 5}$) is picked as an example. There are two different schemes to encode a solution to a binary string, $B = b_1 \ldots b_{50}$ where $b_i \in \{0,1\}$ (see Figure 2). To compare the performance of a simple GA on both encoding schemes, we count the number of subfunctions (F_5) that are solved in the elitist individual observed during a run. The number of subfunctions that are solved is averaged from ten runs. The maximum number of generations is set at 10,000. In Table 1, it can be seen that the first encoding scheme is better than the second encoding scheme. Both encoding schemes share the same optimal solution, but the first encoding scheme gives shorter defining-length BBs. Increasing the population size does not make the second encoding scheme better.

Encoding scheme 1:
$F_{10\times5}(B) = \sum_{i=1}^{10} F_5(b_{5i-4}b_{5i-3}b_{5i-2}b_{5i-1}b_{5i})$
Encoding scheme 1's BB: (defining length = 4)
11111**

Encoding scheme 2:
$F_{10\times5}(B) = \sum_{i=1}^{10} F_5(b_i b_{i+10} b_{i+20} b_{i+30} b_{i+40})$
Encoding scheme 2's BB: (defining length = 40)
1*********1*********1*********1*********1*********

Fig. 2. The performance of simple GA relies on the encoding scheme. An improper encoding scheme reduces the performance dramatically

Table 1. The performance of simple GA on the 10×5-trap function. The second encoding scheme gives long defining length, and therefore resulting in poor performance

Population size	Average subfunctions that are solved	
	Encoding scheme 1	Encoding scheme 2
100	4.0	0.8
1,000	8.4	0.7
10,000	10.0	0.1

Thierens raised the scalability issue of simple GAs [10]. He used the uniform crossover so that solutions are randomly mixed. The fitness function is the sum of 5-bit trap functions ($F_{m\times5}$). The analysis shows that either the computational time grows exponentially with the number of 5-bit trap functions or the population size must be exponentially increased. It is clear that scaling up the problem size requires BB information. In addition, the performance of simple GAs relies on the ordering of solution bits. The ordering may not pack the dependent bits close together. Such an ordering results in poor BB mixing. Therefore the BBs need to be identified to improve the scalability issue.

Many strategies in the literature use bit-reordering approach to pack the dependent bits close together, for example, inversion operator [2], messy GA [2], and linkage learning GA (LLGA) [3].

The inversion operator is shown in Figure 3. First, a chunk of adjacent bits are randomly selected. Next, the chunk is inversed by left-to-right flipping. The bits are moved around, but the meaning (fitness) of the solution does not changed. Only the ordering of solution bits is greatly affected. The bits at positions 4 and 7 are passed together with a higher probability. The inversion operator alters the ordering at random. The tight ordering (dependent bits being close

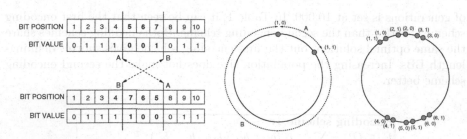

Fig. 3. Inversion operator (left) and linkage learning genetic algorithm (right)

together) are more likely to appear in the final generation. The simple GA enhanced with inversion operator is able to find the optimal solution for additively decomposable functions. In the worst case, the number of function evaluations grows exponentially with the problem size.

The messy GA encodes a solution bit to (p, v) where $p \in \{1, \ldots, \ell\}$ is bit position and $v \in \{0, 1\}$ is bit value. For instance, "01101" is encoded as (1, 0) (2, 1) (3, 1) (4, 0) (5, 1). The bits are tagged with the position numbers so that they can be moved around without losing the meaning. When the solution is mixed, the mixed solution may be over-specified or under-specified. The over-specification is having more than one copy for a bit position. The under-specification is having no copy for a bit position. Several alternatives are proposed for interpreting the over-specified and the under-specified solutions. For example, the over-specification is resolved by majority voting or first-come, first-serve basis. The under-specification is resolved by means of the competitive templates [2]. The messy GA is later developed to fast messy genetic algorithms (FMGA) and gene messy genetic algorithm (GEMGA) [2].

The LLGA encodes ℓ-bit solutions to 2ℓ distinct pieces of (p, v) placed on a circular string where the bit position $p \in \{1, \ldots, \ell\}$ and the bit value $v \in \{0, 1\}$. The 1-bit solution is encoded as it is shown in Figure 3 (left circle). Interpreting the solution is probabilistic. First, a starting point on the circular string is chosen. Second, walking clockwise and picking up (p, v) by first-come, first-serve basis. For instance, if $(1, 0)$ is encountered first, $(1, 1)$ will not be picked up. The 1-bit solution will be interpreted as $(1, 0)$ with probability $\frac{B}{A+B}$, but the interpretation will be $(1, 1)$ with probability $\frac{A}{A+B}$ where A and B are distances on the circular string. In Figure 3 (right circle), the dependent bits come close together. The solution will be interpreted as "111111" with a high probability.

The bit-reordering approach does not explicitly identify BBs, but it successfully delivers the optimal solution. Several papers explicitly analyze the fitness function. The analysis is done on a set of random solutions. Munetomo proposed that bit i and bit j should be in the same BBs if the monotonicity is violated by at least a solution [7]. The monotonicity is defined as follows.

if $\triangle f_i(s) > 0$ and $\triangle f_j(s) > 0$ then $\triangle f_{ij}(s) > \triangle f_i(s)$ and $\triangle f_{ij}(s) > \triangle f_j(s)$

if $\triangle f_i(s) < 0$ and $\triangle f_j(s) < 0$ then $\triangle f_{ij}(s) < \triangle f_i(s)$ and $\triangle f_{ij}(s) < \triangle f_j(s)$

where

$$\triangle f_i(s) = f(\ldots \bar{s}_i \ldots) - f(\ldots s_i \ldots) \tag{3}$$

$$\triangle f_{ij}(s) = f(..\overline{s}_i..\overline{s}_j..) - f(..s_i..s_j..) \tag{4}$$

f denotes fitness function. s denotes binary string. s_i is the ith bit of s. $\overline{s}_i$ denotes $1 - s_i$. In practice, there might be a relaxation of the monotonicity condition.

Another work that explicitly identifies BBs is to reconstruct the fitness function [6]. Any function can be written in terms of Walsh's functions. For example, $f(x_1, x_2, x_3)$, can be written as $f(x_1, x_2, x_3) = w_0 + w_1\Psi_1(x_1) + w_2\Psi_2(x_2) + w_3\Psi_3(x_3) + w_4\Psi_4(x_1, x_2) \ w_5\Psi_5(x_1, x_3) + w_6\Psi_6(x_2, x_3) + w_7\Psi_7(x_1, x_2, x_3)$ where w_i is Walsh's coefficient ($w_i \in R$) and Ψ_i is Walsh's function ($\Psi_i : R \times \ldots \times R \to \{-1, 1\}$). The main algorithm is to compute the Walsh's coefficients. The non-zero Walsh's coefficient indicates the dependency between its associated variables. However, the number of Walsh's coefficients grows exponentially with variables. An underlying assumption is that the function has bounded variable interaction of order-k. Subsequently, the Walsh's coefficients can be calculated in a polynomial time.

Identifying BBs is somewhat related to building a distribution of solutions [4, 8]. The basic concept of optimization by building a distribution is to start with a uniform distribution of solutions. Next, a number of solutions is drawn from the distribution. Some good solutions (winners) are selected. Then the distribution is adjusted toward the winners (the winners-like solutions will be drawn with a higher probability in the next iteration). These steps are repeated until the optimal solution is found or reaching a termination condition. The work in this category is referred to as probabilistic model-building genetic algorithms (PMBGAs).

The Bayesian optimization algorithm (BOA) uses the Bayesian network to represent a distribution [9]. It is shown that if the problem is composed of k-bit trap functions, the network will be fully connected sets of k nodes (see Figure 4) [9, pp. 54]. In addition, the Bayesian network is able to represent joint distributions in the case of overlapped BBs. The BOA can solve the sum of k-bit trap functions ($F_{m \times k}$) in a polynomial relationship between the number of function evaluations and the problem size [9]. The hierarchical BOA (hBOA) is the BOA enhanced with decision tree/graph and a niching method called restricted tournament replacement [9]. The hBOA can solve the hierarchically decomposable functions (HDFs) in a scalable manner. Successful applications for BB identification are financial applications, distributed data mining, cluster optimization, maximum satisfiability of logic formulas (MAXSAT) and Ising spin glass systems [2].

We have present many techniques for identifying BBs. Those techniques have different strength and consume different computational time. The Bayesian network is a powerful tool for identifying BBs, but building the network is time-consuming. Eventually there will be a parallel construction of Bayesian networks. This paper presents a distinctive approach for identifying BBs. Let S be a set of binary solutions and the solution $s = b_1 \ldots b_\ell$, $b_i \in \{0, 1\}$. We construct a symmetric matrix of which the element in row i and column j, denoted by m_{ij}, is the chi-square of variables b_i and b_j. The matrix is called chi-square matrix (CSM). The CSM is further developed from our previous work, the simultaneity

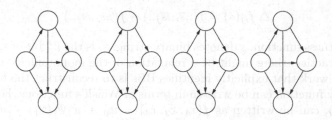

Fig. 4. A final structure of the Bayesian network. An edge indicates dependency between two variables

matrix (SM) [1]. The larger the m_{ij} is, the higher the dependency is between bit i and bit j. The matrix computation is simple and fast. Recently, there is similar work called dependency structure matrix (DSM) [12]. An element of the DSM is only zero (independent) or one (dependent) that is determined by the non-monotonicity [7]. Computing the CSM differs from that of the DSM. However, the papers that are independently developed share some ideas. The remainder of the paper is organized as follows. Section 2 describes the chi-square matrix. Section 3 validates the chi-square matrix with a number of test functions. Section 4 makes a comparison to the BOA and the hBOA. Section 5 concludes the paper.

2   The Chi-Square Matrix

Let $M = (m_{ij})$ be an $\ell \times \ell$ symmetric matrix of numbers. Let S be a set of ℓ-bit binary strings. Let s_i be the ith string, $1 \leq i \leq n$. Let $s_i[j]$ be the jth bit of s_i, $1 \leq j \leq \ell$. The chi-square matrix (CSM) is defined as follows.

$$m_{ij} = \begin{cases} ChiSquare(i,j) & ; \text{ if } i \neq j \\ 0 & ; \text{ otherwise.} \end{cases} \tag{5}$$

The $ChiSquare(i,j)$ is defined as:

$$\sum_{xy} \frac{(C_S^{xy}(i,j) - n/4)^2}{n/4}, \ (x,y) \in \{0,1\}^2 \tag{6}$$

where $C_S^{xy}(i,j)$ counts the number of solutions in which the bit i and the bit j are "00," "01," "10," and "11." The expected frequency of observing "00" or "01" or "10" or "11" is $n/4$ where n is the number of solutions. If the solutions are random, the observed frequency $C_S^{xy}(i,j)$ is close to the expected frequency. The common structures (or building-blocks) appear more often than the expected frequency. Consequently, the chi-square of bit variables that are in the same BB is high. The time complexity of computing the matrix is $O(\ell^2 n)$.

3   A Validation of the CSM

The building-block hypothesis states that the solution quality can be improved by composing BBs. The artificial functions are designed so that the BB hypothe-

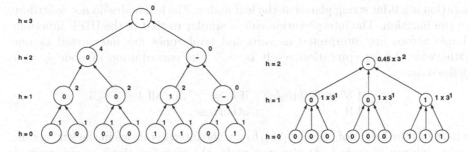

Fig. 5. The HIFF function interprets the solution as a binary tree (left). The 8-bit solution, "00001101," is placed at the lowest level ($h = 0$). The interpretation results are "0," "1," and "-" according to a deterministic rule. Each node excepting the nodes that are "-"contributes to the fitness by 2^h. The fitness is a total of 18. The HTrap1 function interprets the solution as a 3-branch tree (right). The 9-bit solution, "000000111," is placed at the lowest level ($h = 0$). The interpretation results are "0," "1," and "-" according to a deterministic rule. Each node excepting the leaf nodes contributes to the fitness by $3^h \times F_3(b_1b_2b_3)$ where b_i is the interpretation of the child nodes. The fitness of "000000111" is 13.05

sis is true, for example, the additively decomposable functions (ADFs) mentioned in the first section and the hierarchically decomposable functions (HDFs). The HDFs are far more difficult than the ADFs. First, BBs in the lowest level need to be identified. The solution quality is improved by exploiting the identified BBs in solution recombination. Next, the improved population reveals larger BBs. Again the BBs in higher levels need to be identified. Identifying and exploiting BBs are repeated many times until reaching the optimal solution. Commonly used HDFs are hierarchically if-and-only-if (HIFF), hierarchical trap 1 (HTrap1), and hierarchical trap 2 (HTrap2) functions. The original definitions of the HDFs can be found in [11, 9].

To compute the HIFF functions, a solution is interpreted as a binary tree. An example is shown in Figure 5 (left). The solution is an 8-bit string, "00001101." It is placed at the leaf nodes of the binary tree. The leaf nodes are interpreted as the higher levels of the tree. A pair of zeroes and a pair of ones are interpreted as zero and one respectively. Otherwise the interpretation result is "-." The HIFF functions return the sum of values contributed from each node. The contribution of node i, c_i, shown at the upper right of the node, is defined as:

$$c_i = \begin{cases} 2^h & ; \text{ if node } i \text{ is "0" or "1"} \\ 0 & ; \text{ if node } i \text{ is "-,"} \end{cases} \tag{7}$$

where h is the height of node i. In the example, the fitness of "00001101" is $\sum c_i = 18$. The HIFF functions do not bias an optimizer to favor zeroes rather than ones and vice versa. There are two optimal solutions, the string composed of all zeroes and the string composed of all ones.

The HTrap1 functions interpret a solution as a tree in which the number of branches is greater than two. An example is shown in Figure 5 (right). The solution is a 9-bit string placed at the leaf nodes. The leaf nodes do not contribute to the function. The interpretation rule is similar to that of the HIFF functions. Triple zeroes are interpreted as zero and triple ones are interpreted as one. Otherwise the interpretation result is "-." The contribution of node i, c_i, is defined as:

$$c_i = \begin{cases} 3^h \times F_3(b_1 b_2 b_3) & ; \text{ if } b_j \neq \text{"-"} \text{ for all } 1 \leq j \leq 3 \\ 0 & ; \text{ otherwise,} \end{cases} \qquad (8)$$

where h is the height of node i. b_1, b_2, b_3 are the interpretations in the left, middle, right children of node i. At the root node, the trap function's parameters are $f'_{high} = 1$ and $f'_{low} = 0.9$. The other nodes use $f_{high} = 1$ and $f_{low} = 1$. In Figure 5 (right), the HTrap1 function returns $\sum c_i = 13.05$. The optimal solution is composed of all ones.

The HTrap2 functions are similar to the HTrap1 functions. The only difference is the trap function's parameters. In the root node, $f'_{high} = 1$ and $f'_{low} = 0.9$. The other nodes use $f_{high} = 1$ and $f_{low} = 1 + \frac{0.1}{h}$ where h is tree height. The optimal solution is composed of all ones if the following condition is true.

$$f'_{high} - f'_{low} > (h-1)(f_{low} - f_{high}) \qquad (9)$$

The parameter setting ($f'_{high} = 1$, $f'_{low} = 0.9$, $f_{high} = 1$, $f_{low} = 1 + \frac{0.1}{h}$) satisfies the condition. The HTrap2 functions are more deceptive than the HTrap1 functions. Only root node guides an optimizer to favor ones while the other nodes fool the optimizer to favor zeroes by setting $f_{low} > f_{high}$.

To validate the chi-square matrix, an experiment is set as follows. We randomize a population of which the fitness of any individual is greater than a threshold T. Next, the matrix is computed according to the population. Every time step, the threshold T is increased and the matrix is recomputed. The population size is set at 50 for all test functions. The onemax function counts the number of ones. The mixed-trap function is additively composed of 5-bit onemax, 3-bit, 4-bit, 5-bit, 6-bit, and 7-bit trap functions. A sequence of the chi-square matrix is shown in Figure 6-7. A matrix element is represented by a square. The square intensity is proportional to the value of matrix element In the early stage (A), the population is almost random because the threshold T is small. Therefore there are no irregularities in the matrix. The solution quality could be slightly improved without the BB information. That is sufficient to reveal some irregularities or BBs in the next population (B). Improving the solution quality further requires information about BBs (C). Otherwise, randomly mixing disrupts the BBs with a high probability. Finally, the population contains only high-quality solutions. The BBs are clearly seen (D). The correctness of the BBs depends on the quality of solutions observed. An optimization algorithm that exploits the matrix have to extract the BB information from the matrix in order to perform the solution recombination. Hence, moving the population from (A) to (B), (B) to (C), and (C) to (D).

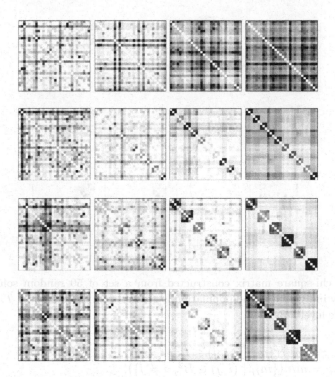

Fig. 6. The chi-square matrix constructed from a set of 50 random solutions. The fitness of any individual in the population is greater than the threshold T. The ADFs have only single-level BBs

4   A Comparison to BOA and hBOA

An exploitation of the chi-square matrix is to compute a partition $\{1, \ldots, \ell\}$ where ℓ is the solution length [1]. The main idea is to put i and j in the same partition subset if the matrix element m_{ij} is significantly high. There are several definitions of the desired partition, for example, the definitions in the senses of non-monotonicity [7], Walsh coefficients [6], and minimal description-length principle [4]. We develop a definition in the sense of chi-square matrix. Algorithm PAR searches for a partition P such that

1. P is a partition.
    1.1 The members of P are disjoint set.
    1.2 The union of all members of P is $\{1, \ldots, \ell\}$.
2. $P \neq \{\{1, \ldots, \ell\}\}$.
3. For all $B \in P$ such that $|B| > 1$,
    3.1 for all $i \in B$, the largest $|B| - 1$ matrix elements in row i are founded in columns of $B \setminus \{i\}$.
4. For all $B \in P$ such that $|B| > 1$,
    4.1 $H_{max} - H_{min} < \alpha(H_{max} - L_{min})$ where

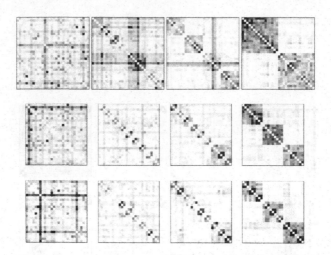

Fig. 7. The chi-square matrix constructed from a set of 50 random solutions. The fitness of any individual in the population is greater than the threshold T. The HDFs have multiple-level BBs

$$H_{max} = max(\{m_{ij} \mid (i,j) \in B^2, \ i \neq j\}),$$
$$H_{min} = min(\{m_{ij} \mid (i,j) \in B^2, \ i \neq j\}),$$
$$L_{min} = min(\{m_{ij} \mid i \in B, j \in \{1,\dots,\ell\} \setminus B\}), \text{ and } \alpha \in [0,1].$$

5. There are no partition P_x such that for some $B \in P$, for some $B_x \in P_x$, P and P_x satisfy the first, the second, the third, and the fourth conditions, $B \subset B_x$.

We assume that the matrix elements $\{m_{ij} \mid i < j\}$ are distinct. In practice, the elements can be made distinct by several techniques [1]. An example of the chi-square matrix is shown in Figure 8. The first condition is obvious. The second condition does not allow the coarsest partition because it is not useful in solution recombination. The third condition makes i and j, in which m_{ij} is significantly high, in the same partition subset. For instance, $P_1 = \{\{1,2,3\}, \{4,5,6\}, \{7,8,9\}, \{10,11,12\}, \{13,14,15\}\}$ satisfies the third condition because the largest two elements in row 1 are found in columns of $\{2,3\}$, the largest two elements in row 2 are found in columns of $\{1,3\}$, the largest two elements in row 3 are found in columns of $\{1,2\}$, and so on. However, there are many partitions that satisfy the third condition, for example, $P_2 = \{\{1,2,3\}, \{4,5,6,7,8,9\}, \{10,11,12\}, \{13,14,15\}\}$. There is a dilemma between choosing the fine partition (P_1) and the coarse partition (P_2). Choosing the fine partition prevents the emergence of large BBs, while the coarse partition results in poor mixing. To overcome the dilemma, the coarse partition will be acceptable if it satisfies the fourth condition. The fifth condition says choosing the coarsest partition that is consistent with the first, the second, the third, and the fourth conditions.

elements governed by {4, 5, 6}

elements governed by {4, 5, 6, 7, 8, 9}

	Col 1	Col 2	Col 3	Col 4	Col 5	Col 6	Col 7	Col 8	Col 9	Col 10	Col 11	Col 12	Col 13	Col 14	Col 15
Row 1	0	7.0220	7.0451	6.1129	6.1841	6.2405	6.3493	6.1560	6.3968	6.0455	6.1065	6.0472	6.2699	6.0534	6.0272
Row 2	7.0220	0	7.0130	6.1115	6.2569	6.1972	6.3075	6.2080	6.1943	6.1290	6.0002	6.1259	6.3515	6.0205	6.1223
Row 3	7.0451	7.0130	0	6.2233	6.3643	6.2571	6.4586	6.4432	6.4146	6.1489	6.1774	6.1260	6.3214	6.1133	6.2010
Row 4	6.1129	6.1115	6.2233	0	7.0999	7.0172	6.8228	6.8722	6.8782	6.1817	6.2222	6.2241	6.3219	6.2016	6.1715
Row 5	6.1841	6.2569	6.3643	7.0999	0	7.1543	6.8738	6.8474	6.8064	6.3443	6.3244	6.2739	6.5128	6.2765	6.2995
Row 6	6.2405	6.1972	6.2571	7.0172	7.1543	0	6.8715	6.8567	6.8727	6.2289	6.2683	6.2613	6.3685	6.2914	6.2791
Row 7	6.3493	6.3075	6.4586	6.8228	6.8738	6.8715	0	7.2764	7.3739	6.3571	6.3877	6.3976	6.5485	6.3230	6.2969
Row 8	6.1560	6.2080	6.4432	6.8722	6.8474	6.8567	7.2764	0	7.3045	6.3215	6.2996	6.3359	6.4957	6.2862	6.2538
Row 9	6.3968	6.1943	6.4146	6.8782	6.8064	6.8727	7.3739	7.3045	0	6.3289	6.3623	6.3590	6.6003	6.3272	6.3170
Row 10	6.0455	6.1290	6.1489	6.1817	6.3443	6.2289	6.3571	6.3215	6.3289	0	7.0259	7.0527	6.2390	6.2794	6.2619
Row 11	6.1065	6.0002	6.1774	6.2222	6.3244	6.2683	6.3877	6.2996	6.3623	7.0259	0	7.0457	6.1318	6.3258	6.1094
Row 12	6.0472	6.1259	6.1260	6.2241	6.2739	6.2613	6.3976	6.3359	6.3590	7.0527	7.0457	0	6.3025	6.1219	6.3465
Row 13	6.2699	6.3515	6.3214	6.3219	6.5128	6.3685	6.5485	6.4957	6.6003	6.2390	6.1318	6.3025	0	7.0316	7.1092
Row 14	6.0534	6.0205	6.1133	6.2016	6.2765	6.2914	6.3230	6.2862	6.3272	6.2794	6.3258	6.1219	7.0316	0	7.0832
Row 15	6.0272	6.1223	6.2010	6.1715	6.2995	6.2791	6.2969	6.2538	6.3170	6.2619	6.1094	6.3465	7.1092	7.0832	0

Fig. 8. An example of the chi-square matrix. The matrix elements in the diagonal are always zero. The matrix is symmetric ($m_{ij} = m_{ji}$)

By condition 4.1, the partition subset $\{4, 5, 6\}$ is acceptable because the values of matrix elements governed by $\{4, 5, 6\}$ are close together (see Figure 8). Being close together is defined by $H_{max} - H_{min}$ where H_{max} and H_{min} is the maximum and the minimum of the nondiagonal matrix elements governed by the partition subset. The $H_{max} - H_{min}$ is a degree of irregularities of the matrix. The main idea is to limit $H_{max} - H_{min}$ to a threshold. The threshold, $\alpha(H_{max} - L_{min})$, is defined relatively to the matrix elements because the threshold cannot be fixed for a problem instance. The partition subset $\{4, 5, 6\}$ gives $H_{max} = 7.1543$, $H_{min} = 7.0172$, and $L_{min} = 6.1115$. L_{min} is the minimum of the nondiagonal matrix elements in rows of $\{4, 5, 6\}$. The fourth condition limits $H_{max} - H_{min}$ to $100 \times \alpha$ percent of the difference between H_{max} and L_{min}. An empirical study showed that α should be set at 0.90 for both ADFs and HDFs. Choosing $\{4, 5, 6, 7, 8, 9\}$ yields ($H_{max} = 7.3739, H_{min} = 6.8064, L_{min} = 6.1115$) which does not violate condition 4.1. The fifth condition prefers a coarse partition $\{\{4, 5, 6, 7, 8, 9\}, \ldots\}$ to a fine partition $\{\{4, 5, 6\}, \ldots\}$ so that the partition subsets can be grown to compose larger BBs in higher levels.

Algorithm PAR is shown in Figure 9. A trace of the algorithm is shown in Table 2. The outer loop processes row 1 to ℓ. In the first step, the columns of the sorted values in row i are stored in $R_{i,1}$ to $R_{i,\ell}$. For $i = 1$, $R_{1,1}$ to $R_{i,\ell}$ are 3, 2, 9, 7, 13, 6, 5, 8, 4, 11, 14, 12, 10, 15, 1, respectively. Next, the inner loop tries a number of partition subsets by enlarging A ($A \leftarrow A \cup \{R_{i,j}\}$). If A satisfies conditions 3.1 and 4.1, A will be saved to B. Finally, P is the partition that satisfies the five conditions. Checking conditions 3.1 and 4.1 is the most time-consuming section. It can be done in $O(\ell^2)$. The checking is done at most ℓ^2 times. Therefore the time complexity of PAR is $O(\ell^4)$.

We customize simple GAs as follows. Every generation, the chi-square matrix is constructed. The PAR algorithm is executed to find the partition. Two parents are chosen by the roulette-wheel method. The solutions are reproduced by a restricted uniform crossover – bits governed by the same partition subset must

$M = (m_{ij})$ denotes $\ell \times \ell$ chi-square matrix.
T_i and $R_{i,j}$ denote arrays of numbers.
A and B are partition subsets.
P denotes a partition.

Algorithm PAR(M, α)
$P \leftarrow \emptyset$;
for $i = 1$ **to** ℓ **do**   // outer loop
   if $i \notin B$ for all $B \in P$ **then**
      $T \leftarrow \{$row i sorted in desc. order$\}$;
      for $j = 1$ **to** ℓ **do**
         $R_{i,j} \leftarrow x$ where $m_{ix} = T_j$;
      endfor
      $A \leftarrow \{i\}$;
      $B \leftarrow \{i\}$;
      for $j = 1$ **to** $\ell - 2$ **do**   // inner loop
         $A \leftarrow A \cup \{R_{i,j}\}$;
         if A satisfies cond. 3.1 and 4.1 **then**
            $B \leftarrow A$;
      endfor
      $P \leftarrow P \cup \{B\}$;
   endif
endfor
return P;

Fig. 9. Algorithm PAR takes an $\ell \times \ell$ symmetric matrix, $M = (m_{ij})$. The output is the partition of $\{1, \ldots, \ell\}$

be passed together. The mutation is turned off. The diversity is maintained by the rank-space method. The population size is determined empirically by the bisection method [9, pp. 64]. The bisection method performs binary search for the minimal population size. There might be 10% different between the population size used in the experiments and the minimal population size that ensures the optimal solution in all independent 10 runs.

The chi-square matrix (CSM) is compared to the Bayesian optimization algorithm (BOA) [9, pp. 115–117]. We also show the results of our previous work, the simultaneity matrix (SM) [1]. Figure 10 shows the number of function evaluations required to reach the optimum. The linear regression in log scale indicates a polynomial relationship between the number of function evaluations and the problem size. The degree of polynomial can be approximated by the slope of linear regression. The maximum number of incoming edges, a parameter of the BOA, limits the number of incoming edges for every vertex in the Bayesian network. The default setting is to set the number of incoming edges to $k - 1$ for $m \times k$-trap functions. It can be seen that the BOA and the CSM can solve the ADFs in a polynomial relationship between the number of function evaluations and the problem size. The BOA performs better than the CSM. However, the performance gap narrows as the problem becomes harder (onemax, $m \times 3$-trap,

Table 2. A trace of the PAR algorithm is shown in the table below. The PAR input is the matrix in Figure 8. The partition subset A is enlarged by $R_{i,j}$. If A satisfies conditions 3.1 and 4.1, A will be saved to B. After finishing the iteration $i = 1$ and $j = 13$, B is added to the partition. Finally, PAR returns the partition $\{\{1,2,3\}, \{4,5,6,7,8,9\}, \{10,11,12\}, \{13,14,15\}\}$

i	j	A	Cond. 3.1	Cond. 4.1	B
1	1	$\{1, 3\}$	True	True	$\{1, 3\}$
1	2	$\{1, 3, 2\}$	True	True	$\{1, 3, 2\}$
1	3	$\{1, 3, 2, 9\}$	False	False	$\{1, 3, 2\}$
1	4	$\{1, 3, 2, 9, 7\}$	False	False	$\{1, 3, 2\}$
1	5	$\{1, 3, 2, 9, 7, 13\}$	False	False	$\{1, 3, 2\}$
1	6	$\{1, 3, 2, 9, 7, 13, 6\}$	False	False	$\{1, 3, 2\}$
1	7	$\{1, 3, 2, 9, 7, 13, 6, 5\}$	False	False	$\{1, 3, 2\}$
1	8	$\{1, 3, 2, 9, 7, 13, 6, 5, 8\}$	False	False	$\{1, 3, 2\}$
1	9	$\{1, 3, 2, 9, 7, 13, 6, 5, 8, 4\}$	False	False	$\{1, 3, 2\}$
1	10	$\{1, 3, 2, 9, 7, 13, 6, 5, 8, 4, 11\}$	False	False	$\{1, 3, 2\}$
1	11	$\{1, 3, 2, 9, 7, 13, 6, 5, 8, 4, 11, 14\}$	False	False	$\{1, 3, 2\}$
1	12	$\{1, 3, 2, 9, 7, 13, 6, 5, 8, 4, 11, 14, 12\}$	False	False	$\{1, 3, 2\}$
1	13	$\{1, 3, 2, 9, 7, 13, 6, 5, 8, 4, 11, 14, 12, 10\}$	False	False	$\{1, 3, 2\}$

and $m \times 5$-trap functions respectively). The difficulty of predetermining the maximum number of incoming edges is resolved in a later version of the BOA called hierarchical BOA (hBOA). We made a comparison between the CSM and the hBOA [9, pp. 164–165] (see Figure 10). The hBOA uses less number of function evaluations than that of the CSM. But in terms of scalability, the hBOA and the CSM are able to solve the HDFs in a polynomial relationship.

Another comparison to the hBOA is made in terms of elapsed time and memory usage. The elapsed time is an execution time of a call on subroutine `constructTheNetwork`. The memory usage is the number of bytes dynamically allocated in the subroutine. The hardware platform is HP NetServer E800, 1GHz Pentium-III, 2GB RAM, and Windows XP. The memory usage in the hBOA is very large because of inefficient memory management in constructing the Bayesian network. A fair implementation of the Bayesian network is the Microsoft WinMine Toolkit. The WinMine is a set of tools that allow you to build statistical models from data. It constructs the Bayesian network with decision tree that is similar to that of the hBOA. The WinMine's elapsed time and memory usage are measured by an execution of `dnet.exe` – a part of the WinMine that constructs the network. All experiments are done with the same biased population that is composed of aligned chunks of zeroes and ones. The parameters of the hBOA and the WinMine Toolkit are set at default. The population size is set at three times greater than the problem size. The elapsed time and memory usage averaged from 10 independent runs are shown in Figure 11. It can be seen that constructing the Bayesian network is time-consuming. In contrast, the matrix computation is 10 times faster and uses 10 times less memory then constructing the network.

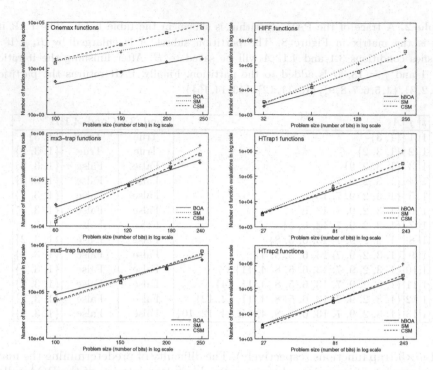

Fig. 10. Performance comparison between the BOA and the CSM on ADFs (left) Performance comparison between the hBOA and the CSM on HDFs (right)

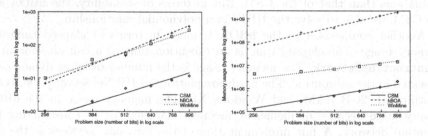

Fig. 11. Elapsed time (left) and memory usage (right) required to construct the Bayesian network and the upper triangle of the matrix (a half of the matrix is needed because the matrix is symmetric)

5   Conclusions

The current BB identification research relies on building a distribution of solutions. The Bayesian network is a powerful representation for the distribution. Nevertheless, building the network is time-and-memory consuming. We have presented a BB identification by the chi-square matrix. The matrix element m_{ij} is the degree of dependency between bit i and bit j. The time complexity of computing the matrix is $O(\ell^2 n)$ where ℓ is the solution length and n is the number of solutions. We put i and j of which m_{ij} is high in the same partition subset. The time complexity of partitioning is $O(\ell^4)$ where ℓ is the solution length. The bits governed by the same partition subsets are passed together when performing solution recombination. The chi-square matrix is able to solve the ADFs and HDFs in a scalable manner. In addition, the matrix computation is efficient in terms of computational time and memory usage.

References

1. Aporntewan, C., and Chongstitvatana, P. (2004). Simultaneity matrix for solving hierarchically decomposable functions. Proceedings of the Genetic and Evolutionary Computation, page 877–888, Springer-Verlag, Heidelberg, Berlin.
2. Goldberg, D. E. (2002). The Design of Innovation: Lessons from and for Competent Genetic Algorithms. Kluwer Academic Publishers, Boston, MA.
3. Harik, G. R. (1997). Learning linkage. Foundation of Genetic Algorithms 4, page 247–262, Morgan Kaufmann, San Francisco, CA.
4. Harik, G. R. (1999). Linkage learning via probabilistic modeling in the ECGA. Technical Report 99010, Illinois Genetic Algorithms Laboratory, University of Illinois at Urbana-Champaign, Champaign, IL.
5. Holland, J. H. (2000). Building blocks, cohort genetic algorithms, and hyperplane-defined functions. Evolutionary Computation, Vol. 8, No. 4, page 373–391, MIT Press, Cambridge, MA.
6. Kargupta, H., and Park, B. (2001). Gene expression and fast construction of distributed evolutionary representation. Evolutionary Computation, Vol. 9, No. 1, page 43–69, MIT Press, Cambridge, MA.
7. Munetomo, M., and Goldberg, D. E. (1999). Linkage identification by non-monotonicity detection for overlapping functions. Evolutionary Computation, Vol. 7, No. 4, page 377–398, MIT Press, Cambridge, MA.
8. Pelikan, M., Goldberg, D. E., and Lobo, F. (1999). A survey of optimization by building and using probabilistic models. Computational Optimization and Applications, Vol. 21, No. 1, page 5–20, Kluwer Academic Publishers.
9. Pelikan, M. (2002). Bayesian optimization algorithm: From single level to hierarchy. Doctoral dissertation, University of Illinois at Urbana-Champaign, Champaign, IL.
10. Thierens, D. (1999). Scalability problems of simple genetic algorithms. Evolutionary Computation, Vol. 7, No. 4, page 331–352, MIT Press, Cambridge, MA.
11. Watson, R. A., and Pollack, J. B. (1999). Hierarchically consistent test problems for genetic algorithms. Proceedings of Congress on Evolutionary Computation, page 1406–1413, IEEE Press, Piscataway, NJ.
12. Yu, T., and Goldberg, D. E. (2004). Dependency structure matrix analysis: off-line utility of the dependency structure matrix genetic algorithm. Proceedings of the Genetic and Evolutionary Computation, page 355–366, Springer-Verlag, Heidelberg, Berlin.

Design Exploration Framework Under Impreciseness Based on Register-Constrained Inclusion Scheduling

Chantana Chantrapornchai[1,*], Wanlop Surakumpolthorn[2], and Edwin Sha[3,**]

[1] Faculty of Science, Silpakorn University, Nakorn Pathom, Thailand
[2] Faculty of Engineering, King Mongkut's Institute of Technology, Ladkrabang, Thailand
[3] Department of Computer Science, University of Texas, Richardson, Texas, USA

Abstract. In this paper, we propose a design exploration framework which consider impreciseness in design specification. In high-level synthesis, imprecise information is often encountered. Two kinds of impreciseness are considered here: imprecise characteristics of functional units and imprecise design constraints. The proposed design exploration framework is based on efficient scheduling algorithm which considers impreciseness, *Register-Constrained Inclusion Scheduling*. We demonstrate the effectiveness of our framework by exploring a design solution for a well-known benchmark, *Voltera filter*. The selected solution meets the acceptability criteria while minimizing the total number of registers.

Keywords: Imprecise Design Exploration, Scheduling/Allocation, Multiple design attributes, Imprecise information, Register constraint, Inclusion Scheduling.

1 Introduction

In architectural level synthesis, imprecise information is almost unavoidable. For instance, an implementation of a particular component in a design may not be known due to several reasons. There may be various choices of modules implementing the functions or the component may have not been completely designed down to the geometry level. Even if it has been designed, variation in fabrication process will likely induce varying area and time measurements. Another kind of impreciseness or vagueness arises from the way a design is considered to be acceptable at architecture level. If a design with latency of 50 cycles is acceptable, what about a design with 51 cycles versus a design with 75 cycles? This even becomes imprecise especially when there are multiple conflicting design criteria. For example, is it worth to expand a latency by two cycles while saving one register and what about expanding 10 more cycles ? Effective treatment of such impreciseness in high level synthesis can undoubtedly play a key role in finding optimal design solutions.

In this paper, we propose a design exploration framework which considers imprecise information underlying in system specification and requirements. Particularly, we

---

* This work was supported in part by the TRF under grant number MRG4680115, Thailand.
** This work was supported in part by TI University Program, NSF EIA 0103709, Texas ARP-009741-0028-2001 and NSF CCR-0309461, USA.

M.J. Maher (Ed.): ASIAN 2004, LNCS 3321, pp. 78–92, 2004.

are interested in the latency and register constraints. However, the approach can be extended to handle other multiple design criteria. The system characteristics are modeled based on the fuzzy set theory. Register count is considered as another dimension of imprecise system requirement. The work in [2, 6] is used as a scheduling core in the iterative design refinement process. The imprecise schedule which minimizes the register usage is generated. If the schedule meets acceptability criteria, the design solution is selected. Otherwise, the resources are adjusted an the process is repeated. Our input system is modeled using a data flow graph with imprecise timing parameters. Such systems can be found in many digital signal processing applications, e.g., communication switches and real-time multimedia rendering systems. Imprecise specification on both system parameters and constraints can have a significant impact on component resource allocation and scheduling for designing these systems. Therefore, it is important to develop synthesis and optimization techniques which incorporate such impreciseness.

Most traditional synthesis tools ignore these vagueness or impreciseness in the specification. In particular, they assume the worst case (or sometimes typical case) execution time of a functional unit. The constraints are usually assumed to be a fixed precise value although in reality some flexibility can be allowed in the constraint due to the individual interpretation of an "acceptable" design. Such assumptions can be misleading, and may result in a longer design process and/or overly expensive design solutions. By properly considering the impreciseness up front in the design process, a good initial design solution can be achieved with provable degree of acceptance. Such a design solution can be used effectively in the iterative process of design refinement, and thus, the number of redesign cycles can be reduced.

Random variables with probability distributions may be used to model such uncertainty. Nevertheless, collecting the probability data is sometimes difficult and time consuming. Furthermore, some imprecise information may not be correctly captured by the probabilistic model. For example, certain inconspicuousness in the design goal/constraint specification, such as the willingness of the user to accept certain designs or the confidence of the engineer towards certain designs, cannot be described by probabilistic distribution.

Many researchers have applied the fuzzy logic approach to various kinds of scheduling problem. In compiler optimization, fuzzy set theory has been used to represent unpredictable real-time events and imprecise knowledge about variables [16]. Lee et.al. applied the fuzzy inference technique to find a feasible real-time schedule where each task satisfies its deadline under resource constraints [20]. In production management area, fuzzy rules were applied to job shop and shop floor scheduling [24, 28]. Kaviani and Vranesic used fuzzy rules to determine the appropriate number of processors for a given set of tasks and deadlines for real-time systems [19]. Soma et.al. considered the schedule optimization based on fuzzy inference engine [27]. These approaches, however, do not take into account the fact that an execution delay of each job can be imprecise and/or multiple attributes of a schedule.

Many research results are available for design space exploration [1, 8, 13, 23]. All of these works differ in the techniques used to generate a design solution as well as the solution justification. These works, however, do not consider the impreciseness in the system attributes such as latency constraints and the execution time of a functional unit.

Recently, Karkowski and Otten introduced a model to handle the imprecise propagation delay of events [17, 18]. In their approach, the fuzzy set theory was employed to model imprecise computation time. Their approach applies possibilistic programming based on the integer linear programming (ILP) formulation to simultaneously schedule and select a functional unit allocation under fuzzy area and time constraints. Nevertheless, the complexity of solving the ILP problem with fuzzy constraints and coefficients can be very high. Furthermore, they do not consider multiple degrees in acceptability of design solutions. Several papers were published on the resource estimation [9, 25, 26]. These approaches, however, neither consider multiple design attributes nor impreciseness in system characteristics.

Many research works related to register allocation exists in high-level synthesis and compiler optimization area for VLIW architecture. For example, Chen et. al. proposed a loop scheduling for timing and memory operation optimization under register constraint [14]. The technique is based on multi-dimensional retiming. Eichenberger et. al. presented an approach for register allocation for VLIW and superscalar code via stage scheduling [11, 12]. Dani et. al. also presented a heuristic which uses stage scheduling to minimize register requirement. They also target at instruction level scheduling [10]. Zalamea et. al. presented hardware and software approach to minimize the register's usage targeting VLIW architecture [21, 22, 30]. On the software side, they proposed an extended version of modulo scheduling which considers register constraint, and register spilling. However, these work focus on loop scheduling and do not consider handling the imprecise system characteristics or specification.

In [2], the inclusion scheduling which takes the imprecise system characteristic was proposed. The algorithm was expanded and used in design exploration under imprecise system requirement as well as the estimation of resource bounds [4, 5, 7]. However, it does not take register criteria in creating a schedule.

In this paper, we particularly consider both imprecise latency and register constraints. We develop a design exploration framework under imprecise specification and constraints. The framework is iterative and based on the developed scheduling core, RCIS, *Register-Constrained Inclusion Scheduling* that takes imprecise information into account. Experimental results show that we can achieve an accpetable design solution with minmized number of registers.

This paper is organized as follows: Section 2 describes our models. It also presents some backgrounds in fuzzy set. Section 3 presents the iterative design framework. Section 4 presents the scheduling core (RCIS) used in the design exploration framework. It also addresses some issues when the register count is calculated during scheduling. Section 5 displays some experimental results. Finally, Section 6 draws a conclusion from our work.

2   Overview and Models

Operations and their dependencies in an application are modeled by a vertex-weighted directed acyclic graph, called a *Data Flow Graph*, $G = (\mathcal{V}, \mathcal{E}, \beta)$, where each vertex in the vertex set $\mathcal{V}$ corresponds to an operation and $\mathcal{E}$ is the set of edges representing data flow between two vertices. Function β defines the type of operation for node $v \in \mathcal{V}$.

Operations in a data flow graph can be mapped to different functional units which in turn can have varying characteristics. Such a system must also satisfy certain design constraints, for instance, power and cost limitations. These specifications are characterized by a tuple $S = (\mathcal{F}, \mathcal{A}, \mathcal{M}, \mathcal{Q})$, where $\mathcal{F}$ is the set of functional unit types available in the system, e.g., {add, mul}. $\mathcal{A}$ is $\{A_f : \forall f \in \mathcal{F}\}$. Each A_f is a set of tuples $(a_1, \ldots, a_k)$, where a_1 to a_k represent attributes of particular f. In this paper, we use only latency as an example attribute. (Note that our approach is readily applicable to include other constraints such as power and area). Hence, $A_f = \{x : \forall x\}$ where x refers to the latency attribute of f. $\mathcal{M}$ is $\{\mu_f : \forall f \in \mathcal{F}\}$ where μ_f is a mapping from A_f to a set of real number in [0,1], representing a possible degree of using the value. Finally, $\mathcal{Q}$ is a function that defines the degree of a system being acceptable for different system attributes. If $\mathcal{Q}(a_1, \ldots, a_k) = 0$ the corresponding design is totally unacceptable while $\mathcal{Q}(a_1, \ldots, a_k) = 1$, the corresponding design is definitely acceptable.

Using a function $\mathcal{Q}$ to define the acceptability of a system is a very powerful model. It can not only define certain constraints but also express certain design goals. For example, one is interested in designing a system with latency under 500 and register count being less than 6 respectively. Also, the smaller latency and register count, the better a system is. The best system would have both latency and register count being less than or equal to 100 and 1 respectively. An acceptability function, $\mathcal{Q}(a_1, a_2)$ for such a specification is formally defined as:

$$\mathcal{Q}(a_1, a_2) = \begin{cases} 0 & \text{if } a_1 > 500 \text{ or } a_2 > 6 \\ 1 & \text{if } a_1 \leq 100 \text{ and } a_2 \leq 1 \\ F(a_1, a_2) & \text{otherwise,} \end{cases} \quad (1)$$

where F is assumed to be linear functions, e.g., $F(a_1, a_2) = 1.249689(a_1 + 2a_2) - 0.001242$ which returns the acceptability between $(0, 1)$. Figures 1(a) and 1(b) illustrates Equation (1) graphically.

Based on the above model, the design solution we would like to find is formulated as following.

Given a specification containing $S = (\mathcal{F}, \mathcal{A}, \mathcal{M}, \mathcal{Q})$, $G = (\mathcal{V}, \mathcal{E}, \beta)$, *and acceptability level* α, *find a design solution whose the acceptability degree is greater than or equal to* α *subject* $\mathcal{Q}$.

Fuzzy sets, proposed by Zadeh, represent a set with imprecise boundary [29]. A fuzzy set x is defined by assigning each element in a universe of discourse its member-

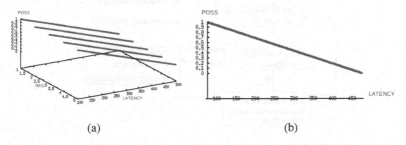

(a)   (b)

Fig. 1. (a) Imprecise constraint Latency : Register $= 1 : 2$ (b) Its projection

ship degree $\mu(x)$ in the unit interval $[0, 1]$, conveying to what degree x is a member in the set. Let A and B be fuzzy numbers with membership functions $\mu_A(x)$ and $\mu_B(y)$, respectively. Let $*$ be a set of binary operations $\{+, -, \times, \div, \min, \max\}$. The arithmetic operations between two fuzzy numbers, defined on $A * B$ with membership function $\mu_{A*B}(z)$, can use the extension principle, by [15]: $\mu_{A*B}(z) = \bigvee_{z=x*y}(\mu_A(x) \wedge \mu_B(y))$ where $\vee$ and $\wedge$ denote max and min operations respectively.

Based on the basic fuzzy set concept, we model the relationship between functional units and possible characteristics such that each functional unit is associated with a fuzzy set of characteristics. Given a functional unit f and its possible characteristic set A_f let $\mu_f(a) \in [0, 1], \forall a \in A_f$, describe a possibility of having attribute a for a functional unit f.

3   Iterative Design Framework

Figure 2 presents an overview of our iterative design process for finding a satisfactory solution. One may estimate the initial design configuration with any heuristic for example using ALAP, and/or ASAP scheduling [7]. The RCIS scheduling and allocation process produces the imprecise schedule attributes which are used to determine whether or not the design configuration is acceptable.

RCIS is a scheduling and allocation process which incorporates varying information of each operation. It takes an application modeled by a directed acyclic graph as well as the number of functional units that can be used to compute this application. Then, the schedule of the application is derived. This schedule shows an execution order of operations in the application based on the available functional units. The total attributes of the application can be derived after the schedule is computed. The given acceptability function is then checked with the derived attributes of the schedule.

In order to determine whether or not the resource configuration is satisfied the objective function, we use the *acceptability threshold*. If the schedule attributes lead to the acceptability level being greater than the threshold, the process stops. Otherwise, the resource configuration is adjusted using a heuristic and this process is repeated until the design solution cannot be improved or the design solution is found.

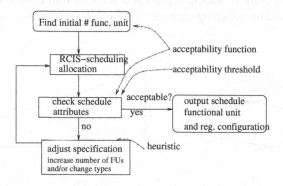

Fig. 2. Design solution finding process using RCIS

4   Register-Constraint Inclusion Scheduling

In this section, we present the register-constraint inclusion scheduling (RCIS) algorithm.The algorithm is based on the *inclusion scheduling* core presented in Algorithm 1. The algorithm evaluates the quality of the schedule by considering imprecise register criteria which will be discussed later subsections

Specifically, inclusion scheduling is a scheduling method which takes into consideration of fuzzy characteristics which in this case is fuzzy set of varying latency values associated with each functional unit. The output schedule, in turn, also consists of fuzzy attributes. In a nutshell, inclusion scheduling simply replaces the computation of accumulated execution times in a traditional scheduling algorithm by the fuzzy arithmetic-based computation. Hence, fuzzy arithmetics is used to compute possible latency from the given functional specification. Then, using a fuzzy scheme, latency of different schedules are compared to select a functional unit for scheduling an operation. Though the concept is simple, the results are very informative. They can be used in many ways such as module selection [3]. Algorithm 1 presents a list-based inclusion scheduling framework.

Algorithm 1 (Register-Constrained Inclusion scheduling)
Input: $G = (\mathcal{V}, \mathcal{E}, \beta)$, $Spec = (F, \mathcal{A}, \mathcal{M}, \mathcal{Q})$, and $N =$ #FUs
Output: A schedule S, with imprecise latency

1	$Q =$ vertices in G with no incoming edges // finding root nodes
2	**while** $Q \neq$ **empty do**
3	$Q = prioritized(Q)$
4	$u = dequeue(Q)$; mark u scheduled
5	good_S = **NULL**;
6	**foreach** $f \in \{f_j :$ where f_j is able to perform $\beta(u), 1 \leq j \leq N\}$ **do**
7	temp_S = $assign\_heuristic(S, u, f)$ // assign u at FU f
8	**if** $Eval\_Schedule\_with\_Reg($good_S, temp_S, $G, Spec)$
9	**then** good_S = temp_S **fi od**
10	$S =$ good_S // keep good schedule
11	**foreach** $v : (u, v) \in E$ **do**
12	$indegree(v) = indegree(v) - 1$
13	**if** $indegree(v) = 0$ **then** $enqueue(Q, v)$ **fi od**
14	**od**
15	**return**(S)

After node u is assigned to f, the imprecise attributes of the intermediate schedule, is computed. *Eval_Schedule_with_Reg* compares the current schedule with the "best" one found in previous iterations. The better one of the two is then chosen and the process is repeated for all nodes in the graph.

In Algorithm 1, fuzzy arithmetic simply takes place in routine *Eval_Schedule_with_Reg* (Line 8). In this routine, we also consider the register count used in the schedule. Since an execution time of a node is imprecise, the life time of a node is imprecise. Traditionally, a life time of a node depends on the location of the node's successors in the schedule. That is the value produced by the node must be

held until its successors have consumed it. For simplicity, let the successors consume
the value right after they start.

Recall that a node's execution time is a fuzzy set, where the membership function
is defined by $\mu(x) = y$. It implies that the node will take x time units with possibility
y. Consequently, a start time and finished time of a node are fuzzy numbers. To be
able to calculate fuzzy start time and finished time, we must assume that all nodes
have been assigned to functional units already. We assume that resource binding and
order of nodes executing in these resources are given based on the modified DFG (i.e,
scheduled DFG). The modified DFG is just the original DFG where extra edges due to
independent nodes executing in the same functional units are inserted (as constructed
in Algorithm 3).

4.1   Imprecise Timing Attributes

In the following, we present basic terminologies used in the algorithm which calculates
register usage under impreciseness.

Definition 1. *For $G = (\mathcal{V}, \mathcal{E}, \beta)$, and a given schedule, a fuzzy start time of node
$u \in V$, $FST(u)$ is a fuzzy set whose membership degree is defined by $\mu_{FST(u)}(x) = y$,
i.e, node u may start at time step x with possibility y.*

For nodes that are executed at time step 0 in each functional unit, $FST(u) = 0$,
which is a crisp value.

Definition 2. *For $G = (\mathcal{V}, \mathcal{E}, \beta)$, and a given schedule, a fuzzy finished time of node
$u \in V$, $FFT(u)$ is a fuzzy set whose membership degree is defined by $\mu_{FFT(u)}(x) =
y$, i.e, node u may finish at time step x with possibility y.*

Hence, $FFT(v) = FST(v) + EXEC(v)$, where $EXEC(v)$ is the fuzzy latency
of v. When considering earliest start time of a node, $FST(v) = \max_i(FFT(u_i)) + 1$,
$\forall u_i \to v$.

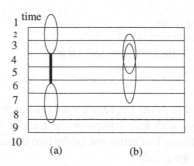

Fig. 3. A view of fuzzy start time and finished time

The general idea of using fuzzy numbers is depicted in Figure 3 for both start time
and finished time. Circles denote the fuzzy boundary which means that the start time

and finished time boundary of a node is unclear. Indeed, they may also be overlapped as shown in Figure 3(b). When a node occupies a resource at a certain time step, a possibility value is associated with the assignment.

Traditionally, when the timing attribute is a crisp value, the start time x and finished time y of a node form an integer interval $[x...y]$, which will be used to compute the register usage in the schedule. In our case, a fuzzy life time for node u contains two fuzzy sets: $FST(u)$ and $MFFT(u)$, the maximum of start time of all its successors.

Definition 3. *For* $G = (\mathcal{V}, \mathcal{E}, \beta)$, *and a given schedule, fuzzy life time of node* u, $FLT(u)$ *is a pair of* $[FST(u), MFFT(u)]$, *where* $\mu_{MFFT(u)} = FFT(u) \underset{\sim}{+}$ $\max(FST(v_i))$, *where* $u \rightarrow v_i \in \mathcal{E}$ *and* $\underset{\sim}{+}$, $\underset{\sim}{\max}$ *are fuzzy addition and fuzzy maximum respectively.*

Given $FLT(u)$, let min_st be the minimum time step from $FST(u)$ whose $\mu_{FST(u)}$ is nonzero, and max_st be the maximum time step from $FST(u)$ whose $\mu_{FST(u)}$ is nonzero. Similarly, let min_fin be the minimum time step from $MFFT(u)$ whose $\mu_{MFFT(u)}$ is nonzero, and let max_fin be the maximum time step from $MFFT(u)$ whose $\mu_{MFFT(u)}$ is nonzero. Without loss of generality, assume that $FST(u)$ and $MFFT(u)$ are sorted in the increasing order of the time step. We create a fuzzy set $IFST(u)$, mapping for a discrete time domain $[min\_st...max\_st]$ to a real value in $[0..1]$, showing the possibility that at time step x, node u will occupy a register for $FST(u)$ and likewise for $IMFFT(u)$ for $MFFT(u)$ as in Definitions 4–5.

Definition 4. $G = (\mathcal{V}, \mathcal{E}, \beta)$, *a given schedule,* $[min\_st...max\_st]$ *and* $FLT(u)$

$$\mu_{IFST(u)}(c) = \begin{cases} 0 & \text{if } c < min\_st \text{ or } c > max\_st \\ \max_{\forall x, min\_st \le x < y}(\mu_{FST(u)}(x)) & \text{otherwise} \\ \quad y = max(FST(u)) \text{ and } y < c \end{cases}$$

Definition 5. $G = (\mathcal{V}, \mathcal{E}, \beta)$, *a given schedule,* $[min\_fin...max\_fin]$ *and* $FLT(u)$

$$\mu_{IMFFT(u)}(c) = \begin{cases} 0 & \text{if } c < min\_fin \text{ or} \\ & c > max\_fin \\ \max_{\forall x, y < x \le max\_fin}(\mu_{MFFT(u)}(x)) & \text{otherwise} \\ \quad y = max(MFFT(u)) \text{ and } y < c \end{cases}$$

From the above calculation, we assume that for any two starting time value $a, b \in FST(u)$ where $a < b$, if node u starts at time a, it will be already started at time b. For $MFFT(u)$, when $a < b$, $a, b \in MFFT(u)$, if the value for node u will not be needed at time a, it will not be needed at time b and vice versa. Thus, Definitions 4–5 give the following properties.

Property 1. The possibility of $IFST(u)$ is in nondecreasing order.

Property 2. The possibility of $IMFFT(u)$ is in non-increasing order.

From $IFST(u)$ and $IMFFT(u)$, we merge the two sets to create a fuzzy interval for a node by defining Definition 6.

Definition 6. $G = (\mathcal{V}, \mathcal{E}, \beta)$, *a given schedule*, $IFST(u)$ *and* $IMFFT(u)$.

$$\mu_{IFLT(u)}(c) = \begin{cases} 0 & \text{if } c < \min(min\_st, min\_fin) \text{ or} \\ & c > \max(max\_st, max\_fin) \\ \max(\mu_{IFST(u)}(c), \mu_{IMFFT(u)}(c)) & \text{if } min\_st \le c \le max\_st \\ & \text{or } min\_fin \le c \le max\_fin \\ 1 & \text{otherwise} \end{cases}$$

After we compute the fuzzy life time interval for each node, we can start compute register usage for each time step.

4.2   Register Usage Calculation

Once a scheduled DFG is created, $FST(u)$ and $FFT(u)$ must be calculated for all $u \in V$. Figure 4 displays the meaning of fuzzy life time implied by Definitions 4–5. SA and FA denote the fuzzy start time and the fuzzy finished time of node A respectively. Similarly, SB and FB denote the start time and the fuzzy finished time of node B The fuzzy life times of A and B are shown in the filled boxes on the right side.

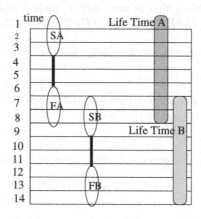

Fig. 4. Relationship between scheduled nodes and life time

In the figure, the life time of A and the life time of B may overlap. Traditionally, when the timing attribute is precise, the overlapped interval implies that two registers are needed during these time steps. In particular, during time steps 7 and 8, two registers are needed.

When an execution time becomes a fuzzy number, each box still implies that one register is needed. However, the derived possibility associated with a time step indicates that that node may not actually exist during the time step. For example, node may start later or finished earlier. In other words, there is a possibility that a node may not use such a register. With this knowledge, the register may be shared with others with high possibility. Consider the overlap interval in Figure 4 at time step 7. One or two registers may be used with some possibility. This depends on whether the dependency between $A \rightarrow B$ exists. If edge $A \rightarrow B$ exists in the original data flow graph, the total register count would be one. Notice that in this case, the intersection of $IFLT(A)$ and $IFLT(B)$ is not empty. On the contrary, if A and B are independent, the total register count would be two although the intersection may not be empty as well. This issue must be considered in calculating register usages.

4.3   Algorithms

Algorithm 2 presents a framework in evaluating fuzzy latency and register counts of a schedule. This algorithm is called after Line 7 in Algorithm 1 which already assigns the start time for each node.

In Algorithm 2, Line 6 invokes Algorithm 3 to calculate the life time of all nodes in schedule and find the maximum register usage. The register usage is then kept in $Reg[S_i]$ for a schedule S_i. Next, the latency of the whole schedule is then calculated. Note that after invoking Algorithm 3, necessary timing attributes for all nodes in G_0 can be obtained. The latency of the schedule is obtained by just fuzzy maximizing the finished time of all leaves in G_0. Line 9 merges latency and register usage attributes of the schedule using some heuristic function. The combined attribute is denoted as a *quality* of the schedule. This quality is then compared in Line 13 to select the best one.

Algorithm 2 (Eval_Schedule_with_Reg)
Input: schedules S_1, S_2, $G = (\mathcal{V}, \mathcal{E}, \beta)$, and $Spec = (F, \mathcal{A}, \mathcal{M}, \mathcal{Q})$
Output: 1 if S_1 is better than S_2, 0 otherwise.

1    $G_0 = (\mathcal{V}_0, \mathcal{E}_0, \beta)$ where $\mathcal{V}_0 = \mathcal{V} - \{\text{unscheduled nodes}\}$, $\mathcal{E}_0 = \emptyset$
2   **foreach** schedule $S_i = S_1$ **to** S_2 **do**
3        $\mathcal{E}_0 = \{(u, v) : u, v \in \mathcal{V}_0$, if u, v in same f.u. in S_i
4   and v is immediately after $u\}$
6   Calculate register usage for G_0 using Algorithm 3
7   Let W is a set of leaves in G_0
8        $latency[S_i] = fuzzymax\_time(W)$
9        $quality[S_i] = Combine(latency[S_i], Reg[S_i])$
10   **od**
12   // comparing the overall attributes of both schedules
13   **return**($compare(quality[S_1], quality[S_2])$)

Algorithm 3 (Calculate_Register_Count)
Input: Scheduled Graph G_0 for schedule S and, original DFG $G = (\mathcal{V}, \mathcal{E}, \beta)$ $Spec = (F, \mathcal{A}, \mathcal{M}, \mathcal{Q})$
Output: $Reg[S]$ contains register counts needed and its possibility

1   Calculate $FLT(u) \ \forall u \in G_0$ by Definition 3
2   Calculate $IFLT(u) \ \forall u \in G_0$ by Definitions 4–5
3   Let max_cs be max. finished time , $\forall u \in G_0$
4   **for** $cs = 1$ **to** max_cs **do**
5        $(RegAt[cs].reg, RegAt[cs].poss) = Count\_Node(IFLT, cs, G_0)$ **od**

```
 6  ∀n, FReg[n] = 0
 7  for cs = 1 to max_cs do
 8      FReg[RegAt[cs].reg].reg = RegAt[cs].reg
 9      FReg[RegAt[cs].reg].poss =
10      max(FReg[RegAt[cs].reg].poss, RegAt[cs].poss) od
12  Reg[S] = FReg
```

In Algorithm 3, $RegAt$ stores maximum number of registers needed at each cs and its associated possibility. The values are obtained by Algorithm $Count\_Node$. Lines 7–10 summarize the overall number of registers needed and its possibility. Algorithm $Count\_Node$ is described in Algorithm 4.

Algorithm 4 ($Count\_Node$)

Input: $IFLT$, G_0, cs
Output: # registers needed and its possibility at cs

```
 1  node_set = {nodes occupy reg at cs}
 2  set G_0 in topological order
 3  Let sorted_node be node_set sorted in by sorted G_0
 4  poss = 0, reg = 0
 5  ∀i ∈ sorted_node, i.ok = FALSE, i.count = FALSE
 6  for every i ∈ sorted_node do
 7      for j = i + 1 to last node in sorted_node do
 8          if i.ok = TRUE and i.count = FALSE
 9          then
10              reg + +; poss = max(poss, μ_{IFLT(i)}(cs))
11              i.count = TRUE fi
13          if FindPath(i, j)
14          then j.ok = FALSE // don't count descendant fi
15      od
16      Let j be the last node in sorted_node
17      if j.ok = TRUE
18      then
19          reg + +; poss = max(poss, μ_{IFLT(j)}(cs))
20          j.count = TRUE fi
21  od
22  return (reg, poss)
```

In Algorithm 4, our heuristic only attempts to consider the ancestor at the current time step. In other words, we assume that the ancestor finishes first and then its descendants can start. Flag ok uses to indicate that the associated node should be counted at the current step or not. If it is a descendant of any of nodes in the current step, the flag will be disable. Since the schedule contains every node, the descendant will be started eventually. reg and $poss$ store the current number of counted nodes and maximum possibility. At Line 3, the nodes currently in this time step indicated by $IFLT$ are sorted in the topological order according to G_0. Then we extract each node in the sorted list to check if any pair are dependent by using $FindPath$ in Line 13. In the loop, it selectively marks descendant nodes in the current step.

Let us consider the complexity of Algorithm 4. The time complexity is dominated by Lines 6–21, which is $O(|V|^2(|V|+|E|))$. Since for DAG, *FindPath* takes $O(|V|+|E|)$.

In Algorithm 3, the calculation for $FLT(u)$ depends on $FST(u)$ and $MFFT(u)$. Let N_1 be the number of discrete points in $FST(u)$ and $MFFT(u)$. Lines 1–2 perform the calculation whose upper bound is of $O(N_1|V||E|)$. The computation for $IFLT(u)$ is simply a double loop for each node. In overall, Algorithm 3 runs in polynomial time.

5   Experimental Results

We present experimental results on the voltera filter benchmark [7], containing 27 nodes, where 10 nodes require adder units and the rest requires multiplier units. Assume that we have two types of functional units: adder and multiplier. whose latencies are as shown according to Table 1. In the figure, an adder may have different latency values with the given possibility. Columns "lat" and "pos" show the latency and its possibility of having the latency value for each adder and multiplier. Thus, if the nodes are executed in the functional unit, the node may have variable latency values as well.

Table 1. Adder and multiplier characteristics

FUs	(lat,poss)		(lat,poss)		(lat,poss)		(lat,poss)	
	lat	poss	lat	poss	lat	poss	lat	poss
adder	5	0.05	10	1	15	0.9	23	0.1
multiplier	7	0.5	12	0.7	17	1	29	0.05

Assume the constraint is depicted in Figure 5 where the register axis is [1..7] and the latency axis is [200..700]. We demonstrate by considering various design configuration of varying the number of functional units using RCIS and original inclusion scheduling as a scheduling core in the design exploration. Due to the characteristics of the filter, increasing the number of multipliers will help reduce the overall latency. Suppose that we set the acceptability threshold to be 0.8. The results are shown in Table 2. In the table, In particular, Columns "RCIS"and "IS" compare the performance of the schedule by RCIS and the original inclusion scheduling (IS) for each functional unit configuration. Row "Avg Latency" shows the weighted sum of latency for each case. Row "Max Reg" displays the maximum number of registers. Row "Acceptability" shows the acceptability value obtained using the "Avg latency" and "Max Reg". Row "Max Latency" presents the maximum latency values for each case. For RCIS, recall that $w_1 = 1$ and $w_2 = 10$. That is we consider register criteria ten times as much as the latency value. RCIS attempts to create a schedule which minimizes the total weighted sum of $w_1x + w_2y$ where x and y are the weighted latency and weighted register counts of the resulting schedule. Figure 6 we depict acceptability values for each design configuration based on RCIS. When we increase the number of functional units the latency decreases while the number of register counts needed increases. However, when the number of multipliers becomes 4 or more, RCIS can create a schedule which gives the maximum acceptability values 0.84 (which is greater than the threshold defined at 0.8).

By inspecting the resulting schedule, we conclude that 4 multipliers would be sufficient and adding more multipliers will be wasteful. Compared this the schedule generated by IS, we found that since IS does not consider the register criteria, IS attempts to utilize all available resources to minimize the overall latency values. Thus, the latency of schedule generated by IS keeps decreasing and the number of register counts keep increasing. This will finally decrease the acceptability value according the constraint.

From the results, we can see that to achieve the acceptability threshold 0.8, using RCIS will give a better design solution using fewer number of registers. Consider the running time. For all the cases, the maximum running time is approximately 2.8 seconds to achieve the results for 1 adder and 5 multipliers under Pentium 4 2.8GHz, 1GB RAM.

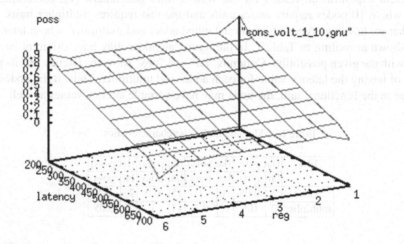

Fig. 5. Constraint for Voltera filer

Table 2. Exploring various number of functional units using RCIS and IS

	1 add 2 muls		1 add 3 muls		1 add 4 muls		1 add 5 muls	
	RCIS	IS	RCIS	IS	RCIS	IS	RCIS	IS
Avg Latency	308	300	267	270	260	260	260	246
Max Reg	2	2	3	3	**4**	4	4	5
Acceptability	0.78	0.80	0.84	0.84	**0.84**	0.84	0.84	0.84
Max Latency	561	561	474	477	445	445	445	416

6   Conclusion

We propose a design exploration framework considering impreciseness. The framework is based on the scheduling core, RCIS which considers impreciseness in the system specification,and constraint and attempts to create a schedule which minimizes both latency and register usages. The framework can be used to generate various design solutions under imprecise system constraints and characteristics and select an acceptable solution under latency and register criteria. The experiments demonstrate the usage of the framework on a well-known benchmark, where the selected design solution can be found with a given acceptability level.

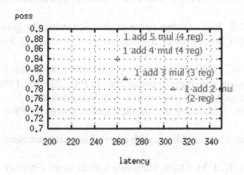

Fig. 6. Acceptability values for each configuration

References

1. I. Ahmad, M. K. Dhodhi, and C.Y.R. Chen. Integrated scheduling, allocation and module selection for design-space exploration in high-level synthesis. *IEEE Proc.-Comput. Digit. Tech.*, 142:65–71, January 1995.
2. C. Chantrapornchai, E. H. Sha, and X. S. Hu. Efficient scheduling for imprecise timing based on fuzzy theory. In *Proc.Midwest Symposium on Circuits and Systems*, pages 272–275, 1998.
3. C. Chantrapornchai, E. II. Sha, and X. S. Hu. Efficient algorithms for finding highly acceptable designs based on module-utility selections. In *Proceedings of the Great Lake Symposium on VLSI*, pages 128–131, 1999.
4. C. Chantrapornchai, E. H-M. Sha, and X. S. Hu. Efficient module selections for finding highly acceptable designs based on inclusion scheduling. *J. of System Architecture*, 11(4):1047–1071, 2000.
5. C. Chantrapornchai, E. H-M. Sha, and Xiaobo S. Hu. Efficient acceptable design exploration based on module utility selection. *IEEE Trans. on Computer Aided Design of Integrated Circuits and Systems*, 19:19–29, Jan. 2000.
6. C. Chantrapornchai, W. Surakumpolthorn, and E.H. Sha. Efficient scheduling for design exploration with imprecise latency and register constraints. In *Lecture Notes in Computer Science: 2004 International Conference on Embedded and Ubiquitous Computing*, 2004.
7. C. Chantrapornchai and S. Tongsima. Resource estimation algorithm under impreciseness using inclusion scheduling. *Intl. J. on Foundation of Computer Science, Special Issue in Scheduling*, 12(5):581–598, 2001.
8. S. Chaudhuri, S. A. Bylthe, and R. A Walker. An exact methodology for scheduling in 3D design space. In *Proceedings of the 1995 International Symposium on System Level Synthesis*, pages 78–83, 1995.
9. S. Chaudhuri and R. Walker. Computing lower bounds on functional units before scheduling. In *Proceedings of the International Symposium on System Level Synthesis*, pages 36–41, 1994.

10. A. Dani, V. Ramanan, and R. Govindarajan. Register-sensitive software pipelining. In *Proceedings. of the Merged 12th International Parallel Processing and 9th International Symposium on Parallel and Distributed Systems*, pages 194–198, April 1998.
11. A. Eichenberger and E. S. Davidson. Register allocation for predicated code. In *Proceeding of MICRO*, 1995.
12. Alexandre E. Eichenberger and Edward S. Davidson. Stage scheduling: A technique to reduce the register requirements of a modulo schedule. In *Proceedings of MICRO-28*, pages 338–349, 1995.
13. H. Esbensen and E. S. Kuh. Design space exploration using the genetic algorithm. In *Proceedings of the 1996 International Symposium on Circuits and Systems*, pages 500–503, 1996.
14. F.Chen, S. Tongsima, and E. H. Sha. Loop scheduling algorithm for timing and memory operation minimization with register constraint. In *Proc. SiP'98*, 1998.
15. K. Gupta. *Introduction to fuzzy arithmetics*. Van Nostrand, 1985.
16. O. Hammami. Fuzzy scheduling in compiler optimizations. In *Proceedings of the ISUMA-NAFIPS*, 1995.
17. I. Karkowski. Architectural synthesis with possibilistic programming. In *HICSS-28*, January 95.
18. I. Karkowski and R. H. J. M. Otten. Retiming synchronous circuitry with imprecise delays. In *Proceedings of the 32nd Design Automation Conference*, pages 322–326, San Francisco, CA, 1995.
19. A. S. Kaviani and Z. G. Vranesic. On scheduling in multiprocess systems using fuzzy logic. In *Proceedings of the International Symposium on Multiple-valued Logic*, pages 141–147, 1994.
20. J. Lee, A. Tiao, and J. Yen. A fuzzy rule-based approach to real-time scheduling. In *Proc. Intl. Conf. FUZZ-94*, volume 2, 1994.
21. Josep Llosa, Eduard Ayguade, Antonio Gonzalez, Mateo Valero, and Jason Eckhardt. Lifetime-sensitive modulo scheduling in a production environment. *IEEE Transactions on Computers*, 50(3):234–249, 2001.
22. Josep Llosa, Mateo Valero, and Eduard Ayguade. Heuristics for register-constrained software pipelining. In *International Symposium on Microarchitecture*, pages 250–261, 1996.
23. C. A. Mandal, P. O. Chakrabarti, and S. Ghose. Design space exploration for data path synthesis. In *Proceedings of the10th International Conference on VLSI Design*, pages 166–170, 1996.
24. K. Mertins et al. Set-up scheduling by fuzzy logic. In *Proceedings of the International conference on computer integrated manufacturing and automation technology*, pages 345–350, 1994.
25. J. Rabaey and M. Potkonjak. Estimating implementation bounds for real time DSP application specific circuits. *IEEE Transactions on Computer-Aided Design of integrated circuits and systems*, 13(6), June 1994.
26. A. Sharma and R. Jain. Estimating architectural resources and performance for high-level synthesis applications. *IEEE Transactions on VLSI systems*, 1(2):175–190, June 1993.
27. H. Soma, M. Hori, and T. Sogou. Schedule optimization using fuzzy inference. In *Proc. FUZZ-95*, pages 1171–1176, 1995.
28. I.B. Turksen et al. Fuzzy expert system shell for scheduling. *SPIE*, pages 308–319, 1993.
29. L. A. Zadeh. The concept of a linguistic variable and its application to approximate reasoning, Part I. *Information Science*, 8:199–249, 1975.
30. J. Zalamea, J. Llosa, E. Ayguade, and M. Valero. Software and hardware techniques to optimize register file utilization in vliw architectures. In *Proceedings of the International Workshop on Advanced Compiler Technology for High Performance and Embedded Systems (IWACT)*, July 2001.

Hiord: A Type-Free Higher-Order Logic Programming Language with Predicate Abstraction

Daniel Cabeza[1], Manuel Hermenegildo[1,2], and James Lipton[1,3]

[1] Technical University of Madrid, Spain
{dcabeza, herme, jlipton}@fi.upm.es
[2] University of New Mexico, USA
herme@unm.edu
[3] Wesleyan University, USA
jlipton@wesleyan.edu

Abstract. A new formalism, called H iord, for defining type-free higher-order logic programming languages with predicate abstraction is introduced. A model theory, based on partial combinatory algebras, is presented, with respect to which the formalism is shown sound. A programming language built on a subset of H iord, and its implementation are discussed. A new proposal for defining modules in this framework is considered, along with several examples.

1 Introduction

This paper presents a new declarative formalism, called Hiord, for logic programming with untyped higher-order logic and predicate abstractions. This is followed by a discussion of various practical restrictions of this logic to make it amenable to speedy translation to WAM-compilable code and static analysis.

A number of proposals have been made over the past two decades to introduce higher-order features into logic programming in a declarative fashion by extending the underlying logic, among them λProlog and Hilog [1–4]. This has proven a very useful way to place on a solid logical ground certain natural steps that, in the original first-order context of pure logic programming, seem to compromise declarative transparency. For example, the simple transformation of code such as the following:

```
all(Prop,[]).
all(Prop,[H|Tl]) :- call(Prop,H), all(Prop,Tl).
```

to

```
all(Prop,[]).
all(Prop,[H|Tl]) :- Prop(H), all(Prop,Tl).
```

or a typed version thereof, turns a Prolog meta-program into a fully declarative program in higher-order logic. This simple example tells only a small part

M.J. Maher (Ed.): ASIAN 2004, LNCS 3321, pp. 93–108, 2004.

of the story, of course, as there are other ways to give a declarative semantics to meta-predicates. However, in our view, higher-order logic with predicate abstraction is an excellent choice for bringing metaprogramming within the scope of declarative programming, especially when management of substitutions in the object language is involved, via higher-order abstract syntax. It is also a natural framework for robust declarative treatment of automated deduction, and code and data specification.

It is also our feeling that translation of explicit higher-order notions into a first-order formalism simply places the original specification at a greater distance from the program semantics, and hence, the programmer a greater distance from the aims of declarative programming.

The work discussed in this paper extends the *untyped* classical first-order Horn clauses of core Prolog to untyped classical Horn clauses in higher-order logic, with predicate abstractions allowed. The main rationale for keeping types out of the picture is compatibility with existing Prolog code and Prolog systems, with all their tools for static analysis and program development, and so as to implement higher-order programming as a package loadable from Prolog. We are proposing an extension to the syntax and semantics of core Prolog, not a compilation into it, as a basis for the syntax and semantics of the input code, (irrespective of whether or not the implementation actually does compile the code to Prolog in the end).

A second, important consideration is that we consider powerful applications, including a new proposal for declarative definition of modules, that make use of self-application and head flex variables that is *not* typable in simply typed lambda-calculus, and, in particular, not legal λProlog code. Many applications may, in fact, be typable in a sufficiently strong type discipline. Even so, our eventual interest is in compile-time type inference through static analysis of type-free code.

There is a type free higher order extension of Prolog, namely Hilog, which has been presented [4] with a proof theory and a semantics based on (the well-known) translation of higher-order logic into first-order logic. It lacks predicate abstraction, however, which for us is an essential feature of Hiord. Although our formalism is inspired by the Hilog work, the semantics requires significant reworking to permit abstraction.

The paper is divided into two parts. The first presents a strong formalism that allows higher-order resolution and term-rewriting with unrestricted abstraction of all terms and goals. The aim is to define a framework within which any number of practical restrictions can be studied. In this formalism we continue along the lines of the "anything goes" philosophy of *Hilog*. All terms can have a truth value, compound terms can be functors, a functor can have multiple arities. We define a model theory based on partial combinatory algebras [5] with certain semilattices serving as an object of truth-values, and show that our completely general resolution is sound.

The formalism defined is, in a sense, too strong to be a useful programming language. Since it contains the full untyped lambda-calculus, it permits untyped higher-order logical-functional programming. Indeed, one could virtually ignore the logic and simply program in the lambda-calculus (which, of course, is not our aim). For this reason we regard the Hiord formalism as more of a blueprint for defining restricted type-free higher order languages, and we have included a second section in which a restriction of the language is discussed along with an implementation, the *Hiord* package included in the latest release (1.11) of the Ciao system.

A serious concern, of course, is that by combining higher-order logic with a type-free function calculus with a fixed point operator one is coming dangerously close to inconsistency. Indeed if one adds unrestricted abstraction to the *full logic* of *Hilog* (as compared to the Horn Clause subset), one has an easy formalization of Curry's paradox, a simple variant of Russell's, by defining a predicate $p = \lambda x.\neg x(x)$. Then we have $p(p) = \neg p(p)$! More subtle paradoxes can be found even in the absence of negation (see e.g. [6, 7]). We steer away from these problems by staying within the Horn fragment of logic with SLD resolution, which is shown sound with respect to the model theory introduced in section 3.

2   The Syntax of Formal Hiord

We initially consider a language with a very liberal syntax, which incorporates the flexibility of Hilog by allowing arity-free functors, and not distinguishing between functors and relators.

A *language* for Hiord is a set S of non-logical parameters (which will contain all names for constants, function and relation symbols). It is also equipped with a set of variables V, as well as the logical parameters "," (comma), = (equality) and "E" (existence).

Hiord terms and formulas are defined by mutual recursion, as shown in the table below:

Terms:
1. A variable is a term.
2. A nonlogical parameter is a term.
3. If $t, t_1, \ldots, t_n$ are terms, then $t(t_1, \ldots, t_n)$ is a term, called a *simple* term if $t \in S \cup V$.
4. If G is a goal and x is a sequence of variables, $\{(x){:}\text{-}G\}$ is a term, known as an *abstraction*.

Atomic Formulas, Goals and clauses

1. $\top$ is an atomic formula, called *true*.
2. If $t, t_1, \ldots, t_n$ are terms, then $t(t_1, \ldots, t_n)$ is an atomic formula. If $t \in S$ the formula is called *rigid*, and if $t \in V$ *flex*.
3. If t_1 and t_2 are terms then $t_1 = t_2$ is an atomic formula.
4. An atomic formula is a goal. If G_1 and G_2 are goals, then the *conjunction* G_1, G_2 is a goal, and for any goal G and variable x, $E(x)G$ is a goal.
5. A clause is a formula of the form $H \leftarrow G$ where H is a rigid atomic formula with no occurrences of equality or abstractions, and G is a goal.

Definition 1. *A* Hiord *logic* **program** *is a finite set of clauses. A* **state** *is a pair* $\langle P|\mathfrak{g}\rangle$ *where P is a program, and $\mathfrak{g}$ is a sequence of goals. The empty sequence, denoted $\square$, is allowed. Goal sequences are defined by the following grammar:*

$$\mathfrak{g} ::= \square \mid G \mid \mathfrak{g} \otimes \mathfrak{g}$$

When the program is understood from context it may be omitted. When we write $\langle P, A \leftarrow Tl \mid \mathfrak{g}\rangle$, it is understood that we are distinguishing one of the clauses $A \leftarrow Tl$ of the program P. All program clauses are treated as tacitly closed by standardizing variables in clauses apart from any other variables appearing in a state or a deduction. For this reason, application of a unifying substitution θ to a state $\langle P|\mathfrak{g}\rangle$ results in the new state $\langle P|\mathfrak{g}\theta\rangle$.

Definition 2. *A* **substitution** θ *is a map from variables in $\mathcal{V}$ to terms over $\mathcal{V} \cup \mathcal{S}$. Such a map lifts to a unique map (also denoted θ) from terms or goals to terms or goals, defined as follows:*

$$t(t_1, \ldots, t_n)\theta = t\theta(t_1\theta, \ldots, t_n\theta)$$
$$\{(\boldsymbol{x}):\text{-}G\}\theta \quad = \{(\boldsymbol{y}):\text{-}G[\boldsymbol{x} := \boldsymbol{y}]\theta\}$$

where $\boldsymbol{y}$ is a sequence of variables (of the same length as $\boldsymbol{x}$) which are disjoint from the domain and range of θ, and $G[\boldsymbol{x} := \boldsymbol{y}]$ is the result of simultaneously replacing, in G, every free occurrence of each variable x_i in the sequence $\boldsymbol{x}$ with the corresponding variable y_i of $\boldsymbol{y}$. For nonatomic goals G_1, G_2 and $\mathrm{E}(x)G$ we define

$$(G_1, G_2)\theta = (G_1\theta, G_2\theta) \text{ and } (\mathrm{E}(x)G)\theta = (\mathrm{E}(y)G[x := y]\theta)$$

with the same conditions as above for x and y.

Substitutions lift to goal sequences in the obvious way. We now give resolution proof rules for Hiord logic programs.

Definition 3. *A* **resolution step** *for* Hiord *is a ternary relation $\rightsquigarrow$ on states $\times$ subs $\times$ states. We write $\mathfrak{s}_1 \stackrel{\theta}{\rightsquigarrow} \mathfrak{s}_2$ instead of $(\mathfrak{s}_1, \theta, \mathfrak{s}_2) \in \rightsquigarrow$.*

There are six kinds of resolution steps. In the rules defining them, below, equality may be taken as one of **strict** *equality (the terms must be identical without any reduction taking place),* **alpha**-*equivalence, or* **beta-eta** *equivalence.*

---

1. **Backchain:** $\langle P, A \leftarrow Tl|\mathfrak{g} \otimes A' \otimes \mathfrak{g}'\rangle \stackrel{\theta}{\rightsquigarrow} \langle P, A \leftarrow Tl|\mathfrak{g}^\theta\rangle \otimes Tl\theta \otimes \mathfrak{g}'\theta$ *where* $\theta A = \theta A'$.
2. **Unify:** $\langle P|\mathfrak{g} \otimes t_1 = t_2 \otimes \mathfrak{g}\rangle \stackrel{\theta}{\rightsquigarrow} \langle P|\mathfrak{g}\theta \otimes \mathfrak{g}'\theta\rangle$ *where* $t_1\theta = t_2\theta$.
3. **Reduce:** $\langle P|\mathfrak{g}\rangle \stackrel{\beta}{\rightsquigarrow} \langle P|\mathfrak{g}'\rangle$. *denoting any α or β reduction (or conversion, their congruence closure) of a term or subterm in a sequence of goals.*
4. **Conjunction:** $\langle P|\mathfrak{g} \otimes G_1, G_2 \otimes \mathfrak{g}'\rangle \stackrel{\otimes}{\rightsquigarrow} \langle P|\mathfrak{g} \otimes G_1 \otimes G_2 \otimes \mathfrak{g}'\rangle$
5. **Existence:** $\langle P|\mathfrak{g} \otimes \exists x.G \otimes \mathfrak{g}'\rangle \stackrel{\exists}{\rightsquigarrow} \langle P|\mathfrak{g} \otimes G \otimes \mathfrak{g}'\rangle$ *where the bound variable x is assumed distinct from any other variable occurring in $\mathfrak{g} \otimes G \otimes \mathfrak{g}'$.*
6. **True:** *The instantiation of head variable in a flex goal $X, X(t_1, \ldots, t_n)$ by $\top$ or $\lambda(x_1, \ldots, x_n).\top$ is a resolution step, $\langle P|\mathfrak{g}\rangle \stackrel{\theta}{\rightsquigarrow} \langle P|\mathfrak{g}\theta\rangle$, with substitution $\theta = [X := \top]$ or $[X := \lambda(x_1, \ldots, x_n).\top]$ as the case may be.*

To fix one formalism, we take the last option for equality, $\beta\eta$ equivalence, as the "official equality" of Hiord. This, of course, requires the implementation of potentially non-terminating higher-order unification. In practice, this is one of the areas where restrictions are of interest [8].

The symbols $\exists$, β and $\otimes$ written over the reduction symbol, a notational convenience, are just different names for the identity substitution. A **convention** we will adopt, except where otherwise indicated, in this paper, is that the unifying substitution displayed over a reduction arrow is understood to be restricted to the free variables in its source.

All bound variables in explicit existential quantifications and lambda-abstractions are assumed to be distinct from each other, all free variables present in each state, and all variables occurring in (the domain or range of) substitutions.

Finally, we define a resolution proof of a state $\langle P|\mathfrak{g}\rangle$ to be a sequence of resolution steps ending with the empty sequence $\square$ of goals.

3   Semantics

We will need the following algebraic notion of an object of truth values, which can be thought of as a limited Boolean or Heyting algebra.

Definition 4. *An LP-algebra* $\boldsymbol{\Omega} = (\Omega, \top, \leq, \wedge)$ *consists of a meet-semilattice* $(\Omega, \leq, \wedge)$ *with a top element* $\top$.

In the presence of a Hiord structure, defined below, LP-agebras will be required to have certain potentially infinite parametrized suprema.

Definition 5. *A* Hiord **Structure**

$$\mathfrak{A} = \langle U, \mathfrak{s}, \mathfrak{k}, \top, \wedge_{\mathfrak{A}}, \exists_{\mathfrak{A}}, \mathsf{eq}_{\mathfrak{A}}, \odot, p, \pi_l, \pi_r, \omega, \Omega \rangle$$

is given by the following data:

1. *A nonempty set U, called the carrier, domain or underlying set of $\mathfrak{A}$, and also denoted $|\mathfrak{A}|$, with $\top, \mathfrak{s}, \mathfrak{k}, \wedge_{\mathfrak{A}}, \exists_{\mathfrak{A}}, p, \pi_l, \pi_r \in U$.*
2. $\langle U, \mathfrak{s}, \mathfrak{k}, \odot \rangle$ *is a partial combinatory algebra with pairing operator p and projections π_l, π_r*
3. *An LP algebra $(\Omega, \top, \leq, \wedge)$ with all (finite and infinite) U-parametrized joins required in the following definition of ω.*
4. *A partial map $\omega : U \to \Omega$.*

This structure must also satisfy the following conditions (where juxtaposition denotes left-associative application $\odot$):

$$\pi_l(puv) = u \qquad \pi_r(puv) = v$$
$$\mathfrak{s}uvw = (uw)(vw)$$
$$\mathfrak{k}uv = u$$

$$\omega(\top) = \top_\Omega$$
$$\omega(\wedge_{\mathfrak{A}} uv) = \wedge_\Omega(\omega(u), \omega(v))$$
$$\omega(\exists_{\mathfrak{A}} u) = \bigvee \{\omega(ud) : d \in U \text{ and } ud \text{ defined in } U\}$$
$$\omega(\mathsf{eq}_{\mathfrak{A}} u_1 u_2 \cdots u_n) = \top_\Omega \text{ iff } u_1 = u_2 = \cdots = u_n$$

In a Hiord structure, the (possibly) infinite meets used in the condition for $\omega(\exists_{\mathfrak{A}} u)$ must exist. This condition is considerably weaker than requiring arbitrary suprema to exist, but the reader may take Ω to be complete lattice without significant loss of generality.

In the presence of a pairing operator and projections, we can assume the existence of the following derived notions of *n-tuples* and *n-ary projections* (where Parenthesized superscripts denote iteration):

$$\langle u \rangle := u$$
$$\langle u_1, \ldots, u_n \rangle := p u_1 \langle u_2, \ldots, u_n \rangle \qquad (n \geq 2)$$
$$\pi_k^n := \begin{cases} \pi_l & \text{if } k = 1 < n \\ \pi_{k-1}^{n-1} \pi_r & \text{if } 1 < k < n \\ \pi_r^{(n-1)} & \text{if } k = n \end{cases}$$

If $u \in \mathfrak{A}$, we write $(u)_i$ for π_i^n when n is clear from context.

With these definitions, we have $\pi_k^n \langle u_1, \ldots, u_n \rangle = u_k$ and if $u = \langle u_1, \ldots, u_n \rangle$ then $(u)_k = u_k$ in $\mathfrak{A}$.

Definition 6. *Let S be a set of parameters. Let $\mathfrak{A}$ be a* Hiord *structure. An* $\mathfrak{A}$-**assignment** *is a map*

$$I : S \to U,$$

and an $\mathfrak{A}$-**environment** *is a map*

$$\nu : \mathcal{V} \to U.$$

A structure $\mathfrak{A}$, an assignment I and an $\mathfrak{A}$-environment ν induce an **interpretation**, that is to say, a map $\nu[\![\_]\!]^{\mathfrak{A},I}$ (abbreviated to $\nu[\![\_]\!]$ when the remaining parameters are clear from context) from terms and formulas to the domain U of $\mathfrak{A}$ as follows:

$$\nu[\![X]\!] = \nu(X) \text{ defined, for } X \in \mathcal{V} \tag{1}$$
$$\nu[\![s]\!] = I(s) \text{ for } s \in S \tag{2}$$
$$\nu[\![t(t_1, \ldots, t_n)]\!] = \nu[\![t]\!] \langle \nu[\![t_1]\!], \ldots, \nu[\![t_n]\!] \rangle \tag{3}$$
$$\nu[\![t_1 = t_2]\!] = \mathsf{eq}_{\mathfrak{A}} \nu[\![t_1]\!] \nu[\![t_2]\!] \tag{4}$$
$$\nu[\![\{(\boldsymbol{x}):\text{-}G\}]\!] = [\mathbf{x}] \nu[\![G[x_i := (\mathbf{x})_i]_{1 \leq i \leq n}]\!] \quad \text{where } \boldsymbol{x} \equiv (x_1, \ldots, x_n) \tag{5}$$
$$\nu[\![G_1, G_2]\!] = \wedge_{\mathfrak{A}} \nu[\![G_1]\!] \nu[\![G_2]\!] \tag{6}$$
$$\nu[\![\exists x G]\!] = \exists_{\mathfrak{A}} [\mathbf{x}] \nu[\![G[x := \mathbf{x}]]\!] \tag{7}$$

and where each $\mathbf{x_i}$ is a fresh variable not in $\mathcal{V}$ (and hence not in the domain of ν) and $[\mathbf{x_i}]u$ denotes so-called *bracket abstraction* in the model, definable in any partial combinatory algebra, and described below. First, we briefly note that in the setting of a combinatory algebra, currying of terms is automatically enforced, since a sequence of applications, allowed in our syntax, has the semantics $\nu[\![(t_1 \cdots t_n)]\!] = \nu[\![t_1]\!] \cdots \nu[\![t_n]\!]$. However we incorporate the conventional syntax of core prolog by treating multiple arity arguments as vectors in clause 3 of the preceding definition.

3.1   Bracket Abstraction Over A

Syntactic translation of closed λ-terms to variable free combinatory logic is well-known, and has been used extensively in compilation of functional programming languages. Here we have to consider the additional wrinkle of doing it with respect to an ambient model, so we give the details.

In order to define *semantic* bracket abstraction and the interpretation of terms and formulas with bound variables rigorously, we will need to make use of several intermediate notions of term: those built up from elements of the carrier U of a Hiord structure and a fresh set variables, and the collection of Hiord terms and goals built up using two sets of variables. Let $\mathcal{W}$ be a fresh set of variables (i.e. disjoint from $\mathcal{S} \cup \mathcal{V}$), in one-to-one correspondence with $\mathcal{V}$, via the mapping $x \mapsto \mathbf{x}$.

Let $\mathfrak{A}$ be a Hiord structure with carrier U, and let $U[\mathcal{W}]$ be the set of terms freely built from U and $\mathcal{W}$ using application in $\mathfrak{A}$:

- if $\mathbf{x} \in \mathcal{W}$ then $\mathbf{x} \in U[\mathcal{W}]$,
- if $u, v \in U[\mathcal{W}]$ then $uv \in U[\mathcal{W}]$.

We extend $U[\mathcal{W}]$ to the set $\lambda U[\mathcal{W}]$ of bracket-abstracted terms as follows:

- if $\mathbf{x} \in \mathcal{W}$ then $\mathbf{x} \in \lambda U[\mathcal{W}]$,
- if $u, v \in \lambda U[\mathcal{W}]$ then $uv \in \lambda U[\mathcal{W}]$.
- if $u \in \lambda U[\mathcal{W}]$ and $\mathbf{x} \in \mathcal{W}$ then $[\mathbf{x}]u \in \lambda U[\mathcal{W}]$.

Let $\wp[\mathcal{W}]$ be the set of Hiord terms and goals containing occurrences (possibly bound) of variables in $\mathcal{V}$ and only free occurrences of variables in $\mathcal{W}$. Let $\nu : \mathcal{V} \to U$ be a $\mathfrak{A}$-environment. Extend ν to a function $\nu : \mathcal{V} \cup \mathcal{W} \to U[\mathcal{W}]$ by defining it to be the identity function on $\mathcal{W}$, and then to a function $\nu : \wp[\mathcal{W}] \to U[\mathcal{W}]$ in the usual way, i.e. according to the equations (1-7). Then, the result of applying ν to goals using these equations is a bracket abstraction, i.e. a term in $\lambda U[\mathcal{W}]$. These expressions denote members of U according to the following rules:

Definition 7. *We define* **bracket abstraction** *with respect to* $(\mathfrak{A}, \nu)$. *Terms in* $\lambda U[\mathcal{W}]$ *denote the following unique members of* U:

$$[\mathbf{x}]\mathbf{x} = \mathfrak{s}\mathfrak{k}\mathfrak{k} \tag{8}$$

$$[\mathbf{x}]u = \mathfrak{k}u \qquad \textit{if } \mathbf{x} \textit{ does not occur freely in } u, \tag{9}$$

$$[\mathbf{x}]uv = \mathfrak{s}([\mathbf{x}]u)([\mathbf{x}]v) \tag{10}$$

Note that these rules define the denotation of nested abstractions (such as $[\mathbf{x}][\mathbf{y}]u$) by first replacing $[\mathbf{y}]u$ by the member of U it denotes.[1]

3.2   Truth in an Interpretation

A structure together with an assignment $(\mathfrak{A}, I)$ will be called a *model*. It induces, in the presence of an environment ν a mapping from goals to truth values:

$$[\![\_]\!]^{\mathfrak{A}}_{\nu} : Goals \to \Omega$$

given by:

$$[\![G]\!] = \omega(\nu[\![G]\!]) \qquad \text{for all goals } G.$$

The mapping is independent of the environment if G is closed.

We say a clause $Tl \to Hd$ is true in a model if, for every environment ν we have $[\![Tl]\!]_{\nu} \leq_{\Omega} [\![Hd]\!]_{\nu}$. A program is true in a model (or $(\mathfrak{A}, I)$ is a model of a program) if its clauses are.

We can lift interpretations to sequences of goals in the obvious way:

$$[\![G_1 \otimes \cdots \otimes G_n]\!] = [\![G_1]\!] \wedge \cdots \wedge [\![G_n]\!].$$

Theorem 1 (Soundness). *If* $\langle P|\mathfrak{g}_1 \rangle \overset{\theta}{\leadsto} \langle P|\mathfrak{g}_2 \rangle$ *then in every model of P and for any environment ν,* $[\![\mathfrak{g}_2]\!]_{\nu} \leq [\![\mathfrak{g}_1\theta]\!]_{\nu}$. *Therefore, in particular, if* $\langle P|\mathfrak{g} \rangle \overset{\theta}{\leadsto} \square$ *then* $[\![\mathfrak{g}\theta]\!]_{\nu} = \top_{\Omega}$.

To prove this theorem, we need several technical lemmas.

Lemma 1 (α and β Soundness). *Renaming of bound variables is sound in* Hiord *semantics. In particular, suppose y is a variable distinct from x and not occurring freely in G. Then*

$$[\![\lambda x.G]\!]_{\nu} = [\![\lambda y.G[x := y]]\!]_{\nu} \qquad y \text{ fresh}$$
$$[\![\exists x.G]\!]_{\nu} = [\![\exists y.G[x := y]]\!]_{\nu} \qquad y \text{ fresh}$$

β-*reduction is sound. In particular*

$$[\![(\lambda x.G)t]\!]_{\nu} = [\mathbf{x}] [\![G[x := \mathbf{x}]]\!]_{\nu} \cdot [\![t]\!]_{\nu}$$

The proof of soundness of α, by induction on the cases of the definition of bracket abstraction, is straightforward and left to the reader.

The proof of the soundness of β, given the combinatory nature of our models, is a straightforward adaptation of Curry's combinatory completeness arguments [9] to structures with environments. We prove one step of contraction is sound for

---

[1] The denotation of bracket abstraction can be defined in terms of an evaluation map $()^*$ from $\lambda U[\mathcal{W}]$ to U given by transition rules imitating the equations just given. The extra notational step does not seem to add any clarity to the definition. In either case one must show that for any (closed) bracket abstraction u, there is a unique normal form, i.e. a unique abstraction-free member of U denoted by u. This is left to the reader.

the last two of the three cases of the bracket abstraction definition, by showing $([\mathbf{x}]u)v = u[\mathbf{x} := v]$, the first case amounting to a verification of the fact that $\mathfrak{see}$ is the identity function in a combinatory algebra.

If $\mathbf{x}$ does not occur freely in u, then $([\mathbf{x}]u)v = \mathfrak{k}uv = u$ which agrees with $u[\mathbf{x} := v]$. If u is $u_1 u_2$, then $([\mathbf{x}]u)v = \mathfrak{s}([\mathbf{x}]u_1)([\mathbf{x}]u_2)v$ which in turn gives $(([\mathbf{x}]u_1)v)(([\mathbf{x}]u_2)v)$. By induction, the result is immediate.

The reader can easily check full β conversion is sound.

Lemma 2 (Substitution Lemma). *Let G be a goal (or a sequence of goals), θ a substitution, ν an environment into a structure $\mathfrak{A}$. Let ν_θ be the modified environment induced by θ, that is to say, for each variable x, $\nu_\theta(x) = \nu(\theta(x))$. Then $\nu[\![G\theta]\!] = \nu_\theta[\![G]\!]$, and hence $[\![G\theta]\!]_\nu^{\mathfrak{A}} = [\![G]\!]_{\nu_\theta}^{\mathfrak{A}}$.*

We now prove the substitution lemma, first for *terms* t, then for formulas G by structural induction.

Proof. Suppose

t is a parameter in S:
Then $\nu[\![t\theta]\!] = I(s)$ and also $\nu_\theta[\![t]\!] = I(s)$.
t is a variable X:
Then $\nu[\![X\theta]\!] = \nu_\theta[\![X]\!]$ by definition.
t is of the form $u \cdot v$:
A special case is $t = X(t_1 \ldots t_n)$ for a variable X or $t = r(t_1 \ldots t_n)$ for some parameter r. Then $\nu[\![u \cdot v\theta]\!] = \nu[\![u\theta \cdot v\theta]\!] = \nu[\![u\theta]\!] \cdot \nu[\![v\theta]\!]$. By induction hypothesis, this is equal to $\nu_\theta[\![u]\!] \cdot \nu_\theta[\![v]\!]$ and hence $\nu_\theta[\![u \cdot v]\!]$.
t is $\lambda x.G$:

Then $\nu[\![t\theta]\!]$ is $\nu[\![\lambda y.G[x := y]\theta]\!]$ for any y distinct from x and disjoint from the variables in the domain or range of θ, which gives $[\mathbf{y}]\nu[\![G[x := y]\theta]\!]$. By the induction hypothesis, this is equal to $[\mathbf{y}]\nu_\theta[\![G[x := y]]\!]$. and, by soundness of α conversion, to $[\mathbf{x}]\nu_\theta[\![G]\!]$, which is precisely $\nu_\theta[\![\lambda x.G]\!]$.

The cases $t_1 = t_2$ and G_1, G_2 follow immediately from the induction hypothesis, and the $\exists x.G$ case is similar to abstraction, and left to the reader.

We now prove the soundness theorem.

Proof (Soundness). The result is shown by induction on the length of the given resolution deduction. The length 0 case gives the conclusion trivially. Suppose the claim holds for all deductions length smaller than sone natural number $n > 0$, and that we are given a deduction of length n whose first step is *backchain*:

$$\langle P, A \leftarrow Tl|\mathfrak{g}_0 \otimes A' \otimes \mathfrak{g}_1\rangle \overset{\theta_1}{\rightsquigarrow} \langle P, A \leftarrow Tl|\mathfrak{g}_0\theta_1 \otimes Tl\theta_1 \otimes \mathfrak{g}_1\theta_1\rangle \overset{\theta_1'}{\rightsquigarrow} \cdots \rightsquigarrow \langle P, A \leftarrow Tl|\mathfrak{g}'\rangle$$

where $\theta = \theta_1\theta_1'$.

By the induction hypothesis, in any model of $P, A \leftarrow Tl$ we have

$$[\![\mathfrak{g}']\!]_\nu \leq [\![(\mathfrak{g}_0\theta_1 \otimes Tl\theta_1 \otimes \mathfrak{g}_1\theta_1)\theta']\!]_\nu$$

the latter truth value being equal to $[\![\mathfrak{g}_0\theta]\!]_\nu \wedge [\![Tl\theta]\!]_\nu \wedge [\![\mathfrak{g}_1\theta]\!]_\nu$. Now, since $[\![]$ is assumed a model of P, for any environment ν we have $[\![Tl]\!]_\nu \leq [\![A]\!]_\nu$, so, in particular, for any substitution θ we have $[\![Tl]\!]_{\nu_\theta} \leq [\![A]\!]_{\nu_\theta}$. By the substitution lemma $[\![Tl\theta]\!]_\nu \leq [\![A\theta]\!]_\nu$. Since $[\![A\theta]\!]_\nu = [\![A'\theta]\!]_\nu$, we have

$$
\begin{aligned}
[\![\mathfrak{g}']\!]_\nu &\leq [\![\mathfrak{g}_0\theta]\!]_\nu \wedge [\![Tl\theta]\!]_\nu \wedge [\![\mathfrak{g}_1\theta]\!]_\nu \\
&\leq [\![\mathfrak{g}_0\theta]\!]_\nu \wedge [\![A'\theta]\!]_\nu \wedge [\![\mathfrak{g}_1\theta]\!]_\nu \\
&\leq [\![(\mathfrak{g}_0 \otimes A' \otimes \mathfrak{g}_1)\theta]\!]_\nu
\end{aligned}
$$

as we wanted to show.

Now we consider the case where the first step in the deduction is an occurrence of the *unify* rule:

$$
\langle P | \mathfrak{g}_0 \otimes t_1 = t_2 \otimes \mathfrak{g}_1 \rangle \overset{\theta}{\rightsquigarrow} \langle P | \mathfrak{g}_0\theta \otimes \mathfrak{g}_1\theta \rangle \rightsquigarrow \cdots \rightsquigarrow .
$$

where $t_1\theta = t_2\theta$.

It suffices to show that $[\![(\mathfrak{g}_0 \otimes \mathfrak{g}_1)\theta]\!]_\nu \leq [\![(\mathfrak{g}_0 \otimes t_1 = t_2 \otimes \mathfrak{g}_1)\theta]\!]_\nu$ in any model of P. But this requirement is equivalent to

$$
\begin{aligned}
[\![(\mathfrak{g}_0 \otimes t_1 = t_2 \otimes \mathfrak{g}_1)\theta]\!]_\nu &= [\![\mathfrak{g}_0\theta]\!]_\nu \wedge [\![t_1\theta = t_2\theta]\!]_\nu \wedge [\![\mathfrak{g}_1\theta]\!]_\nu \\
&= \mathsf{eq}_{\mathfrak{A}} \, [\![t_1\theta]\!]_\nu \, [\![t_2\theta]\!]_\nu = \top
\end{aligned}
$$

which always holds.

Now suppose the first step in the deduction is an occurrence of the *exists* rule:

$$
\langle P | \mathfrak{g}_0 \otimes \exists x G \otimes \mathfrak{g}_1 \rangle \overset{\exists}{\rightsquigarrow} \langle P | \mathfrak{g}_0 \otimes G\theta \otimes \mathfrak{g} \rangle \rightsquigarrow \cdots \rightsquigarrow ,
$$

with x fresh. It suffices to show that for any environment $[\![G\theta]\!]_\nu \leq [\![\exists x G]\!]_\nu$ which is straightforward, and left to the reader.

4   Restricting the Formalism

We have so far defined a very general formalism intended to capture essentially all the higher-order features of Hilog, together with full-blown abstraction and β rewriting. As mentioned in the introduction, the formalism should be viewed as a framework for defining higher-order declarative languages, by suitably restricting the calculus, imposing type disciplines, and making use of abstract interpretation for type inference and specialization.

The aim of this section is to give examples of the use of the higher order features described, and suggest some interesting restrictions of the formalism.

4.1   The Hiord-1 Language

A language for Hiord-1 is composed by a set $\mathcal{F}$ of names for constants and functions, a set $\mathcal{R}$ of names of relations, and a set $\mathcal{V}$ of variables, such that the three are nonempty and disjoint pairwise.

Data (terms) and predicates are distinguished. Terms are restricted to those formed using parameters in $\mathcal{F}$ at the head, and Atomic goals are restricted to

---

Definition of terms:

1. A variable is a term.
2. A name in $\mathcal{F}$ is a term.
3. If $t_1, \ldots, t_n$ are terms, and $s \in \mathcal{F}$, then $s(t_1, \ldots, t_n)$ is a term.
4. If G is a goal and $\overline{x}$ is a sequence of variables, then $\{(\overline{x}) :\text{-}\, G\}$ is a term, known as an abstraction.

Definitions of atomic formulas, Goals and Clauses:

1. $\top$ is an atomic formula.
2. A name $r \in \mathcal{R}$ is an atomic formula, and if $t_1, \ldots, t_n$ are terms, then $r(t_1, \ldots, t_n)$ is an atomic formula. This kind of atomic formulas are called *rigid*.
3. If X is a variable and $t_1, \ldots, t_n$ are terms, then X and $X(t_1, \ldots, t_n)$ are atomic formulas.
4. If t_1 and t_2 are terms, then $t_1 = t_2$ is an atomic formula.
5. An atomic formula is a goal. If G_1 and G_2 are goals, then $G_1 \& G_2$ is a goal, and if x is a variable, $\mathrm{E}(x)G_1$ is a goal.
6. A clause is a formula of the form $H \leftarrow G$, where H is a rigid atomic formula and G is a goal.

---

terms formed with relational parameters or variables at the head. The table below summarizes the formal abstract syntax of Hiord-1, a restriction of the Hiord syntax.

Formally, we take Hiord-1 resolution rules to be a subset of Hiord, using strict first-order equality of terms in unification, which is, of course, a subset of $\beta\eta$-conversion. Thus our model theory and soundness results provide a semantic base for this fragment. In practice, the language is sufficiently restricted to permit some obvious compile-time transformations that produce WAM-ready Prolog code. These are discussed in [10] and in the documentation for its implementation as the Hiord-1 *package* in Ciao [11].

4.2   Concrete Syntax of Higher-Order Data and Examples

We now propose a concrete syntax for higher-order data in Hiord-1. Our proposal aims at syntactically differentiating higher-order data from ordinary terms. Thus, in modules using the `hiord` package, all terms to be considered higher-order data are surrounded by {}. The most general syntax for *predicate abstractions* follows the pattern:

$$\{ \; sharedvars \; \text{->} \; {}^{\prime\prime}(absvars) \; :\text{-} \; \mathrm{G} \; \}$$

which represents the term in the formalism $\{(\overline{x}) :\text{-}\, \mathrm{E}(y)\,\mathrm{G}\}$ and where *absvars* is a comma-separated sequence of distinct variables representing the vector of abstracted variables listed in $\overline{x}$, and $\langle sharedvars \rangle$ is a comma-separated sequence listing all *exported* variables, i.e. all variables that are not existentially quantified. These variables are shared with the rest of the clause. Variables not appearing in *sharedvars* or in *absvars* are existentially quantified (i.e., correspond to those in y), i.e., they are local to the predicate abstraction, even if their names happen to coincide with variables outside the predicate abstraction.

When *sharedvars* is empty the arrow is omitted. Also, when *absvars* is empty no surrounding parenthesis are written (i.e., only '' is used). Finally, when G is true the ":- true" part can also be omitted. Note that the functor name in the head is the void atom ''.

Conjunction is written with a comma and disjunction, treated in the theory as a defined symbol, is written with semicolon.

Some simple examples of higher-order predicates and uses of predicate abstractions are:

```
% list(List,Pred) : Pred is true for all elements of List
list([], _).
list([X|Xs], P):- P(X), list(Xs, P).

% map(List1,Rel,List2) : Rel is true for all pairs of elements of
%                        List1 and List2 in the same position
map([], _, []).
map([X|Xs], P, [Y|Ys]) :- P(X,Y), map(Xs, P, Ys).

all_less(L1, L2) :- map(L1, {''(X,Y) :- X < Y}, L2).

% child_of(Person, Mother, Father) : Family database
child_of(tom, mary, john).
...

same_mother(L) :- list(L, {M -> ''(S) :- child_of(S,M,_)}).

same_parents(L) :- list(L, {M,F -> ''(S) :- child_of(S,M,F)}).
```

The decision of marking shared variables instead of existential variables is based on the following considerations:

– This approach makes clear which variables can affect the predicate abstraction "from outside."
– Unique existential variables in the predicate abstraction can be written simply as anonymous variables (_).
– Compile-time code transformations are simplified, since new variables introduced by expansions should be existential (there is no need to add them to the head).
– The compilation of the predicate abstraction is also simplified.

An additional syntactic form is provided: *closures*. Closures are "syntactic sugar" for certain predicate abstractions, and can always be written as predicate abstractions instead (but they are more compact). All higher-order data (surrounded by {}) not adhering to predicate abstraction syntax is a closure. In a closure, each occurrence of the atom # corresponds to a parameter of the predicate abstraction that it represents. All variables in a closure are shared with the rest of the clause (for compatibility with meta-programming). As an example, the following definition of same_parents/2 using a closure is equivalent to the previous one:

```
same_parents(L) :- list(L, {child_of(#,_M,_F)}).
```

If there are several #'s in the closure, they each correspond to a successive element of the sequence of abstracted variables in the corresponding predicate abstraction (i.e., in the same order). For example, the following definition of all_less/2 is equivalent to the one above:

```
all_less(L1, L2) :- map(L1, {# < #}, L2).
```

Note that closures are simply a compact but limited abbreviation. If a different order is required, then a predicate abstraction should be used instead.

Some Examples and a Comparison of Programming Style with Hilog.
We start by showing the higher-order predicate which defines the transitive closure of a given relation:

```
closure(R,X,Y) :- R(X,Y).
closure(R,X,Y) :- R(X,Z), closure(R,Z,Y).
```

Assume now that we have the family database defined in a previous example child_of(Person, Mother, Father). Then, given a list of people, to verify that all have the same father one could do:

```
same_father(L) :- list(L, {F -> ''(S) :- child_of(S,_,F)}).
```

This would be expressed in Hilog as:

```
father(F)(S) :- child_of(S,_,F).
same_father(L) :- list(L, father(_)).
```

Which is more laborious. Admittedly, the explicit definition of a predicate sometimes is more clear, but this can also be done in Hiord-1, of course.

```
father(F, S) :- child_of(S,_,F).
same_father(L) :- list(L, {father(_, #)}).
```

But assume now that we want to define a predicate to enumerate the descendents of someone which may share a Y-chromosome feature (that is, only the father relation is taken into account). In Hiord-1, given the father/2 relation above one could say:

```
descendent_Y(X,Y) :- closure({father(#,#)},X,Y).
```

and if father/2 were not defined, one would say:

```
descendent_Y(X,Y) :- closure({''(F,S) :- child_of(S,_,F)},X,Y).
```

In Hilog one would think that, as a father relation was already defined for same_father, it could be used for this new predicate. But note that the father relation defined above (Hilog version) is not a binary relation and thus one would need to define another father2/2 relation:

```
father2(F,S) :- child_of(S,_,F).
descendent_Y(X,Y) :- closure(father2, X, Y).
```

That father2 relation is semantically equivalent to the father relation, but their different uses in higher-order predicates forces one to make several versions.

4.3 Formalizing Module Structure in Hiord

We now consider an example that exploits the fact that we are working with a higher-order function calculus that allows explicit recursion in our case, a fragment of the untyped λ-calculus. The reader should note that although flex terms are not explicitly allowed as heads of clauses, such a generalized notion of clause is expressible via bindings to lambda-terms, in an essential way, below.

In our example we define modules using predicates and higher-order variables and viewing a module as a predicate which returns a series of predicate abstractions. The module in question, list_mod, defines the Member, List, and Reverse predicates, using some auxiliary predicates.

```
lists_mod(Member, List, Reverse) :-
      Member =   {Member -> ''(X, L) :- L = [X|_],
                          ;  L = [_|Xs], Member(X, Xs)
                  },
      List =     {List -> ''(L, P) :- L = []
                                    ;  L = [X|Xs], P(X), List(Xs, P)
                  }, %% Higher-Order predicate
      Rev3 =     {Rev3 -> ''(L, R1, R2) :- L = [], R1 = R2
                                    ;  L = [E|Es], Rev3(Es,[E|L],R)
                  }, %% Internal predicate
      Reverse = {Rev3 ->  ''(L, R) :- Rev3(L, [], R) }.
```

Note that the definitions of the predicate abstractions which are recursive (Member, List, and Rev3) involve unifications which do not pass the occur-check. While this may be considered a problem, note that the compiler can easily detect such cases (a variable is unified to a predicate abstraction which uses this variable in a flex goal) and translate them defining an auxiliary higher-order extension of the predicate, as the following code shows:

```
lists_mod(Member, List, Reverse) :-
      Me =       { ''(X, L, R) :- L = [X|_]
                              ;  L = [_|Xs], R(X, Xs, R)
                 }, %% Internal predicate
      Member = { Me ->
                   ''(X,L) :- Me(X, L, Me)
                 },
      Li =       { ''(L, P, R) :- L = []
                              ;  L = [X|Xs], P(X), R(Xs, P, R)
                 }, %% Internal predicate
      List =    { Li ->
                   ''(L,P) :- Li(L, P, Li)
                 },
      ...
```

This module can then be used in another module for example as follows:

```
main(X) :-
    lists_mod(Member,_,_), %% Import Member from lists_mod.
    Member(X,[1,3,5]).      %% Call Member.
```

or simply called from the top level for example as follows:

```
?- lists_mod(Member,_,_), Member(X,[1,3,5]).
```

5   Conclusions and Further Work

This paper studies a framework for defining the syntax and semantics of a type-free higher-order extensions to core Prolog with predicate abstractions. Some of the uses of type-free predicate abstraction and higher-order features are underscored, including ways to capture metaprogramming and to formalize module structure. The formalism (and the various subsets considered) is shown sound with respect to a model theory based on partial combinatory algebras with an object of truth-values. Practical restrictions of this framework are then discussed, along with an implementation included as a package in Ciao-prolog. Examples are given showing the use of the notions introduced to define various higher-order predicates, and databases applications. These include programs to compare programming in this framework with code in other Higher-order formalisms.

The framework proposed gives rise to many questions the authors hope to address in future research. In particular, a rigorous treatment must be developed for comparison with other higher-order formal systems (Hilog, Lambda-Prolog). For example, it is reasonably straightforward to conservatively translate the Higher-order Horn fragment of λProlog into Hiord by erasing types, as the resolution rules are essentially the same (assuming a type-safe higher-order unification procedure).

Clearly, the formalisms presented need a more thoroughgoing semantical analysis –declarative and operational– as well as completeness theorems for various typed and type-free restrictions, and a with abstract interpretation taken into account. Also, a formal treatment is needed for the new proposal for module definition given in this paper.

References

1. Miller, D., Nadathur, G., Pfenning, F., Scedrov, A.: Uniform proofs as a foundation for logic programming. Annals of Pure and Applied Logic **51** (1991) 125–157
2. Nadathur, G., Miller, D.: Higher-order horn clauses. Journal of the ACM **37** (1990) 777–814
3. Miller, D.: λprolog: an introduction to the language and its logic. Manuscript available at http://www.cse.psu.edu/~dale (2002)
4. Chen, W., Kifer, M., Warren, D.S.: Hilog: A Foundation for Higher-Order Logic Programming. Journal of Logic Programming (1989)

5. Beeson, M.: Foundations of Constructive Mathematics. Springer-Verlag (1985)
6. Lawvere, F.W.: Diagonal arguments and cartesian-closed categories. In: Category Theory, Homology Theory and their Applications. Springer (1969)
7. Huwig, H., Poigné, A.: A Note on Inconsistencies Caused by Fixpoints in a Cartesian Closed Categories. Theoretical Computer Science **73** (1990) 101–112
8. Miller, D.: A logic programming language with lambda-abstraction, function variables, and simple unification. Journal of Logic and Computation **1** (1991) 497 – 536
9. Curry, H.B., Feys, R.: Combinatory Logic. North-Holland, Amsterdam (1958)
10. Cabeza, D.: An Extensible, Global Analysis Friendly Logic Programming System. PhD thesis, Dept. of Computer Science, Technical University of Madrid, Spain, Madrid, Spain (2004)
11. Bueno, F., Cabeza, D., Carro, M., Hermenegildo, M., López-García, P., (Eds.), G.P.: The Ciao System. Reference Manual (v1.10). The ciao system documentation series–TR, School of Computer Science, Technical University of Madrid (UPM) (2002) System and on-line version of the manual available at http://clip.dia.fi.upm.es/Software/Ciao/.
12. Warren, D.H.: Higher-order extensions to prolog: are they needed? In Hayes, J., Michie, D., Pao, Y.H., eds.: Machine Intelligence 10. Ellis Horwood Ltd., Chicester, England (1982) 441–454
13. Naish, L.: Higher-order logic programming. Technical Report 96/2, Department of Computer Science, University of Melbourne, Melbourne, Australia (1996)
14. McDowell, R., Miller, D.: A logic for reasoning with higher-order abstract syntax. In: Proceedings of the 12th Annual IEEE Symposium on Logic in Computer Science, IEEE Computer Society (1997) 434
15. Miller, D.: Abstract syntax and logic programming. In: Proceedings of the Second Russian Conference on Logic Programming, Springer Verlag (1991)

Assessment Aggregation in the Evidential Reasoning Approach to MADM Under Uncertainty: Orthogonal Versus Weighted Sum

Van-Nam Huynh, Yoshiteru Nakamori, and Tu-Bao Ho

School of Knowledge Science,
Japan Advanced Institute of Science and Technology,
Tatsunokuchi, Ishikawa, 923-1292, Japan
{huynh, nakamori, bao}@jaist.ac.jp

Abstract. In this paper, we revisit the evidential reasoning (ER) approach to multiple-attribute decision making (MADM) with uncertainty. The attribute aggregation problem in MADM under uncertainty is generally formulated as a problem of evidence combination. Then several new aggregation schemes are proposed and simultaneously their theoretical features are explored. A numerical example traditionally examined in published sources on the ER approach is used to illustrate the proposed techniques.

1 Introduction

So far, many attempts have been made to integrate techniques from artificial intelligence (AI) and operational research (OR) for handling uncertain information, e.g., [1, 4, 5, 8, 9, 11, 19]. During the last decade, an evidential reasoning (ER) approach has been proposed and developed for MADM under uncertainty in [20, 21, 23–25]. Essentially, this approach is based on an evaluation analysis model [26] and the evidence combination rule of the Dempster-Shafer (D-S) theory [14]. The ER approach has been applied to a range of MADM problems in engineering and management, including motorcycle assessment [21], general cargo ship design [13], system safety analysis and synthesis [17], retro-fit ferry design [22] among others.

Recently, due to a need of developing theoretically sound methods and tools for dealing with MADM problems under uncertainty, Yang and Xu [25] have proposed a system of four synthesis axioms within the ER assessment framework with which a rational aggregation process needs to satisfy. It has also been shown that the original ER algorithm only satisfies these axioms approximately. At the same time, guided by the aim exactly, the authors have proposed a new ER algorithm that satisfies all the synthesis axioms precisely.

It is worth emphasizing that the underlying basis of using Dempster's rule of combination is the independent assumption of information sources to be combined. However, in situations of multiple attribute assessment based on a multi-level structure of attributes, assumptions regarding the independence of

M.J. Maher (Ed.): ASIAN 2004, LNCS 3321, pp. 109–127, 2004.

attributes' uncertain evaluations may not be appropriate in general. In this paper, we reanalysis the previous ER approach in terms of D-S theory so that the attribute aggregation problem in MADM under uncertainty can be generally formulated as a problem of evidence combination. Then we propose a new aggregation scheme and simultaneously examine its theoretical features. For the purpose of the present paper, we take only qualitative attributes of an MADM problem with uncertainty into account, though quantitative attributes would be also included in a similar way as considered in [20, 21].

2 Background

2.1 Problem Description

This subsection describes an MADM problem with uncertainty through a tutorial example taken from [25].

Let us consider a problem of motorcycle evaluation [6]. To evaluate the quality of the *operation* of a motorcycle, the following set of distinct evaluation grades is defined

$$\mathcal{H} = \{poor\ (H_1), indifferent\ (H_2), average\ (H_3), good\ (H_4), excellent\ (H_5)\} \quad (1)$$

Because *operation* is a general technical concept and is not easy to evaluate directly, it needs to be decomposed into detailed concepts such as *handling*, *transmission*, and *brakes*. Again, if a detailed concept is still too general to assess directly, it may be further decomposed into more detailed concepts. For example, the concept of *brakes* is measured by *stopping power*, *braking stability*, and *feel at control*, which can probably be directly evaluated by an expert and therefore referred to as basic attributes (or basic factors).

Generally, a qualitative attribute y may be evaluated through a hierarchical structure of its subattributes. For instance, the hierarchy for evaluation of the *operation* of a motorcycle is depicted as in Fig. 1.

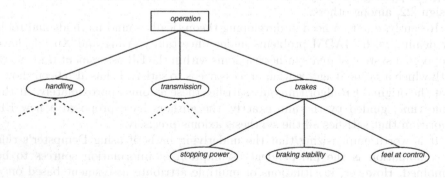

Fig. 1. Evaluation hierarchy for *operation* [25]

In evaluation of qualitative attributes, judgments could be uncertain. For example, in the problem of evaluating different types of motorcycles, the following type of uncertain subjective judgments for the *brakes* of a motorcycle, say "Yamaha", was frequently used [6, 25]:

1. Its *stopping power* is *average* with a confidence degree of 0.3 and it is *good* with a confidence degree of 0.6.
2. Its *braking stability* is *good* with a confidence degree of 1.
3. Its *feel at control* is evaluated to be *good* with a confidence degree of 0.5 and to be *excellent* with a confidence degree of 0.5.

In the above statements, the confidence degrees represent the uncertainty in the evaluation. Note that the total confidence degree in each statement may be smaller than 1 as the case of the first statement. This may be due to incomplete of available information.

In a similar fashion, all basic attributes in question could be evaluated. Then the problem is to generate an overall assessment of the *operation* of a motorcycle by aggregating all uncertain judgments of its basic attributes in a rational way.

2.2   Evaluation Analysis Model

The evaluation analysis model was proposed in [26] to represent uncertain subjective judgments, such as statements specified in preceding subsection, in a hierarchical structure of attributes.

To begin with, let us suppose a simple hierarchical structure consisting of two levels with a general attribute, denoted by y, at the top level and a finite set E of its basic attributes at the bottom level. Let $E = \{e_1, \ldots, e_i, \ldots, e_L\}$ and assume the weights of basic attributes are given by $W = (w_1, \ldots, w_i, \ldots, w_L)$, where w_i is the relative weight of the ith basic attribute (e_i) with $0 \leq w_i \leq 1$.

Given the following set of evaluation grades

$$\mathcal{H} = \{H_1, \ldots, H_n, \ldots, H_N\}$$

designed as distinct standards for assessing an attribute, then an assessment for e_i of an alternative can be mathematically represented in terms of the following distribution [25]

$$S(e_i) = \{(H_n, \beta_{n,i}) \mid n = 1, \ldots, N\}, \text{ for } i = 1, \ldots, L \qquad (2)$$

where $\beta_{n,i}$ denotes a degree of belief satisfying $\beta_{n,i} \geq 0$, and $\sum_{n=1}^{N} \beta_{n,i} \leq 1$. An assessment $S(e_i)$ is called *complete* (respectively, *incomplete*) if $\sum_{n=1}^{N} \beta_{n,i} = 1$ (respectively, $\sum_{n=1}^{N} \beta_{n,i} < 1$).

For example, the three assessments 1.–3. given in preceding subsection can be represented in the form of distributions defined by (2) as

$$
\begin{aligned}
S(stopping\ power) &= \{(H_3, 0.3), (H_4, 0.6)\} \\
S(braking\ stability) &= \{(H_4, 1)\} \\
S(feel\ at\ control) &= \{(H_4, 0.5), (H_5, 0.5)\}
\end{aligned}
$$

where only grades with nonzero degrees of belief are listed in the distributions.

Let us denote β_n the degree of belief to which the general attribute y is assessed to the evaluation grade of H_n. The problem now is to how to generate β_n, for $n = 1, \ldots, N$, by combinating the assessments for all associated basic attributes e_i $(i = 1, \ldots, L)$ as given in (2). However, before continuing the discussion, it is necessary to briefly review the basis of D-S theory of evidence in the next subsection.

2.3   Dempster-Shafer Theory of Evidence

In D-S theory, a problem domain is represented by a finite set Θ of mutually exclusive and exhaustive hypotheses, called *frame of discernment* [14]. Formally, a basic probability assignment (BPA, for short) is a function $m : 2^\Theta \rightarrow [0, 1]$ verifying

$$m(\emptyset) = 0, \text{ and } \sum_{A \in 2^\Theta} m(A) = 1$$

The quantity $m(A)$ can be interpreted as a measure of the belief that is committed exactly to A, given the available evidence. A subset $A \in 2^\Theta$ with $m(A) > 0$ is called a *focal element* of m.

Two useful operations that play a central role in the manipulation of belief functions are *discounting* and *Dempster's rule of combination* [14]. The discounting operation is used when a source of information provides a BPA m, but one knows that this source has probability α of reliable. Then m is discounted by a factor of $(1 - \alpha)$, resulting in a new BPA m^α defined by

$$m^\alpha(A) = \alpha m(A), \text{ for any } A \subset \Theta \tag{3}$$
$$m^\alpha(\Theta) = (1 - \alpha) + \alpha m(\Theta) \tag{4}$$

Consider now two pieces of evidence on the same frame Θ represented by two BPAs m_1 and m_2. Dempster's rule of combination is then used to generate a new BPA, denoted by $(m_1 \oplus m_2)$ (also called the orthogonal sum of m_1 and m_2), defined as follows

$$(m_1 \oplus m_2)(\emptyset) = 0, (m_1 \oplus m_2)(A) = \frac{1}{K} \sum_{B, C \subseteq \Theta : B \cap C = A} m_1(B) m_2(C) \tag{5}$$

where

$$K = 1 - \sum_{B, C \subseteq \Theta : B \cap C = \emptyset} m_1(B) m_2(C) \tag{6}$$

Note that the orthogonal sum combination is only applicable to such two BPAs that verify the condition $K > 0$.

As we will see in the following sections, these two operations essentially play an important role in the ER approach to MADM under uncertainty developed in, e.g., [20, 21, 25]. Although the discounting operation has not been mentioned explicitly in these published sources.

3   The Evidential Reasoning Approach

Let us return to the two-level hierarchical structure with a general attribute y at the top level and a finite set $E = \{e_1, \ldots, e_i, \ldots, e_L\}$ of its basic attributes at the bottom level. Denote β_n the degree of belief to which the general attribute y is assessed to the evaluation grade of H_n, for $n = 1, \ldots, N$.

3.1   The Original ER Algorithm

The original ER algorithm proposed in [20] has been used for the purpose of obtaining β_n $(n = 1, \ldots, N)$ by aggregating the assessments of basic attributes given in (2). The summary of the algorithm in this subsection is taken from [25].

Given the assessment $S(e_i)$ of a basic attribute e_i $(i = 1, \ldots, L)$, let $m_{n,i}$ be a basic probability mass representing the belief degree to which the basic attribute e_i supports the hypothesis that the attribute y is assessed to the evaluation grade H_n. Let $m_{\mathcal{H},i}$ be the remaining probability mass unassigned to any individual grade after all the N grades have been considered for assessing the general attribute y as far as e_i is concerned. These quantities are defined as follows

$$m_{n,i} = w_i \beta_{n,i}, \text{ for } n = 1, \ldots, N \tag{7}$$

$$m_{\mathcal{H},i} = 1 - \sum_{n=1}^{N} m_{n,i} = 1 - w_i \sum_{n=1}^{N} \beta_{n,i} \tag{8}$$

Let $E_{I(i)} = \{e_1, \ldots, e_i\}$ be the subset of first i basic attributes. Let $m_{n,I(i)}$ be a probability mass defined as the belief degree to which all the basic attributes in $E_{I(i)}$ supports the hypothesis that y is assessed to H_n. Let $m_{\mathcal{H},I(i)}$ be the remaining probability mass unassigned to individual grades after all the basic attributes in $E_{I(i)}$ have been assessed. The quantities $m_{n,I(i)}$ and $m_{\mathcal{H},I(i)}$ can be generated by combining the basic probability masses $m_{n,j}$ and $m_{\mathcal{H},j}$ for all $n = 1, \ldots, N$, and $j = 1, \ldots, i$.

With these notations, the key step in the original ER algorithm is to inductively calculate $m_{n,I(i+1)}$ and $m_{\mathcal{H},I(i+1)}$ as follows

$$m_{n,I(i+1)} = K_{I(i+1)}(m_{n,I(i)}m_{n,i+1} + m_{n,I(i)}m_{\mathcal{H},i+1} + m_{\mathcal{H},I(i)}m_{n,i+1}) \tag{9}$$

$$m_{\mathcal{H},I(i+1)} = K_{I(i+1)}(m_{\mathcal{H},I(i)}m_{\mathcal{H},i+1}) \tag{10}$$

for $n = 1, \ldots, N$, $i = 1, \ldots, L-1$, and $K_{I(i+1)}$ is a normalizing factor defined by

$$K_{I(i+1)} = \left[1 - \sum_{t=1}^{N} \sum_{\substack{j=1 \\ j \neq t}}^{N} m_{t,I(i)}m_{j,i+1} \right]^{-1} \tag{11}$$

Then we obtain

$$\beta_n = m_{n,I(L)}, \text{ for } n = 1, \ldots, N$$

$$\beta_{\mathcal{H}} = m_{\mathcal{H},I(L)} = 1 - \sum_{n=1}^{N} \beta_n \tag{12}$$

3.2   Synthesis Axioms and the Modified ER Algorithm

Inclined to developing theoretically sound methods and tools for dealing with MADM problems under uncertainty, Yang and Xu [25] have recently proposed a system of four synthesis axioms with which a rational aggregation process needs to satisfy. These axioms are symbolically stated as below.

Axiom 1. (*Independency*) If $\beta_{n,i} = 0$ for all $i = 1, \ldots, L$, then $\beta_n = 0$.

Axiom 2. (*Consensus*) If $\beta_{k,i} = 1$ and $\beta_{n,i} = 0$, for all $i = 1, \ldots, L$, and $n = 1, \ldots, N$, $n \neq k$, then $\beta_k = 1$, $\beta_n = 0$, for $n = 1, \ldots, N$, $n \neq k$.

Axiom 3. (*Completeness*) Assume $\mathcal{H}^+ \subset \mathcal{H}$ and denote $I^+ = \{n | h_n \in \mathcal{H}^+\}$. If $\sum_{n \in I^+} \beta_{n,i}(> 0) = 1$, for all $i = 1, \ldots, L$, then $\sum_{n \in I^+} \beta_n(> 0) = 1$ as well.

Axiom 4. (*Incompleteness*) If there exists $i \in \{1, \ldots, L\}$ such that $\sum_{n=1}^{N} \beta_{n,i} < 1$, then $\sum_{n=1}^{N} \beta_n < 1$.

It is easily seen from (9–12) that the original ER algorithm naturally follows the independency axiom. However, it has been shown in [25] that the original ER algorithm only satisfies the consensus axiom approximately, and does not satisfy the completeness axiom.

In [25], Yang and Xu proposed a new ER algorithm that satisfies all the synthesis axioms. Its main features are summarized as follows

1) *Weight Normalization*. In the new ER algorithm, the weights w_i ($i = 1, \ldots, L$) of basic attributes are normalized such that: $0 \leq w_i \leq 1$ and $\sum_{i=1}^{L} w_i = 1$.

2) *Aggregation Process*. First, the probability mass $m_{\mathcal{H},i}$ given in (8) is decomposed into two parts: $m_{\mathcal{H},i} = \tilde{m}_{\mathcal{H},i} + \overline{m}_{\mathcal{H},i}$, where

$$\overline{m}_{\mathcal{H},i} = 1 - w_i, \text{ and } \tilde{m}_{\mathcal{H},i} = w_i \left(1 - \sum_{n=1}^{N} \beta_{n,i} \right) \tag{13}$$

Then, with the notations as in preceding section, the process of aggregating the first i assessments with the $(i+1)$th assessment is recursively carried out as follows

$$m_{n,I(i+1)} = K_{I(i+1)}[m_{n,I(i)} m_{n,i+1} + m_{n,I(i)} m_{\mathcal{H},i+1}$$
$$+ m_{\mathcal{H},I(i)} m_{n,i+1}] \tag{14}$$

$$m_{\mathcal{H},I(i)} = \tilde{m}_{\mathcal{H},I(i)} + \overline{m}_{\mathcal{H},I(i)}, \ n = 1, \ldots, N$$

$$\tilde{m}_{\mathcal{H},I(i+1)} = K_{I(i+1)}[\tilde{m}_{\mathcal{H},I(i)} \tilde{m}_{\mathcal{H},i+1}$$
$$+ \overline{m}_{\mathcal{H},I(i)} \tilde{m}_{\mathcal{H},i+1} + \tilde{m}_{\mathcal{H},I(i)} \overline{m}_{\mathcal{H},i+1}] \tag{15}$$

$$\overline{m}_{\mathcal{H},I(i+1)} = K_{I(i+1)}[\overline{m}_{\mathcal{H},I(i)} + \overline{m}_{\mathcal{H},i+1}] \tag{16}$$

where $K_{I(i+1)}$ is defined as same as in (11).

For assigning the assessment $S(y)$ for the general attribute y, after all L assessments of basic attributes have been aggregated, the algorithm finally defines

$$\beta_n = \frac{m_{n,I(L)}}{1 - \overline{m}_{\mathcal{H},I(L)}}, \text{ for } n = 1, \ldots, N \tag{17}$$

$$\beta_{\mathcal{H}} = \frac{\tilde{m}_{\mathcal{H},I(L)}}{1 - \overline{m}_{\mathcal{H},I(L)}} \tag{18}$$

and then

$$S(y) = \{(H_n, \beta_n), n = 1, \ldots, N\} \tag{19}$$

The following theorems are due to Yang and Xu [25] that are taken for granted to develop the new ER algorithm above.

Theorem 1. *The degrees of belief defined by (17) and (18) satisfy the following*

$$0 \le \beta_n, \beta_{\mathcal{H}} \le 1, n = 1, \ldots, N$$

$$\sum_{n=1}^{N} \beta_n + \beta_{\mathcal{H}} = 1$$

Theorem 2. *The aggregated assessment for y defined by (19) exactly satisfies all four synthesis axioms.*

Although proofs of these theorems given in [25] are somehow complicated, however, by analyzing the ER approach in terms of D-S theory in the next section, we show that these theorems follow quite simply.

4   A Reanalysis of the ER Approach

Let us remind ourselves the available information given to an assessment problem in the two-level hierarchical structure:

– the assessments $S(e_i)$ for basic attributes e_i $(i = 1, \ldots, L)$, and
– the weights w_i of the basic attributes e_i $(i = 1, \ldots, L)$.

Given the assessment $S(e_i)$ of a basic attribute e_i $(i = 1, \ldots, L)$, we now define a corresponding BPA, denoted by m_i, which quantifies the belief about the performance of e_i as follows

$$m_i(H_n) \triangleq \beta_{n,i}, \text{ for } n = 1, \ldots, N \tag{20}$$

$$m_i(\mathcal{H}) \triangleq 1 - \sum_{n=1}^{N} m_i(H_n) = 1 - \sum_{n=1}^{N} \beta_{n,i} \tag{21}$$

The quantity $m_i(H_n)$ represents the belief degree that supports for the hypothesis that e_i is assessed to the evaluation grade H_n. While $m_i(\mathcal{H})$ is the remaining probability mass unassigned to any individual grade after all evaluation grades have been considered for assessing e_i. If $S(e_i)$ is a complete assessment,

m_i is a probability distribution. Otherwise, m_i quantifies the ignorance resulted in $m_i(\mathcal{H}) > 0$.

As such with L basic attributes e_i, we obtain L corresponding BPAs m_i as quantified beliefs of the assessments for basic attributes. The problem now is how to generate an assessment for y, i.e. $S(y)$, represented by a BPA m from m_i and w_i $(i = 1, \ldots, L)$. Formally, we aim at obtaining the BPA m that combines all m_i's with taking weights w_i's into account in the form of the following

$$m = \bigoplus_{i=1}^{L} (w_i \otimes m_i) \tag{22}$$

where $\otimes$ is a product-type operation and $\oplus$ is a sum-type operation in general.

Under such a reformulation, we may have different schemes for obtaining the BPA m represented the generated assessment $S(y)$.

4.1   The Discounting-and-Orthogonal Sum Scheme

Let us first consider $\otimes$ as the discounting operation and $\oplus$ as the orthogonal sum in D-S theory. Then, for each $i = 1, \ldots, L$, we have $(w_i \otimes m_i)$ is a BPA (refer to (3–4)) defined by

$$(w_i \otimes m_i)(H_n) \overset{\triangle}{=} m_i^{w_i}(H_n) = w_i m_i(H_n) = w_i \beta_{n,i}, \text{ for } i = 1, \ldots, L \tag{23}$$

$$(w_i \otimes m_i)(\mathcal{H}) \overset{\triangle}{=} m_i^{w_i}(\mathcal{H}) = (1 - w_i) + w_i m_i(\mathcal{H})$$

$$= (1 - w_i) + w_i(1 - \sum_{n=1}^{N} \beta_{n,i}) = 1 - w_i \sum_{n=1}^{N} \beta_{n,i} \tag{24}$$

With this formulation, we consider each m_i as the belief quantified from the information source $S(e_i)$ and the weight w_i as the "probability" of $S(e_i)$ supporting the assessment of y.

Now Dempster's rule of combination allows us to combine BPAs $m_i^{w_i}$ $(i = 1, \ldots, L)$ under the independent assumption of information sources for generating the BPA m for the assessment of y. Namely,

$$m = \bigoplus_{i=1}^{L} m_i^{w_i} \tag{25}$$

where, with an abuse of the notation, $\oplus$ stands for the orthogonal sum.

It would be worth noting that two BPAs $m_i^{w_i}$ and $m_j^{w_j}$ are combinable, i.e. $(m_i^{w_i} \oplus m_j^{w_j})$ does exist, if and only if

$$\sum_{t=1}^{N} \sum_{\substack{n=1 \\ n \neq t}}^{N} m_i^{w_i}(H_n) m_j^{w_j}(H_t) < 1$$

For example, assume that we have two basic attributes e_1 and e_2 with

$$S(e_1) = \{(H_1, 0), (H_2, 0), (H_3, 0), (H_4, 1), (H_5, 0)\}$$
$$S(e_2) = \{(H_1, 0), (H_2, 0), (H_3, 1), (H_4, 0), (H_5, 0)\}$$

and both are equally important, or $w_1 = w_2$. If the weights w_1 and w_2 are normalized so that $w_1 = w_2 = 1$, then $(m_1^{w_1} \oplus m_2^{w_2})$ does not exist.

Note further that, by definition, focal elements of each $m_i^{w_i}$ are either singleton sets or the whole set $\mathcal{H}$. It is easy to see that m also verifies this property if applicable. Interestingly, the commutative and associative properties of Dempster's rule of combination with respect to a combinable collection of BPAs $m_i^{w_i}$ $(i = 1, \ldots, L)$ and the mentioned property essentially form the basis for the ER algorithms developed in [20, 25]. More particularly, with the same notations as in preceding section, we have

$$m(H_n) = m_{n,I(L)}, \text{ for } n = 1, \ldots, N \tag{26}$$

$$m(\mathcal{H}) = m_{\mathcal{H}, I(L)} \tag{27}$$

Further, by a simple induction, we easily see that the following holds

Lemma 1. *With the quantity $\overline{m}_{\mathcal{H}, I(L)}$ inductively defined by (16), we have*

$$\overline{m}_{\mathcal{H}, I(L)} = K_{I(L)} \prod_{i=1}^{L} (1 - w_i) \tag{28}$$

where $K_{I(L)}$ is inductively defined by (11).

Except the weight normalization, the key difference between the original ER algorithm and the modified ER algorithm is nothing but the way of assignment of β_n $(n = 1, \ldots, N)$ and $\beta_{\mathcal{H}}$ after obtained m. That is, in the original ER algorithm, the BPA m is directly used to define the assessment for y by assigning

$$\beta_n = m(H_n) = m_{n,I(L)}, \text{ for } n = 1, \ldots, N \tag{29}$$

$$\beta_{\mathcal{H}} = m(\mathcal{H}) = m_{\mathcal{H}, I(L)} \tag{30}$$

While in the modified ER algorithm, after obtained the BPA m, instead of using m to define the assessment for y as in the original ER algorithm, it defines a BPA m' derived from m as follows

$$m'(H_n) = \frac{m(H_n)}{1 - \overline{m}_{\mathcal{H}, I(L)}}, \text{ for } n = 1, \ldots, N \tag{31}$$

$$m'(\mathcal{H}) = \frac{(m(\mathcal{H}) - \overline{m}_{\mathcal{H}, I(L)})}{1 - \overline{m}_{\mathcal{H}, I(L)}} = \frac{\tilde{m}_{\mathcal{H}, I(L)}}{1 - \overline{m}_{\mathcal{H}, I(L)}} \tag{32}$$

Then the assessment for y is defined by assigning

$$\beta_n = m'(H_n), \text{ for } n = 1, \ldots, N \tag{33}$$

$$\beta_{\mathcal{H}} = m'(\mathcal{H}) \tag{34}$$

By (31)–(32), Theorem 1 straightforwardly follows as m is a BPA.

Lemma 2. *If all assessments $S(e_i)$ $(i = 1, \ldots, L)$ are complete, we have*

$$m(\mathcal{H}) = \overline{m}_{\mathcal{H}, I(L)} = K_{I(L)} \prod_{i=1}^{L} (1 - w_i) \tag{35}$$

i.e., $\tilde{m}_{\mathcal{H}, I(L)} = 0$; and, consequently, $S(y)$ defined by (33) is also complete.

As if $w_i = 0$ then the BPA $m_i^{w_i}$ immediately becomes the *vacuous* BPA, and, consequently, plays no role in the aggregation. Thus, without any loss of generality, we assume that $0 < w_i < 1$ for all $i = 1, \ldots, L$. Under this assumption, we are easily to see that if the assumption of the completeness axiom holds, then

$$\mathcal{F}_{m_i^{w_i}} = \{\{h_n\}|n \in I^+\} \cup \{\mathcal{H}\}, \text{ for } i = 1, \ldots, L \qquad (36)$$

where $\mathcal{F}_{m_i^{w_i}}$ denotes the family of focal elements of $m_i^{w_i}$. Hence, by a simple induction, we also have

$$\mathcal{F}_m = \{\{h_n\}|n \in I^+\} \cup \{\mathcal{H}\} \qquad (37)$$

Note that the assumption of the consensus axiom is the same as that of the completeness axiom with $|I^+| = 1$.

Therefore, the consensus and completeness axioms immediately follow from Lemma 2 along with (31)–(34) and (37).

It is also easily seen that

$$m(\mathcal{H}) = K_{I(L)} \prod_{i=1}^{L} m_i^{w_i}(\mathcal{H}) = K_{I(L)} \prod_{i=1}^{L} [w_i m_i(\mathcal{H}) + (1 - w_i)] \qquad (38)$$

and in addition, if there is an incomplete assessment $S(e_j)$ then $w_j m_j(\mathcal{H}) > 0$, resulting in

$$w_j m_j(\mathcal{H}) \prod_{\substack{i=1 \\ i \neq j}}^{L} (1 - w_i) > 0$$

This directly implies $m'(\mathcal{H}) > 0$. Consequently, the incompleteness axiom follows as (33)–(34).

4.2   The Discounting-and-Averaging Scheme

In this subsection, instead of applying the the orthogonal sum operation after discounting m_i's, we apply the averaging operation over L BPAs $m_i^{w_i}$ ($i = 1, \ldots, L$) to obtain a BPA $\overline{m}$ defined by

$$\overline{m}(H) = \frac{1}{L} \sum_{i=1}^{L} m_i^{w_i}(H), \text{ for any } H \subseteq \mathcal{H} \qquad (39)$$

Therefore, we have

$$\overline{m}(H) = \begin{cases} \frac{1}{L} \sum_{i=1}^{L} w_i \beta_{n,i}, & \text{if } H = \{H_n\} \\ \frac{1}{L} \sum_{i=1}^{L} \left(1 - w_i \sum_{n=1}^{N} \beta_{n,i}\right), & \text{if } H = \mathcal{H} \\ 0, & \text{otherwise} \end{cases} \qquad (40)$$

After obtaining the aggregated BPA $\overline{m}$, the problem now is to use $\overline{m}$ for generating the aggregated assessment for the general attribute y. Naturally, we can assign

$$\beta_n = \overline{m}(H_n) = \frac{1}{L} \sum_{i=1}^{L} w_i \beta_{n,i}, \text{ for } n = 1, \ldots, N \tag{41}$$

$$\beta_{\mathcal{H}} = \overline{m}(\mathcal{H}) = \frac{1}{L} \sum_{i=1}^{L} \left(1 - w_i \sum_{n=1}^{N} \beta_{n,i} \right) \tag{42}$$

Then the assessment for y is defined by

$$S(y) = \{(H_n, \beta_n) | n = 1, \ldots, N\} \tag{43}$$

Regarding the synthesis axioms, we easily see that the first axiom holds for the assessment (43). For the next two axioms, we have the following

Theorem 3. *The assessment (43) defined via (41)–(42) satisfies the consensus axiom and/or the completeness axiom if and only if $w_i = 1$ for all $i = 1, \ldots, L$.*

The assessment for y according to this aggregation scheme also satisfies the incompleteness axiom trivially due to the nature of discounting-and-averaging.

Unfortunately, the requirement of $w_i = 1$ for all i to satisfy the consensus axiom and the completeness axiom would not be appropriate in general. This is due to the allocation of the average of discount rates

$$\overline{\alpha} \triangleq \left(1 - \frac{\sum_{i=1}^{L} w_i}{L} \right)$$

to $\mathcal{H}$ as a part of unassigned probability mass. This dilemma can be resolved in a similar way as in the modified algorithms above. Interestingly, this modification leads to the weighted sum scheme as shown in the following.

4.3   Weighted Sum as the Modified Discounting-and-Averaging Scheme

By applying the discounting-and-averaging scheme, we obtain the BPA $\overline{m}$ as defined by (40). Now, guided by the synthesis axioms, instead of making direct use of $\overline{m}$ in defining the generated assessment $S(y)$ (i.e., allocating the average discount rate $\overline{\alpha}$ to $\beta_{\mathcal{H}}$ as a part of unassigned probability mass) as above, we define a new BPA denoted by $\overline{m}'$ derived from $\overline{m}$ by making use of $(1 - \overline{\alpha})$ as a normalization factor. More particularly, we define

$$\overline{m}'(H_n) = \frac{\overline{m}(H_n)}{1 - \overline{\alpha}}, \text{ for } n = 1, \ldots, N \tag{44}$$

$$\overline{m}'(\mathcal{H}) = \frac{\overline{m}(\mathcal{H}) - \overline{\alpha}}{1 - \overline{\alpha}} \tag{45}$$

Then by (40) and a simple transformation, we easily obtain

$$\overline{m}'(H_n) = \sum_{i=1}^{L} \overline{w}_i \beta_{n,i}, \text{ for } n = 1, \ldots, N \tag{46}$$

$$\overline{m}'(\mathcal{H}) = \sum_{i=1}^{L} \overline{w}_i \left(1 - \sum_{n=1}^{N} \beta_{n,i} \right) \tag{47}$$

where

$$\overline{w}_i = \frac{w_i}{\sum_{i=1}^{L} w_i}, \text{ for } i = 1, \ldots, L$$

Let us turn back to the general scheme of combination given in (22). Under the view of this general scheme, the above BPA $\overline{m}'$ is nothing but an instance of it by simply considering $\otimes$ as the multiplication and $\oplus$ as the weighted sum. Namely, we have

$$\overline{m}'(H_n) = \sum_{i=1}^{L} \overline{w}_i m_i(H_n), \text{ for } n = 1, \ldots, N \tag{48}$$

$$\overline{m}'(\mathcal{H}) = \sum_{i=1}^{L} \overline{w}_i m_i(\mathcal{H}) \tag{49}$$

where relative weights $\overline{w}_i$ are normalized as above so that $\sum_i \overline{w}_i = 1$. It is of interest to note that the possibility of using such an operation has previously been mentioned in, for example, [18]. Especially, the weighted sum operation of two BPAs has been used for the integration of distributed databases for purposes of data mining [10].

Now we quite naturally define the assessment for y by assigning

$$\beta_n = \overline{m}'(H_n) = \sum_{i=1}^{L} \overline{w}_i m_i(H_n), \text{ for } n = 1, \ldots, N \tag{50}$$

$$\beta_{\mathcal{H}} = \overline{m}'(\mathcal{H}) = \sum_{i=1}^{L} \overline{w}_i m_i(\mathcal{H}) \tag{51}$$

Appealingly simple as it is, we can see quite straightforwardly that the following holds.

Proposition 1. *The degrees of belief generated using (50)–(51) satisfy the following*

$$0 \leq \beta_n, \beta_{\mathcal{H}} \leq 1, \text{ for } n = 1, \ldots, N$$

$$\sum_{n=1}^{N} \beta_n + \beta_{\mathcal{H}} = 1$$

Furthermore, we have the following theorem.

Theorem 4. *The aggregated assessment for y defined as in (50)–(51) exactly satisfies all four synthesis axioms.*

4.4   Expected Utility in the ER Approaches

In the tradition of decision making under uncertainty [12], the notion of expected utility has been mainly used to rank alternatives in a particular problem. That is one can represent the preference relation $\succeq$ on a set of alternatives X with a single-valued function $u(x)$ on X, called *expected utility*, such that for any $x, y \in X$, $x \succeq y$ if and only if $u(x) \geq u(y)$. Maximization of $u(x)$ over X provides the solution to the problem of selecting x.

In the ER approach, we assume a utility function $u' : \mathcal{H} \to [0, 1]$ satisfying

$$u'(H_{n+1}) > u'(H_n) \text{ if } H_{n+1} \text{ is preferred to } H_n.$$

This utility function u' may be determined using the probability assignment method [8] or using other methods as in [20, 25].

If all assessments for basic attributes are complete, Lemma 2 shows that the assessment for y is also complete, i.e. $\beta_{\mathcal{H}} = 0$. Then the expected utility of an alternative on the attribute y is defined by

$$u(y) = \sum_{n=1}^{N} \beta_n u'(H_n) \tag{52}$$

An alternative a is strictly preferred to another alternative b if and only if $u(y(a)) > u(y(b))$.

Due to incompleteness, in general, in basic assessments, the assessment for y may result in incomplete. In such a case, in [25] the authors defined three measures, called minimum, maximum and average expected utilities, and proposed a ranking scheme based on these measures (see, e.g., [25] for more details).

In this paper, based on the *Generalized Insufficient Reason Principle*, we define a probability function P_m on $\mathcal{H}$ derived from m for the purpose of making decisions via the *pignistic transformation* [15]. Namely,

$$P_m(H_n) = m(H_n) + \frac{1}{N} m(\mathcal{H}) \text{ for } n = 1, \ldots, N \tag{53}$$

That is, as in the two-level language of the so-called *transferable belief model* [15], the aggregated BPA m itself represented the belief is entertained based on the available evidence at the *credal level*, and when a decision must be made, the belief at the credal level induces the probability function P_m defined by (53) for decision making. Particularly, the approximately assessment for y for the purpose of decision making is then defined as

$$\beta'_n = P_m(H_n) = \beta_n + \frac{1}{N} \beta_{\mathcal{H}}, \text{ for } n = 1, \ldots, N \tag{54}$$

Therefore, the expected utility of an alternative on the attribute y is straightforwardly defined by

$$u(y) = \sum_{n=1}^{N} \beta'_n u'(H_n) = \sum_{n=1}^{N} (\beta_n + \frac{1}{N} \beta_{\mathcal{H}}) u'(H_n) \tag{55}$$

In fact, while the amount of belief $\beta_{\mathcal{H}}$ (due to ignorance) is allocated either to the least preferred grade H_1 or to the most preferred grade H_N to define the expected utility interval in Yang's approach [25], it is uniformly allocated to every evaluation grade H_n, guided by the Generalized Insufficient Reason Principle [15], to define an approximately assessment for y and, hence, a single-valued expected utility function.

5   An Example: Motorcycle Assessment Problem

The problem is to evaluate the performance of four types of motorcycles, namely *Kawasaki, Yamaha, Honda,* and *BMW.*

The overall performance of each motorcycle is evaluated based on three major attributes which are *quality of engine, operation, general finish*. The process of attribute decomposition for the evaluation problem of motorcycles results in a hierarchy graphically depicted in Fig. 2, where the relative weights of attributes at a single level associated with the same upper level attribute are defined by w_i, w_{ij}, and w_{ijk}, respectively.

Using the five-grade evaluation scale as given in (1), the assessment problem of motorcycles is given in Table 1, where P, I, A, G, and E are the abbreviations

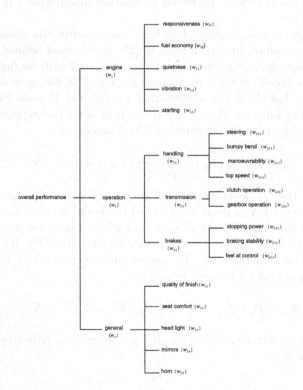

Fig. 2. Evaluation hierarchy for motorcycle performance assessment [25]

Table 1. Generalized Decision Matrix for Motorcycle Assessment [25]

General attributes		Basic attributes	types of motor cycle (alternatives)			
			Kawasaki (a_1)	Yamaha (a_2)	Honda (a_3)	BMW (a_4)
Overall performance	engine	responsiveness	E (0.8)	G (0.3) E (0.6)	G (1.0)	I (1.0)
		fuel economy	A(1.0)	I (1.0)	I(0.5) A(0.5)	E(1.0)
		quietness	I(0.5) A(0.5)	A(1.0)	G(0.5) E(0.3)	E(1.0)
		vibration	G (1.0)	I (1.0)	G(0.5) E(0.5)	P(1.0)
		starting	G (1.0)	A(0.6) G(0.3)	G (1.0)	A(1.0)
		steering	E(0.9)	G(1.0)	A(1.0)	A(0.6)
	handling	bumpy bends	A(0.5) G(0.5)	G (1.0)	G(0.8) E(0.1)	P(0.5) I(0.5)
		maneuverability	A(1.0)	E(0.9)	I (1.0)	P(1.0)
		top speed stability	E(1.0)	G(1.0)	G(1.0)	G(0.6) E(0.4)
	transmission	clutch operation	A(0.8)	G(1.0)	E(0.85)	I(0.2) A(0.8)
		gearbox operation	A(0.5) G(0.5)	I(0.5) A(0.5)	E(1.0)	P(1.0)
	brakes	stopping power	G(1.0)	A(0.3) G(0.6)	G(0.6)	E(1.0)
		braking stability	G(0.5) E(0.5)	G(1.0)	A(0.5) G(0.5)	E(1.0)
		feel at control	P(1.0)	G(0.5) E(0.5)	G(1.0)	G(0.5) E(0.5)
	general	quality of finish	P(0.5) I(0.5)	G(1.0)	E(1.0)	G(0.5) E(0.5)
		seat comfort	G(1.0)	G(0.5) E(0.5)	G(0.6)	E(1.0)
		headlight	G(1.0)	A(1.0)	E(1.0)	G(0.5) E(0.5)
		mirrors	A(0.5) G(0.5)	G(0.5) E(0.5)	E(1.0)	G(1.0)
		horn	A(1.0)	G(1.0)	G(0.5) E(0.5)	E(1.0)

of *poor*, *indifferent*, *average*, *good*, and *excellent*, respectively, and a number in bracket denoted the degree of belief to which an attribute is assessed to a grade. For example, $E(0.8)$ means "*excellent* to a degree of 0.8".

Further, all relevant attributes are assumed to be of equal relative important [25]. That is

$$w_1 = w_2 = w_3 = 0.3333$$
$$w_{11} = w_{12} = w_{13} = w_{14} = w_{15} = 0.2$$
$$w_{21} = w_{22} = w_{23} = 0.3333$$
$$w_{211} = w_{212} = w_{213} = w_{214} = 0.25$$
$$w_{221} = w_{222} = 0.5$$
$$w_{231} = w_{232} = w_{233} = 0.3333$$
$$w_{31} = w_{32} = w_{33} = w_{34} = w_{35} = 0.2$$

In the sequent, for the purpose of comparison, we generate two different results of aggregation via the modified ER approach (refer to (33)–(34)), and the new approach taken in this paper (refer to (50)–(51)).

By applying the modified ER approach, the distributed assessments for overall performance of four types of motorcycles are given in Table 2. These four distributions are graphically shown as in Fig. 3 (a).

Table 2. Aggregated assessments for four types of motorcycles obtained by using the modified ER approach [25]

	$Poor(P)$	$Indifference(I)$	$Average(A)$	$Good(G)$	$Excellent(E)$	$Unknown(U)$
Kawasaki	0.0547	0.0541	0.3216	0.4452	0.1058	0.0186
Yamaha	0.0	0.1447	0.1832	0.5435	0.1148	0.0138
Honda	0.0	0.0474	0.0621	0.4437	0.4068	0.0399
BMW	0.1576	0.0792	0.1124	0.1404	0.5026	0.0078

At the same time, by applying the weighted sum aggregation scheme, we obtain the distributed assessments for overall performance of four types of motorcycles as shown in Table 3 (graphically depicted in Fig. 3 (b)).

As we can easily see, it is not much difference between the result obtained by the modified ER algorithm and that obtained by our method, especially the behavior of correspondingly assessment distributions as Fig. 3 has shown.

Now, as mentioned above, for the purpose of making decisions we apply the *pignistic transformation* (refer to (53)) to obtain the approximately assessment for overall performance of motorcycles given in Table 4 below.

Assume the same utility function $u' : \mathcal{H} \to [0,1]$ as in [25] defined by

$$u'(P) = 0, u'(I) = 0.35, u'(A) = 0.55, u'(G) = 0.85, u'(E) = 1$$

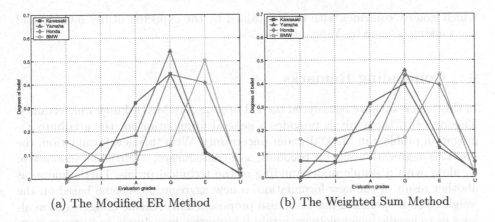

(a) The Modified ER Method   (b) The Weighted Sum Method

Fig. 3. Overall Evaluation of Motorcycles

Table 3. Aggregated assessments for four types of motorcycles obtained by using the weighted sum aggregation scheme

	Poor(P)	Indifference(I)	Average(A)	Good(G)	Excellent(E)	Unknown(U)
Kawasaki	0.0703	0.0667	0.3139	0.3972	0.1247	0.0272
Yamaha	0.0	0.1611	0.2122	0.4567	0.1501	0.0198
Honda	0.0	0.0611	0.0796	0.4344	0.3922	0.0659
BMW	0.1639	0.0917	0.1278	0.1685	0.437	0.0111

Table 4. Approximately assessments for four types of motorcycles obtained by using the pignistic transformation

	Poor(P)	Indifference(I)	Average(A)	Good(G)	Excellent(E)
Kawasaki	0.07574	0.07214	0.31934	0.40264	0.13014
Yamaha	0.00396	0.16506	0.21616	0.46066	0.15406
Honda	0.01318	0.07428	0.09278	0.44758	0.40538
BMW	0.16612	0.09392	0.13	0.17072	0.43922

Using (55), we easily obtain the expected utility of motorcycles given as

$$u(Kawasaki) = 0.6733, \ u(Yamaha) = 0.7223$$
$$u(Honda) = 0.8628, \ u(BMW) = 0.6887$$

Consequently, the ranking of the four types of motorcycles is given by

$$Honda \succ Yamaha \succ BMW \succ Kawasaki$$

which exactly coincides with that obtained by the expected utility interval and the ranking scheme by Yang and Xu [25].

6 Concluding Remarks

In this paper, we have reanalyzed the ER approach to MADM under uncertainty. Theoretically, the analysis provides a general formulation for the attribute aggregation problem in MADM under uncertainty. With this new formulation, the previous aggregation scheme becomes, as a consequence, a particular instance of it, along with a simple understanding of the technical proofs. Furthermore, as another result of the new formulation, a new aggregation scheme based on the weighted sum operation has been also proposed. This scheme of aggregation allows us to handle incomplete uncertain information in a simple and proper manner when the assumption regarding the independence of attributes' uncertain evaluations is not appropriate. For the purpose of decision making, an approximate method of uncertain assessments based on the so-called pignistic transformation [15] has been applied to define the expected utility function, instead of using the expected utility interval proposed previously. A tutorial example has been examined to illustrate the discussed techniques.

References

1. G. Balestra, A. Tsoukias, Multicriteria analysis represented by artificial intelligence techniques, *European Journal of Operational Research* **41** (1990) 419–430.
2. V. Belton, Comparison of the analytic hierarchy process and a simple multi-attribute value function, *European Journal of Operational Research* **26** (1986) 7–21.
3. V. Belton, T. J. Stewart, *Multiple Criteria Decision Analysis: An Integrated Approach* (Kluwer, Norwell, MA, 2002).
4. B. G. Buchanan, E. H. Shortliffe, *Rule-Based Expert Systems* (Addison-Wesley, Reading, MA, 1984).
5. S. J. Chen, C. L. Hwang, *Fuzzy Multiple Attribute Decision Making-Methods and Applications.* Berlin: Springer, 1992.
6. T. Isitt, The Sports Tourers, *Motor Cycle International* **64** (1990) 19–27.
7. C. L. Hwang and K. Yoon, *Multiple Attribute Decision Making Methods and Applications* (Springer-Verlag, New York, 1981).
8. R. L. Keeney, H. Raiffa, *Decisions with Multiple Objectives: Preferences and Value Tradeoffs* (John Wiley & Sons, New York, 1976).
9. L. Levrat, A. Voisin, S. Bombardier, J. Bremont, Subjective evaluation of car seat comfort with fuzzy set techniques, *International Journal of Intelligent Systems* **12** (1997) 891–913.
10. S. McClean, B. Scotney, Using evidence theory for the integration of distributed databases, *International Journal of Intelligent Systems* **12** (1997) 763–776.
11. M. Roubens, Fuzzy sets and decision analysis, *Fuzzy Sets and Systems* **90** (1997) 199–206.
12. L. J. Savage, *The Foundations of Statistics* (John Wiley and Sons, New York, 1954).
13. P. Sen, J. B. Yang, Multiple criteria decision making in design selection and synthesis, *Journal of Engineering Design* **6** (1995) 207–230.

14. G. Shafer, *A Mathematical Theory of Evidence* (Princeton University Press, Princeton, 1976)

15. P. Smets, R. Kennes, The transferable belief model, *Artificial Intelligence* **66** (1994) 191–234.

16. T. J. Stewart, A critical survey on the status of multiple criteria decision making theory and practice, *OMEGA International Journal of Management Science* **20** (5–6) (1992) 569–586.

17. J. Wang, J. B. Yang, P. Sen, Multi-person and multi-attribute design evaluations using evidential reasoning based on subjective safety and cost analysis, *Reliability Engineering and System Safety* **52** (1996) 113–127.

18. R. R. Yager, Non-monotonic reasoning with belief structures, in R. R. Yager, M. Fedrizzi, J. Kacprzyk (Eds.), *Advances in the Dempster-Shafer Theory of Evidence* (Wiley, New York, 1994) pp. 534–554.

19. R. R. Yager, Decision-making under various types of uncertainties, *Journal of Intelligent & Fuzzy Systems* **3** (1995) 317–323.

20. J. B. Yang, M. G. Singh, An evidential reasoning approach for multiple attribute decision making with uncertainty, *IEEE Transactions on Systems, Man, and Cybernetics* **24** (1994) 1–18.

21. J. B. Yang, P. Sen, A general multi-level evaluation process for hyprid multiple attribute decision making with uncertainty, *IEEE Transactions on Systems, Man, and Cybernetics* **24** (1994) 1458–1473.

22. J. B. Yang, P. Sen, Multiple attribute design evaluation of large engineering products using the evidential reasoning approach, *Journal of Engineering Design* **8** (1997) 211–230.

23. J. B. Yang, Rule and utility based evidential reasoning approach for multiple attribute decision analysis under uncertainty, *European Journal of Operational Research* **131** (1) (2001) 31–61.

24. J. B. Yang, D. L. Xu, Nonlinear information aggregation via evidential reasoning in multiattribute decision analysis under uncertainty, *IEEE Transactions on Systems, Man, and Cybernetics* **32** (2002) 376–393.

25. J. B. Yang, D. L. Xu, On the evidential reasoning algorithm for multiple attribute decision analysis under uncertainty, *IEEE Transactions on Systems, Man, and Cybernetics* **32** (2002) 289–304.

26. Z. J. Zhang, J. B. Yang, D. L. Xu, A hierarchical analysis model for multiobjective decision making, in *Analysis, Design and Evaluation of Man-Machine Systems*, selected papers from the 4th IFAC/IFIP/ IFORS/IEA Conference, Xian, P.R. China, September 1989. Pergamon, Oxford, UK, 1990, pp. 13–18.

Learnability of Simply-Moded Logic Programs from Entailment

M. R. K. Krishna Rao

Information and Computer Science Department,
King Fahd University of Petroleum and Minerals,
Dhahran 31261, Saudi Arabia
krishna@ccse.kfupm.edu.sa

Abstract. In this paper, we study exact learning of logic programs from entailment queries and present a polynomial time algorithm to learn a rich class of logic programs that allow local variables and include many standard programs like addition, multiplication, exponentiation, member, prefix, suffix, length, append, merge, split, delete, insert, insertion-sort, quick-sort, merge-sort, preorder and inorder traversal of binary trees, polynomial recognition, derivatives, sum of a list of naturals. Our algorithm asks at most polynomial number of queries and our class is the largest of all the known classes of programs learnable from entailment.

1   Introduction

Logic programs with elegant and simple declarative semantics have become very common in many areas of artificial intelligence such as knowledge acquisition, knowledge representation and common sense and legal reasoning. The problem of learning logic programs from examples and queries [21, 22] has attracted a lot of attention in the last fifteen years. Several techniques and systems for learning logic programs are developed and used in many applications. See [15] for a survey. In this paper, we consider the framework of *learning from entailment* [1,2,4-8,12,13,17-19] and present a polynomial time algorithm to learn a rich class of logic programs.

The framework of *learning from entailment* has been introduced by Angluin [1] and Franzier and Pitt [7] to study learnability of propositional Horn sentences. In the last few years, this framework (with a few modifications) has been used in learning first order Horn programs and many results have been published in [17, 4, 13, 12, 19]. In [4, 13, 12, 19], the learner is allowed to ask the following types of queries in learning a concept (logic program) from a teacher. Through an *entailment equivalence query* $EQUIV(H)$, the learner asks the teacher whether his program H is logically equivalent to the target program H^* or not. The teacher answers 'yes' to this query if H and H^* are equivalent, i.e., $H \models H^*$ and $H^* \models H$. Otherwise, the teacher produces a clause C such that $H^* \models C$ but $H \not\models C$ or $H^* \not\models C$ but $H \models C$. A *subsumption query* $SUBSUME(C)$ produces an answer 'yes' if the clause C is subsumed by a clause in H^*, otherwise answer

M.J. Maher (Ed.): ASIAN 2004, LNCS 3321, pp. 128–141, 2004.

'no'. Besides, equivalence and subsumption queries, the learner is also *allowed to ask for hints*. If C is a clause $A \leftarrow B_1, \cdots, B_n$ such that $H^* \models C$, the request-for-hint query $REQ(C)$ returns (1) an answer 'subsumed' if $A \leftarrow B_1, \cdots, B_n$ is subsumed by a clause in H^*, otherwise returns (2) an atom (hint) B in the proof of $H^* \models C$.

In this paper, we identify a class of programs and present an algorithm to learn them using entailment queries. This class is a major fragment of the well-known class of well-moded programs and is the largest of all the known classes of programs learnable from entailment. Furthermore, our algorithm takes at most polynomial (in the size of the largest counterexample/hint provided by the teacher) number of queries.

The rest of the paper is organized as follows. The next section gives preliminary definitions and section 3 defines the class of simply-moded programs. Section 4 presents a few results about subsumption and entailment and section 5 presents the learning algorithm for simply-moded programs. Section 6 provides correctness proof of the learning algorithm and section 7 compares our results with the existing results.

2   Preliminaries

Assuming that the reader is familiar with the basic terminology of first order logic and logic programming [14], we use the first order logic language with a finite set Σ of function symbols and a finite set Π of predicate symbols including Prolog builtins $=, <, >, \leq, \geq, \neq$. The arity of a predicate/function symbol f is denoted by $arity(f)$. Function symbols of arity zero are also called constants. The size of a term/atom/clause/program is defined as the number of (occurrences of) variables, predicate and function symbols in it. For a sequence $\mathbf{s}$ of terms, $[\![\mathbf{s}]\!]$ denotes the sum of the sizes of all the terms in $\mathbf{s}$.

Definition 1. A *mode* m of an n-ary predicate p is a function from $\{1, \cdots, n\}$ to the set $\{in, out\}$. The sets $in(p) = \{j \mid m(j) = in\}$ and $out(p) = \{j \mid m(j) = out\}$ are the sets of input and output positions of p respectively.

A moded program is a logic program with each predicate having a unique mode associated with it. In the following, $p(\mathbf{s}; \mathbf{t})$ denotes an atom with input terms $\mathbf{s}$ and output terms $\mathbf{t}$. The set of variables occurring in a sequence of terms $\mathbf{t}$ is denoted by $Var(\mathbf{t})$.

Definition 2. A definite clause

$$p_0(\mathbf{s_0}; \mathbf{t_0}) \leftarrow p_1(\mathbf{s_1}; \mathbf{t_1}), \cdots, p_k(\mathbf{s_k}; \mathbf{t_k})$$

$k \geq 0$ is *well-moded* if (a) $Var(\mathbf{t_0}) \subseteq Var(\mathbf{s_0}, \mathbf{t_1}, \cdots, \mathbf{t_k})$ and (b) $Var(\mathbf{s_i}) \subseteq Var(\mathbf{s_0}, \mathbf{t_1}, \cdots, \mathbf{t_{i-1}})$ for each $\mathbf{i} \in [1, k]$. A logic program is *well-moded* if each clause in it is well-moded.

The class of well-moded programs is extensively studied in the literature and the following lemma is one of the well-known facts about well-moded programs.

Lemma 1. Let P be a well-moded program and Q be the query $\leftarrow p(\mathbf{s}; \mathbf{t})$ with ground input terms $\mathbf{s}$. *If there is an SLD-refutation of $P \cup \{Q\}$ with θ as computed answer substitution then $\mathbf{t}\theta$ is ground as well.*

Definition 3. A well-moded program P is *deterministic* if every predicate p defined in P is *deterministic*, i.e., $\mathbf{t}_1 \equiv \mathbf{t}_2$ whenever $P \models p(\mathbf{s}; \mathbf{t}_1)$ and $P \models p(\mathbf{s}; \mathbf{t}_2)$ for any sequence of ground input terms $\mathbf{s}$.

In this paper, we only consider deterministic well-moded programs.

3   Simply-Moded Programs

In this section, we introduce the class of simply-moded programs, for which a learning algorithm is presented in a later section.

Definition 4. A well-moded program P is *simply-moded* [3] if each non-unit clause

$$p_0(\mathbf{s}_0; \mathbf{t}_0) \leftarrow p_1(\mathbf{s}_1; \mathbf{t}_1), \cdots, p_k(\mathbf{s}_k; \mathbf{t}_k)$$

$k \geq 1$, in P satisfies the following: for each $i \in [1, k]$, every term in $\mathbf{s}_i$ is either (1) a term in $\mathbf{t}_1, \cdots, \mathbf{t}_{i-1}$ or a subterm of a term in $\mathbf{s}_0$ or (2) of the form $f(u_1, \ldots, u_n)$ such that each u_j is either (a) a term in $\mathbf{t}_1, \cdots, \mathbf{t}_{i-1}$ or (b) a subterm of a term in $\mathbf{s}_0$.

Remark 1. It may be noted that u_j cannot be a proper subterm of a term in $\mathbf{t}_1, \cdots, \mathbf{t}_{i-1}$, though it can be proper subterm of a term in $\mathbf{s}_0$. This is a crucial restriction needed to obtain the *polynomial-time* learnability, as shown in the sequel.

Example 1. Consider the following quick-sort program.

> Moding: qs (in, out), app (in, in, out) and
> part (in, in, out, out).
>
> app([ ], Ys, Ys) $\leftarrow$
> app([X|Xs], Ys, [X|Zs]) $\leftarrow$ app(Xs, Ys, Zs)
>
> part([ ], H, [ ], [ ]) $\leftarrow$
> part([X|Xs], H, [X|Ls], Bs) $\leftarrow$ X $\leq$ H, part(Xs, H, Ls, Bs)
> part([X|Xs], H, Ls, [X|Bs]) $\leftarrow$ X $>$ H, part(Xs, H, Ls, Bs)
>
> qs([ ], [ ]) $\leftarrow$
> qs([H|L], S) $\leftarrow$ part(L, H, A, B), qs(A, A1), qs(B, B1), app(A1, [H|B1], S)

Input terms of each atom (except app(A1, [H|B1], S) in the last clause) are subterms of input terms of head and output terms of the atoms before that. For atom app(A1, [H|B1], S) in the last clause, the term [H|B1] is of the form $f(u_1, \ldots, u_n)$ such that $u_1, \ldots, u_n$ are subterms of input terms of head and output terms of the earlier atoms. Hence, this program is simply-moded.   □

Example 2. Consider the following program for exponentiation.

> Moding: add(in,in, out), mult(in,in, out) and
>   power(in,in, out).

> add(0, Y, Y) ←
> add(s(X), Y, s(Z)) ← add(X, Y, Z)

> mult(0, Y, 0) ←
> mult(s(X), Y, Z) ← mult(X, Y, Z1), add(Y, Z1, Z)

> power(0, Y, 1) ←
> power(s(X), Y, Z) ← power(X, Y, Z1), mult(Y, Z1, Z)

It is easy to see that this program is simply-moded.   □

4   Subsumption and Entailment

Definition 5. Let C_1 and C_2 be clauses $H_1 \leftarrow Body_1$ and $H_2 \leftarrow Body_2$ respectively. We say C_1 *subsumes* C_2 and write $C_1 \succeq C_2$ if there exists a substitution θ such that $H_1\theta \equiv H_2$ and $Body_1\theta \subseteq Body_2$.

Definition 6. A program P *entails* a clause C, denoted by $P \models C$, if C is a logical consequence of P.

The relation between subsumption and entailment is discussed below.

Definition 7. A *derivation* of a clause C from a program P is a finite sequence of clauses $C_1, \ldots, C_k = C$ such that each C_i is either an instance of a clause in P or a resolvent of two clauses in $C_1, \ldots, C_{i-1}$. If such a derivation exists, we write $P \vdash_d C$.

The following theorem is proved in Nienhuys-Cheng and de Wolf [16].

Theorem 1. (Subsumption Theorem)
Let P be a program and C be a clause. Then $P \models C$ *if and only if one of the following holds:*
(1) C is a tautology or
(2) there exists a clause D such that $P \vdash_d D$ and D subsumes C.

When C is ground, the above theorem can be reformulated as the following theorem.

Definition 8. An SLD-refutation is *minimal* if selected atoms are resolved with unit clauses whenever possible.

Theorem 2. Let P be a program and C be a ground clause $A \leftarrow B_1, \cdots, B_n$. Then $P \models C$ *if and only if one of the following holds.*
(1) C is a tautology.
(2) C is subsumed by a clause in P.
(3) There is a minimal SLD-refutation of $P \cup \{B_i \leftarrow \ | \ i \in [1,n]\} \cup \{\leftarrow A\}$.

Even though (2) is covered by (3) in the above theorem, we explicitly mention (2) in view of its importance in our learning algorithm.

Lemma 2. *If C_1 and C_2 are two simply-moded clauses, $C_1 \succeq C_2$ is decidable in polynomial time over the sizes of C_1 and C_2.*
Proof : Easy. ☐

5   Learning Algorithm

In this section, we present an algorithm **Learn-SM$_l$** for exact learning of **terminating** simply-moded programs, whose clauses have no more than l atoms in the body, from entailment using equivalence, subsumption and request-for-hint queries. When C is a ground clause $A \leftarrow B_1, \cdots, B_n$ such that $H^* \models C$, the request-for-hint query $REQ(C)$ returns (1) an answer 'subsumed' if $A \leftarrow B_1, \cdots, B_n$ is subsumed by a clause in H^*, otherwise returns (2) an atom (hint) $B\theta$ in a minimal SLD-refutation of $H^* \cup \{\leftarrow A\}$ with answer substitution θ such that $B\theta \notin \{B_1, \cdots, B_n\}$. As we are interested in polynomial time learnability, we consider programs for which the above three types of queries can be answered by the teacher in polynomial time[1].

Algorithm **Learn-SM$_l$** uses the notions of saturation [9, 20] and least general generalization (lgg).

Definition 9. A clause C is a saturation of an example E w.r.t. a theory (program) H if and only if C is a reformulation of E w.r.t. H and $C' \Rightarrow C$ for every reformulation C' of E w.r.t. H. A clause D is a reformulation of E w.r.t. H if and only if $H \wedge E \Leftrightarrow H \wedge D$.

We are concerned with simply-moded programs and define saturation of an example E w.r.t. a program H as $E \leftarrow Closure_H(E)$, where $Closure_H(E)$ is the following set of ground atoms.

---

[1] This requirement is similar to the restriction that $|D_B(t)|$ is bounded by a polynomial in the size of t, placed in [10, 13]. In fact, if the above three types of queries cannot be answered by the teacher in polynomial time (which essentially means that some SLD-derivations are of exponential –in the size of the initial query– length), it is not possible to learn from entailment in polynomial time, as the teacher in a bad mood keeps giving hints from such long derivations.

Definition 10. The closure of an example $E \equiv p_0(\mathbf{s}_0; \mathbf{t}_0)$ w.r.t. a program H is defined as $Closure_H(E) = S_1 \cup S_2 \cup \cdots \cup S_l$, where

1. $T_1 = \{s \mid s$ is a subterm of a term in $\mathbf{s}_0\}$ and
2. for $1 \le i \le l$, $S_i = \{p(\mathbf{u}; \mathbf{v}) \mid H \models p(\mathbf{u}; \mathbf{v})$ and each term in $\mathbf{u}$ is either a term in T_i or of the form $f(\mathbf{w})$ such that each term in $\mathbf{w}$ is a term in T_i, f is a function symbol in Σ and p is a predicate symbol in Π with at least one output position$\}$,
3. for $1 \le i < l$, $T_{i+1} = T_i \cup \{s \mid s$ is an output term of an atom in $S_i\}$.

Definition 11. Let C_1 and C_2 be two simply-moded clauses $A_1 \leftarrow Body_1$ and $A_2 \leftarrow Body_2$ respectively. The *least general generalization* $C_1 \sqcup C_2$ of C_1 and C_2 is defined as a simply-moded clause $A \leftarrow Body$ if σ_1 and σ_2 are two least general substitutions such that (1) A is the least general generalization of $A_1 \equiv A\sigma_1$ and $A_2 \equiv A\sigma_2$ and (2) $Body = \{B \mid B\sigma_i \in Body_i, i \in [1,2]\}$.

Procedure **Learn-SM**$_l$;
begin $H := \phi$;
while $EQUIV(H) \ne$ 'yes' do
  begin $A := EQUIV(H)$; $C := A \leftarrow Closure_H(A)$;
    while $REQ(C)$ returns a hint B do $C := B \leftarrow Closure_H(B)$;
    % This **while** loop exits when C is subsumed by a clause in H^*. %
    $C := \mathbf{Reduce}(C)$;
    if $SUBSUME(C \sqcup D)$ returns 'yes' for some clause $D \in H$
        then $H := H \cup \{\mathbf{Reduce}(C \sqcup D)\} - \{D\}$ else $H := H \cup \{C\}$
  end;
Return(H)
end **Learn-SM**$_l$;

Function **Reduce**$(A \leftarrow Body)$;
begin
    for each atom $B \in Body$ do
        if $SUBSUME(A \leftarrow (Body - \{B\}))$ then
            $Body := (Body - \{B\})$;
    Return$(A \leftarrow Body)$
end **Reduce**;

Remark 2. It may be noted that the application of the above function **Reduce** is not mandatory for the correctness of the algorithm **Learn-SM**$_l$, but it improves the efficiency. In particular, checking subsumption of reduced clauses is easier than that of non-reduced clauses.

Remark 3. The structure of **Learn-SM**$_l$ is very similar to that of the learning algorithms given in [4, 13, 12, 19]. However, the construction of new clauses from the hints, in particular $Closure_H(B)$, is different from the existing works, and facilitates learning the largest known class of programs, i.e., simply-moded programs.

Example 3. We illustrate the working of **Learn-SM**$_l$ by outlining the trace of **Learn-SM**$_l$ run on the standard quick-sort program given in Example 1. For this example, we take $l = 4$. **Learn-SM**$_l$ starts with $H = \phi$ as the initial hypothesis and query $EQUIV(H)$ returns a counterexample, say $A = qs([3, 1, 5], [1, 3, 5])$. The inner **while** loop asks $REQ(A \leftarrow Closure_H(A))$, which results in a hint, say $B = part([1, 5], 3, [1], [5])$. Then, it asks $REQ(B \leftarrow Closure_H(B))$ and continue. This loop terminates with a unit clause

$$part([\ ], 3, [\ ], [\ ]) \leftarrow$$

which will be added to H in the **if** statement.

The outer **while** loop asks $EQUIV(H)$ and gets a counterexample, say $A_1 = qs([5, 2, 7], [2, 5, 7])$. The inner **while** loop terminates with a clause

$$part([\ ], 5, [\ ], [\ ]) \leftarrow .$$

The **if** statement replaces the clause $part([\ ], 3, [\ ], [\ ]) \leftarrow$ in H by the lgg of this clause and the above clause, i.e., $part([\ ], H, [\ ], [\ ]) \leftarrow$.

The outer **while** loop asks $EQUIV(H)$ and gets a counterexample, say $A_2 = qs([5, 2, 7, 3], [2, 3, 5, 7])$. The inner **while** loop asks $REQ(A_2 \leftarrow Closure_H(A_2))$, which results in a hint, say $B_2 = part([3], 5, [3], [\ ])$. In the next iteration, it asks $REQ(B_2 \leftarrow Closure_H(B_2))$ and gets answer 'subsumed' as $Closure_H(B_2)$ includes $part([\ ], 5, [\ ], [\ ])$ as well as $3 \leq 5$. The clause

$$part([3], 5, [3], [\ ]) \leftarrow 3 \leq 5, part([\ ], 5, [\ ], [\ ])$$

will be added to H.

The outer **while** loop asks $EQUIV(H)$ and gets a counterexample, say $A_3 = qs([3, 1, 8, 2], [1, 2, 3, 8])$. The inner **while** loop asks $REQ(A_3 \leftarrow Closure_H(A_3))$, which results in a hint, say $B_3 = part([2], 3, [2], [\ ])$. In the next iteration, it asks $REQ(B_3 \leftarrow Closure_H(B_3))$ and gets answer 'subsumed' as $Closure_H(B_3)$ includes $part([\ ], 3, [\ ], [\ ])$ as well as $2 \leq 3$. Now, **Reduce**(C) is the clause

$$part([2], 3, [2], [\ ]) \leftarrow 2 \leq 3, part([\ ], 3, [\ ], [\ ]).$$

The **if** statement replaces the clause $part([\ ], H, [\ ], [\ ]) \leftarrow$ in H by the lgg of these two clauses, i.e., $part([X], H, [X], [\ ]) \leftarrow X \leq H, part([\ ], H, [\ ], [\ ])$.

Similarly, a clause $part([X], H, [\ ], [X]) \leftarrow X > H, part([\ ], H, [\ ], [\ ])$ will be added to H. At some later stages, these two clauses will be replaced by their generalizations:

$$part([X|Xs], H, [X|Ls], Bs) \leftarrow X \leq H, part(Xs, H, Ls, Bs)$$
$$part([X|Xs], H, Ls, [X|Bs]) \leftarrow X > H, part(Xs, H, Ls, Bs)$$

The clauses for *app* will be learnt in a similar fashion. Now, we explain the generation of clauses for *qs*.

The outer **while** loop asks $EQUIV(H)$ and gets a counterexample, say $A_4 = qs([3, 1, 8, 6], [1, 3, 6, 8])$. The inner **while** loop terminates (after asking some REQ queries) with a unit clause $qs([\ ], [\ ]) \leftarrow$ adding it to H.

The outer **while** loop asks $EQUIV(H)$ and gets a counterexample, say $A_5 = qs([6], [6])$. The inner **while** loop asks $REQ(A_5 \leftarrow Closure_H(A_5))$ and gets answer 'subsumed' as $Closure_H(A_5)$ includes the atoms $part([\ ], 6, [\ ], [\ ])$, $qs([\ ], [\ ])$, $app([], [6], [6])$ and the clause

$$\text{qs}([6], [6]) \leftarrow \text{part}([\ ], 6, [\ ], [\ ]), \text{qs}([\ ], [\ ]), app([\ ], [6], [6])$$

will be added to H. This clause will be later replaced by a generalization

$$\text{qs}([X], [X]) \leftarrow \text{part}([\ ], X, [\ ], [\ ]), \text{qs}([\ ], [\ ]), app([\ ], [X], [X]).$$

The outer **while** loop asks $EQUIV(H)$ and gets a counterexample, say $A_6 = qs([2, 1, 3], [1, 2, 3])$. The inner **while** loop asks $REQ(A_6 \leftarrow Closure_H(A_6))$ and gets answer 'subsumed' as $Closure_H(A_6)$ includes the atoms $part([1, 3], 2, [1], [3])$, $qs([1], [1])$, $qs([3], [3])$, $app([1], [2, 3], [1, 2, 3])$. Now, **Reduce**$(C)$ is the clause

$$\text{qs}([2, 1, 3], [1, 2, 3]) \leftarrow \text{part}([1, 3], 2, [1], [3]), \text{qs}([1], [1]), \text{qs}([3], [3]),$$
$$app([1], [2, 3], [1, 2, 3]).$$

The **if** statement replaces the clause $\text{qs}([X], [X]) \leftarrow \text{part}([\ ], X, [\ ], [\ ])$, $\text{qs}([\ ], [\ ]), app([\ ], [X], [X])$ by the lgg of these two clauses, i.e.,

$$\text{qs}([H|L], S) \leftarrow \text{part}(L, H, A, B), \text{qs}(A, A1), \text{qs}(B, B1), app(A1, [H|B1], S). \quad \square$$

6   Correctness of the Learning Algorithm

In this section, we prove that the learning algorithm **Learn-SM**$_l$ exactly identifies simply-moded programs and takes at most polynomial number of queries. Let H^* be the target program, $H_0, H_1, \cdots$ be the sequence of hypotheses proposed in the equivalence queries, and $A_1, A_2, \cdots$ be the sequence of counterexamples returned by those queries. The following theorem states an important property of the hints returned by the teacher.

Theorem 3. Let A be a positive example and $B_1, \cdots, B_n$ be the sequence of hints returned in the inner **while** loop of **Learn-SM**$_l$. Then *there is an SLD-refutation* $G_0, G_1, \cdots, G_{n1}$ of $H^* \cup \{\leftarrow A\}$ *with answer substitution* θ *such that for each* $i \in [1, n]$, $B_i = B_i'\theta$, *where* B_i' *is the selected atom in goal* G_{k_i} *and* $1 < k_1 < k_2 < \cdots < k_n \leq n1$. *Furthermore,* $H^* \models B_i$ *for each* $i \in [1, n]$.

Proof: Induction on i. Since A is a positive example, there is a minimal SLD-refutation, say, $G_0, G_1, \cdots, G_{n1}$ of $H^* \cup \{\leftarrow A\}$ with an answer substitution θ. By the definition of hint, there is a selected atom B_1' in goal G_{k_1} such that $B_1 \equiv B_1'\theta$. It is obvious that $H^* \models B_1$.

Now, assuming that the theorem holds for each $1 \leq i < j$, we prove that it holds for $i = j$ as well. In j^{th} iteration of the inner **while** loop, the query $REQ(B_{j-1} \leftarrow Closure_H(B_{j-1}))$ is asked and hint B_j is obtained. By induction hypothesis, $H^* \models B_{j-1}$ and hence there is a minimal SLD-refutation of $H^* \cup \{\leftarrow B_{j-1}\}$ with answer substitution θ and $B_j \equiv B_j'\theta$, where B_j' is the selected atom of a goal in it. Therefore, $H^* \models B_j$. It is easy to see that the SLD-refutation of

$H^* \cup \{\leftarrow B_{j-1}\}$ is a sub-derivation of the derivation $G_{k_{j-1}}, \cdots, G_{n1}$ in the sense that all atoms in each goal in the SLD-refutation of $H^* \cup \{\leftarrow B_{j-1}\}$ are atoms in a goal in $G_{k_{j-1}}, \cdots, G_{n1}$. Therefore, B_j is the selected atom of a goal (say, G_{k_j}) in $G_{k_{j-1}}, \cdots, G_{n1}$ and $k_j > k_{j-1}$.   □

We need the following definition.

Definition 12. A program P_1 is a *refinement* of program P_2, denoted by $P_1 \sqsubseteq P_2$ if $(\forall C_1 \in P_1)(\exists C_2 \in P_2) C_2 \succeq C_1$. Furthermore, P_1 is a *conservative refinement* of P_2 if there exists at most one $C_1 \in P_1$ such $C_2 \succeq C_1$ for any $C_2 \in P_2$.

Theorem 4. For each $i \geq 0$, hypothesis H_i is a conservative refinement of H^* and counterexample A_{i+1} is positive.

Proof : Proof by induction on i. For $i = 0$, H_i is the empty program and the theorem obviously holds. We prove the theorem holds for $i = m$ if it holds for $i = m - 1$. Consider m^{th} iteration of the main **while** loop. Since A_i is a positive counterexample, $H^* \models A_i$ and hence $H^* \models A_i \leftarrow Closure_{H_{m-1}}(A_i)$. By the above theorem, $H^* \models B$ for each hint B and hence $H^* \models B \leftarrow Closure_{H_{m-1}}(B)$ as well. By definition, $H_{m-1} \not\models B$ for any hint B. That is, $H^* \models C$ and $H_{m-1} \not\models C$ for each clause C considered in this iteration, in particular for the clause C at the exit of the inner **while** loop. We have two cases: (a) there is a clause $D \in H_{m-1}$ such that $C \sqcup D$ is subsumed by a clause $C^* \in H^*$ and $H_m = H_{m-1} \cup \{\textbf{Reduce}(C \sqcup D)\} - D$ or (b) there is no such clause D and $H_m = H_{m-1} \cup \{C\}$.

By hypothesis, H_{m-1} is a conservative refinement of H^* and it is easy to see that H_m is a conservative refinement of H^* in case (b). Consider case (a) now. Since H_{m-1} is a conservative refinement of H^*, D is the unique clause in H_{m-1} subsumed by C^*. As H_m is obtained from H_{m-1} by replacing D with $\textbf{Reduce}(C \sqcup D)$, it is clear that H_m is a conservative refinement of H^* in this case also. Since each hypothesis is a refinement, each counterexample is positive.

□

It is easy to see that any conservative refinement H of a deterministic program H^* is deterministic and $H \models A$ is decidable in polynomial time if $H^* \models A$ is decidable in polynomial time.

Now, we establish polynomial time complexity of the learning algorithm **Learn-SM**$_l$.

Theorem 5. For any counterexample A of size n, the number of iterations of the inner **while** loop of **Learn-SM**$_l$ is bounded by a polynomial in n.

Proof : The request query $REQ(A \leftarrow Closure_H(A))$ is answered in polynomial (in n) time and hence the length of the minimal SLD-refutation of $H^* \cup \{\leftarrow A\}$ is bounded by a polynomial in n. By Theorem 3, the number of hints returned (and hence the number of iterations of the inner **while** loop of **Learn-SM**$_l$) is bounded by a polynomial in n.   □

Now, we prove that each iteration of the inner **while** loop of **Learn-SM**$_l$ takes polynomial time. The main work done in each iteration is the computation of $Closure_H(B)$ and the following theorem establishes that the size of

$Closure_H(B)$ is bounded by a polynomial in the size of B. We fix the following constants for the discussions in the sequel: $k_1 = |\Pi|$, $k_2 = |\Sigma|$, k_3 is the maximum number of input positions of predicates in Π and k_4 is the maximum arity of functions in Σ.

Theorem 6. If $B \equiv p_0(\mathbf{s}_0; \mathbf{t}_0)$ is a ground atom and $n = [\![\mathbf{s}_0]\!]$ then $|Closure_H(B)|$ is bounded by a polynomial in n.

Proof : $Closure_H(B)$ is defined as $S_1 \cup S_2 \cup \cdots \cup S_l$. We show that each $|S_i|$ is bounded by a polynomial in n, by induction on i.

Basis : $i = 1$. S_1 is defined as the set $\{q(\mathbf{u}; \mathbf{v}) \mid H \models q(\mathbf{u}; \mathbf{v})$ and each term in $\mathbf{u}$ is either a subterm of a term in $\mathbf{s}_0$ or of the form $f(\mathbf{w})$ such that each term in $\mathbf{w}$ is a subterm of a term in $\mathbf{s}_0$ and f is a function symbol in $\Sigma\}$. Since H is a deterministic program, input terms $\mathbf{u}$ uniquely determine the output terms $\mathbf{v}$ of the atom $q(\mathbf{u}; \mathbf{v})$. There are at most k_3 input positions for q and hence at most k_3 terms of the form $f(\mathbf{w})$ in $\mathbf{u}$. Therefore, there are at most $k_3 k_4$ subterms of $\mathbf{s}_0$ in $\mathbf{u}$. There are at most n distinct subterms of $\mathbf{s}_0$ (and hence $|T_1|$ is bounded by n) as $n = [\![\mathbf{s}_0]\!]$. Therefore, there are at most $k_1 k_2^{k_3} n^{k_3 k_4}$ atoms in S_1. That is, $|S_1|$ is bounded by a polynomial in n.

Induction Step : Assuming $|S_i|$ is bounded by a polynomial in n for each $i < j$, we prove that $|S_j|$ is also bounded by a polynomial in n. S_j is defined as $\{q(\mathbf{u}; \mathbf{v}) \mid H \models q(\mathbf{u}; \mathbf{v})$ and each term in $\mathbf{u}$ is either a term in T_j or of the form $f(\mathbf{w})$ such that each term in $\mathbf{w}$ is a term in T_j and f is a function symbol in $\Sigma\}$, where $T_j = T_{j-1} \cup \{s \mid s$ is an output term of an atom in $S_{j-1}\}$. Using an argument similar to the one in the basis step, we can prove that there are at most $k_1 k_2^{k_3} |T_j|^{k_3 k_4}$ atoms in S_j. Since $|S_{j-1}|$ is bounded by a polynomial in n, $|T_j|$ is also bounded by a polynomial in n. Hence, $|S_j|$ is bounded by a polynomial in n. □

Remark 4. As remarked earlier, our restriction that *an input term of an atom in the body is either (1) a subterm of an input term of the head or (2) an output term of an atom before that atom or (3) of the form $f(\mathbf{w})$ such that each term in $\mathbf{w}$ is either (a) a subterm of an input term of the head or (b) an output term of an atom before that atom* plays a crucial role in the above proof (condition (2) in particular). If we allow the input term of an atom in the body (or an argument of f) to be a subterm of an output term of an atom before that atom, $|T_j|$ is no longer guaranteed to be bounded by a polynomial in n as the size of an output term can be exponential in the size of input and hence there are exponentially many subterms of such output term (it may also be noted that the definition of T_j also needs to be changed to account for all the subterms of output terms of atoms in S_{j-1}).

The following lemma is needed in proving polynomial time complexity of the learning algorithm **Learn-SM$_l$**.

Lemma 3. If C is a clause of size n, then the sequence $C = C_0, \prec C_1 \prec C_2$ is of length no more than $2n$.

Proof : When $C_i \prec C_{i+1}$, one of the following holds: (1) $size(C_{i+1}) = size(C_i)$ and $|Var(C_{i+1})| > |Var(C_i)|$, i.e., a constant or an occurrence of a variable (which occurs in C_i more than once) is replaced by a new variable or (2) $size(C_{i+1}) < size(C_i)$. The change (1) can occur at most n times as the number of variables in a clause is less than its size. The change (2) can occur at most n times as the size of any clause is positive. $\qquad\square$

Now, we can prove the main result of the paper.

Theorem 7. The above algorithm **Learn-SM$_l$** exactly identifies any deterministic simply-moded program with m clauses of length at most l, in a polynomial time over m and n, where n is the size of the largest counterexample/hint provided.

Proof : Termination condition of the main **while** loop is $EQUIV(H)$. Therefore **Learn-SM$_l$** exactly identifies the target program H^* if **Learn-SM$_l$** terminates. Now, we prove that the number of iterations of the main **while** loop is bounded by a polynomial over m and n.

By Theorem 4, H is always a conservative refinement of H^* and hence H has at most m clauses. The size of each clause in H is bounded by a polynomial in n by Theorem 6. Each iteration of the main **while** loop either adds a clause to H or generalizes a clause in H. By Lemma 3, the number of times a clause can be generalized is bounded by twice the size of the clause. Therefore, the number of iterations of the main **while** loop is bounded by $m.poly(n)$, where $poly(n)$ is a polynomial in n. Each iteration takes polynomial time as (1) saturation and lgg are polynomial time computable, (2) each query is answered in polynomial time and (3) by Theorem 5, the number of iterations of the inner **while** loop (and hence the number of queries asked) is bounded by a polynomial in n and by Theorem 6, each iteration takes polynomial time. Therefore, **Learn-SM$_l$** exactly identifies any simply-moded program with m clauses in a polynomial time over m and n. $\qquad\square$

7   Comparison

We compare the results of this paper with four recent works [4, 13, 12, 19] on learning first order acyclic (terminating) deterministic Horn programs in polynomial time. As already remarked, all these works [4, 13, 12, 19] essentially use the same learning algorithm but use different ways of constructing new clauses from the hints given by the teacher. In other words, the structure of the learning algorithm is a constant factor and the significance of the results is reflected by the class of programs they learn and the way new clauses are constructed. In the following, we compare our class of simply-moded programs with the classes of programs considered in [4, 13, 12, 19].

In [4], Arimura presents the class of acyclic constrained Horn (ACH) programs. The main restriction ACH-programs is that each term occurring in the body of a clause is a subterm of a term in the head. This excludes all the programs using local variables. However, local variables play an important role of *sideways information passing* in the paradigm of logic programming and hence it is important to extend the results for classes of programs which allow local variables.

In [13], Krishna Rao and Sattar present the class of finely-moded programs, that properly contains the class of ACH programs and allow local variables. The main restriction of finely-moded programs is the following: if a local variable occurs in an output position of an atom A in the body, then each input term of A is a subterm of an input term in the head. This excludes programs like *quick-sort* and *merge-sort*. In fact, the class of simply-moded programs properly contains the class of finely-moded programs.

Theorem 8. *The class of acyclic constrained Horn (ACH) programs is a subclass of the class of finely-moded programs. The class of finely-moded programs is a subclass of the class of simply-moded programs.*

In [19], Reddy and Prasad present the class of acyclic Horn (AH) programs. The main restriction AH-programs is that each term occurring in the head of a clause is a subterm of a term in the body. This is a strong restriction and excludes even simple programs like *member, append* and *prefix*. However, Reddy and Tadepalli [19] argue that the class of acyclic Horn (AH) programs is quite useful for representing planning knowledge.

In [12], Krishna Rao and Sattar present the class of simple linearly-moded programs. The main restriction of simple linearly-moded programs is that they cannot capture non-linear relationships between inputs and outputs. Hence programs like multiplication and exponentiation are beyond the scope of the results from [12]. Though the class of simple linearly-moded programs is not a subclass of the class of simply-moded programs (as simple linearly-moded programs allow the input term of a body atom to be a *proper subterm* of an output term of an atom before that atom; see Remarks 1 and 4), all the simple linearly-moded programs from Sterling and Shapiro's book [23] are simply-moded.

The relationship between various classes discussed above is depicted by the following diagram. It is clear that **the class of simply-moded programs is by far the largest known class of programs learnable from entailment**.

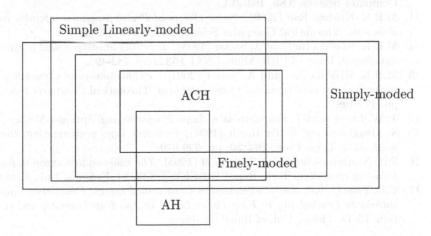

8   Conclusion

In this paper, we considered exact learning of logic programs from entailment and presented a polynomial time algorithm to learn a rich class of logic programs that allow local variables. Our learning algorithm asks at most polynomial number of queries. Unlike the class of linearly-moded programs considered in [11,12], our class allows non-linear relationship between inputs and outputs. The naturality of our class and applicability of the results in practice can be infered from the fact that our class contains a substantial portion of the programs given in standard books on Prolog.

Acknowledgement. The author thanks King Fahd University of Petroleum and Minerals for the generous support provided in conducting this research.

References

1. D. Angluin (1988), *Learning with hints*, Proc. COLT'88, pp. 223-237.
2. D. Angluin (1988), *Queries and concept learning*, Machine Learning **2**, pp. 319-342.
3. K. R. Apt and I. Luitjes. 1995. Verification of logic programs with delay declarations. Proc. of AMAST'95, LNCS **936**, 66-90. Springer-Verlag.
4. H. Arimura (1997), *Learning acyclic first-order Horn sentences from entailment*, Proc. of Algorithmic Learning Theory, ALT'97, LNAI **1316**, pp. 432-445.
5. W. Cohen and H. Hirsh (1992), *Learnability of description logics*, Proc. COLT'92, pp. 116-127.
6. S. Dzeroski, S. Muggleton and S. Russel (1992), *PAC-learnability of determinate logic programs*, Proc. of COLT'92, pp. 128-135.
7. M. Frazier and L. Pitt (1993), *Learning from entailment: an application to propositional Horn sentences*, Proc. ICML'93, pp. 120-127.
8. M. Frazier and L. Pitt (1994), *CLASSIC learning*, Proc. COLT'94, pp. 23-34.
9. P. Idestam-Almquist (1996), *Efficient induction of recursive definitions by structural analysis of saturations*, pp. 192-205 in L. De Raedt (ed.), *Advances in inductive logic programming*, IOS Press.
10. M.R.K. Krishna Rao (1998), *Incremental learning of logic programs*, Theoretical Computer Science, **185**, 193-213.
11. M.R.K. Krishna Rao (2000), *Some classes of Prolog programs inferable from positive data*, Theoretical Computer Science **241**, pp. 211-234.
12. M.R.K. Krishna Rao and A. Sattar (1998), *Learning linearly-moded programs from entailment*, Proc. of PRICAI'98, LNAI **1531**, pp. 482-493.
13. M.R.K. Krishna Rao and A. Sattar (2001), *Polynomial-time earnability of logic programs with local variables from entailment*, Theoretical Computer Science **268**, pp. 179-198.
14. J. W. Lloyd (1987), *Foundations of Logic Programming*, Springer-Verlag.
15. S. Muggleton and L. De Raedt (1994), *Inductive logic programming: theory and methods*, J. Logic Prog. **19&20**, pp. 629-679.
16. S.H. Nienhuys-Cheng and R. de Wolf (1995), *The subsumption theorem for several forms of resolution*, Tech. Rep. EUR-FEW-CS-96-14, Erasmus Uni., Rotterdam.
17. C.D. Page (1993), *Anti-Unification in Constrained Logics: Foundations and Applications to Learnability in First-Order Logic, to Speed-up Learning and to Deduction*, Ph.D. Thesis, Uni. of Illinois, Urbana.

18. C.D. Page and A.M. Frish (1992), *Generalization and learnability: a study of constrained atoms*, in Muggleton (ed.) **Inductive Logic programming**, pp. 29-61.
19. C. Reddy and P. Tadepalli (1998), *Learning first order acyclic Horn programs from entailment*, Proc. of Inductive Logic Programming, ILP'98.
20. C. Rouveirol (1992), *Extensions of inversion of resolution applied to theory completion*, in Muggleton (ed.) **Inductive Logic programming**, pp. 63-92.
21. E. Shapiro (1981), *Inductive inference of theories from facts*, Tech. Rep., Yale Univ.
22. E. Shapiro (1983), *Algorithmic Program Debugging*, MIT Press.
23. L. Sterling and E. Shapiro (1994), *The Art of Prolog*, MIT Press.

A Temporalised Belief Logic for Specifying the Dynamics of Trust for Multi-agent Systems

Chuchang Liu[1], Maris A. Ozols[1], and Mehmet Orgun[2]

[1] Information Networks Division,
Defence Science and Technology Organisation,
PO Box 1500, Edinburgh, SA 5111, Australia
{Chuchang.Liu, Maris.Ozols}@dsto.defence.gov.au
[2] Department of Computing, Macquarie University,
Sydney, NSW 2109, Australia
mehmet@ics.mq.edu.au

Abstract. Temporalisation is a methodology for combining logics whereby a given logic system can be enriched with temporal features to create a new logic system. TML (Typed Modal Logic) extends classical first-order logic with typed variables and multiple belief modal operators; it can be applied to the description of, and reasoning about, trust for multi-agent systems. Without the introduction of a temporal dimension, this logic may not be able to express the dynamics of trust. In this paper, adopting the temporalisation method, we combine TML with a temporal logic to obtain a new logic, so that the users can specify the dynamics of trust and model evolving theories of trust for multi-agent systems.

1   Introduction

Multi-agent systems (MASs for short) consist of a collection of agents that interact with each other in dynamic and unpredictable environments. Agents can be human beings, organizations, machines, and other entities. They may have their own goals, intentions, beliefs, obligations, and perform actions (cooperatively sometimes) in a society of other agents. A very basic problem regarding security properties of a MAS is that whether a message sent by an agent through the system is reliably received by other agents and whether the message received is regarded as reliable in the view of receivers. The problem generally depends on trust-based security mechanisms of a system and the trust that agents would put in the mechanisms. That is, agents should gain their beliefs regarding whether messages they received are reliable based on their trust in the security mechanisms of a system. Therefore, it is important to provide a formal method for specifying the trust that agents have in the security mechanisms of the system, so as to support reasoning about agent beliefs as well as the security properties that the system may satisfy.

M.J. Maher (Ed.): ASIAN 2004, LNCS 3321, pp. 142–156, 2004.

Several logics have been proposed for describing MASs, for example, logics for knowledge and belief [8], and logics for goals and intentions [20]. There are also a number of logics and approaches, such as the BAN logic and related extensions [1, 2, 19], axiomatic methods [16, 18], proof methods based on higher-order logics [17], and the compositional logic [3], that are specifically used for analysing secure communication systems where trust is essential. Many of these formalisms have successfully been used for dealing with some particular aspect of agents that they address, but they generally ignore the other aspects. It is clear that any logical system modeling active agents should be a combined system of logics of knowledge, belief, time and norms. Combining logics is therefore emerging as an active research area in formal methods. Investigating new techniques for combining logics, studying the properties of combined logics, and evaluating the expressive power and computational aspects of these new logic systems with respect to practical applications are a challenging work.

Temporalisation [5] is a methodology for combining logics whereby an arbitrary logic system can be enriched with temporal features to generate a new logic system. In the previous work, Liu [13] proposed a belief logic, called TML (Typed Modal Logic), which extends classical first-order logic with typed variables and multiple belief modal operators. Based on TML, system-specific theories of trust can be constructed, and they provide a basis for analysing and reasoning about trust in particular environments and systems. TML seems to be more suitable to express static properties of trust. However, trust is the outcome of observations leading to the belief that the actions of another may be relied upon, without explicit guarantee, to achieve a goal in a risky situation. Trust can therefore be developed over time as the outcome of a series of confirming observations [4]. An agent may lose its trust or gain new trust at any moment in time due to some reasons such as recommendations from other agents. Without the introduction of a temporal dimension, TML is unable to express the dynamics of trust. The motivation of our work is to build a logical framework, in which we are able to describe both the "statics" and "dynamics" of trust. In TML we have had the ability to describe the "statics" of trust, but not to describe the "dynamics". In contrast, temporal logics have the potential to deal with "dynamics", i.e., the changes in time of a set of properties. In this paper, adopting the temporalisation method, we enrich TML by combining it with a temporal logic, so that users can specify the dynamics of trust and also model evolving theories of trust. The new logic system we develop is constructed by combining TML with a simple linear-time temporal logic (SLTL) that offers two operators, **first** and **next**, which refer to the initial moment and the next moment in time respectively. As we will see, the new logic can express time-dependent properties regarding agent beliefs in a natural manner. It allows users to specify the dynamics of trust and timely update the theory of trust for a given system based on trust changes. We choose SLTL to temporalise TML, not only because SLTL is very simple to handle, but also due to it being executable [14]. As the combination of TML with SLTL, the resulting logic is therefore in particular suitable for applications involving the description of, and reasoning about, the dynamics of trust for multi-agent systems. Furthermore, the executable aspects may be suited for use within security agents that need to consider their changing environment and interactions.

The rest of this paper is organized as follows. Section 2 is a brief introduction to the logic TML. Section 3 discusses TML$^+$, the temporalised logic. In Section 4, we propose a framework for describing the dynamics of trust. In Section 5, we show how to construct a practical theory of trust for a given system based on the initial trust of agents through an example, and present a principle regarding the revision of a theory when agents lose their trust. The last section concludes the paper with a short discussion about future research directions.

2   The Logic TML

Assume we have n agents $a_1, \ldots, a_n$ and, correspondingly, n modal operators $\mathbf{B}_{a_1}, \ldots, \mathbf{B}_{a_n}$ in the logic, where $\mathbf{B}_{a_i}$ ($1 \leq i \leq n$) stands for "agent a_i believes that". Let $\varPhi_0$ be a set of countably many atomic formulae of the (typed) first-order logic. We define $\mathcal{L}_{tml}$ as the smallest set of well-formed formulae (wffs) of the logic TML such that:

- $\varPhi_0 \subset \mathcal{L}_{tml}$;
- if φ is in $\mathcal{L}_{tml}$, so are $\neg\varphi$ and $\mathbf{B}_{a_i} \varphi$ for all i ($1 \leq i \leq n$);
- if φ and ψ are in $\mathcal{L}_{tml}$, then $\varphi \wedge \psi$ is in $\mathcal{L}_{tml}$; and
- if $\varphi(X)$ is in $\mathcal{L}_{tml}$ where X is a free variable, then $\forall X \varphi(X)$ is, too.

Connectives $\vee, \rightarrow$ and $\leftrightarrow$, and the quantifier $\exists$ are defined in the usual manner. Furthermore, we assume that in TML all the wffs are correctly typed.

A formal definition of the semantics for TML is defined below, using the notion of possible worlds for the semantics interpretation of belief [12]. A *Kripke frame* is denoted by $(S, R_1, \ldots, R_n)$, where S is the set of states and each R_i, $i = 1, \ldots, n$, is a relation over S. R_i, called the *possibility relation* according to agent a_i, is defined as follows: R_i is a non-empty set consisting of state pairs (s, t) such that $(s, t) \in R_i$ iff, at state s, agent a_i considers the state t possible. A *model* is a tuple $\langle S, R_1, \ldots, R_n, \pi \rangle$, where $(S, R_1, \ldots, R_n)$ is a Kripke frame and π is the *assignment function*, which gives a value $\pi(s, q) \in \{true, false\}$ for any $s \in S$ and atomic formula q. We write $\mathcal{M}, s \models \varphi$ to stand for "φ is true at s in the model $\mathcal{M}$" or "φ holds at s in $\mathcal{M}$". Thus, the semantics of formulae in the logic TML is inductively defined as follows:

1. For any atomic formula $p(e_1, \ldots, e_k)$, $\mathcal{M}, s \models p(e_1, \ldots, e_k)$ iff in $\mathcal{M}$ we have $\pi(s, p(e_1, \ldots, e_k)) = true$.
2. $\mathcal{M}, s \models \neg\varphi$ iff it is not the case that $\mathcal{M}, s \models \varphi$.

3. $\mathcal{M}, s \models (\varphi \wedge \psi)$ iff $\mathcal{M}, s \models \varphi$ and $\mathcal{M}, s \models \psi$.
4. $\mathcal{M}, s \models \forall X \varphi(X)$ iff $\mathcal{M}[X/d], s \models \varphi(X)$ for all $d \in \mathcal{T}$, where $\mathcal{T}$ is the type of X and $\mathcal{M}[X/d]$ is just like $\mathcal{M}$ except that it assigns d to X.
5. $\mathcal{M}, s \models \mathbf{B}_{a_i} \varphi$ iff, for all t such that $(s, t) \in R_i$, $\mathcal{M}, t \models \varphi$.

Furthermore, we say that φ is valid in the model $\mathcal{M}$, and write $\mathcal{M} \models \varphi$ if $\mathcal{M}, s \models \varphi$ for all states $s \in S$; we say that φ is satisfiable in $\mathcal{M}$ if $\mathcal{M}, s \models \varphi$ for some state $s \in S$. If, for all models $\mathcal{M}$, $\mathcal{M} \models \varphi$ then we have $\models \varphi$.

The axiom set of TML consists of the following axiom schemata:

B0.   all axioms of the classical first-order logic.
B1. $\mathbf{B}_{a_i}(\varphi \rightarrow \psi) \wedge \mathbf{B}_{a_i} \varphi \rightarrow \mathbf{B}_{a_i} \psi$ for all i $(1 \leq i \leq n)$.
B2. $\mathbf{B}_{a_i}(\neg \varphi) \rightarrow \neg(\mathbf{B}_{a_i} \varphi)$ for all i $(1 \leq i \leq n)$.

The rules of inference in this logic include:

R1.   From φ and $\varphi \rightarrow \psi$ infer ψ.   (Modus Ponens)
R2.   From $\forall X \varphi(X)$ infer $\varphi(Y)$.   (Instantiation)
R3.   From $\varphi(X)$ infer $\forall X \varphi(X)$.   (Generalisation)
R4.   From φ infer $\mathbf{B}_{a_i} \varphi$ for all i $(1 \leq i \leq n)$.   (Necessitation)

The soundness and completeness of the axiomatisation system for TML, consisting of axioms B0-B2 and rules of inference R1-R4, can be proved in a standard pattern [9].

3   TML$^+$: A Temporalised Logic

In this section, we first introduce the temporal logic SLTL, then discuss the approach to temporalising TML and present some properties of the new logic.

3.1   SLTL

In SLTL, the collection of moments in time is the set of natural numbers with its usual ordering relation $<$. Let $\mathcal{P}$ be a set of countably many propositional symbols. We define $\mathcal{L}_{sltl}$ as the smallest set of temporal propositional formulae of SLTL such that:

- $\mathcal{P} \subset \mathcal{L}_{sltl}$;
- If A is in $\mathcal{L}_{sltl}$, then $\neg A$, **first** A and **next** A are in $\mathcal{L}_{sltl}$; and
- If A and B are in $\mathcal{L}_{sltl}$, so $A \wedge B$ is, too.

Connectives $\vee, \rightarrow$ and $\leftrightarrow$ are derived from the primitive connectives as usually.

We define the global clock as the increasing sequence of natural numbers, i.e., $\langle 0, 1, 2, \ldots \rangle$, and a local clock is an infinite subsequence of the global clock. A *time frame* is denoted by $(\mathbf{C}, <)$, where $\mathbf{C} = (t_0, t_1, t_2, \ldots)$ is a local clock and $<$ is the ordinary binary relation over $\mathbf{C}$. A *model* $\mathcal{M}$ is a triple $\langle \mathbf{C}, <, v \rangle$, where $(\mathbf{C}, <)$ is a time frame and v is a total function, called a valuation, from $\mathbf{C}$ to the power set of $\mathcal{P}$, that is, for any moment in time $t_i \in \mathbf{C}$, $v(t_i) \subseteq \mathcal{P}$. $\mathcal{M}, t_i \models A$ means that the formula A holds over model $\mathcal{M}$ at time t_i, and it is defined recursively as follows:

- $\mathcal{M}, t_i \models p$, $p \in \mathcal{P}$ iff $p \in v(t_i)$.
- $\mathcal{M}, t_i \models \neg A$ iff it is not the case that $\mathcal{M}, t_i \models A$.
- $\mathcal{M}, t_i \models A \wedge B$ iff $\mathcal{M}, t_i \models A$ and $\mathcal{M}, t_i \models B$.
- $\mathcal{M}, t_i \models \mathbf{first}\ A$ iff $\mathcal{M}, t_0 \models A$.
- $\mathcal{M}, t_i \models \mathbf{next}\ A$ iff $\mathcal{M}, t_{i+1} \models A$.

A formula A is *valid* over a class $\mathcal{C}$ of clocks, indicated by $\mathcal{C} \models A$, if for any model $\mathcal{M} = \langle \mathbf{C}, <, v \rangle$ where $\mathbf{C} \in \mathcal{C}$ and for any $t \in \mathbf{C}$, we have $\mathcal{M}, t \models A$. If Σ is a set of formulae, we write $\mathcal{C} \models \Sigma$ to indicate that $\mathcal{C} \models A$ for every $A \in \Sigma$. Therefore, for different classes we may have different sets of valid formulae.

Let $\mathcal{C}_0$ be the class of all local clocks. A minimal axiomatic system for the propositional temporal logic over the class $\mathcal{C}_0$ contains the following axioms (axiom schemata):

A0. all classical tautologies.   A1. $\nabla(\mathbf{first}\ A) \leftrightarrow \mathbf{first}\ A$.

A2. $\nabla(\neg A) \leftrightarrow \neg(\nabla A)$.   A3. $\nabla(A \wedge B) \leftrightarrow (\nabla A) \wedge (\nabla B)$.

where ∇ stands for **first** or **next** in any axiom schema. The rules of inference are:

US.   From $\vdash A(p)$ infer $\vdash A(p \backslash B)$, where p is the propositional symbol and B any formula and $A(p \backslash B)$ is a formula resulting from substituting all appearances of p in A by B.   (Uniform Substitution)

MP.   From $\vdash A$ and $\vdash A \rightarrow B$ infer $\vdash B$.   (Modus Ponens)

TG.   From $\vdash A$ infer $\vdash \mathbf{first}\ A$ and $\vdash \mathbf{next}\ A$.   (Temporal Generalisation)

The soundness and completeness of the axiomatisation system for SLTL with respect the class $\mathcal{C}_0$ are straightforward [14].

3.2   Temporalising TML

We write $\mathrm{TML}^+ = \mathrm{TML} + \mathrm{SLTL}$ to indicate that the new logic TML^+ is constructed by combining TML with SLTL. The symbol "$+$" in TML $+$ SLTL can be viewed as a combination operator applied for the two logics; while, in TML^+, the "$+$" is used to indicate that the new logic is an extension of the logic TML or TML^+ has a stronger expressive power than TML does.

To define the syntax of the new logic TML^+, we first consider $\mathcal{L}_{tml}^{(m)}$, a proper subset of $\mathcal{L}_{tml}$, which is recursively defined as follows:

(1) If φ is an atomic formula in $\mathcal{L}_{tml}$, then $\varphi \in \mathcal{L}_{tml}^{(m)}$;

(2) If $\varphi \in \mathcal{L}_{tml}^{(m)}$, so $\mathbf{B}_{a_i} \varphi \in \mathcal{L}_{tml}^{(m)}$, for all i ($1 \leq i \leq n$);

(3) If $\varphi(X) \in \mathcal{L}_{tml}^{(m)}$ where X is a free variable, then $\forall X \varphi(X) \in \mathcal{L}_{tml}^{(m)}$, too; and

(4) Only those formulae that are created by (1), (2) and (3) belong to $\mathcal{L}_{tml}^{(m)}$.

This definition suggests that we may partition $\mathcal{L}_{tml}$ into two sets, $\mathcal{L}_{tml}^{(b)}$ and $\mathcal{L}_{tml}^{(m)}$. A formula $\varphi \in \mathcal{L}_{tml}^{(b)}$ is called a *boolean combination* iff it is built up from other formulae with the use of the boolean connectives $\neg$ or $\wedge$ or any other connectives defined only in terms of those; it is called a monolithic formula and belongs to $\mathcal{L}_{tml}^{(m)}$ otherwise. Thus, for any formula $\varphi \in \mathcal{L}_{tml}$, we must have $\varphi \in \mathcal{L}_{tml}^{(b)}$ or $\varphi \in \mathcal{L}_{tml}^{(m)}$, but not both.

In the following, we use $\alpha, \beta, \gamma, \ldots$ to range over formulae of TML$^+$. Let $\mathcal{L}_{tml+}$ be the set of formulae of the logic system TML$^+$. Then $\mathcal{L}_{tml+}$ is defined as the smallest set such that:

- If $\alpha \in \mathcal{L}_{tml}^{(m)}$, then $\alpha \in \mathcal{L}_{tml+}$;
- If $\alpha \in \mathcal{L}_{tml+}$ and $\beta \in \mathcal{L}_{tml+}$ then $\neg\alpha \in \mathcal{L}_{tml+}$ and $(\alpha \wedge \beta) \in \mathcal{L}_{tml+}$;
- If $\alpha \in \mathcal{L}_{tml+}$, then **first** $\alpha \in \mathcal{L}_{tml+}$ and **next** $\alpha \in \mathcal{L}_{tml+}$.

From the definition above, **first** and **next** never appear within the scope of a modal operator $\mathbf{B}_{a_i}$ or a quantifier $\forall$ or $\exists$.

Let $\mathcal{K}$ be a class of (Kripke) models of the logic TML. Consider a time frame $(\mathbf{C}, <)$, where $\mathbf{C} = (t_0, t_1, t_2, \ldots)$, and a function $v : \mathbf{C} \to \mathcal{K}$, mapping moments in time on the clock $\mathbf{C}$ to a model in the class $\mathcal{K}$. Then we define a model of TML$^+$ as a triple $\langle \mathbf{C}, <, v \rangle$, denoted by $\mathcal{M}_{tml+}$, and use $\mathcal{M}_{tml+}, t_i \models \alpha$ to mean that α is true at the time t_i in the model $\mathcal{M}_{tml+}$. Thus, the semantics of the logic TML$^+$ can recursively be defined as follows:

- $\mathcal{M}_{tml+}, t_i \models \alpha$, $\alpha \in \mathcal{L}_{tml}^{(m)}$ iff $v(t_i) = \mathcal{M}_{tml}$, $\mathcal{M}_{tml} \models \alpha$.
- $\mathcal{M}_{tml+}, t_i \models \neg\alpha$ iff it is not the case that $\mathcal{M}_{tml+}, t_i \models \alpha$.
- $\mathcal{M}_{tml+}, t_i \models \alpha \wedge \beta$ iff $\mathcal{M}_{tml+}, t_i \models \alpha$ and $\mathcal{M}_{tml+}, t_i \models \beta$.
- $\mathcal{M}_{tml+}, t_i \models$ **first** α iff $\mathcal{M}_{tml+}, t_0 \models \alpha$.
- $\mathcal{M}_{tml+}, t_i \models$ **next** α iff $\mathcal{M}_{tml+}, t_{i+1} \models \alpha$.

We write $\mathcal{K} \models \alpha$ if, for every model $\mathcal{M}_{tml+}$ in $\mathcal{K}$ and the relevant clock $\mathbf{C} \in \mathcal{C}$ and for every time $t \in \mathbf{C}$, it is the case $\mathcal{M}_{tml+}, t \models \alpha$.

3.3   Axiomatisation for TML$^+$

We say that a formula α in $\mathcal{L}_{tml}$ is a standard-monolithic boolean combination if $\alpha \in \mathcal{L}_{tml}^{(m)}$ or $\alpha = \neg\beta$ where β is a standard-monolithic boolean combination or $\alpha = \beta \wedge \gamma$ where both β and γ are standard-monolithic boolean combinations. Thus the following theorem can inductively be proved based on structures of formulae in $\mathcal{L}_{tml}$.

Theorem 1. *For any formula* $\alpha \in \mathcal{L}_{tml}$, *there exists some standard-monolithic boolean combination* α', *such that* α *and* α' *are equivalent, i.e.* α *is valid in TML iff* α' *is valid in TML.*

For example, formula $\forall X(p(X) \;\wedge\; q(X))$ is equivalent to the standard-monolithic boolean combination $\forall X p(X) \wedge \forall X q(X)$, and $\mathbf{B}_{a_1}(\varphi \wedge \psi)$ is equivalent to $\mathbf{B}_{a_1}\varphi \wedge \mathbf{B}_{a_1}\psi$. Note that a formula in TML, such as $\forall X(p(X) \wedge q(X))$, may not be a formula in TML$^+$, but a standard-monolithic boolean combination equivalent to it must be a formula of TML$^+$. According to the formation rule of formulae of TML$^+$ and the definition of standard-monolithic boolean combinations, this assertion holds true.

The proof system of TML$^+$ is given by the following axioms and rules of inference:

- The axioms of SLTL;
- The rules of inference for SLTL;
- For every formula α in $\mathcal{L}_{tml}$, if $\vdash_{tml} \alpha$ and α' is a standard-monolithic boolean combination equivalent to α, then $\vdash_{tml+} \alpha'$.

where $\vdash_{tml} \alpha$ means that α is a theorem of the system TML and $\vdash_{tml+} \alpha$ means that α is a theorem of the system TML$^+$.

The third item above is a new rule of inference that is applied to preserve the theoremhood of formulae of the logic TML. We call it the *rule of theoremhood preservation* (or **TP**). Therefore, the axiomatisation for TML$^+$ consists of all axioms of the linear-time temporal logic SLTL and 4 rules of inference, **US**, **MP**, **TG** and **TP**.

By the method of Finger and Gabbay [5], which is used to show theorems 2.2 and 2.3 of their paper, the following theorem can be proved.

Theorem 2. *The logic system TML$^+$ is sound and complete with respect the class* $\mathcal{C}_0$.

Based on the new logic TML$^+$, we can express time-dependent properties of trust regarding agent beliefs in a natural manner. As an example, let us consider the assertion:

– On day 0 (initially) John believes that Mary has the secure key k but, on day 5, he does not believe that Mary still has that key, i.e., it is not the case that John believes Mary has that key (assuming the reason, for example, that he suspects the key to be stolen by an attacker).

This assertion can be formalized as

$$\textbf{first } \mathbf{B}_{john} \; has(mary, k) \wedge \textbf{first next}^{(5)} \; \neg \mathbf{B}_{john} \; has(mary, k),$$

where **next**$^{(n)}$ denotes n applications of **next** and **next**$^{(0)}\varphi = \varphi$.

Another example is given as follows. We use S, R, and I to stand for the sender, the receiver, and the intruder respectively. Then the formula

$sends(S, R, M) \wedge intercepts(I, M) \rightarrow$
  $(\textbf{next } sends(I(S), R, M') \wedge M \neq M' \rightarrow$
  $\textbf{next next } (receives(R, M') \wedge \textbf{B}_R \; comes\_from(M', S)))$

describes a rule that the intruder may likely follow for setting up an attack. It means that, if the intruder I intercepts a message M sent to R by S, then, if he masquerades as S to send a fake message M' to R at the immediately next moment in time, then at the next next moment in time R will receive M' and believe that M' comes from S. Correspondingly, a successful attack may be formalized by the formula as follows:

$sends(S, R, M) \wedge intercepts(I, M) \wedge \textbf{next } sends(I(S), R, M') \wedge M \neq M' \wedge$
  $\textbf{next next } (receives(R, M') \wedge \textbf{B}_R \; comes\_from(M', S)).$

4   Describing the Dynamics of Trust

Trust is based on many factors. Any event related to these factors may cause agents to lose or gain trust in other agents. Jonker and Treur [10] called an event that can influence the degree of trust of an agent a *trust-positive experience* or a *trust-negative experience* of the agent. If the event an agent experiences is a trust-negative experience then the agent may loose his trust to some degree; if it is a trust-positive experience then the agent may gain his trust to some degree. With these concepts, they further proposed two functions, *trust evolution function* and *trust update function*, to model the dynamics of trust, and claimed that their formal models within this framework could be applied for the specification of the dynamics of trust for software agents as part of their design.

We discuss the dynamics of trust for multi-agent systems in a different way. Let τ be a multi-agent system and $\mathcal{A}$ the set of agents. The trust relation over the set $\mathcal{A}$, denoted by $\mathcal{R}$, is a set consisting of pairs of agents, which is defined as: $(a, b) \in \mathcal{R}$ if and only if a has *trust* in b. Therefore, our model is based on binary values, which restricts the expression of trust to a certain degree (*trust* or *no trust*): $(a, b) \in \mathcal{R}$ means that a has *trust* in b, $\neg((a, b) \in \mathcal{R})$ or $(a, b) \notin \mathcal{R}$ will mean that a has *no trust* in b. We define that a *trust state* is a function (ρ) from $\mathcal{A}$ to $2^{\mathcal{A}}$. Given a trust state ρ, for any agent $a \in \mathcal{A}$, $\rho(a) \subseteq \mathcal{A}$ is a set, associated with a, that contains all the agents whom agent a trusts. Furthermore, we define a predicate $is\_the\_trust\_state\_of(\rho, \tau)$ meaning that ρ is the trust state of the system τ. Then, in particular, if we have $\textbf{first } is\_the\_trust\_state\_of(\rho_0, \tau)$, ρ_0 is called the *initial trust* of τ.

Given the initial trust ρ_0 for the system τ, if $\bigcap_{a \in \mathcal{A}} \rho_0(a) \neq \emptyset$, the empty set, then initially there are some agents whom all agents trust. For example, we may have that initially all agents trust the security mechanism of the system (this may mean that all agents would believe that the mechanism can guarantee the reliability of messages they received through the system). Assume that we have

$is\_the\_trust\_state\_of(\rho, \tau) \wedge \textbf{next } is\_the\_trust\_state\_of(\rho', \tau),$

then it is obvious that

- $\rho = \rho'$ iff at the next moment in time there is no agent who changes its trust;
- $\rho \neq \rho'$ iff at the next moment in time there is at least one agent who changes his trust.

In the case when $\rho = \rho'$, the trust state at the next moment is the same one at the current moment, so there are no changes to the trust of the system; while in the case when $\rho \neq \rho'$, there must be some changes to the trust of the system. Changes may come from many reasons, but the actions for making these changes can simply be classified into precisely two types: adding, for example, trust of an agent a in an agent b to the trust relation of the system, or deleting trust from the trust relation. For instance, assume that for two weeks John trusted Peter, but this morning John found that Peter lied to him, so John now no longer trusts Peter, thus, if $(john, peter) \in \mathcal{R}$, the trust relation of the system, then $(john, peter)$ should now be deleted from $\mathcal{R}$.

Borrowing the denotations regarding experiences of agents from [10], we assume that there are two sets, one is the set of trust-negative experiences, say $\{e_1^{(n)}, e_2^{(n)}, \ldots\}$, and the other is the set of trust-positive experiences, say $\{e_1^{(p)}, e_2^{(p)}, \ldots\}$. We also assume that, if currently an agent, say a, has trust in an agent, say b, and at the next moment in time a has any trust-negative experience with b, then at that moment a loses trust in b and (a, b) will therefore be deleted. The change with deleting trust of agent a in agent b can formally be expressed as follows:

$$is\_the\_trust\_state\_of(\rho, \tau) \wedge \mathbf{next}\ is\_the\_trust\_state\_of(\rho', \tau) \rightarrow$$
$$(b \in \rho(a) \wedge \mathbf{next}\ has(a, e_i^{(n)}, b) \rightarrow \mathbf{next}\ b \notin \rho'(a)),$$

where $has(a, e_i^{(n)}, b)$ means that a has the experience $e_i^{(n)}$ ($i = 1, 2 \ldots$) with b.

Similarly, the change with adding trust of a in b can formally be expressed as follows:

$$is\_the\_trust\_state\_of(\rho, \tau) \wedge \mathbf{next}\ is\_the\_trust\_state\_of(\rho', \tau) \rightarrow$$
$$(b \notin \rho(a) \wedge \mathbf{next}\ has(a, e_i^{(p)}, b) \rightarrow \mathbf{next}\ b \in \rho'(a)).$$

5   Theory of Trust, and Theory Revision

In this section, we discuss how to construct a system-specific theory of trust supporting reasoning about the dynamics of trust and agent beliefs through a simple example, and present a principle regarding theory revision.

5.1   Constructing a Theory

In formalizing trust for multi-agent systems, Liu and Ozols [15] proposed a simple trust model, in which it is assumed that an agent only trusts the security

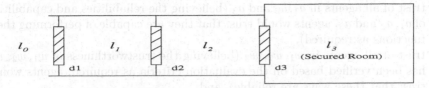

Fig. 1. Four locations with the secured room

mechanisms (as special agents) in the system whose trustworthiness has been verified based on required evaluation criteria. This model is appropriate to express initial trust of an agent in the system. The use of the simple trust model is reasonable in the case when considering those systems that are critical and require a high assurance [15]. According to this model, beliefs of an agent can only be obtained based on his/her trust in the security mechanisms of the system. Therefore, for reasoning about agent beliefs, the key is to obtain the rules specifying trust of agents in the system. That is, we should first specify the initial trust that agents have in the security mechanisms of the system. In general, under a logic framework, the assumptions describing trust of agents in the security mechanisms for a given system are encapsulated in a notion of trust and represented as a set of axioms (rules). Thus, these axioms with the logic together form a theory, so called the *trust theory* (or *theory of trust*) for the system [13].

Let us consider the case: to enter a secured room, one must pass through three successive doors, d_1, d_2 and d_3. Assume that agents a_1, a_2 and a_3 control doors d_1, d_2 and d_3, respectively, by checking the identity of the person who wants to enter the room in different ways (for example, verifying photos or fingerprint *etc.*; formally, we may say authentication checks w_1, w_2 and w_3) and an agent allows the person pass through the door it controls at a moment in time only if at that moment the agent believes that the identity is authenticated. We also assume that, if the person has passed door d_1, he may pass door d_2, go back out of the door d_1, or stay where he is at the next moment in time; similarly, if the person has passed door d_2, he may pass door d_3, go back out of the door d_2, or stay where he is at the next moment in time. This is a multi-agent authentication system.

We identify four locations, l_0, l_1, l_2 and l_3, divided by walls and doors, as shown in Figure 1. According to the simple trust model, initially an agent does not trust anyone, but only the security mechanisms. The security mechanisms for the system consists of agents, a_1, a_2 and a_3, the ways, w_1, w_2 and w_3, applied for checking identity of a person, and the physical security environment (walls and doors), denoted by *pse*. Formally, assume ρ_0 is the initial trust of the system, then the security mechanisms of the system is the set $\bigcap_{a \in \mathcal{A}} \rho_0(a) = \{w_1, w_2, w_3, a_1, a_2, a_3, pse\}$. Therefore, the initial trust of the system include

- trust of all agents in a_1, a_2 and a_3 (believing the reliabilities and capabilities of a_1, a_2 and a_3, agents would trust that they are capable of performing their functions as required),
- trust of all agents in w_1, w_2, w_3 (believing the trustworthiness of w_1, w_2, w_3 has been verified based on the evaluation criteria as required, agents would trust that these ways are reliable), and
- trust of all agents in *pse* (believing the *pse* satisfies security properties as required, agents would trust that there is no problem with it on the security objective).

We formalize the initial trust through a number of rules that are expressed as the axiom schemata as follows:

I1. $at(X, l_{i-1}) \wedge \mathbf{B}_{a_i} \ is\_authenticated(X, w_i) \rightarrow \mathbf{next} \ at(X, l_i)$, for $i = 1, 2, 3$.

I2. $\neg \mathbf{B}_{a_i} \ is\_authenticated(X, w_i) \rightarrow \mathbf{next} \ \neg at(X, l_i)$, for $i = 1, 2, 3$.

I3. $at(X, l_0) \rightarrow \mathbf{next} \ (at(X, l_0) \vee at(X, l_1))$.

I4. $at(X, l_i) \rightarrow \mathbf{next} \ (at(X, l_{i-1}) \vee at(X, l_i) \vee at(X, l_{i+1}))$, for $i = 1, 2$.

where rules I1 and I2 specify the behavioural (functional) properties of the authentication system, and rules I3 and I4 specify the physical security properties enforced by *pse* (walls and doors) regarding locations where the user may stay at a certain moment in time. These are essentially topological axioms.

Combining the physical security and door access security, the logic TML$^+$ together with the rules I1, I2, I3 and I4 will forms a theory of trust for the system. Denoting the theory by $\mathbf{T}$, without confusion, we write $\mathbf{T} = \{I1, I2, I3, I4\}$.

A theory of trust can be applied to reasoning about agent beliefs. For instance, having the theory $\mathbf{T}$, we can show that any person cannot get into the security room if anyone of the agents a_1, a_2 and a_3 does not believe the person's identity is authenticated. We outline a proof procedure as follows: Suppose that initially the person , say X, is at l_0, then if agent a_1 does not believe the person's identity is authenticated, i.e., if $\neg \mathbf{B}_{a_1} \ is\_authenticated(X, w_1)$, then we have $\neg \mathbf{next} \ at(X, l_1)$ (by I2), further by I3 we have $\mathbf{next} \ at(X, l_0)$, i.e., at next moment in time, the person is still at the location l_0. Similarly, we can prove that, if the person is at l_1 but we have $\neg \mathbf{B}_{a_2} \ is\_authenticated(X, w_2)$, then at the next moment, the person must be at l_1 or l_0; and if the person is at l_2 but we have $\neg \mathbf{B}_{a_3} \ is\_authenticated(X, w_3)$, then at the next moment, the person must be at l_2 or l_1. Therefore, the person can never be in the security room (l_3) if any of these agents does not believe his identity is authenticated.

5.2   Theory Revision

The discussion above indicates that, if these agents (a_1, a_2 and a_3), the ways (w_1, w_2 and w_3) for checking the identity of a person, and the physical security environment (*pse*) can be trusted, the authentication system (as a whole) would also be trusted. It also indicates that, if agents have trust in the security mechanisms for the system, then the theory describing trust of agents in the system can be used for reasoning about agent beliefs, in particular, from the theory, agent may obtain new trust and beliefs.

Trust changes dynamically, an agent may lose its trust or gain new trust at any moment in time after it has gained the initial trust. Suppose from the initial moment in time until some time t, there are no agents losing their initial trust, then the theory could still be valid. Regarding the soundness of trust theory, we have the following property:

- Given a system τ, suppose we have **first next**$^{(i)}$ $is\_the\_trust\_state\_of(\rho_i, \tau)$, for $i = 0, 1, 2, \ldots$, and T is a sound theory of trust for τ that specifies the initial trust of agents in the security mechanisms expressed by $\bigcap_{a \in \mathcal{A}} \rho_0(a)$, then, if from time 0 until some time t, $\bigcap_{a \in \mathcal{A}} \rho_0(a) \subseteq \bigcap_{a \in \mathcal{A}} \rho_i(a)$, for any $0 \leq i \leq t$, then the theory T is valid for the whole time interval $[0, t]$ and, therefore, any (time-dependent related to $[0, t]$) statement derived from the theory would be regarded as valid.

However, once agents lose their trust in the security mechanisms of a system, the theory based on the initial trust of the system is no longer valid, so, in that case, we need to revise the theory or stop using it. Regarding theory revision, we have the following principle:

- Given a system τ, suppose we have **first next**$^{(i)}$ $is\_the\_trust\_state\_of(\rho_i, \tau)$, for $i = 0, 1, 2, \ldots$, and T is a sound theory of trust for τ that specifies the initial trust of agents in the security mechanisms, $\bigcap_{a \in \mathcal{A}} \rho_0(a)$. If at some time, say t, $\bigcap_{a \in \mathcal{A}} \rho_0(a) \subseteq \bigcap_{a \in \mathcal{A}} \rho_t(a)$ does not hold, i.e., there is some agent losing its trust in the mechanisms, then the theory must be revised or not be used for any security purpose.

For example, with the secured room, assume agents lose their trust in the physical security environment (pse) due to a reason that they have found a hole in the ceiling of the room from which one can enter it. In the case, rules I3 and I4 are no longer correct to specify the physical security properties: someone at the locations l_0, l_1 or l_2 may directly enter the room through the hole. Therefore, we need to revise the theory **T**. To do it, we first need to improve the mechanisms, for example, by adding a new trusted agent, say a_4, into it. Agent a_4 is to check whether there is a hole and decides where a person should stay: if a_4 believes that there is no hole with the room, there are no changes to the controls; if it is not the case he believes that there is no hole with the room, then at the next moment in time all agents will stay where they are currently. Let p stand for an assertion that there is no hole with the room. We may revise axioms I1, I3 and I4 of the theory **T** as follows:

I1'. $\mathbf{B}_{a_4} p \rightarrow (at(X, l_{i-1}) \wedge \mathbf{B}_{a_i} is\_authenticated(X, w_i) \rightarrow \mathbf{next} \ at(X, l_i))$, for $i = 1, 2, 3$.

I3'. $\mathbf{B}_{a_4} p \rightarrow (at(X, l_0) \rightarrow \mathbf{next} \ (at(X, l_0) \vee at(X, l_1)))$.

I4'. $\mathbf{B}_{a_4} p \rightarrow (at(X, l_i) \rightarrow \mathbf{next} \ (at(X, l_{i-1}) \vee at(X, l_i) \vee at(X, l_{i+1})))$, for $i = 1, 2$.

We also add a new axiom to the theory:

I5 $\neg \mathbf{B}_{a_4}\, p \to (at(X, l_i) \to \mathbf{next}\; at(X, l_i))$, for $i = 0, 1, 2, 3$.

Thus, we obtain a revised theory, $\mathbf{T}'=\{I1', I2, I3', I4', I5\}$, which is based on trust of agents in the extended security mechanisms $\{w_1, w_2, w_3, a_1, a_2, a_3, a_4, pse\}$. From the theory, we can show that a person who want to enter the secured room must pass through the three doors and no one can enter it through a hole.

6   Conclusion

A temporalised belief logic has been presented. It provides a logical framework for users to specify the dynamics of trust and model evolving theories of trust for multi-agent systems.

Combined logics have been attracting interest, and there are many applications of logic that concern composite domains. The techniques for combining logics recently under investigation contain the fusion of two mono-modal logics [11], fibring technique [7], and the product of modal logics [6] *etc.* In simple terms, the problem of combining logic is: given two logics $\mathbf{L}_1$ and $\mathbf{L}_2$, how may we combine them into a single logic $\mathbf{L}_1 + \mathbf{L}_2$ that extends the expressive power of each. Different choices (definitions) of the combining operator may give different logic systems with different expressive power. The method we adopt for TML + SLTL can be viewed as a hierarchy combination technique: when applying "+" to the logics TML and SLTL, we in fact combine SLTL only on the above of TML. Defining "+" in a different way, we would have another new logic system. Generally, given the logics TML and SLTL, we may have a sequence of operators, $+_1, +_2, \ldots$, which can be applied for combining them to obtain a sequence of new logic systems: TML^{+_1}, TML^{+_2}, *etc.* As future work, we may investigate different techniques that can be applied for combining the two logics and compare the properties of those resulting logic systems and their expressive power.

Another research direction is to investigate what will happen if we choose different belief logics or different temporal logics while using the same combining technique. We may extend TML^+ with the introduction of operators $\Diamond$ and $\Box$ to SLTL. We define the semantics of the formula $\Diamond \alpha$ by the following

$$\models \Diamond \alpha \text{ iff } \models \mathbf{first}\; \mathbf{next}^{(n)}\, \alpha, \text{ for some } n \in \mathcal{N},$$

and $\Box \alpha$ as $\neg \Diamond \neg \alpha$. As usually, $\Box$ can be read as "Always" and $\Diamond$ as "Sometime". The introduction of these temporal operators would be useful in expressing security properties that a system may satisfy.

Investigating applications of the combining techniques is also an important research direction. In this paper, we have proposed a framework for describing the dynamics of trust for multi-agent systems, and through a practical example shown that our combined logic TML^+ can be applied to construct system-specific theories of trust supporting reasoning about time-dependent beliefs of agents. With the understanding we gain from the combined logic systems, we could consider the richer applicable logics, and investigate how they provide more expressive theories for describing, and reasoning about, security properties within trusted communication systems.

Acknowledgements

The work presented in this article has been supported in part by an Australian Research Council (ARC) Discovery Project grant.

References

1. M. Burrows, M. Abadi, and R. M. Needham. A logic of authentication. *ACM Transactions on Computer Systems*, 8(1):18–36, 1990.
2. A. Dekker. C3PO: A tool for automatic sound cryptographic protocol analysis. In *Proceedings of the 13th IEEE Computer Security Foundations Workshop*, pages 77–87, Cambridge, UK, 3-5 July 2000. IEEE Computer Society.
3. N. Durgin, J. Mitchell, and D. Pavlovic. A compositional logic for proving security properties of protocols. *Journal of Computer Security*, 11(2003):677–721.
4. G. Elofson. Developing trust with intelligent agents: An exploratory study. In *Proceedings of the first International Workshop on Deception, Fraud and Trust in Agent Societies*, pages 125–139, Minneapolis/St Paul, USA, 1998.
5. M. Finger and D. M. Gabbay. Adding a temporal dimension to a logic system. *Journal of Logic, Language and Information*, 1:203–233, 1992.
6. D. M. Gabbay and V. Shehtman. Products of modal logics, part 1. *Logic Journal of the IGPL*, 6(1):71–146, 1998.
7. Dov M. Gabbay. *Fibring Logics*. Oxford University Press, Oxford, 1999.
8. J. Y. Halpern and Y. Moses. A guide to completeness and complexity for modal logics of knowledge and belief. *Artificial Intelligence* **54**, pages 319–379, 1992.
9. G. E. Hughes and M. J. Cresswell. *A New Introduction to Modal Logic*. Routledge, 1996.
10. C. M. Jonker and J. Treur. Formal analysis of models for the dynamics of trust based on experiences. In *Multi-Agent System Engineering, Proc. of MAAMAW'99*, Lecture Notes in Artificial Intelligence, Vol. 1647, pages 221–231. Springer, 1999.
11. M. Kracht and F. Wolter. Properties of independently axiomatizable bimodal logics. *The Journal of Symbolic Logic*, 56(4):1469–1485, 1991.
12. S. Kripke. Semantical considerations on modal logic. *Acta Philosophica Fennica*, 16:83–94, 1963.
13. C. Liu. Logical foundations for reasoning about trust in secure digital communication. In *AI2001: Advances in Artificial Intelligence*, Lecture Notes in Artificial Intelligence, Vol. 2256, pages 333–344. Springer, 2001.
14. C. Liu and M. A. Orgun. Dealing with multiple granularity of time in temporal logic programming. *Journal of Symbolic Computation*, 22:699–720, 1996.
15. C. Liu and M. A. Ozols. Trust in secure communication systems – the concept, representations, and reasoning techniques. In *AI2002: Advances in Artificial Intelligence*, Lecture Notes in Artificial Intelligence, Vol. 2257, pages 60–70. Springer, 2002.

16. C. Liu, M. A. Ozols, and T. Cant. An axiomatic basis for reasoning about trust in PKIs. In *Proceedings of the 6th Australasian Conference on Information Security and Privacy (ACISP 2001)*, Lecture Notes in Computer Science, Vol. 2119, pages 274–291. Springer, 2001.
17. L. C. Paulson. The inductive approach to verifying cryptographic protocols. *Journal of Computer Security*, 6(1-2):85–128, 1998.
18. P. V. Rangan. An axiomatic basis of trust in distributed systems. In *Proceedings of the 1988 IEEE Computer Society Symposium on Research in Security and Privacy*, pages 204–211, 1988.
19. P. F. Syverson and P. C. van Oorschot. On unifying some cryptographic protocol logics. In *Proceedings of the IEEE Society Symposium on Research in Security and Privacy*, pages 234–248, Oakland, CA USA, 1994. IEEE Computer Society Press.
20. M. Wooldridge. Coherent social action. In *Proceedings of the Eleventh European Conference on Artificial Intelligence (ECAI-94)*, pages 279–283, Amsterdam, The Netherlands, 1994.

Using Optimal Golomb Rulers for Minimizing Collisions in Closed Hashing

Lars Lundberg, Håkan Lennerstad, Kam illa Klonowska, and Göran Gustafsson

School of Engineering, Blekinge Institute of Technology,
S-37225 Ronneby, Sweden
{Lars.Lundberg, hakan.lennerstad, kamilla.klonowska,
goran.gustafsson}@bth.se

Abstract. We give conditions for hash table probing which minimize the expected number of collisions. A probing algorithm is determined by a sequence of numbers denoting jumps for an item during multiple collisions. In linear probing, this sequence consists of only ones – for each collision we jump to the next location. To minimize the collisions, it turns out that one should use the Golomb ruler conditions: consecutive partial sums of the jump sequence should be distinct. The commonly used quadratic probing scheme fulfils the Golomb condition for some cases. We define a new probing scheme – Golomb probing - that fulfills the Golomb conditions for a much larger set of cases. Simulations show that Golomb probing is always better than quadratic and linear and in some cases the collisions can be reduced with 25% compared to quadratic and with more than 50% compared to linear.

1 Introduction

For a long time, hash tables have been used to optimize the speed with which a stored data item can be accessed. A hashing table data structure is usually implemented as a large array. Whenever an item must be added to the hash table it is not simply added at the end of the structure. Instead a hash function determines the location at which a given item is stored based, somehow, on some attribute of the item to be stored. This computed location is called an items hash value. The value being hashed is sometimes called the *pre-image*.

In order to insert, delete or access an item in a hash table it is necessary to compute the hash value of the item. Once the hash value is known, we can look in the table at the location specified by the hash value. If it is an insert operation and if this table location is empty, the new item can be inserted without problems. However if this location in the table is already in use by some other data item an alternate strategy must be used. This occurrence is known as a *collision*.

One way of dealing with collisions is to make each location in the hash table the head of a linked-list data structure. If a collision has occurred, you traverse down the linked list at the hash value and add the new data item to the lists tail. This method is known as *open hashing* because multiple data items sharing the same hash value are stored "outside" the hash table.

M. J. Maher (Ed.): ASIAN 2004, LNCS 3321, pp. 157–168, 2004.
© Springer-Verlag Berlin Heidelberg 2004

An alternative approach is to store collided data inside the hash table at different locations than their computed hash value. This is known as *closed hashing*. The classic way to deal with collisions is to simply increment an item's hash value by one until finding an unoccupied hash table location. This method is known as *linear probing*. Linear probing, while simple to understand and implement, leads to data clumping and is not the ideal way of handling collisions.

A different approach is *quadratic probing*. In quadratic probing, instead of simply moving one address down in the hash table, the number of spaces moved is dependent on the number of times that we have moved so far. The advantage of quadratic probing compared to linear probing is that the expected number of collisions is lower (we will assume a hash function that distributes the hashed values evenly). Here we present a new probing scheme that provides even less collisions than traditional quadratic probing. The results presented here benefit from previous results on optimal recovery schemes in fault tolerant distributed and cluster computing [7,8].

2 Problem Definition

We consider closed hashing with a fixed hash table size m. When the hash function (*HF*) maps two items (pre-image values) to the same entry in the hash table a collision occurs. In that case we need to find another entry in the hash table for the value that cannot be inserted in the already occupied entry in the hash table. The new entry after i collisions is defined by a probing function $o(i)$ ($0 \leq i < m$).

   entry_after_i_collisions $= HF(item) + o(i) \bmod m$

For all probing functions we have $o(0) = 0$. It is also clear that for $0 < i \leq m$, all $o(i)$ should be unique and $o(i) < m$. For linear probing we have $o(i) = i$. For quadratic probing we have a more complex function. One example is $o(i) = 2^i - 1$ [5], as long as $2^i - 1 < m$, another example is $o(i) = i^2$, as long as $i^2 < m$. When $2^i - 1 < m$ or $i^2 \geq m$ the function $o(i)$ is defined as the largest value that has not yet been used. These are the two most common approaches to quadratic probing. In the discussion below we will use the function $o(i) = 2^i - 1$ as an example, but in the performance evaluations in Section 5 we compared both of these methods with the approach that we suggest (defined in Section 4), and it turns out that the performance of these two approaches are almost identical.

Let v be a vector of length m, and let entry i in this vector have value $o(i)$. For $m = 10$ we get: $v = (0,1,2,3,4,5,6,7,8,9)$ for linear probing, and $v = (0,1,3,7,2,4,5,6,8,9)$ for quadratic probing (using $o(i) = 2^i - 1$, when $2^i - 1 < m$, and the largest value that has not yet been used when $2^i - 1 \geq m$).

We also define the difference sequence r as an alternative way of describing the probing algorithm. Here $r_i = o(i) - o(i - 1)$. We require for a difference sequence that all $r_i \neq 0$ and that the numbers $r_1, r_1 + r_2, \ldots, r_1 + \ldots + r_m$ are distinct.

In order to understand why quadratic probing results in less collisions than linear probing we consider a small example where we insert four items in a hash table with $m = 10$. We assume a hash function where the probability of mapping an item to entry x in the hash table is simply $1/10$, i.e. the reduction of collisions due to using quadratic instead of linear probing does not depend on the use of suboptimal hash functions that spread the items unevenly in the hash table (this is sometimes a misconception). We will now discuss the expected number of collisions, using linear and quadratic probing respectively, when we insert four items in the hash table:

1. When we insert item one the table is empty so there will be zero collisions for both the linear and the quadratic cases.
2. When we insert item two there will be on average 1/10 collisions for both the linear and the quadratic cases.
3. When we insert item three, there is a 2/10 chance that we will hit an occupied entry. If we hit an occupied entry there can be either one or two collisions. Two collisions may occur in the following cases:
   a. $HF(item3) = HF(item2) = HF(item1)- o(1) \bmod m$
   b. $HF(item3) = HF(item1) = HF(item2)- o(1) \bmod m$
   c. $HF(item3) = HF(item1) = HF(item2)$

   This means that the expected number of collision when we insert item three is the same for the linear and the quadratic cases.
4. When we insert item four there is a 3/10 chance that we will hit an occupied entry. If we hit an occupied entry there can be either one, two or three collisions. Three collisions may occur in the following cases:
   a. $HF(item4) = HF(item1) = HF(item2) = HF(item3)$
   b. $HF(item4) = HF(item1) = HF(item2)-o(1) \bmod\ m = HF(item3)+ o(2)$ mod m
   c. $HF(item4) = HF(item1) = HF(item3)- o(1) \bmod m = HF(item2)+ o(2)$ mod m
   d. $HF(item4) = HF(item2) = HF(item1)- o(1) \bmod m = HF(item3)+ o(2)$ mod m
   e. $HF(item4) = HF(item2) = HF(item3)- o(1) \bmod m = HF(item1)- o(2)$ mod m
   f. $HF(item4) = HF(item3) = HF(item1)- o(1) \bmod m = HF(item2)+ o(2)$ mod m
   g. $HF(item4) = HF(item3) = HF(item2)- o(1) \bmod m = HF(item1)- o(2)$ mod m
   h. $HF(item4) = HF(item1) = HF(item2) = HF(item3)- o(2) \bmod m$
   i. $HF(item4) = HF(item3) = HF(item2) = HF(item1)- o(2) \bmod m$
   j. $HF(item4) = HF(item1) = HF(item3) = HF(item2)- o(2) \bmod m$

   The cases a-j above are the only possibilities for three collisions for the quadratic case, but for the linear case the following scenarios will also result in three collisions (for the quadratic case these scenarios will only result in two collisions).

   k. HF(item4) = HF(item1) = HF(item2) = HF(item3)-o(1) mod m
   l. HF(item4) = HF(item1) = HF(item3) = HF(item2)-o(1) mod m
   m. HF(item4) = HF(item3) = HF(item2) = HF(item1)-o(1) mod m

   Besides cases k-m above and quadratic probing, two and one collisions will occur for the same cases for linear and quadratic probing. This means that the average number of collisions is larger using linear probing compared to quadratic probing.

The advantages with quadratic probing increases as m and the number of items in the hash table increases. In order to fully understand the reasons why quadratic probing is better than linear probing we will, in Section 3, introduce a formal system model. Based on this model we will, in Section 4, define an alternative probing scheme called Golomb probing, and in Section 5 we will evaluate this scheme using simulations. Section 6 discusses some assumptions, practical aspects and related techniques. Section 7 concludes the paper.

3 Formal Model of a Hash Table

Consider a hash table of m locations, where some locations are filled with items. A location attempt at the j:th location will in general result in a chain of collisions, defined by $HF(item\_j) + o(0)$, $HF(item\_j) + o(1)$,... $HF(item\_j) + o(k-1)$, or in terms of jumps: $HF(item\_j)$, $HF(item\_j) + r_1$, $HF(item\_j) + r_1 + r_2$,... $HF(item\_j) + r_1 + ... + r_{k-1}$. We suppose that $HF(item\_j) + o(k)$ is not occupied, but all k entries in the chain are. This is the definition of a *chain*.

Hence, a chain of length k results in k collisions if an allocation is made at the first entry $HF(item\_j)$ in the chain. Denote by $C(i)$ the length of the chain starting at location i. An empty location is a chain of length 0. Since we assume a hash function that distributes the items with uniform probability for items arriving to all m locations, the expected value for the number of collisions is simply

$$\frac{1}{m}\sum_{i=1}^{m}C(i).$$

For example, if k locations are occupied, and no chain has length more than one, the expected value of the number of collisions is simply k/m.

The performance of a probing algorithm is determined by the expected number of collisions of different configurations, and the probability that such configurations appear. We characterize a configuration by the set of chains that are present.

Conditions of Golomb type

For optimal behavior, the chains should not overlap. Consider a chain $HF(item\_j)$, $HF(item\_j)+r_1$, $HF(item\_j)+r_1+r_2$,... $HF(item\_j)+r_1+...+r_{k-1}$ of occupied entries. Assume that $HF(item\_j)+r_1$ is addressed. Since this entry is occupied, the item is sent to $HF(item\_j)+2r_1$. If this entry coincides with $HF(item\_j)+r_1+r_2$, hence if $r_1 = r_2$, we are certain that we obtain a new collision. Otherwise we may or may not obtain a collision. Thus, if the condition $r_1 \neq r_2$ is fulfilled, we will have a lower number of collisions. Similarly, the possibility of collisions with $HF(item\_j)+r_1+...+r_{i-1}$ for any $i \leq k$, give a condition $r_1 \neq r_2 + ... + r_{j-1}$ for any $i \leq k$.

A similar argument when not starting to allocate $HF(item\_j)+r_1$ but $HF(item\_j)+r_1 + ... + r_{i-1}$ give the conditions $r_1 \neq r_i + ... + r_{j-1}$ for any $i,j \leq k$.

We may also obtain collisions after more than one jump. This corresponds to the condition $r_1 + \ldots + r_{i-1} \neq r_j + \ldots + r_{k-1}$, for certain integers i, j and k. Hence, as many as possible of the partial sums $r_i + \ldots + r_j$ should be distinct. This is a subset of the Golomb conditions that we will discuss in the next section.

Finally, a chain may be entered not as the first allocation attempt. Hence, we obtain the requirement that all consecutive partial sums $r_j + r_{j+1} + \ldots + r_{k-1}$ should be distinct, which is the Golomb condition (see next section).

One may also consider how a sequence $r_1, r_2, \ldots$ affects the probability that chains will occur. We remark that the expectation of the number of collisions at a certain point while constructing the hash table is $\dfrac{1}{m} \sum_{i=1}^{m} C(i)$, where $C(i)$ is the length of the chain starting at the i:th location in the hash table. Note that if the condition $r_1 = r_2$ is fulfilled, two chains have possibly grown since the event of a later allocation to $HF(item\_j) + r_1$ will give at least two collisions with certainty. Chains are growing in the event of collisions, so avoiding collisions is also favorable for minimizing the probability that chains will occur.

One may summarize this argument in that if chains overlap, several chains of different length may grow simultaneously, thus increasing the expected number of collisions. The Golomb conditions avoid these overlaps. The Golomb conditions avoid more overlaps than quadratic probing, since minimization of overlaps is how these are constructed (see Section 4).

We thus have proved:

Theorem 1 Minimizing Conditions are (Golomb Conditions) $r_i + \ldots + r_j \neq r_k + \ldots + r_l$, for All $1 \leq i \leq j < k \leq l \leq m - 1$.

One effect of the Golomb conditions is that there is no overlap between the entries i visited before an item ends up at entry X after i collisions, and the j entries visited before an item ends up on entry X after j collision ($i \neq j$). This is illustrated for quadratic probing in Fig. 1. The entries visited before item D ends up on entry X are marked with d, the entries visited before item C ends up on entry X are marked with c, and so on. Fig. 1 shows that the quadratic probing scheme meets the Golomb conditions as long as $2^i - 1 < m$, i.e. as long as the number of items in the hash table is less than $\log_2(m+1)$ (for the other quadratic scheme is, $o(i) = i^2$, this is not true, e.g. $r_1 + r_2 + r_3 = 3^2 = r_5 = 5^2 - 4^2$). For linear probing there are a lot of overlaps between the entries that have been visited before the collided item ends up on entry X and that probing scheme is thus not optimal (does not meet the Golomb condition).

We would like to find a probing scheme that could guarantee optimal behavior (i.e. minimal probability for collisions) even when there are more than $\log_2(m+1)$ items in the hash table. We expect that such schemes will be very efficient even when the number of items in the hash table is too large to guarantee optimality. We will, therefore, look at the problem of finding probing schemes that guarantee optimal behavior for as large j as possible (j is the number of items in the hash table). This problem is identical to a previously studied problem dealing with static recovery schemes in fault tolerant cluster computing. The initial solution for that problem was in fact the quadratic probing scheme 2^i -1 (although called something else [9]). However, later work showed that there are other (probing) functions that will obtain optimal behavior also when the number of items in the hash table exceeds $\log_2(m+1)$ [7,8]. These results are based on so called Golomb rules, and we will now apply these to hash table probing.

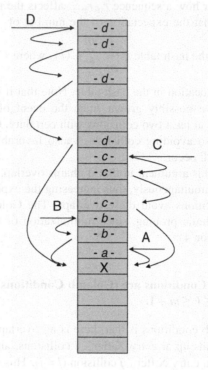

Fig. 1. Ways an item can end up on entry X using quadratic probing (2^i-1)

4   Golomb Probing

The Golomb ruler (the spanning ruler) is a sequence of non-negative integers such that no two distinct pairs of numbers from the set have the same difference. These numbers are called *marks* and correspond to positions on a linear scale. The difference between the values of any two marks is called the *distance*. The shortest Golomb rulers for a given number of marks are called the Optimal Golomb Rulers (OGRs) [157,4,5,10]. An example of the representation of OGR with four marks is shown in Fig. 2. It is possible to measure the distances: 1, 2, 3, 4 as 1+3, 5, 7 as 5+2, 8 as 3+5, 9 as 1+3+5, 10 as 3+5+2 and 11 as 1+3+5+2, but we cannot measure the distance 6 and no distance can be measured in more than one way.

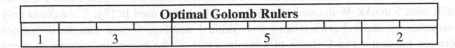

Optimal Golomb Rulers			
1	3	5	2

Fig. 2. The ruler presentation of the OGR with four marks

One property of the Golomb ruler is that if we use a Golomb ruler with j marks as a probing function, there will be no overlaps between a, b, c,... (see discussion above), i.e. the probing function will be optimal. The length (m) of an OGR with j marks defines the minimal hash table size for which we can guarantee optimal behavior (i.e. minimal probability for collisions) when there are at most j items in the table.

Fig. 1 showed a hash table with 21 entries and quadratic probing. In that case we could guarantee optimal behavior for cases A-D, i.e. as long as there are at most four items in the hash table. Table 1 shows the first 10 OGR sequences, and if we have a hash table with $m = 21$ we select the longest sequence that will fit within the table. In this case sequence five (1,4,10,15,17) and use this one as the probing function, i.e. we have $o(1) = 1$, $o(2) = 4$, $o(3) = 10$, $o(4) = 14$, and $o(5) = 17$. The rest of the function $o(i)$ is defined as the largest value that has not yet been used. Let v be a vector of length 21, and let entry i in this vector have value $o(i)$. In that case we get: $v =$ (0,1,4,10,15,17,2,3,5,6,7,8,9,11,12,13,14,16,18,19,20) for Golomb probing, and $v =$ (0,1,3,7,15,2,4,5,6,8,9,10,11,12,13,14,16,17,18,19,20) for quadratic probing.

Table 1. Optimal Golomb sequences and quadratic probing sequences

Length of sequence	Optimal Golomb Sequence	Quadratic Sequence ($2^i - 1$)
1	1	1
2	1,3	1,3
3	1,4,6	1,3,7
4	1,4,9,11	1,3,7,15
5	1,4,10,15,17	1,3,7,15,31
6	1,4,10,18,23,25	1,3,7,15,31,63
7	1,4,9,15,22,32,34	1,3,7,15,31,63,127
8	1,5,12,25,27,35,41,44	1,3,7,15,31,63,127,255
9	1,6,10,23,26,34,41,53,55	1,3,7,15,31,63,127,255,511
10	1,4,13,28,33,47,54,64,70,72	1,3,7,15,31,63,127,255,511,1023

Fig. 3 shows the same hash table as Fig. 1, but now we use Golomb probing. The figure shows that there are no overlaps and that we are now able to fill the table denser with a, b, c,... thus resulting in guaranteed optimal behavior when $j \leq 5$ (i.e. we can guarantee optimality for cases A-E). The figure shows that only entries 5 and

9 are unmarked. This means that it is not possible to measure lengths 5 and 9 with an OGR with 5 marks. With quadratic probing, which was used in Fig. 1, we could only guarantee optimal behavior when $j \leq 4$. Table 1 and Fig. 4 show that the difference between quadratic and Golomb probing in terms of the maximum number of items for which optimal behavior can be guaranteed increases rapidly when the hash table size increases, e.g. for $m = 373$, Golomb probing guarantees optimal behavior when $j \leq 22$, whereas quadratic probing only guarantees optimal behavior when $j \leq 8$.

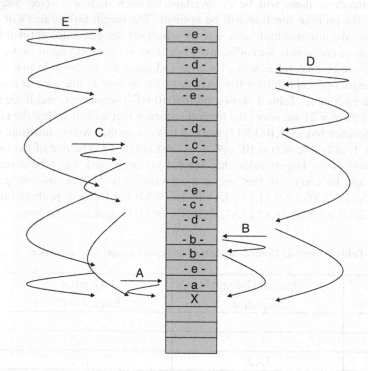

Fig. 3. Different ways an item can end up on entry X using Golomb probing

5 Performance Evaluation

In order to compare the performance of Golomb probing with the performance of quadratic and linear probing we wrote a simulation program. We looked at five different hash table sizes: 100, 200, 400, 800, and 1600. For each of these sizes we looked at different utilization levels ranging from 10% to 95%. When we measured utilization level X we counted the numbe r of collisions when we insert X m items in an empty hash table of size m. For each combination of utilization level and hash table size we generated 10,000,000 random insertion sequences with uniform distribution over the hash table size. We counted the number of collisions using the three schemes: linear, quadratic and Golomb for each case. For the quadratic scheme we used the best value of the two commonly used schemes: is $o(i) = 2^i - 1$ and $o(i) = i^2$ (it turned out that the difference between these schemes was very marginal, even if $o(i) = 2^i - 1$ was slightly better in most cases). The performance degradation due to using the linear scheme instead of Golomb is shown in Fig. 5 and the performance degradation due to using a quadratic scheme instead of Golomb is shown in Fig. 6.

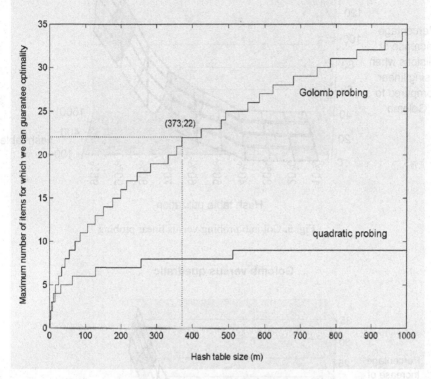

Fig. 4. The maximum number of items in the hash table for which we can guarantee minimal probability for collisions for Golomb and quadratic probing ($2^i - 1$)

The figures show that the advantage of using Golomb increases rapidly as the utilization increases. This is hardly surprising since one might expect that improved collision handling (such as Golomb) pays off when the utilization is high and there are a large number of collisions. The diagrams also show that the advantage of using the Golomb scheme increases somewhat when the hash table size increases, even if this effect is rather limited. It is clear that the improvement compared to linear is much larger than the improvement compared to quadratic. However, Golomb is clearly better than quadratic particularly for high utilization values. By looking at Fig. 6 we see that for 95% utilization there is almost 35% more collisions using quadratic probing compared to Golomb probing.

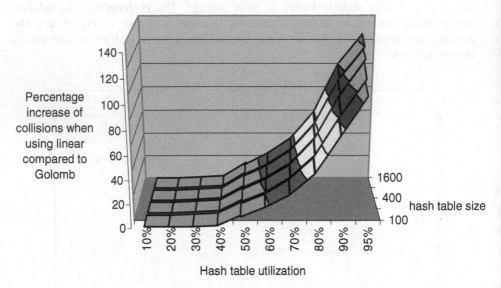

Golomb versus linear

Percentage increase of collisions when using linear compared to Golomb

hash table size

Hash table utilization

Fig. 5. Golomb probing versus linear probing

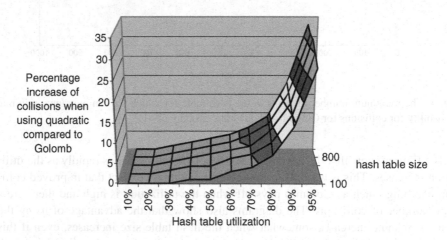

Golomb versus quadratic

Percentage increase of collisions when using quadratic compared to Golomb

hash table size

Hash table utilization

Fig. 6. Golomb probing versus quadratic probing

6   Discussion and Related Work

There are alternative approaches to restarting with the unused entries when $2^i - 1 \geq m$ or $i^2 \geq m$. One common approach is to simply continue to use the probing function

$o(i) = 2^i - 1$ or $o(i) = i^2$ also when $2^i -1 \geq m$ (or $i^2 \geq m$). It is well known that when we continue to use the probing function $o(i) = 2^i - 1$ or $o(i) = i^2$ also when $2^i -1 \geq m$ or $i^2 \geq m$, we should select a hash table size that is a prime number [1]. The advantage with continuing to use the probing function is that it is easier to implement since we can use a simple probing function in order to obtain all values. The disadvantage is that the probing function may examine the same entry more than once (thus resulting in a sure collision the second time) and some entries may not be examined at all. We did some tests where we compared these approaches with the approach considered here for a number of prime number hash table sizes. The performance difference (in terms of the number of collision) was very marginal. For some cases, the approach that we used here was slightly better.

An alternative to quadratic probing is to use a secondary hash function when the primary hash function results in a collision. This is referred to as double hashing. The performance gain (in terms of reduced collisions) of using double hashing compared to linear probing is somewhat better than the performance gain of using our quadratic scheme, but not as good as the Golomb scheme. For instance, for 90% hash table utilization we get 87% more collisions with linear probing compared to double hashing (see Table 9.22 in [11]). The corresponding values (for 90% hash table utilization) measured by us are 80% for quadratic and 100% for Golomb (see figures 5 and 6).

A practical problem with the Golomb scheme (and with the quadratic probing approach discussed here) is that they use large lists for generating the next value in the sequence. One way of reducing this problem is to only include the values in the actual Golomb sequence in the list and then restart with a linear search. The disadvantage of this is that some entries may be examined more than once. We did some test with this and the number of additional collisions using this approach was extremely small (less than 0.1%).

Another problem is how to generate the Golomb sequences for large m. Finding optimal Golomb sequences is an NP-hard problem and the known ones stop at about $m = 22,000$. However, we have in previous work shown that it is possible to define a heuristic algorithm that can generate large (non-optimal) Golomb sequences very fast (and in polynomial time) [7]. The difference in terms of the number of marks for a certain m is very small between the optimal Golomb sequences and the ones generated by the heuristic algorithm, so this aspect is not a real problem.

In some cases there is a static set of known values that we want to map to a hash table. In those cases it is possible to define a so called perfect hash function that maps each of the items to a unique entry in the hash table [3], i.e. there are no collisions. A perfect hash function that, without any collisions, maps a set of m known values to a hash table with m entries is called a minimal perfect hash function [3].

7 Conclusion

We have presented a new probing scheme – Golomb probing - for handling collisions in closed hashing. Performance evaluations based on simulation show that

Golomb probing results in significantly less collisions than linear and quadratic probing.

The rational behind the Golomb probing scheme is based on a formal model of the hash table, which defines Golomb condition for minimizing the number of collisions. Quadratic probing sequences fulfill the Golomb conditions when the number of items in the hash table is small. The Golomb probing scheme defined in this paper meets the Golomb conditions also for a much larger set of cases.

We have also discussed how a Golomb probing scheme can be implemented in a practical way, and how Golomb sequences can be generated also for very large hash tables.

References

1. L.V. Atkinson, A.J. Cornah, *Full period quadratic hashing*, International Journal of Computer Mathematics, Vol. 4, No. 2, 1974. pp. 177-189.
2. G.S.Bloom and S.W.Golomb, *Applications of Numbered, Undirected Graphs*, Proceedings of the IEEE, vol.65, No.4, April, 1977, pp.562-571.
3. Z.J. Czech, G. Havas, B.S. Majewski, *Perfect hashing*, Theoretical Computer Science, 1997, pp. 1-143.
4. A. Dimitromanolakis, *Analysis of the Golomb Ruler and the Sidon Set Problems, and Determination of large, near-optimal Golomb Rulers*, Dept. of Electronic and Computer Engineering Technical University of Crete, June, 2002.
5. Scott Gasch, *Dealing with Collisions*, http://www.fearme.com/misc/alg/node30.html
6. B.Hayes, *Computing Science: Collective Wisdom*, American Scientist, vol. 98, no. 2, March/April, 1998, pp.118-122.
7. K. Klonowska, L. Lundberg, and H. Lennerstad, *Using Golomb Rulers for Optimal Recovery Schemes in Fault Tolerant Distributed Computing*, in proceedings of the 17th International Parallel and Distributed Processing Symposium (IPDPS 2003), Nice, France April 2003.
8. L. Lundberg, D. Haggander, K. Klonowska, and C. Svahnberg, *Recovery schemes for high availability and high performance distributed real-time computing*, in proceedings of the 17th International Parallel and Distributed Processing Symposium (IPDPS 2003), Nice, France April 2003.
9. Lars Lundberg and C. Svahnberg: *Optimal Recovery Schemes for High-Availability Cluster and Distributed Computing*. J. Parallel Distrib. Comput. 61(11): 1680-1691 (2001)
10. Project OGR, http://www1.distributed.net/ogr/index.php.en
11. Thomas A. Standish, *Data structures in Java*, Addison-Wesley, 1997.

Identity-Based Authenticated Broadcast Encryption and Distributed Authenticated Encryption

Yi Mu[1], Willy Susilo[1], Yan-Xia Lin[2], and Chun Ruan[3]

[1]School of IT and Computer Science
[2]School of Mathematics and Applied Statistics,
University of Wollongong, Wollongong, NSW 2522, Australia
{ymu, wsusilo, yanxia}@uow.edu.au
[3]School of Computing and IT, University of Western Sydney,
South Penrith DC, NSW 1797, Australia
chun@cit.uws.edu.au

Abstract. Since its introduction, broadcast encryption has attracted
many useful applications. In this paper, we propose two identity-based
schemes for authenticated broadcasting and distributed message authentication. The first scheme supports multiple broadcasters and allows each
broadcaster to dynamically broadcast messages into an arbitrary group
of receivers determined by the broadcaster. The receivers can obtain the
broadcasted message using the identity of the broadcaster and his own
secret decryption key; hence it ensures both *confidentiality* and *authenticity* of the message. The second scheme allows users (receivers) to send
messages back to the broadcaster where the authentication of messages
is done with the identity of the user. We also provide security proofs for
our schemes under the random oracle model.

1   Introduction

Broadcast encryption is useful for distributing digital contents to Internet users
over a broadcast channel. It allows a server to deliver information to a set of
users that can be dynamically formed. Namely, the broadcaster can determine
which users will receive the information with the pre-defined user information.
Each authorized user can recover the information by using the corresponding
secret key.

The broadcast encryption was introduced by Fiat and Naor [5]. Since then,
there have been a number of schemes in the literature (e.g., [12, 6, 7, 14, 8]). Those
schemes vary from bounded to unbounded number of broadcasts. They may be
composed of either fixed user groups or variable (or dynamic) user groups. Some
schemes allow a special system setup; namely, if users make any new secret
decryption keys with their own secret information, they can be detected by the
broadcaster. This is called traitor tracing (e.g., [4, 1]).

One of important issues in various broadcast schemes is whether they are
symmetric-key based or asymmetric key based. Symmetric-key based schemes

M.J. Maher (Ed.): ASIAN 2004, LNCS 3321, pp. 169–181, 2004.

has their advantage of efficiency and are sufficient for many applications; however, it has disadvantage in managing users (lack of dynamic user management). In contrast, in an asymmetric key setting, the broadcaster encrypts a message (or a session key) using a unique key that is unknown to receivers. Then, each authorized receiver can then decrypt the message (or the session key) using his or her secret decryption key. Therefore, it allows a dynamic user management, i.e., only the authorized users can decrypt the message. To ensure computational efficiency, usually, asymmetric-key encryptions are used to transmit session keys only and messages are encrypted with a session key.

When a broadcast system is composed of multiple broadcasters, it raises an issue of message or broadcaster authentication. This issue is especially important for mutlicasting in computer networks. Actually, there are two distinct issues in this context, i.e., how does a receiver authenticate a message and how does a broadcaster authenticate a message sent back from a receiver? We notice that these issues differ from regular public-key cryptography where a certification authority can been employed. For example, in some settings, the public key becomes a secret broadcast encryption key that can be formed during a run time[1, 14]. The methods based on certification authority are not sufficient to accommodate this situation.

It is noted that Boneh and Franklin [2] recently introduced a new type of cryptography using the Weil pairing (a type of bilinear pairings). The Weil pairing was originally considered to be a bad tool, since it was mainly used for attacking elliptic curves [13]. Recently, it has been showed that the Weil pairing can be used to construct a protocol for three party one round Diffie-Hellman key exchange[10]. Based on pairings, Boneh and Franklin gave a concrete identity based encryption protocol [2], where instead of using a public key of the receiver, the ID of a receiver is used for encryption. The pairings have been shown to have many applications, for example short signature schemes [3] and many ID-based applications.

Our Contribution. We resolve the authentication problem in broadcast system by non-trivially utilizing the concepts of ID-based encryption [2]. In particular, we show how a receiver *authenticates* a broadcasted message using the ID of a broadcaster and how a broadcaster *authenticates* a message sent back from a receiver using the ID of the receiver. These issues are important in networked environment since ID's, such as an IP number and an email address, can conveniently be used for authentication. We note that in the previous constructions in the literature, these issues have never been addressed. Furthermore, the message authentication by broadcasters in our system is *deniable*, in the sense that the verification associated with an authentication process is *not* transferable [9, 15]. No other verifier can be convinced with the authenticity of the broadcasted message other than the intended verifier.

The rest of this paper is organized as follows. In Section 2, we provide the definitions of our schemes and the security requirements. In Section 3, we describe the ID-based authenticated broadcast encryption scheme. We also provide the security proofs for the proposed schemes. In Section 4, we present the ID-based distributed authentication scheme, where the broadcaster can authenticate the message sent back by a receiver via the ID of the receiver. We also prove that the proposed scheme is secure against outsider attacks. Section 5 concludes the paper. The detail of security proofs are provided in the appendix.

2 Definitions

Let $\mathbb{B}$ be a group of broadcasters and $\mathbb{R}$ be a group of receivers. For simplicity, we employ Bob $\in \mathbb{B}$ and Ray $\in \mathbb{R}$ as representatives in the presentation. Let Alice be a trusted authority who sets up the system.

We now define two schemes in our system: ID-Based Authenticated Broadcast Encryption (IBABE) and ID-Based Designated Distributed Authentication (IBDDA).

2.1 The ID-Based Authenticated Broadcast Encryption

The IBABE is a protocol that allows multiple broadcasters to securely broadcast messages to authorized receivers. All broadcasters share an common key construction. Receivers can identify or authenticate the broadcasted message in terms of the identity of a broadcaster and some public information. Using a pay TV system as an example, we may refer a broadcaster to a program or channel which has been subscribed by an user.

Definition 1. (ID-Based Authenticated Broadcast Encryption) *An ID-Based Authenticated Broadcast Encryption is a triple of polynomial-time algorithms (*KeyGen, Encrypt, Decrypt*), where*

- KeyGen, *the key generation algorithm, is a polynomial algorithm used by Alice to set up all the parameters of the system including the public parameters* params. *It takes as input a security parameter ℓ and generates the master secret key K_i for a broadcaster (say, K_B for Bob $\in \mathbb{B}$) and the secret decryption keys x_i for users or receivers ($i = 1, \cdots, n$).*
- Encrypt, *the encryption algorithm, is a probabilistic algorithm used by a broadcaster (say, Bob $\in \mathbb{B}$) to securely send a message $M \in \{0,1\}^l$ to authorized users within the system with the master secret key K_B. It takes as input* params, M, K_B, *and $\{x_i\}_{i=1,\cdots,t}$ for all authorized receivers and returns a ciphertext C of authenticated encryption and the public parameters* params2.
- Decrypt, *the decryption algorithm, is a deterministic algorithm used by receivers to recover the message M from a ciphertext C. It takes one of $\{x_i\}_{i=1,\cdots,t}$, the ID of the broadcaster,* params2, *and C as input and returns M and True or $\perp$, where True or $\perp$ indicates success or failure in the authentication.*

2.2 ID-Based Designated Distributed Authentication

In the IBDDA protocol, a broadcaster can authenticate a message sent by a user $\in \mathbb{R}$. The authentication relies on the ID of the user and no other public information on the user is required. The authentication is designed to be deniable; therefore, the authentication token is not transferable to others [9, 15].

Definition 2. (ID-based Designated Distributed Authentication) *An ID-based Designated Distributed Authentication is a triple of polynomial-time algorithms (KeyGen, Signcrypt, Decrypt), where*

- KeyGen, *the key generation algorithm, is a polynomial algorithm used by Alice to set up all the parameters of the system including the public parameters* params. *It takes as input a security parameter ℓ and generates the master secret key K_i for a broadcaster (say, K_B for Bob $\in \mathbb{B}$) and the secret keys A_i, $i = 1, \cdots, n$ for users. (Note: the setup for broadcasters is the same as that of the IBABE scheme).*
- Signcrypt, *the encryption algorithm, is a probabilistic algorithm used by a receiver (Say, Ray $\in \mathbb{R}$) to securely send a message $M \in \{0,1\}^l$ to a broadcaster (say, Bob) within the system with the secret key A_R. It takes the system parameters* params, A_R, *and M as input and returns a ciphertext C of signcryption and the associated parameters* params2.
- Decrypt, *the decryption algorithm, is a deterministic algorithm used by receivers to recover the message M from a ciphertext C. It takes K_B, the ID of the sender Ray, the public parameters* params2, *and C as input and returns M and True or $\perp$, where True or $\perp$ indicates success or failure in the authentication.*

We note that IBDDA is particularly useful in practice, for example it allows the pay TV subscribers to submit their vote to the broadcaster. This is a novel application which has never been considered in the literature before.

2.3   Bilinear Pairings and Security Definitions

We set up our systems using bilinear pairings due to Boneh and Franklin [2]. Define two cyclic groups $\mathbb{G}_1, \mathbb{G}_2$. $\mathbb{G}_1$ is an additive group and $\mathbb{G}_2$ is multiplicative group. Both have a prime order q. Let e be a computable bilinear map $e : \mathbb{G}_1 \times \mathbb{G}_1 \to \mathbb{G}_2$. For $a, b \in \mathbb{Z}_q$ and $P, Q \in \mathbb{G}_1$, we have $e(aP, bQ) = e(P, Q)^{ab}$.

A decisional Diffie-Hellman problem (DDHP) is easy [11], since $e(aP, bP) = e(P, P)^{ab}$. The security of a pairing based algorithm is based on the computational Diffie-Hellman problem (CDHP), which is given below.

Definition 3. (Computational Diffie-Hellman Problem) *Let a, b be chosen from $\mathbb{Z}_q$ at random and P be a generator chosen from $\mathbb{G}_1$ at random. Given (P, aP, bP), compute $abP \in \mathbb{G}_1$.*

$\mathbb{G}_1$ is referred to as a Gap Diffie-Hellman (GDH) group and CDHP can be referred to as a Gap Diffie-Hellman problem, if DDHP can be solved in polynomial time and no polynomial algorithm can solve CDHP with non-negligible advantage in polynomial time. The other hard problem used in this paper is called Bilinear Diffie-Hellman problem (BDHP) [2]:

Definition 4. (Bilinear Diffie-Hellman Problem) *Let a, b, c be parameters chosen from $\mathbb{Z}_q$ and $P \in \mathbb{G}_1$ be a generator. Given (P, aP, bP, cP), compute $e(P, P)^{abc} \in \mathbb{G}_2$.*

We will also utilize a factorization problem, i.e., Elliptic Curve Factorization Problem (ECFP), defined as follows.

Definition 5. (Elliptic Curve Factorization Problem) *Let P be a generator in $\mathbb{G}_1$ and x, y be members chosen from $\mathbb{Z}_q$. Given $xP + yP$ and P, find xP and yP.*

This type of problems have been used for many cryptographic systems such as the ElGamal encryption scheme.

2.4   Security Notion

The security of the IBABE scheme is considered to be of two types: security against *outsiders* who have public information only and security against *insiders* or users who hold legitimate private keys but are malicious to broadcasters Definitions 6 and 7 define these cases respectively.

Definition 6. *Suppose there exists a polynomial time adversary $\mathcal{A}_1$. Given a ciphertext C of the message M and the corresponding public parameters* params2, *$\mathcal{A}_1$ has the following advantage in solving and finding M.*

$$\mathsf{Succ}_{\mathcal{A}_1}(\ell) = \Pr\left[\begin{array}{l}\mathsf{params2} \to K, \\ \mathsf{Decrypt}(ID, K, C) \to (M, True)\end{array}\right] < negl(\ell).$$

for some negligible function negl and security parameter ℓ.

Definition 7. *Suppose there exists a polynomial time adversary $\mathcal{A}_2$. Given a ciphertext C of the message M, the public parameters* params2, *and one of decryption keys x_i, $\mathcal{A}_2$ has the following advantage in solving and finding the targeted master secret key K_T of a broadcaster.*

$$\mathsf{Succ}_{\mathcal{A}_2}(\ell) = \Pr\left[\begin{array}{l}(C, \mathsf{params2}, x_i) \to K_T, \\ \mathsf{Encrypt}(M', K_T, \mathsf{params}) \to (C', \mathsf{params2'}), \\ \mathsf{Decrypt}(ID, C', \mathsf{params2'}, x_i) \to (M', True)\end{array}\right] < negl(\ell).$$

for some negligible function negl and security parameter ℓ.

The security of the IBDDA scheme is defined as follows.

Definition 8. *Suppose there exists a polynomial time adversary $\mathcal{A}_3$. Given* params *and its own decryption key $A_{\mathcal{A}_3}$, $\mathcal{A}_3$ has the following advantage in forging a signcryption of another user j on a message M'.*

$$\mathsf{Succ}_{\mathcal{A}_3}(\ell) = \Pr\left[\begin{array}{l}(A_{\mathcal{A}_3}, \mathsf{params}) \to A_j, \\ \mathsf{Signcrypt}(A_j, M', \mathsf{params}) \to (C, S, \mathsf{params2}), \\ \mathsf{Decrypt}(ID_j, C, \mathsf{params2}, K_B) \to (M', True)\end{array}\right] < negl(\ell),$$

for some negligible function negl and the security parameter ℓ.

The IBDDA is a designated signcryption scheme and the signcrypted message is not transferable; therefore it is not required to discuss the case against broadcasters.

Definition 9. (Deniability) *Given a legitimate IBDDA signcryption tuple* (params2, C) *on message M, the designated receiver can verify the validity of the signcryption with overwhelming probability. Other parties have negligible advantage to verify the authenticity of the signcryption.*

3   The Identity-Based Authenticated Broadcast Encryption

We now describe the IBABE protocol defined in the preceding section. The notations of bilinear pairings provided in Section 2.3 are applicable to the rest of this paper.

– KeyGen (by Alice).
  Broadcaster setup:
  - Taking ℓ as input, select master private keys $s_i \in \mathbb{Z}_q$ for all broadcasters ($i = 1, \cdots, n_b$).
  - Select three strong public one-way hash functions: $H_1 : \{0,1\}^* \to \mathbb{G}_1$, $H_2 : \{0,1\}^* \to \{0,1\}^l$, $H_3 : \{0,1\}^* \to \mathbb{Z}_q$.
  - Extract $Q_i \leftarrow H_1(ID_i)$, where ID_i are identifiers of broadcasters.
  - Compute the private key of broadcasters: $K_i \leftarrow s_i Q_i$ and $\bar{K}_i \leftarrow s_i P$. Without losing generality, we assume Bob is the broadcaster used in the presentation, therefore the index i is replaced by B.
  User setup:
  - Select $x_i \in \mathbb{Z}_q, i = 1, \cdots, n$ and assign each of $\{x_i\}$ to a user. Assign $\{x_i\}$ to all broadcasters. Again, without losing generality, we assume Ray is a user who has $x_j \in \{x_i\}$.
– Encrypt (by Bob).
  - Select a subset of $\{x_i\}$ and denote it by $\{\bar{x}_i\}$, for $i = 1, \cdots, t$, where the elements are associated with the t users who entitle to receive the message.
  - Compute
    * $\prod_{i=1}^{t}(x - \bar{x}_i)$ with given $\bar{x}_i$ to generate a polynomial function $f(x) = \sum_{i=0}^{t} c_i x^i$. Notice that $f(\bar{x}_i) = 0$ for $i = 1, \cdots, t$,
    * $P_0 \leftarrow r(c_0 P + \bar{K}_B), P_1 \leftarrow c_1 r P, \cdots, P_t \leftarrow c_t r P$,
    * $k \leftarrow e(P, r^2 K_B)$,
    * $C \leftarrow M \oplus H_3(k)$, where M denotes the message,
    * $S \leftarrow r(r - H_2(C)) K_B$.
  - Broadcast $(C, S, ID_B, P_0, \cdots, P_t)$.

- Decrypt (by Ray)
  - Compute

$$D \leftarrow \sum_{i=0}^{t} \bar{x}_j^i P_i,$$

$$e(P, S)e(D, H_3(C)H_1(ID_B)) \rightarrow k,$$

$$H_2(k) \oplus C \rightarrow M.$$

S can be considered as a digital signature on C; however, unlike a normal signature, the value of M is not provided for the receivers. The actual encryption key k is constructed with S and one of the secret key x_i. The ID of the broadcaster is the only unique public information in the authentication. This scheme can be referred to as the notion of digital signature with message recovery (e.g., [16]). It can also be considered as a variant of signcryption.

The completeness of the protocol is easy to verify. If Bob has correctly follow the protocol, Ray can decrypt the message and know that Bob indeed is the broadcaster.

$$e(P, S)e(D, H_2(C)H_1(ID_B)))$$

$$= e(P, r(r - H_2(C))K_B)e(\sum_{i=0}^{t} \bar{x}_j^i P_i, H_2(C)Q_B))$$

$$= e(P, r(r - H_2(C))s_B Q_B)e(rs_B P, H_2(C)Q_B))$$

$$= e(P, r(r - H_2(C))s_B Q_B)e(P, rs_B H_2(C)Q_B))$$

$$= e(P, r^2 K_B))$$

$$= k.$$

We now study the security of the IBABE. As defined earlier, the IBABE scheme should be secure against two types of attack: attacks by an outsider (Adversary $\mathcal{A}_1$) and attacks by a receiver (Adversary $\mathcal{A}_2$).

In the former type, $\mathcal{A}_1$ attempts to find the message broadcasted by Bob. The security can be referred to as the following problem: given $(C, S, P, Q_B, P_0, \cdots, P_t)$ or $(C, S, P, uP, r(c_0 + s_B)P, rP_1, \cdots, rP_t)$, find $e(P, P)^{r^2 u s_B}$ or k. This problem can be reduced to as the ECFP on $r(c_0 + s_B)P \rightarrow U + V$. If $V = rs_B P$ were found, then k can be computed with known S. Finding the correct factors is entirely probabilistic. In this case, the probability of finding $rs_B P$ is $1/q$, which is negligible if q is large.

It is interesting to see that how the ECFP is solved in the random oracle model, where H_3 is considered to be a random oracle. Lemma 1 shows the probability of a success.

Lemma 1. *Let H_3 be a random oracle from G_2 to $\{0,1\}^l$. Let $\mathcal{A}$ be an adversary that has advantage ϵ against IBABE after $\mathcal{A}$ has made a total of $q_h > 0$ queries to H_3. Then there is an algorithm $\mathcal{B}$ that solves the ECFP with advantage at least $\frac{q_h \epsilon}{2^l}(1 - \frac{2^l - q_h}{2^l q_h}) + \frac{2^l - q_h}{2^{2l}}$ and the running time $\mathcal{O}(time\ (\mathcal{A}))$.*

The proof of Lemma 1 is given in the appendix.

In the second type of attacks, $\mathcal{A}_2$ is one of users and targets on Bob in order to find his private key.

Lemma 2. *If there exists an inside attacker $\mathcal{A}_2$, who holds the private key x_2 and can output a secret key of the broadcaster, $(K_B, \bar{K}_B)$, given a valid encrypted broadcast $(C, S, ID_B, P_0, \cdots, P_t)$, then we can use this attacker to solve an Elliptic Curve Discrete Logarithm problem (ECDLP).*

Proof. Assume the attacker $\mathcal{A}_2$ exists. The attacker holds a valid private key x_2. Given a valid decrypted broadcast $(C, S, ID_B, P_0, \cdots, P_t)$, he will output the secret key of the broadcaster, $(K_B, \bar{K}_B)$. We will show that we can create an algorithm $\mathcal{B}$ that will use this attacker to solve an instance of ECDLP. The algorithm works as follows.

It sets $rQ = \sum_{i=0}^{t} x_2^i P_i$, and sets all the required parameters. Then, from the computed values, it constructs a valid encrypted broadcast $(C, S, ID_B, P_0, \cdots, P_t)$, by executing the signing oracle. We omit the detail of the signing oracle in this description, but the signing oracle will basically produce a valid encrypted broadcast $(C, S, ID_B, P_0, \cdots, P_t)$, for a given $(P_0, \cdots, P_t)$. Then, this encrypted broadcast is given to the attacker $\mathcal{A}_2$, who will return the secret key of the broadcaster, $(K_B, \bar{K}_B)$. Now, we will show how the algorithm solves the ECDLP.

The algorithm then construct the following equation.

$$S = (r - H_2(C))(rK_B)$$

$$r(r - H_2(C)) = S/K_B$$

$$r^2 - rH_2(C) - S/K_B = 0$$

We note that this equation can be solved to obtain a valid r, which is the solution of ECDLP, given rQ. Hence, it contradicts with the ECDLP assumption. □

The security in this attack model is reduced to an attack against the signature 4-tuple (S, h, C, k), where $h \leftarrow H_3(C)$ and C is the signed message. If we treat H_3 to be a random oracle, then the security proof against an adaptively chosen message attack is associated with the random oracle model given by Pointcheval and Stern [17] using Forking Lemma. Clearly, Forking Lemma can be applied to our scheme, because if the same r (or the same k) has been used to two different signatures, the secure signing key K_B can then be found. For example, given (S, h, C, k) and (S', h', C', k), we can compute:

$$S - S' = r(r - h)K - r(r - h')K = (h' - h)K.$$

$$K = (h' - h)^{-1}(S - S').$$

The complexity for the attack is described in Lemma 3.

Lemma 3. *Let $\mathcal{A}$ be a probabilistic polynomial time algorithm for an adaptively chosen message attack to the signer (say, Bob). Let H_3 be a random oracle* *from G_2 to $\{0,1\}^l$. $\mathcal{A}$ asks the random oracle q_h queries and Bob q_b queries. Assume that $\mathcal{A}$ has the running time t_1 and advantage $\epsilon \geq 10(q_b+1)(q_b+q_h)/2^\ell$, then the CDHP can be solved with probability $\epsilon \geq 1/9$ with the running time $t_2 \leq 23q_h t_1/\epsilon_1$.*

The proof can be deduced directly from [17] and therefore, it is omited it. $\square$

4   ID-Based Designated Distributed Authentication

In this section, we present the ID-Based Designated Distributed Authentication (IBDDA) protocol used for receivers to send messages to a broadcaster who can verify the authenticity of the messages. This method has potential applicability for user/message authentication in distributed systems.

- KeyGen. (By Alice) Alice generates keys for broadcasters and users.
  Key generation for broadcasters (same as the IBABE):
  - Taking ℓ as input, select master private keys $s_i \in \mathbb{Z}_q$ $(i = 1, \cdots, n_b)$.
  - Select three strong one-way hash functions: $H_1 : \{0,1\}^* \to \mathbb{G}_1$, $H_2 : \{0,1\}^* \to \mathbb{Z}_q$, $H_3 : \{0,1\}^* \to \{0,1\}^l$.
  - Extract $Q_i \leftarrow H_1(ID_i)$ for all broadcasters $(i = 1, \cdots, n_b)$.
  - Compute the private key of broadcasters: $K_i \leftarrow s_i Q_i$. Again, we will consider Bob as a representative and replace the index i with B for Bob.
  Key generation for users:
  - Extract $Q_i \leftarrow H_1(ID_i)$ for all receivers $(i = 1, \cdots, n)$.
  - Compute private keys for users: $A_i \leftarrow s_B Q_i$, $i = 1, \cdots, n$. We will consider Ray as a representative and replace i with R.
- Signcrypt. Ray sends a message, $M \in \{0,1\}^*$, signed by him to Bob. Only Bob can decrypt and verify the message.
  - Select random $r \in \mathbb{Z}_q$ and compute $R \leftarrow rQ_R$.
  - Compute $h \leftarrow H_2(M, R)$.
  - Compute $S \leftarrow (h + r)A_R$.
  - Extract $Q_B \leftarrow H_1(ID_B)$.
  - Compute $b \leftarrow e(S, Q_B)$ and $C \leftarrow M \oplus H_3(b)$.
  - Send the 4-tuple (ID_R, C, h, R) to Bob.
- Decrypt. Bob decrypts C and verifies the validity of C.
  - Compute $e(hH_1(ID_R) + R, K_B) \to b$.
  - Compute $C \oplus H_3(b) \to M$.
  - Check $h \overset{?}{=} H_2(M, R)$.

The correctness of the verification is obvious:

$$e(hH_1(ID_R) + R, K_B) = e(s_B(hQ_R + rQ_R), Q_B) = e(S, Q_B) = b.$$

The Signcrypt is associated with the concept of authenticated encryption. The message M is encrypted and also signed by Bob. The corresponding verification results in the recovery and validity checking of the message. Notice that although we call the second phase "Signcrypt", the signature part (S) is not made public. Only the legitimate users can obtain it with their secret key. It is not revealed to outsiders.

The merit of this scheme lies in the fact that the identifier of the sender is the only public information required for verification. In addition, the signcryption is deniable, as shown in Lemma 4.

Lemma 4. *The signcryption is designated to Bob; therefore it is not transferable.*

Proof: The verification of the signcryption requires Bob's secret key K_B; therefore only Bob can verify it and obtain the message. The signcryption is designated to Bob, because upon receiving the message Bob can alter it. In order to change h to $h' \leftarrow H_2(M', R)$, Bob can compute

$$e(hH_1(ID_R), K_B)^{h^{-1}h'} e(R, K_B) \rightarrow b',$$

$$C' \leftarrow M' \oplus H_3(b').$$

C and C' are indistinguishable to other parties.   □

We now study the security of this scheme. The security mechanism against outsiders is considered to be the following problem: given (h, C, P, Q_R, Q_B, R), find b. Setting the tuple as (h, C, P, uP, vP, ruP), the problem becomes: given (h, C, P, uP, vP, ruP), find $e(P, P)^{uvs_B(h+r)}$. Observe that $e(P, P)^{uvs_B(h+r)} = e(P, P)^{uvs_B} e(P, P)^{ruvs_B}$. It could be considered as two BDHP's, i.e., given $(P, uP, vP, s_B P)$, find $e(P, P)^{uvs_B}$; and given $(P, ruP, vP, s_B P)$, find $e(P, P)^{ruvs_B}$. However, since $s_B P$ is not given, we believe that it is harder then the BDHP. We omit the detailed proof in this paper.

One user might try to frame another user. The security mechanism against such attacks is associated with the CDHP. Namely, for an attacker j, given $A_j = s_B Q_j$ and $Q_i \equiv w Q_j$, find $s_B w Q_j$ or $A_i \leftarrow s_B Q_i$. We can also consider the security to be associated with an ID attack, i.e., given $Q_i \equiv uP$, find u. When we treat H_1 as a random oracle, we can borrow the lemma given by Boneh-Franklin [2]:

Lemma 5. *Let H_1 be a random oracle from $\{0, 1\}^*$ to $\mathbb{G}_1$. Let $\mathcal{A}$ be an ID adversary that has advantage ϵ against our scheme. Suppose $\mathcal{A}$ makes at most q_{H_1} private key extraction queries. Then there is an adversary $\mathcal{B}$ that has advantage at least $\epsilon/e(1 + q_{H_1})$ against our scheme with the running time $\mathcal{O}(time(\mathcal{A}))$.*

5   Conclusion

We have described two useful schemes for dynamic interactive broadcast. They are referred to as ID-based authenticated broadcast encryption and ID-based distributed authentication. In the first scheme, broadcaster can dynamically broadcast messages into an arbitrary receiver group, where the ID's of broadcasters are used to authenticate broadcasters. The second scheme is used for broadcasters to authenticate users/receivers who send messages back to broadcasters. Again, the ID's of users are used to authenticated users and no other public information (public key) is required. Our scheme has potential application in multicasting and digital content distribution.

References

1. D. Boneh and M. Franklin. An efficient public key traitor tracing scheme. In *Adances in cryptology - CRYPTO '99, Lecture Notes in Computer Secience 1666*. Springer Verlag, 1999.
2. D. Boneh and M. Franklin. Identity-based encryption from the Weil pairing. In J. Kilian, editor, *Advances in Cryptology, Proc. CRYPTO 2001*, LNCS 2139, pages 213–229. Springer Verlag, 2001.
3. D. Boneh, B. Lynn, and H. Shacham. Short signatures from the weil pairing. In *Advances in Cryptology–ASIACRYPT 2001*, LNCS 2248, pages 514–532. Springer Verlag, 2001.
4. B. Chor, A. Fiat, and M. Naor. Tracing traitors. In Y. G. Desmedt, editor, *Advances in Cryptology, Proc. CRYPTO 94*, LNCS 839, pages 257–270. Springer, 1994. Lecture Notes in Computer Science No. 839.
5. A. Fiat and M. Naor. Broadcast encryption. In D. R. Stinson, editor, *Advances in Cryptology, Proc. CRYPTO 93*, LNCS 773, pages 480–491. Springer Verlag, 1994.
6. E. Gafni, J. Staddon, and Y. Yin. Efficient methods for integrating traceability and broadcast encryption. In *Advances in Cryptology, Proc. CRYPTO 99*, LNCS 1666, pages 372–352. Springer Verlag, 1999.
7. A. Garay, J. Staddon, and A. Wool. Long-lived broadcast encryption. In *Advances in Cryptology, Proc. CRYPTO 2000*, LNCS 1880, pages 333–352. Springer Verlag, 2000.
8. D. Halevy and A. Shamir. The LSD broadcast encryption scheme. In *Advances in Cryptology, Proc. CRYPTO 2002*, LNCS 2442, pages 47–60. Springer Verlag, 2002.
9. M. Jakobsson, K.Sako, and R. Impagliazzo. Designated verifier proofs and their applications. In *Advances in Cryptology, Proc. EUROCRYPT 96*, LNCS 1070, pages 143–154. Springer-Verlag, Berlin, 1996.
10. A. Joux. A one round protocol for tripartite deffie-hellman. In W. Bosma, editor, *Proc. of the ANTS-IV conference, Lecture Notes in Computer Science*, pages 385–394. Springer Verlag, 2000.
11. A. Joux and K. Nguyen. Separate decision deffie-hellman from deffie-hellman in cryptographic groups. available from eprint.iacr.org.
12. M. Luby and J. Staddon. Combinatorial bounds for broadcast encryption. In *Advances in Cryptology, Proc. EUROCRYPT 98*, LNCS 1403, pages 512–526. Springer Verlag, 1998.
13. A. Menezes, T. Okamoto, and S. Vanstone. Reducing elliptic curve logarithms to logarithms in a finite field. *IEEE Transaction on Information Theory*, 39:1639–1646, 1993.
14. Y. Mu and V. Varadharajan. Robust and secure broadcasting. In *Proc. of Indocrypt 2001, LNCS 2247*, pages 223–231. Springer Verlag, 2001.
15. M. Naor. Deniable ring authentication. In *Advances in Cryptology, Proc. CRYPTO 2002*, LNCS 2442, pages 481–498. Springer-Verlag, Berlin, 2002.

16. K. Nyberg and R. A. Rueppel. Message recovery for signature schemes based on the discrete logarithm problem. In *Advances in Cryptology, Proc. EUROCRYPT 94*, LNCS 950, pages 182–193. Springer-Verlag, 1994.

17. D. Pointcheval and J. Stern. Security arguments for digital signatures and blind signatures. *Journal of Cryptology*, 13(3):361–396, 2000.

A. Proof of Lemma 1

Algorithm $\mathcal{B}$ is given as input the parameters associated with the ECFP: Given $(C, S, P, Q_B, P_0, \cdots, P_t)$ or $(C, S, P, uP, r(c_0 + s)P, rP_1, \cdots, rP_t)$, find $e(P, P)^{r^2 u s_B}$.

Challenge: Define the ciphertext as $\langle \mathsf{params2}, C \rangle$, where C is a random string picked from $\{0, 1\}^l$ and $\mathsf{params2}$ denotes all public parameters associated with the scheme. $\mathcal{B}$ gives $\langle \mathsf{params2}, C \rangle$ to $\mathcal{A}$ as the challenge. Imagine that there is a true message M and a true k such that $M = C \oplus H_3(k)$, where k varies with the scheme.

H_3**-Queries:** $\mathcal{A}$ issues queries X_j to H_3. Assume the total number of queries is q_h and all X_j are different. For each query, $\mathcal{B}$ maintains a list of tuples called H_{list}. Each entry in the list has the form $\langle X_j, H_j \rangle$. This set is denoted by H_{list}.

Guess: After q_h queries have been issued, $\mathcal{A}$ will output its guess M_A to the decryption of C. At this point, if there exists $X_j \in H_{list}$ such that $M_A = C \oplus H_3(X_j)$ (we say that M_A is in the list), $\mathcal{B}$ will output X_j as the solution of the associated hard problem. If M_A is not in the list, $\mathcal{B}$ will randomly pick an element from H_{list} as the solution. The output of $\mathcal{B}$ is denoted as M_B.

In the following, we calculate the probability that $\mathcal{B}$ outputs M.

$$
\begin{aligned}
\Pr(M_B = M) &= \Pr(M_B = M, M_A = M) + \Pr(M_B = M, M_A \neq M) \\
&= \Pr(M_B = M, M_A = M, M_A \in H_{list}) \\
&\quad + \Pr(M_B = M, M_A = M, M_A \notin H_{list}) \\
&\quad + \Pr(M_B = M, M_A \neq M, M_A \in H_{list}) \\
&\quad + \Pr(M_B = M, M_A \neq M, M_A \notin H_{list}) \\
&= \Pr(M_B = M, M_A = M, M_A \in H_{list}) \\
&\quad + \Pr(M_B = M, M_A = M, M_A \notin H_{list}) \\
&\quad + \Pr(M_B = M, M_A \neq M, M_A \in H_{list}) \\
&\quad + \Pr(M_B = M, M_A \neq M, M_A \notin H_{list}, M \in H_{list}) \\
&\quad + \Pr(M_B = M, M_A \neq M, M_A \notin H_{list}, M \notin H_{list}) \\
&= \Pr(M_A = M)\Pr(M_A \in H_{list}|M_A = M) \\
&\quad \times \Pr(M_B = M|M_A = M, M_A \in list) + 0 + 0 \\
&\quad + \Pr(M_A \neq M)\Pr(M_A \notin H_{list}|M_A \neq M) \\
&\quad \times \Pr(M_B = M, M \in H_{list}|M_A \neq M, M_A \notin H_{list}) + 0
\end{aligned}
$$

Whether M_A is in H_{list} is independent of whether M_A is the same as M. Therefore,

$$
\begin{aligned}
\Pr(M_B = M) &= \Pr(M_A = M)\Pr(M_A \in H_{list}) \\
&\quad \times \Pr(M_B = M | M_A = M, M_A \in H_{list}) \\
&\quad + \Pr(M_A \neq M)\Pr(M_A \notin H_{list}) \\
&\quad \times \Pr(M_B = M, M \in H_{list} | M_A \neq M, M_A \notin H_{list}) \\
&= \Pr(M_A = M)\Pr(M_A \in H_{list}) \\
&\quad \times \Pr(M_B = M | M_A = M, M_A \in H_{list}) \\
&\quad + \Pr(M_A \neq M)\Pr(M_A \notin H_{list}) \\
&\quad \times \Pr(M_B = M | M \in H_{list}, M_A \neq M, M_A \notin H_{list}) \\
&\quad \times \Pr(M \in H_{list} | M_A \neq M, M_A \notin H_{list}) \\
&= \Pr(M_A = M)\Pr(M_A \in H_{list}) \\
&\quad \times \Pr(M_B = M | M_A = M, M_A \in H_{list}) \\
&\quad + \Pr(M_A \neq M)\Pr(M_A \notin H_{list}) \\
&\quad \times \Pr(M_B = M | M \in H_{list})\Pr(M \in H_{list}) \\
&= \epsilon \frac{C_{2^l-1}^{q_h-1}}{C_{2^l}^{q_h}} + (1-\epsilon)\frac{C_{2^l-1}^{q_h}}{C_{2^l}^{q_h}}\frac{1}{q_h}\frac{C_{2^l-1}^{q_h-1}}{C_{2^l}^{q_h}} \\
&= \epsilon \frac{q_h}{2^l} + (1-\epsilon)\frac{2^l - q_h}{2^l}\frac{1}{q_h}\frac{q_h}{2^l} \\
&= \frac{q_h\epsilon}{2^l}\left(1 - \frac{2^l - q_h}{2^l q_h}\right) + \frac{2^l - q_h}{2^{2l}}.
\end{aligned}
$$

It follows that $\mathcal{B}$ produces the correct answer with probability at least $\frac{q_h\epsilon}{2^l}\left(1 - \frac{2^l-q_h}{2^l q_h}\right) + \frac{2^l-q_h}{2^{2l}}$ as required.

Deniable Partial Proxy Signatures

Yi Mu[1], Fangguo Zhang[2], and Willy Susilo[1]

[1] School of IT and Computer Science,
University of Wollongong, Wollongong, NSW 2522, Australia
{ymu, wsusilo}@uow.edu.au
[2] Department of Electronics and Communication Engineering,
Sun Yat-Sen University, Guangzhou 510275, P. R. China
isdzhfg@zsu.edu.cn

Abstract. This paper describes a proxy signature scheme where a signer can delegate a partial signing right to a party who can then sign on behalf of the original signer to generate a partial proxy signature. A partial proxy signature can be converted into a full signature with the aid of the original signer. Our proxy signature scheme has the feature of deniability, i.e., only the designated receiver can verify the partial proxy signature and the full signature associated to him, while they are not transferable. This paper also describes an application of our scheme in a deniable optimistic fair exchange.

1 Introduction

We look at the scenario where a signing process is divided into two levels. At the first level, a message is partially signed by an authorized party (proxy) and the partially-signed message is verifiable by the designated receiver. At the second level, the partial signature is converted into a full signature with the aid of the original signer for the designated receiver. Only the designated receiver can verify and believe the signature; therefore the real message signed in the signature is ambiguous or deniable to other parties. In other words, the signatures are not transferable.

This scenario can be associated with proxy signature schemes. The notion of proxy signature was introduced in [10]. The basic idea is to allow a signer to delegate his full signing right to a proxy and the original signer cannot sign on behalf of the proxy signer. The receiver of a proxy signature can verify its validity with a proxy public key which is normally made of the public keys of both the original signer and the proxy signer. There are a number of works associated proxy signing, for example, threshold proxy signature [20], one-time proxy signature [18], and ID-based proxy signature [19]. In a threshold proxy signature scheme, the proxy consists of a group of proxy signers. A proxy signature can be computed only by a predefined number of signers. In a one-time proxy signature scheme, a proxy right is made one-time, in the sense that one one proxy signature can be constructed with a given signing key.

M.J. Maher (Ed.): ASIAN 2004, LNCS 3321, pp. 182–194, 2004.

A traditional proxy signature scheme cannot be applied to our scenario, since it does not accommodate the features of deniability and partially-signing.

A digital signature can be used to achieve non-repudiation of a message. However, to achieve user privacy, non-repudiation might not be always desirable [12]. When Alice sends a message signed by herself to Bob, it could be sufficient that only Bob can verify the validity of the signature. It may not be important whether or not other parties believe the authentication of the message. This is why deniability is introduced into digital signatures. The concept of deniable signature was proposed by Jakobsson, Sako and Implagliazzo in [7]. It is noted in [15] that ring signature scheme can be used to provide such schemes if the number of users is limited to two users. Naor recently proposed a deniable signature scheme for ring authentication [12].

The concept of partial signature is often associated with multisignature [5] and threshold signature (e.g., [17]). We should differentiate the concept of "partial signature" from "multisignature." In a typical multisignature scheme, the secret signing key is split into multiple pieces; each is owned by a participating signer. Each signer can sign a message with the corresponding secret key to generate a portion of the signature which cannot be verified. A combination of all signature portions gives the full signature that can then be verified with the given public key. The concept of partial signature differs from that, because it allows the partially signed message to be verified. Therefore, the partial signature itself is meaningful.

Threshold signatures are similar to multisignatures, but it allows a predefined number of signers from a signer group to construct a signature. There are a number of different constructions on threshold signature. Primary constructions of threshold signatures are based on Shamir's secret sharing scheme [16], where the shared secret is the primary signing key. A piece of threshold signature can be constructed by using a share and can be made verifiable to other parties by using Pedersen's verificable secret sharing scheme [14].

Our Contribution. In this paper, we introduce a novel notion for proxy signing: Deniable Partial Proxy Signature (DPPS). It comprises of two features: partially-signing and strong deniability. By "strong" we mean that the deniability is not reversible even if the secret key of the receiver is revealed. We will also provide security proofs for our scheme and show that the proxy signer is fairly treated in the sense that the original signer cannot farm the proxy signer by generating a full signature with the involvement of the proxy signer. We also give a concrete example showing the applicability of the proposed scheme in optimistic fair exchange (e.g., [1, 2]).

Our scheme is based on the bilinear pairings (e.g., Weil pairing). The Weil pairing was originally considered to be a bad thing, since it can be used for attacking elliptic curves[11]. Recently, it has been showed that the Weil pairing can be used to construct a protocol for three party one round Diffie-Hellman key exchange[8]. Boneh-Franklin have recently proposed a concrete identity based encryption protocol[3] and a short signature scheme based on the Weil pairing[4]. A number of schemes based on bilinear pairings have been proposed. The security our scheme is based on so-called Computational Diffie-Hellman problem and Bilinear Diffie-Hellman problem.

The rest of this paper is organized as follows. Section 2 gives the formal model and definitions of the novel DPPS scheme. Section 3 describes the DPPS protocol in detail. Section 4 considers the security of the DPPS protocol, including the fairness to proxy signer and the hard problems associated with the protocol. Section 5 presents a deniable optimistic fair exchange scheme as a concrete application for our scheme.

2   Definitions

We start by precisely defining what is a Designated Partial Proxy Signature (DPPS) scheme. Suppose the original signer Oliver delegates his partial signing right to Penny who can then partially sign on behalf of Oliver for the designated receiver Don. Oliver sends the following

$$K_O = f(w, k_o),$$

where w is the proxy warrant, k_o is the secret key of Oliver, and $f()$ is a suitable function. Upon receiving K_O, Penny constructs the proxy signing key

$$K_P = f_2(w, K_O, k_p),$$

where the k_p is the secret key of Penny and $f_2()$ is another suitable function. The K_P can be used to construct a partial proxy signature by Penny:

$$\sigma_p = f_3(K_P, m, PK_D),$$

where m is the message, PK_D is the public key of Don, and $f_3()$ is a suitable function. σ_p is designated to Don. In order to convert σ_p into a full signature, Oliver constructs a converter $\sigma_o = f_4(k_o, m)$. $f_4()$ is a suitable function. With σ_o, Don can construct the full signature

$$\sigma = f_5(\sigma_p, \sigma_o),$$

where $f_5()$ is a suitable function and σ is designated to Don. σ can be converted into a non-designated signature by Oliver and Penny, when it is necessary.

Definition 1. (DPPS) *A DPPS scheme is a septuple of polynomial algorithms* (Setup, KeyGen, Sign, Verify, Convert, VerifyFull, Open), *where*

- Setup *comprises of three probabilistic algorithms* (Setup1, Setup2, Setup3) *used by Oliver, Penny, and Don, respectively to set up the parameters of the system.*
  - Setup1, *used by Oliver, taking as input a security parameter ℓ, generates his secret key tuple and the corresponding public key PK_O. It also generates some public parameters.*

- **Setup2**, *used by Penny, taking as input a security parameter ℓ, generates her private key k_p and the corresponding public key PK_P.*
- **Setup3**, *used by Don, taking as input a security parameter ℓ, generates his private key k_d and the corresponding public key PK_D.*

- **KeyGen**, *the proxy key generation algorithm, is a three-move probabilistic algorithm used by Oliver and Penny to set up all parameters of the proxy signing phase. KeyGen takes as input a security parameter ℓ and generates a private proxy signing key K_P for Penny.*

- **Sign**, *the partial proxy signing algorithm, is a probabilistic algorithm used by Penny to construct a DPPS. It takes as input the proxy signing key K_P and a message m and returns a DPPS σ_p.*

- **Verify**, *the DPPS verification algorithm, is a deterministic algorithm used by Don to verify a DPPS. Verify takes as input σ_p and k_d and returns True or $\perp$.*

- **Convert**, *the full signature generation algorithm, comprises of two probabilistic algorithms used by Oliver and Don, respectively. Algorithm 1, used by Oliver, takes Oliver's secret key tuple and public parameters associated with σ_p as input and generates a converter σ_o. Algorithm 2, used by Don, takes as input σ_o and σ_p and returns the full designated signature σ.*

- **VerifyFull**, *the verification algorithm of a full designated signature, is a probabilistic algorithm used by Don. It takes as input σ, m, k_d, PK_F and returns True or $\perp$, where PK_F denotes the public verification key. VerifyFull can be used by other parties when k_d is made public.*

- **Open**, *the full signature conversion algorithm, is a deterministic algorithm. It*
  - *takes as input S_O, PK_O and returns True or $\perp$;*
  - *takes as input S_P, PK_P and returns True or $\perp$;*
  - *checks if S_O is associated with σ_o;*
  - *checks if S_P is associated with σ_p.*

  S_O and S_P are an non-designated or "open" version of σ_o and σ_p, respectively.

The DPPS protocol should satisfy the completeness and soundness:

- **Completeness.** If Oliver and Penny have correctly followed the protocol for a DPPS, then Don will accept the signature with high probability.
- **Soundness.** The DPPS scheme is not forgeable (Definitions 2-7).

Definition 2. *There exists a polynomial time adversary $\mathcal{A}_1$. Given a deniable proxy signature σ_p, $\mathcal{A}_1$ has the following advantage in finding the proxy signing key K_P over a random oracle.*

$$\overline{\mathrm{Succ}}_{\mathcal{A}_1}(\ell) = \Pr\left[\sigma_p \to K_P, \sigma_p' = f_3(K_P, m', PK_D), \mathsf{Verify}(\sigma_p') \to True\right].$$

We require $\overline{\mathrm{Succ}}_{\mathcal{A}_1}(\ell) < negl(\ell)$ for some negligible function negl.

Definition 3. *There exists a polynomial time adversary $\mathcal{A}_2$. Given a deniable full signature converter σ_o, $\mathcal{A}_2$ has the following advantage in finding Oliver's secret key k_o over a random oracle.*

$$\overline{\mathsf{Succ}}_{\mathcal{A}_2}(\ell) = \Pr\left[\sigma_o \to k_o, \sigma'_o = f_4(k_o, m), \mathsf{VerifyFull}(\sigma = \sigma_p \sigma'_o) \to True\right].$$

We require $\overline{\mathsf{Succ}}_{\mathcal{A}_2}(\ell) < negl(\ell)$ for some negligible function negl.

Polynomially-bounded Oliver cannot farm Penny by gaining the proxy signing key of Penny.

Definition 4. *There exists a polynomial time Oliver $\mathcal{A}_3$. Given a partial proxy signature σ_p and $\mathsf{Open}(\sigma_p)$, $\mathcal{A}_3$ has the following advantage in finding the proxy key over a random oracle.*

$$\overline{\mathsf{Succ}}_{\mathcal{A}_3}(\ell) = \Pr\left[(\sigma_p, \mathsf{Open}(\sigma_p)) \to K_P, \sigma'_p = f_3(K_P, m, PK_D), \mathsf{Verify}(\sigma'_p) \to True\right].$$

We require $\overline{\mathsf{Succ}}_{\mathcal{A}_3}(\ell) < negl(\ell)$ for some negligible function negl.

Polynomially-bounded Penny cannot gain the secret key of Oliver from K_O.

Definition 5. *There exists a polynomial time Penny $\mathcal{A}_4$. Given a K_O, $\mathcal{A}_4$ has the following advantage in finding the secret key of Oliver.*

$$\overline{\mathsf{Succ}}_{\mathcal{A}_4}(\ell) = \Pr\left[\begin{matrix} K_O \to k_o, K'_O = f(w', k_o), K'_P = f_2(w, K'_O, k_p), \\ \sigma'_p = f_3(K'_P, m, PK_D), \mathsf{Verify}(\sigma'_p) \to True \end{matrix}\right].$$

We require $\overline{\mathsf{Succ}}_{\mathcal{A}_4}(\ell) < negl(\ell)$ for some negligible function negl.

Polynomially-bounded Oliver cannot farm Penny by creating a full signature without the involvement of Penny.

Definition 6. *There exists a polynomial time Oliver $\mathcal{A}_5$ that has the following advantage in creating a full signature without the involvement of Penny.*

$$\overline{\mathsf{Succ}}_{\mathcal{A}_5}(\ell) = \Pr\left[\begin{matrix} \sigma' = \sigma'_p \sigma'_o \\ \mathsf{Verify}(\sigma') \to True \end{matrix}\right] < negl(\ell).$$

Polynomially-bounded Oliver and Don cannot collude to farm Penny by generating a full signature without the involvement of Penny.

Definition 7. *There exist a polynomial time Oliver $\mathcal{A}_6$ and Don $\mathcal{A}_7$ that has the following advantage in creating a full signature without the involvement of Penny.*

$$\overline{\mathsf{Succ}}_{\mathcal{A}_6, \mathcal{A}_7}(\ell) = \Pr\left[\begin{matrix} \sigma' = \sigma'_p \sigma'_o \\ \mathsf{Verify}(\sigma') \to True \end{matrix}\right] < negl(\ell).$$

We now define the deniability of our proxy signature scheme. In the DPPS scheme, any other parties, other than the designated receiver himself can be convinced about the validity of the signature (a proxy signature or a full signature). This is because the receiver can always alter the message in a signature. It is indistinguishable to others about the real signature and the altered signature. We call this property "Deniability" defined as follows.

Definition 8. (Deniability) *There is a polynomial adversary $\mathcal{A}_8$. Given a designated partial proxy signatures $\sigma_p(m)$ on message m (or a designated full signature $\sigma(m)$ on m) and $\sigma_p(m')$ on message m' (or a designated full signature $\sigma(m')$ on m'), $\mathcal{A}_8$ has a distinguishing advantage ϵ, where ϵ is negligible.*

2.1   Bilinear Pairings

We set up our systems using bilinear pairings due to Boneh and Franklin [3]. Define two cyclic groups $\mathbb{G}_1, \mathbb{G}_2$, where $\mathbb{G}_1$ is an additive group and $\mathbb{G}_2$ is a multiplicative group. Both have a prime order q. Let e be a computable bilinear map $e : \mathbb{G}_1 \times \mathbb{G}_1 \rightarrow \mathbb{G}_2$. For $a, b \in \mathbb{Z}_q$ and $P, Q \in \mathbb{G}_1$, we have $e(aP, bQ) = e(P, Q)^{ab}$.

Definition 9. (Decisional Diffie-Hellman Problem) *Given $P, aP, bP, cP \in \mathbb{G}_1$ and $a, b, c \in Z_q$, decide whether $c = ab \in \mathbb{Z}_q$.*

A decisional Diffie-Hellman problem (DDHP) is easy [9], since $e(aP, bP) = e(P, P)^{ab}$. The security of a pairing based algorithm is based on the computational Diffie-Hellman problem (CDHP), which is described as follows.

Definition 10. (Computational Diffie-Hellman Problem) *Let a, b be chosen from $\mathbb{Z}_q$ at random and P be a generator chosen from $\mathbb{G}_1$ at random. Given (P, aP, bP), compute $abP \in \mathbb{G}$.*

$\mathbb{G}_1$ is referred to as a Gap Diffie-Hellman (GDH) group and CDHP can be referred to as a Gap Diffie-Hellman problem, if DDHP can be solved in polynomial time and no polynomial algorithm can solve CDHP with non-negligible advantage within polynomial time [13, 4, 6].

3   Construction for DPPS

In this section, we construct a designated partial proxy signature scheme, based on the definitions given in the preceding section.

The protocol is described as follows.

- Setup.
    - Setup1: Oliver
        * selects $x_1, x_2 \in \mathbb{Z}_q$ as his private key double.
        * selects $P \in \mathbb{G}_1$, a generator of $\mathbb{G}_1$, as a public parameter.
        * sets the public keys $PK_O = x_1 P$ and $PK_F = x_2 P$.
        * selects two public one-way hash functions $H_1 : \{0,1\}^* \rightarrow \mathbb{G}_1$ and $H_2 : \{0,1\}^* \rightarrow \mathbb{G}_2$.
    - Setup2: Penny selects $k_p \in \mathbb{Z}_q$ and sets the corresponding public key $PK_P = k_p P$.
    - Setup3: Don selects $k_d \in \mathbb{Z}_q$ and sets the corresponding public key $PK_D = k_d P$.

- KeyGen.
  - Oliver:
    * sets $K_O = x_1 W$, where $H_1(w)$ and w is the warrant,
    * sends K_O, W, w to Penny via a secure authenticated channel.
  - Penny:
    * computes the blind proxy key,

$$K_P = K_O + k_p W = (x_1 + k_p)W.$$

    Clearly, K_P is only known to Penny; therefore only Penny can construct a partial proxy signature.
- Sign. Penny:
  - selects a random $r \in_R \mathbb{Z}_q$ and sets $R = rW$;
  - computes $S_P = (H_2(m, R) + r)K_P$;
  - sets the signature 4-tuple on m as (σ_p, w, m, R), where $\sigma_p = e(PK_D, S_P)$;
  - sends (σ_p, w, m, R) to Don;
  - sends (σ_p, w, m, R, S_P) to the original signer via an authenticated secure channel. This step can be delayed until the partial proxy signature needs to be coverted into the full signature.
- Verify. Don verifies:

$$W \overset{?}{=} H_1(w),$$

$$\sigma_p \overset{?}{=} e(PK_O + PK_P, H_2(m, R)W + R)^{k_d}.$$

The verification can be done only by Don, since it requires the secret key k_d. There is no way for other parties to believe the partial proxy signature, even if Don reveals the value of k_d. The proof is given in Claim 1. The correctness of the verification is shown as follows.

$$\begin{aligned}
\sigma_p &= e(PK_D, S_P) \\
&= e(k_d P, (H_2(m, R) + r)(x_1 + k_p)W) \\
&= e((x_1 + k_p)P, (H_2(m, R) + r)W)^{k_d} \\
&= e(PK_O + PK_P, H_2(m, R)W + R)^{k_d}.
\end{aligned}$$

- Convert. To convert it into a full proxy signature, Oliver:
  - does nothing if (σ_p, w, m, R, S_P) has not been received from Penny.
  - otherwise, verifies

$$W \overset{?}{=} H_1(w),$$

$$\hat{e}(P, S_P) \overset{?}{=} e(PK_O + PK_P, H_2(m, R)W + R),$$

$$\sigma_p \overset{?}{=} e(PK_D, S_P);$$

  - computes the full signature converter:

$$S_O = (x_2 - x_1)PK_D,$$

$$\sigma_o = e(S_O, H_2(m, R)W + R);$$

  - sends σ_o to Don.

- To convert it into a full signature, Don computes

$$\sigma = \sigma_p \sigma_o.$$

– To verify the full proxy signature, Don checks

$$\sigma \stackrel{?}{=} e(PK_F + PK_P, H_2(m, R)W + R)^{k_d}.$$

The full signature is also designated to Don, since it requires k_d for verification. In Claim 1, we show that it is deniable even if k_d is revealed by Don. The correctness of the full signature is due to

$$\begin{aligned}
\sigma &= e(PK_D, S_P)e(S_O, H_2(m, R)W + R) \\
&= e(PK_D, (H_2(m, R) + r)(x_1 + k_p)W)e((x_2 - x_1)PK_D, H_2(m, R)W + R) \\
&= e((x_1 + k_p)PK_D, (H_2(m, R)W + R))e((x_2 - x_1)PK_D, H_2(m, R)W + R) \\
&= e((x_2 + k_p)PK_D, H_2(m, R)W + R) \\
&= e((PK_F + PK_P, H_2(m, R)W + R)^{k_d}.
\end{aligned}$$

Claim 1. *The partial proxy signature and the full signature are both deniable.*

Proof: Don, as the designated receiver, can alter the partial signed message. Let m' be the altered message, the receiver can compute

$$\begin{aligned}
\sigma'_p &= e(PK_P + PK_O, k_dW)^{H_2(m', R)}e(PK_P + PK_O, R)^{k_d} \\
&= e(PK_P + PK_O, H_2(m', R)W + R)^{k_d}.
\end{aligned}$$

He then claims that m' is the signed message. Obviously, it can still be verified with given k_d and m'. Similar to that, the deniability of the full signature is due to the fact that Don can alter the signed message:

$$\sigma' = e(PK_F, H_2(m', R)W + R)^{k_d}.$$

σ' and σ'_p are indistinguishable to other parties. □

We notice that the sole purpose of a deniable signature is for the designated receiver to believe a signature. Of course, Don can arbitrary construct fake partial proxy signatures and full signatures designated to himself without the involvement of Oliver and Penny, because he knows k_d. However, due to the deniability, no one will believe them.

Claim 2. *The deniability (or non-transferability) of the DPPS scheme is universal. That is, given any two partial proxy signatures or two full signatures, any one can construct partial proxy signatures or full signatures.*

Proof: Without losing generality, we use partial proxy signatures as an example. Two partial proxy signatures (legitimate or fake one(s)) give

$$\sigma = e(PK_P + PK_O, k_d W)^{H_2(m,R)} e(PK_P + PK_O, R)^{k_d},$$

$$\sigma' = e(PK_P + PK_O, k_d W)^{H_2(m',R)} e(PK_P + PK_O, R)^{k_d}.$$

or

$$\sigma = A^{H_2(m,R)} B, \tag{1}$$

$$\sigma' = A^{H_2(m',R)} B. \tag{2}$$

where $A = e(PK_P + PK_O, k_d K_p)$ and $B = e(PK_P + PK_O, R)^{k_d}$. By (1)/(2), we obtain

$$\sigma \sigma'^{-1} = A^{H_2(m,R) - H_2(m',R)}.$$

or

$$A = (\sigma \sigma'^{-1})^{(H_2(m,R) - H_2(m',R))^{-1}}. \tag{3}$$

From Equations (1) and (3), we have

$$B = \sigma \left((\sigma \sigma'^{-1})^{(H_2(m,R) - H_2(m',R))^{-1} H_2(m,R)} \right)^{-1}.$$

With A and B, any one can construct signatures: pick a forged message m'' and compute the fake signature as $\sigma'' = A^{H_2(m'',R)} B$. □

The full designated proxy signature can be converted into a normal signature by the collaboration of Oliver and Penny using the Open algorithm: Oliver reveals S_O and Penny reveals S_P. The full signature can then be verified with

$$e(P, S_P) \overset{?}{=} e(PK_O + PK_P, H_2(m, R) W + R),$$

$$\sigma \overset{?}{=} e(PK_D, S_P) e(S_O, H_2(m, R) W + R).$$

4   Security Consideration

Theorem 1. *In the DPPS scheme, Penny is fairly treated against dishonest Oliver.*

Refer to the following attacking scenarios and Lemmas 1 and 2 for the proof.

We first take a look at the scenario where Oliver is dishonest. He can set $PK_F = xP - PK_P$ for a random x and then generates a full signature without the involvement of Penny:

$$\sigma' = e(xPK_D, H_2(m', R) W + R).$$

This is obviously targeted on Don, because no one else will believe it. The equality obviously holds: $\sigma' = e(PK_F + PK_P, H_2(m', R) W + R)$. However, it is shown in Lemma 1 that the fraud can be detected by Don.

Lemma 1. *In the DPPS scheme, Oliver cannot generate a valid full signature on a message and claims that Penny has been involved in the construction of the signature, while she has not been involved.*

Proof: In order to prove that σ' is genuine, Oliver needs to provide σ'_o and σ'_p such that

$$\sigma' = \sigma'_p \sigma'_o, \tag{4}$$

where $\sigma'_p = e(PK_D, S'_P)$ and $\sigma'_o = e(S'_O, H_2(m', R)W + R)$. The equality shown in Equation (4) is achievable, provided S'_O and S'_P are chosen prior to computing σ'. However, Oliver cannot compute S'_P such that

$$\sigma'_p = e(PK_O + PK_P, H_2(m, R)W + R)^{k_d}, \tag{5}$$

because he does not know k_p (notice that $\sigma'_P = \hat{e}(PK_D, S'_P)$). Therefore, the validity of σ' is denied by Don, when the equality shown in Equation (5) does not hold. Notice that the verification of Equation (5) does not require S'_P. $\square$

Let us take a look at another attack scenario against the DPPS scheme. As the designated receiver, Don gives a fake DPPS, σ'_p, on m' to Oliver, who then generates σ'_o wrt m'. With σ'_p and σ_o, they can get a full designated proxy signature $\sigma' = o'_p \sigma'_o$. Then, Oliver and Don can claim that Penny has signed m'. However, according to Lemma 2, this is infeasible.

Lemma 2. *The DPPS scheme is secure against the collusion of Don and Oliver.*

Proof: Assume that Don has generated a fake signature

$$\sigma'_p = e(PK_O + PK_P, H_2(m', R)W + R)^{k_d}.$$

He then sends it to Oliver, who in turn computes

$$\sigma'_o = e(S_O, H_2(m', R)W + R)$$

and sends σ'_o to Don. Don computes the product, $\sigma' = \sigma'_p \sigma'_o$, as the full signature. Then, they claim that Penny has signed m'. We show in the following that Oliver cannot prove the validity of σ'_p, since he cannot provide a legitimate S'_P such that $\sigma'_p = e(PK_D, S'_P)$. If σ'_p had been computed by Penny, she should have given the corresponding S'_P to Oliver as part of the protocol. When a dispute happens, Penny can ask Oliver to provide the related S'_P and prove $\sigma'_p = e(PK_D, S'_P)$. Given that computing S'_P from σ'_p is a hard problem, Oliver cannot work out the value of S'_P from given σ'_p. Therefore, he cannot provide S'_P that meets the equality $\sigma'_p = e(PK_D, S'_P)$. It therefore shows that Penny has not signed m'.

The malicious Oliver might make a false statement that he has not received S_P from Penny and claims that Penny is a cheater. However, The protocol states that Oliver should not generate the converter without receiving the 5-tuple $(\sigma_p, S_P, R, \sigma_p, m)$ for Penny. Therefore, Oliver fails. $\square$

The security of the DPPS scheme is based on the Computational Diffie-Hellman Problem. Our scheme is secure as long as the CDHP is hard. This statement is justified in Lemma 3.

Lemma 3. *The security of DPPS is equivalent to the CDHP.*

Proof (Sketch): Set $W = aP$ for $a \in \mathbb{Z}_q$. There are two CDH problems: (1) Given W and $PK_p = k_pP$, compute ak_pP; (2) Given W and $PK_O = x_1P$, compute ax_1P (the original key). Having solved the CDH problems, the proxy key can computed: $K_P = ak_pP + ak_oP$. Since the original key is also obtained, a fake proxy signature can be converted into a full signature.   □

5   Deniable Optimistic Fair Exchange

The problem of fair exchange is one of the fundamental problems in secure electronic transactions. Intuitively, it allows two parties to exchange items in a fair way, so that either party gets the other's item, or neither party does. The nature of this problem often causes the situation where Alice waits for an item signed by Bob before she sends her signed item to Bob and Bob waits for the item signed by Alice before he sends his one to Alice. Optimistic fair exchange schemes are normally applied to solve this problem.

Both Alice and Bob register themselves with a semi-trusted third party. Alice sends her partially signed item to Bob and Bob sends his partially signed item to Alice. Alice then converts her partial signature into a full signature. Bob does the same. If one of the parties fails to do so, the semi-trusted third party can do it for them. The DPPS scheme can be used in designated optimistic fair exchange (DOFE) with the feature of deniability. That is, the exchanged items are designated to Alice and Bob respectively, while no one else will believe the exchange.

We now assume that Alice and Bob are "proxy signers" who have registered with Oliver the semi-trusted third party. In the meanwhile, Bob is also the designated receiver of Alice's partial signature and Alice is also the designated receiver of Bob's partial signature. In order to fit the DPPS into the DOFE, we need to slightly modify the scheme, so that a "proxy signer" can convert the associated partial signature into a full signature.

– Setup of the system.
  • Both Alice and Bob get registered with Oliver.
  • Oliver: similar to that of the DSSP protocol. The only difference is that he needs to handle two "proxies." He possesses two full signature converters, one for each "proxy."
  • Alice and Bob have registered with Oliver. Each has a partial "proxy" signing key. Each has a private key and its corresponding public key.
    * Alice possesses a private key k_a, the corresponding public key $PK_A = aP$, the partial "proxy" signing key K_A, and her full signature converter $\sigma_o(\text{Alice})$ given by Oliver.
    * Bob possesses a private key k_b, the corresponding public key $PK_A = bP$, and the partial "proxy" signing key K_B, and his full signature converter $\sigma_o(\text{Bob})$ given by Oliver.

- Partial signatures.
  - Alice partially signs an item using K_A to construct $\sigma_p(\text{Alice} \xrightarrow{m_1} \text{Bob})$, where $X \xrightarrow{m_i} Y$ denotes that the designated signature receiver is Y and the signer is X. m_i represents a partially signed item.
  - Bob partially signs an item using K_B to construct $\sigma_p(\text{Bob} \xrightarrow{m_2} \text{Alice})$.
- Full signatures.
  - Alice sends her converter $\sigma_o(\text{Alice})$ to Bob.
  - Bob then converts $\sigma_p(\text{Alice} \xrightarrow{m_1} \text{Bob})$ into a full signature $\sigma(\text{Alice} \xrightarrow{m_1} \text{Bob}) = \sigma_p(\text{Alice} \xrightarrow{m_1} \text{Bob})\sigma_o(\text{Alice})$.
  - Bob sends his converter $\sigma_o(\text{Bob})$ to Bob.
  - Alice then converts $\sigma_p(\text{Bob} \xrightarrow{m_2} \text{Alice})$ into a full signature $\sigma(\text{Bob} \xrightarrow{m_2} \text{Alice}) = \sigma_p(\text{Bob} \xrightarrow{m_2} \text{Alice})\sigma_o(\text{Bob})$
- Open full signatures (in case of dispute).
  - Oliver reveals $S_P(\text{Alice})$, $S_P(\text{Bob})$, $S_O(\text{Alice})$, and $S_O(\text{Bob})$, where S_P and S_O were defined earlier and the names represent the association.

If either Alice or Bob refuses to give the associated converter, Oliver sends the corresponding converter to the associated party.

6   Conclusion

We have described a novel proxy signature scheme that has two distinct features, partial-signing and deniability. The authorized signing right to the proxy signer is limited into partially signing; it provides the first level of signing. A partial signature can be converted into a full signature by the original signer and the proxy signer. Both partial signature and full signature in our scheme are deniable, since only the designated receiver can verify the validity of signatures. The proposed partial proxy signature scheme has applicability in many aspects such as contract signing and electronic commerce. In this paper, we have given an example of optimistic fair exchange.

References

1. N. Asokan, M. Schunter, and M. Waidner. Optimistic protocols for fair exchange. In *Proc. 4th ACM Conference on Computer and Communications Security*, pages 8–17, 1997.
2. N. Asokan, V. Shoup, and M. Waidner. Optimistic fair exchange of digital signatures. *IEEE Journal on Selected Areas in Communications*, 18, 2000.
3. D. Boneh and M. Franklin. Identity-based encryption from the Weil pairing. In J. Kilian, editor, *Advances in Cryptology, Proc. CRYPTO 2001*, LNCS 2139, pages 213–229. Springer Verlag, 2001.
4. D. Boneh, B. Lynn, and H. Shacham. Short signatures from the weil pairing. In *Advances in Cryptology–ASIACRYPT 2001*, LNCS 2248, pages 514–532. Springer Verlag, 2001.
5. C. Boyd. Digital multisignatures. In *in Proceedings of the Cryptography and Coding*, pages 241–256. Oxford University Press, 1989.

6. J. C. Cha and J. H. Cheon. An identity-based signature from gap Diffie-Hellman groups. In *PKC 2003, Lecture Notes in Computer Science.* Springer Verlag, 2003.

7. M. Jakobsson, K.Sako, and R. Impagliazzo. Designated verifier proofs and their applications. In *Advances in Cryptology, Proc. EUROCRYPT 96,* LNCS 1070, pages 143–154. Springer-Verlag, Berlin, 1996.

8. A. Joux. A one round protocol for tripartite deffie-hellman. In W. Bosma, editor, *Proc. of the ANTS-IV conference, Lecture Notes in Computer Science,* pages 385–394. Springer Verlag, 2000.

9. A. Joux and K. Nguyen. Separate decision deffie-hellman from deffie-hellman in cryptographic groups. available from eprint.iacr.org.

10. M. Mambo, K. Usuda, and E. Okamoto. Proxy signatures for delegating signing operation. In *Proc. of the Third ACM Conf. on Computer and Communications Security,* pages 48–57, 1996.

11. A. Menezes, T. Okamoto, and S. Vanstone. Reducing elliptic curve logarithms to logarithms in a finite field. *IEEE Transaction on Information Theory,* 39:1639–1646, 1993.

12. M. Naor. Deniable ring authentication. In *Advances in Cryptology, Proc. CRYPTO 2002,* LNCS 2442, pages 481–498. Springer-Verlag, Berlin, 2002.

13. T. Okamoto and D. Pointcheval. The gap problems: a new class of problems for the security of cryptographic schemes. In *PKC 2001, Lecture Notes in Computer Science 1992,* pages 104–118. Springer Verlag, Berlin, 2001.

14. T. P. Pedersen. Non-interactive and information-theoretical secure verifiable secret sharing. In *Advances in Cryptology, Proc. CRYPTO 91,* LNCS 576, pages 130–140, 1991.

15. R. Rivest, A. Shamir, and Y. Tauman. How to leak a secret. In *Adances in cryptology - Asiacrypto 2001, Lecture Notes in Computer Secience 2248,* pages 552–565. Springer-Verlag, Berlin, 2001.

16. A. Shamir. How to share a secret. *Communications of the ACM,* 22:612–613, November 1979.

17. V. Shoup. Practial threshold signatures. In *Adances in cryptology - EUROCRYPT'96, Lecture Notes in Computer Secience 1807.* Springer-Verlag, Berlin, 2000.

18. H. Wang and J. Pieprzyk. Efficient one-time proxy signature. In *Aisacrypt 2003, LNCS.* Springer-Verlag, 2003.

19. F. Zhang and K. Kim. Id-based blind signature and proxy signature from bilinear pairings. In *In: Information Security and Privacy (ACISP'03), LNCS 2727,* pages 312–323. Springer-Verlag, 2003.

20. K. Zhang. Threshold proxy signature schemes. In *In Proc. Information Security (ISW'97), LNCS 1396,* pages 282–290. Springer-Verlag, 1997.

Formal Concept Mining: A Statistic-Based Approach for Pertinent Concept Lattice Construction

Taweechai Ouypornkochagorn and Kitsana Waiyamai

Knowledge Discovery from very Large database research group: KDL,
Computer Engineering Department, Kasetsart University, Thailand
o_taweechai@kdl.cpe.ku.ac.th, fengknw@ku.ac.th

Abstract. In this paper, we define formal concept mining, a method for generating and evaluating all the pertinent concepts from large transaction databases. We propose a novel efficient formal concept mining algorithm, called Distribution Curve Self-Evaluation (DCSEA). Attempting repeatedly to self-adjust the normal distribution curve to be as close as the symmetry curve, DCSEA automatically identifies all the pertinent concepts by deleting and masking non-pertinent concepts. Instead of using the global support threshold, DCSEA allows users to specify the interestingness of the output concepts by using a more understandable statistic-based threshold, called minimum significance threshold. Such threshold measures the level of significance of the concept extent size (the number of objects) from all the concept extent sizes. Experimental results showed that the proposed algorithm gives high concept retrieval performance, and efficient concept focusing, especially on large databases.

Keywords: Data Mining, Concept Lattice, Formal Concept Mining, DESEA, Statistic-based, Pertinent Concept.

1 Introduction

Introduced in the early 80ies, concept lattice [14] is a form of classification that provides an identification of object clustering according to the mutual similarity between the objects. The lattice structure offers an algebraic support for representing maximal sets of objects and maximal sets of properties (or items) which have been shown to provide a theoretical framework for a number of practical problems in machine learning [2,3], information retrieval [1,10] and knowledge engineering [16,17]. Recently, research in concept lattice theory has shifted from theoretical foundations to practical applications, with a particular emphasis on knowledge discovery and decision-making tasks. There are already several data mining techniques that have been developed based on concept lattice such as association rule discovery [4,12,18], data classification and [8,11,13], data clustering [7,9].

Concept lattice construction is an important issue of Conceptual Knowledge Discovery in Databases (CKDD) [5]. However, almost existing concept lattice construction methods are not suitable in the real-world data. Indeed, these methods

M. J. Maher (Ed.): ASIAN 2004, LNCS 3321, pp. 195–211, 2004.

perform an exhaustive search of all possible concepts which result in an enormous number of concepts generated, especially in the large size of database. Among the generated concepts, there are lots of non useful and uninteresting concepts from the user point of view. Efficient methods that do not carry out an exhaustive search are needed to generate only the *pertinent concepts*. In this paper, we define *pertinent concept* a concept that has a significance threshold greater than a user-specified minimum significance. Such threshold measures the level of significance of a given concept extent size (the number of objects in the concept extent) from the whole concepts extent sizes. That is to say in another way, a pertinent concept is a concept with extent size (the number of objects in the concept extent) similar to almost of the other concept extent sizes. Inversely, a *non-pertinent concept* is a concept with extent size very different to almost of the other concept extent sizes. Then, we define *formal concept mining* a method for generating and evaluating all the pertinent concepts from a given database (context). Formal concept mining result is the set of pertinent concepts organized into a lattice structure called *pertinent concept lattice*.

Several concept lattice construction methods that do not carry out an exhaustive search of all the possible concepts have been proposed [4,6,9,12]. Almost of them use the notion of support threshold to prune out non-frequent concepts. A *frequent concept* is a concept that covers at least some minimum number of objects of the database, that say the items (in the concept's intent) shared by those objects in the concept's extent must have a support greater or equal the user-specified minimum support. Disadvantage of such approach is that the use of support threshold cannot effectively prune out non-pertinent concepts. The resulting frequent concept lattice still contains lot of non-pertinent concepts with high support value, especially at the general level of the lattice. Further, the support threshold has to be specified by the user. This support value must be defined based on the number of objects in the database (context), and this value may be changed when news objects are inserted. Generally, the maximum extent size (the maximum number of objects in concept extent) of all the concepts is very small compared to the number of database transactions. Thus the value of the minimum support is quite difficult to be defined by the user.

In this paper, we propose a formal concept mining algorithm, called *Distribution Curve Self-Evaluation Algorithm (DCSEA)*. DCSEA generates all the pertinent concepts from a given context. The proposed algorithm makes use of the descriptive statistics [15] to evaluate the pertinent concepts. DCSEA iteratively self-adjusts the normal distribution curve of concept extent sizes to be as close as the symmetry curve by deleting and masking non-pertinent concepts. Automatically, concepts that have extent size (number of objects) very different from the other concept extent sizes of the normal distribution curve are deleted or masked. DCSEA allows users to specify the interestingness of the pertinent concepts by using a statistic-based significance threshold. The proposed significance threshold is easily understandable in statistic sense. Empirical evaluations comparing DCSEA with a support threshold-based lattice construction methods show that DCSEA gives better concept retrieval performance and lower computation time. Further, the experimental results show that DCSEA can be applicable to real-world databases.

The rest of the paper is organized as follows. Section 2 gives basic definitions of formal concept mining and its related works. Section 3 describes our DCSEA formal concept mining algorithm. Section 4 gives the empirical evaluations comparing DCSEA with the frequent concept lattice construction algorithms. Finally, we conclude the paper in Section 5.

2 Formal Concept Mining

In this section, we start by briefly recall the basic definitions of formal concept analysis. Then, we introduce the framework of formal concept mining with the definitions of pertinent concept and pertinent concept lattice. Finally, we compare formal concept mining with the frequent concept lattice construction.

2.1 Formal Concept Analysis

Concepts are necessary for expressing human knowledge which can be activated to communicatively represent knowledge coded in database. *Formal Concept Analysis* [14] offers such a formalization by mathematizing concepts that are understood as units of thought constituted by their extension and intension. Following are the definitions of context, concept and concept lattice which compose formal concept analysis theory.

Definition 2.1. A *formal context* (or database) is a triple (G, M, R) where G and M are sets and $R \subseteq G \times M$ is a relation. The elements if G and M are called objects and items, respectively, and $(g, m) \in R$ if object g has the item m. For $A \subseteq G$ and $B \subseteq M$, we define:

$$A' := \{ m \in M \,|\, \forall g \in A : gRm \}$$

$$B' := \{ g \in G \,|\, \forall m \in B : gRm \}$$

A *(formal) concept* is a pair (A, B) such that $A \subseteq G$, $B \subseteq M$, $A' = B$ and $B' = A$; A and B called the extent and the intent of the concept (A, B), respectively. The *subconcept-superconcept-relation* is formalized by

$$(A_1, B_1) \leq (A_2, B_2) :\Leftrightarrow A_1 \subseteq A_2 \quad (\Leftrightarrow B_1 \supseteq B_2)$$

The set of all concepts of a context (G, M, R) together with the order relation $\leq$ is always a complete lattice, called the *concept lattice* of (G, M, R) and denoted by $B(G, M, R)$. The lattice structure imposes a partial order on concepts and any concept subset must have one greatest common sub-concept and one smallest common super-concept. Figure 1 shows the concept lattice generated from our context example in table 1. Due to the space limitation, concept extents are represented by their number of objects given in parenthesis. Noticed that the lattice example is organized into several levels ordering from most general level or lowest level (level 1) to most specific level or highest level (level 5). A level n contains the set of all concepts having n items in their intent part.

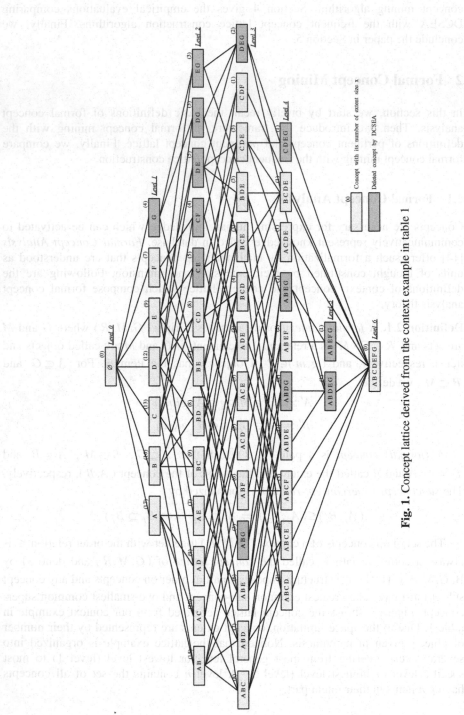

Fig. 1. Concept lattice derived from the context example in table 1

Table 1. Context (transaction database)

ObjectID	Items	ObjectID	Items
1	A B D E G	11	A B D E
2	A B E F G	12	A B D G
3	A B C D	13	A B E F
4	A B C D	14	A C D E
5	A B C D	15	B C D E
6	A B C E	16	C D E G
7	A B C E	17	A B D
8	A B C F	18	A B F
9	A B C F	19	A C D
10	A B C F	20	C D F

Principal disadvantage of the formal concept analysis is its enormous number of concepts that are generated. The number of items, number of transactions, and distribution of items can influence directly the number of concepts generated. With i items context, it can generate at most 2^i concepts, this means the fact that the big oh induced from this characteristic is 2^i. Further, real-world transaction databases may contain large number of non-pertinent items (out of user's point of view). Constructing a concept lattice from a database containing non-pertinent items can produce very large set of non-pertinent concepts to at most 2^{i-1} concepts per non-pertinent item. We called such behavior: *Non-pertinent Concept Propagation.*

2.1   Formal Concept Mining

Formal Concept Mining is a method for generating and evaluating all the pertinent concepts from a given context (transaction database), organized into a lattice structure. A *pertinent concept* is a concept with extent size (the number of objects in the concept extent) similar to almost of the other concept extent sizes. Inversely, a *non-pertinent concept* is a concept with extent size very different from almost of the other concept extent sizes. Following is the formal definition:

Definition 2.2. Let L be a set of all concepts derived from context (G, M, R), $K \subseteq L$ be a set of concepts, and $k \in K$ a concept. Let *min-significance$_K$* be a user-specified minimum level of significance of K. A *significance* threshold associated to a concept k and a set of concepts K, denoted as $los(k, K)$ is the level of significance of extent size of concept k from the set of extent sizes of all the concepts in K. Significance value is ranging from 0% to 100%. A *pertinent concept of K* is a concept that has $los(k, K) \geq$ *min-significance$_K$*.

Given a significance threshold, a *pertinent concept* is a concept that has a significance threshold greater than a user-specified minimum significance. Such threshold measures the level of significance of a given concept extent size (the number of objects) from the whole concepts extent sizes. The set of all generated pertinent concepts of a context (G, M, R) together with the order relation $\leq$ forms a complete lattice, called *pertinent concept lattice* of (G, M, R). Pertinent concept lattice can be defined as follows:

Definition 2.3. Let L be a set of all concepts derived from context (G,M,R), c is a concept of L, and I a set of all pertinent concepts, i.e. $I = \{c \in L \setminus los(c,L) \geq min\text{-}significance_L\}$. A *Pertinent Concept Lattice* $\underline{P}(G,M,R) = (I \cup \{\perp\}, \leq)$ of a context (G,M,R), is a complete lattice of pertinent concepts derived from the context (G,M,R).

Based the above definition, the pertinent concept lattice with 80%- minimum significance derived from the context example (cf. table 1) is given in figure 2. Compared with the concept lattice constructed with traditional method, we observe the following concept characteristics: number of concepts is 46, mean and standard deviation of the extent sizes are, respectively 5.41 and 4.12. Notice that the *min-significance* is 0.8 and the z-scores is 0.8418.

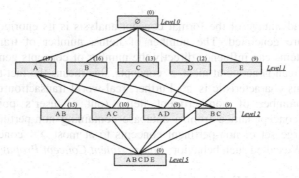

Fig. 2. Pertinent concept lattice derived from the context example in table 1 with 80%-minimum significance

Notice that, our significance threshold has been defined based on mean and standard deviation of the whole concept extent sizes. Such threshold implicitly determines an approximate number of pertinent concepts to be evaluated.

In the pertinent concepts mining process, users can specify range of pertinence by using significance threshold. We call that specification operation, *pertinent concept focusing*. Output of the pertinent concept focusing is the set of pertinent concepts in different lattice levels. Different concept generation methods can generate different sets of pertinent concepts or *different concept focusing*. At the equal number of generated pertinent concepts, concept focusing difference between two concept generation methods can be calculated based on the average level of lattice that can represent how specific or general of the generated pertinent concepts.

We define the notion of *concept covering* as the ratio between the number of pertinent concepts to be evaluated and the total number of concepts. Considering the example lattice given in figure 2, the lattice is composed of 9 pertinent concepts (excluding the empty concepts in level 0 and 5), it follows that the concept covering is equal to 19.57% (9 concepts from the total of 46 concepts, excluding the empty concepts). With 80% significance, the concept covering is around 20%, which conforms to the expected concept covering result.

3   Distribution Curve Self-Evaluation Algorithm (DCSEA) for Formal Concept Mining

DCSEA iteratively self-adjusts the concept extent size normal distribution curve level by level in the lattice to be as close as the symmetry curve. With respect to that distribution curve, DCSEA applies standard deviation (σ), kurtosis and skewness of data for masking or deleting those concepts having the extent size very different from almost dataset. Standard deviation has to be greater or equal to 1, in order to protect too low data distribution. Kurtosis has to be smaller than 0, in order to make sure that there are no data that are too much different from most of data. Skewness has be as near as 0, for non-pertinent concepts filtering.

The steps in DCSEA is composed of masking and deleting operations for pertinent concepts evaluation. The too large concept extent sizes will highly effect kurtosis and skewness. These concepts are too general, uninteresting, however they can be used later to generate more specific pertinent concepts. They will be temporarily *denoted* as masked concepts and will be used later for generating concepts at the more specific level of the lattice. In the similar way, concepts with too small extent size that highly effect both kurtosis and skewness will be deleted from the lattice. Notice that if the intent of a non-pertinent concept is included in the intent of another concept c', then c' becomes automatically non-pertinent. Further, pertinent concepts can be also generated by other pertinent concepts.

DCSEA notations are described in table 2. Main steps of the algorithm consist in computing iteratively pertinent concepts, level by level, in ascending order of the intent size (the number of items in concept intent).

DCSEA starts at level-0 by generating candidate concept having empty intent part, by calling the Candidate-Concept-Gen function (line 1). Then, by calling the Concept-Verification function (line 2), it verifies whether a given set of candidate concepts preserves all the lattice properties (cf. 2.1 for lattice properties). At the next step, level-1 candidate concepts having one item in their intent part are generated (line 3). Then, DCSEA-Evaluation function is called for identifying and evaluating all the level-1 generated concepts (line 4). Resulting candidate concept intents are then used for determining itemsets (line 5), denoted as Y_1, for generating candidates concepts at any lattice level. Again, the Concept-Verification function is called for lattice properties preservation check.

For each level k, all the pertinent concepts with intent size k are determined (line 8-10). One database scan is performed for each lattice level. Based on the lattice property - a concept can be generated from concept generator at the lower lattice level, all the candidate concept intents can be generated from Y_1, in a bottom-up manner (line 8). Next, candidate concepts are checked for lattice properties preservation (line 9), and then are evaluated for their pertinence (line 10). Notice that, the maximum number of iterations (or the highest level of lattice), can be determined based on the number of 1-itemset: $\|Y_1\|$ (in line 7). Different steps of the Candidate-Concept-Gen function are similar to the Apriori-Gen function. In the following, Concept-Verification and DCSEA-Evaluation functions are respectively explained.

Algorithm DCSEA

1) $L_0 \leftarrow$ Candidate-Concept-Gen($\varnothing$) ;
2) $L_0 \leftarrow$ Concept-Verification(L_0) ;
3) $L_1 \leftarrow$ Candidate-Concept-Gen({1-itemsets}) ;
4) $L_1 \leftarrow$ DCSEA-Evaluation(L_1) ;
5) insertinto Y_1 select c.intent from L_1 ;
6) $L_1 \leftarrow$ Concept-Verification(L_1) ;
7) for($k \leftarrow 2$; $k \leq \|Y_1\|$; k ++) do
8) $L_k \leftarrow$ Candidate-Concept-Gen(Candidate-Gen(k,Y_1)) ;
9) $L_k \leftarrow$ Concept-Verification(L_k) ;
10) $L_k \leftarrow$ DCSEA-Evaluation(L_k) ;
11) Answer $\leftarrow \bigcup_k L_k$;

Function Candidate-Gen (i , Y_1)

1) insertinto Y_i select $y_1, y_2, ..., y_i$ from Y_1 ;
2) Answer $\leftarrow Y_i$;

Function Candidate-Concept-Gen (Y_i)

1) forall transaction $t \in T$ do
2) $Y_t \leftarrow$ Subset(Y_i,t) ;
3) forall candidate $y \in Y_t$ do
4) c.intent $\leftarrow y$;
5) c.extent $\leftarrow \bigcup \{ t.tid \}$;
6) Answer $\leftarrow \bigcup \{ c \in L_i \}$;

Function Concept-Verification(L_i)

1) forall candidate concept $c \in L_i$ do
2) $t_c \leftarrow$ {1-itemsets} ;
3) forall transaction t do
4) forall candidate concept $c \in L_i$ do
5) if $t.tid \in c$.extent do
6) $t_c \leftarrow t_c \cap t$;
7) forall candidate concept $c \in L_i$ do
8) if c.intent $\neq t_c$ do
9) delete c ;
10) Answer $\leftarrow \bigcup \{ c \in L_i \}$;

Table 2. Notations

C	Concept of a lattice L
Y_1	Set of items to be used for generating candidates concepts at any lattice level
Y_i	Set of candidate i-itemsets, $y \in Y_i$
L_k	Set of concepts at level k of a lattice
H_i	Set of concept extent sizes (number of objects) of the concepts in L_i
C	Concept of a lattice L
Y_1	Set of items to be used for generating candidates concepts at any lattice level

Function DCSEA-Evaluation($c \in L_i$)

1) $H_i \leftarrow \{ \|c_1.\text{extent}\|$, $\|c_2.\text{extent}\|$, ..., $\|c_n.\text{extent}\| \}$;
2) if $\sigma(H_i) > 0$ do
3) if $Kurt(H_i) > 0$ do
4) delete $Min(H_i)$ or mask $Max(H_i)$ that greatest decreases $Kurt(H_i)$;
5) goto 2) ;
6) if $Skew(H_i) > 0$ do
7) mask $Max(H_i)$;
8) if $Skew(H_i)$ decreases do
9) goto End ;
10) else goto 2) ;
11) elseif $Skew(H_i) < 0$ do
12) delete $Min(H_i)$;
13) if $Skew(H_i)$ increases do
14) goto End ;
15) else goto 2) ;
16) elseif $Skew(H_i) = 0$ do goto End ;
17) End: Answer $\leftarrow$ L_i | $\|c.\text{extent}\| \in H_i$;

Table 2. Notations (continued)

Y_i	Set of candidate i-itemsets, $y \in Y_i$	$\|c.\text{extent}\|$	Number of objects in the extent of c
L_k	Set of concepts at level k of a lattice		
H_i	Set of concept extent sizes (number of objects) of the concepts in L_i	$\sigma(H_i)$	Standard deviation of H_i
		$Kurt(H_i)$	Kurtosis of H_i
t	Transaction of a database T	$Skew(H_i)$	Skewness of H_i
$\|Y_I\|$	Number of items in Y_I		

Concept-Verification function takes as input, the set of candidate concepts generated by the Candidate-Concept-Gen function, then it checks against each candidate concept for lattice property preservation. A concept preserves lattice properties if and only if the functions A' and B' are verified (cf. 2.1 for lattice properties). Said in another way, concept's intent must be equal to the intersection all of transactions contained in concept's extent.

DCSEA-Evaluation function applies statistic theory for concept evaluation. In order to adjust statistical curve, DCSEA-Evaluation tries repeatedly to mask or delete concepts having very different extent sizes from the other concept extent sizes. The standard deviation is verified first, then the kurtosis and finally the skewness, until no further concepts are deleted or masked. In the following, different steps of DCSEA-Evaluation are described using the context example given in table 1. Due to the space limit, we only show how masking and deleting operations are performed in level-2 of the output lattice.

Let us recall that concept "G" of the previous level (level 1) was deleted from the output lattice. Thus, in level-2, concepts which include "G" in their intent: "DG" and "EG" are not taken into account. Then, the concepts "AB", "AC", "AD", "AE", "BC", "BD", "BE", "CD", "CE", "CF" and "DE" having extent sizes (the number of objects) 15, 10, 9, 7, 9, 8, 7, 8, 5, 4 and 5, respectively, are evaluated in the level-2. The evaluation result is shown in table 3.

The DCSEA-Evaluation function starts by analyzing characteristics of concept extent sizes in the level-2. Figure 3(a), step 1, shows the result of that operation where the kurtosis of the concept extent sizes is greater than 0. Then, the function searches for the greatest data distribution by comparing the results between masking largest extent size ("15") and deleting smallest extent size ("4"). The comparison shows that masking largest extent size can greatest decrease the kurtosis (-1.15 for masking and 1.13 for deleting). Thereby, the largest extent size is masked, and the result is shown figure 3(b), step2. The skewness is below 0 after the masking operation. Consequently, DCSEA deletes smallest extent size "4" at this step (cf. figure 3(c) at step 3), then checks the skewness value. Based on the skewness value, it continues to delete next smallest extent size "5" (cf. figure 3(d) at step 4). Since, the deletion operation increases the skewness value, then the evaluation terminates at this step.

Different steps of the DCSEA-Evaluation function can be justified as follows. In step 1, greatest extent size concepts are masked because they are common concepts which are uninteresting. Objective of the smallest concept extent sizes deletion operation, in step 2, is to adjust the normal distribution curve to be symmetry by deleting non-pertinent concepts. Notice that, such operation can decrease both the number of non-pertinent concepts and the skewness value because the resulting number of pertinent concepts can be much more different from the number of non-pertinent concepts. This means that some non-pertinent concepts still remain. Hence, further deletion of smallest extent size concepts is needed, in step 3.

Table 3. The operations of DCSEA at level 2 of the output lattice

Step	Action	σ	Kurt	Skew	Comment
1	**Start**	2.88	0.90	0.99	Kurt. > 0
2	Mask "15"	1.89	-1.15	-0.29	Skew to left
3	Delete "4"	1.64	-1.01	-0.33	Skew Decreases
4	Delete "5"	1.03	-1.14	0.19	Skew Increases
5	**End**	1.03	-1.14	0.19	

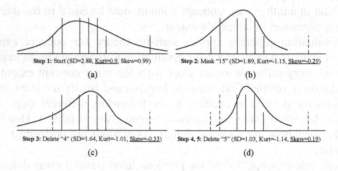

Step 1: Start (SD=2.88, Kurt=0.9, Skew=0.99)   Step 2: Mask "15" (SD=1.89, Kurt=-1.15, Skew=-0.29)

(a)   (b)

Step 3: Delete "4" (SD=1.64, Kurt=-1.01, Skew=-0.33)   Step 4, 5: Delete "5" (SD=1.03, Kurt=-1.14, Skew=0.19)

(c)   (d)

Fig. 3. Distribution curves associated with DCSEA operations in table 3

Comparing DCSEA with the number of concepts generated (cf. figure 1, with boxes in gray color) using traditional global support based concept generation method, DCSEA evaluates 35 pertinent concepts from 48 concepts or 72.92%. Notice that the support value of the deleted concepts is greater than the support value of the frequent concepts obtained from traditional global support based method. For example, deleted concepts in the level 2 - "CE", "CF" and "DE" have greater support value compared to the 15 pertinent concepts (which represents 42.86% of the total number of pertinent concepts) in higher levels. Notice also that equal support value between two concepts does not mean that the two concepts have equal level of pertinence. Probability of concept occurrence is not the same in each lattice level, common or general concepts at the very low lattice level have very high support value, and specific concepts at the very high lattice level have very small support value. While traditional concepts generation methods use the all lattice levels global support threshold for concept evaluation, DCSEA evaluates the pertinent concepts based on distribution of concept extent sizes in each lattice level.

Though DCSEA can automatically evaluate pertinent concepts from a given context (database), it is inconvenient for users to specify the user pertinence level. To alleviate that problem, we allow the user to specify statistic-based range of pertinence, called *significance* threshold. And we call that operation, *pertinent concepts focusing*.

Formal definition of the significance threshold is given in section 2. Significance threshold is defined based on mean and standard deviation of all concepts' extent sizes. Our pertinent concept lattice (cf. figure 1) has the following characteristics: number of pertinent concepts: 33 (excluding the empty concepts in level 0 and in level 5), concept extent size mean: 6.45, standard deviation: 4.37, kurtosis: -0.18, skewness: 0.79. Table 4 shows the result of our formal concepts mining method with the use of the significance threshold, where the concept covering column, the number of pertinent concepts with minimum significance is shown. Notice that the smallest number of objects (smallest extent size) and the support threshold can de derived from the significance threshold.

Table 4. DCSEA formal concept mining results with minimum significance varying from 20% to 90%

No.	Min. significance	Min. number of objects	Support	Concept covering
1	20%	2.78	0.084	27 (81.82%)
2	40%	5.35	0.162	18 (54.55%)
3	60%	7.56	0.229	12 (36.36%)
4	80%	10.13	0.307	5 (15.15%)
5	90%	12.06	0.365	4 (12.12%)

Based on the above example, we show that the significance threshold can be easily understandable by the end-user. Since, definitions of DCSEA algorithm and significance threshold are based on statistic distribution curve, it is highly recommended that the number of items should be greater than 30 in order to obtain convincing result.

4   Experimental Results

Two data sets: synthetic data and real-world data are used for testing the proposed algorithm. The synthetic data was generated using a randomizing function with 11 predefined pertinent items and 12 non-pertinent items (used for generating, respectively pertinent and non-pertinent concepts). The total number of transactions is 100,000 with average 8 items per transaction and the maximum transaction size is 10 items. We define *Noise* as the proportion of non-pertinent item occurrences to the total number of item occurrences. In our experiments, we vary that proportion from *low level of noise* to *high level of noise*. With *low level of noise* data, most of database transactions are pertinent. With *high level of noise* data, the database contains a lot of non-pertinent transactions.

The real-world data consists of name and address fields (country, county, region, state, city and street sub-field) of aircraft registration database of the Federal Aviation Administration (FAA), US (http://registry.faa.gov). Due to memory limitation and computation time acceptance, we select only 288,226 transactions from the total of 343,682 transactions. The resulting dataset 110,491 items at average 26.39 items per transaction. We consider *pertinent items*, all words that are present in all the address fields, and *non-pertinent items*, all words that are not present in the address fields. Notice that some US cities share similar names in several states. From this experiment, we expect that the pertinent concept lattice should clarify this name-sharing problem. Our experiments are performed on a Duron CPU at clock rate of 1 GHz, 256 MB of main memory. Computation time, retrieval efficiency and pertinent concepts focusing are studied. Results obtained using our algorithm are compared with the global support threshold-based approach for concept lattice construction.

4.1 Synthetic Dataset

4.1.1. Time Performance

The time performance is shown in figure 4. We notice that DCSEA computation time on low level of noise data (average time for all testing minimum significance) data is very small compared with the global support method. This can be explained by the fact that the filtering performance on low level of noise data is better that the one obtained on high level of noise data, especially at the level 1 of the lattice. With high level of noise data, the performance of DCSEA decreases with respect to the filtering performance. It follows that the computation time on high level of noise is not much different between the two approaches. In summary, DCSEA stops the non-pertinent concept propagation better on low level of noise data.

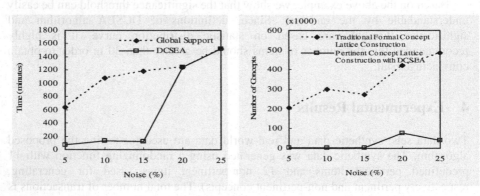

Fig. 4. Time performance on synthetic data   **Fig. 5.** Retrieval performance using synthetic data

4.1.2. Concept Retrieval Efficiency

Experimental results of concept retrieval efficiency between traditional concept lattice construction method and DCSEA are shown in figure 5. Notice that the number of retrieved concepts varies directly with level of noise. Traditional method generates very large number of concepts compared to the number of concepts generated with DCSEA. DCSEA generates only 6.31% of the number of concepts generated by the traditional method.

Figure 6 compares retrieval efficiency in terms of precision, recall and f-measure between global support-based approach and DCSEA with the use of significance

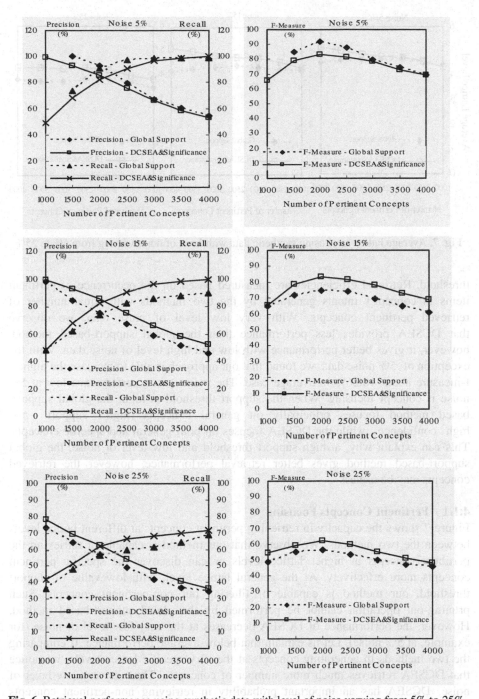

Fig. 6. Retrieval performance using synthetic data with level of noise varying from 5% to 25%

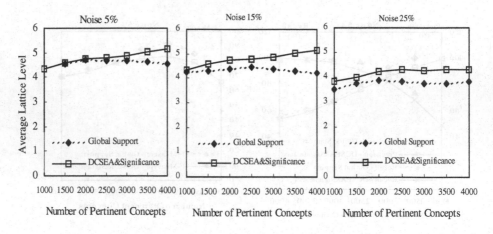

Fig. 7. Average lattice level using synthetic data with level of noise varying from 5% to 25%

threshold. Retrieval efficiencies are measured based on the occurrence of pertinent items in concept's intents generated by the two methods with equal number of retrieved pertinent concepts. With very low level of noise data, we observe that DCSEA provides less performance than the global support-based method, however, it gives better performance with low and high level of noise data. With the exception of 5% noise data, we found that our approach is able to increase the highest f-measure about 10.26% in average case. The main reason performance low at 5% noise is concept focusing. When the support threshold is high, the global support-based method focuses principally on general concepts (high support and high confidence), while the DCSEA focuses on both general and specific concepts. This can explain why, at high support threshold and low level of noise, the global support-based method gives better retrieval performance, however the retrieved concepts may be useless.

4.1.1 Pertinent Concepts Focusing
Figure 7 shows the capacity in retrieving pertinent concepts at different lattice levels between the two methods. We observe that our method is capable of retrieving the pertinent concepts at higher lattice levels or can discover the specific pertinent concepts more effectively. At the general lattice level, with low value of support threshold, our method is capable of filtering out non-pertinent concepts. Such pruning-out operation cannot be performed by the global support-based method. However, the performance of DCSEA decreases at the very low level of noise (for example, at 5% noise). Let us explain that below-average performance. If comparing the two methods in generating concepts at the very specific lattice levels, we notice that DCSEA retrieves much more number of concepts. Thus, at very low level of noise, DCSEA has an important probability in retrieving non-pertinent concepts which contain non-pertinent items.

4.1 Real-World Dataset

Due to memory limitation and too long computation time, it is not appropriate to construct full concept lattice without using any filtering technique. Experimental results using our method show that, with 505 items, 6,659 minutes are needed to generate all the possible concepts. When using the traditional concept lattice construction method with the same number of pertinent items (505), 6,650 minutes are needed to generate partially the concept lattice. It must be noticed that the highest level of lattice corresponds to the greatest transaction size, by updating the transaction database with remaining items. At the final state, the average transaction size is 5.66, maximal transaction size is 7 with 0.91% noise.

Using the traditional lattice construction method, the number of generated concepts, with minimum support equal to 0% is 284,912. With DCSEA, the number of generated pertinent concepts, with minimum significance equal to 0% is 2,619. Figure 8 and 9 show, respectively precision and concept focusing, between DCSEA and global support-based method.

In figure 8, we observe that DCSEA is less accurate than the global support-based method. The same observation has been already mentioned from our previous experiments with very low level of noise data (cf. figure 6). With respect to the number of retrieved pertinent concepts, different average levels of the output lattice are shown in Figure 9. With real-world data, DCSEA generates higher average lattice level compared to the global support-based method. Notice also that, DCSEA is capable of retrieving specific pertinent concepts more effectively, and it clearly represents the city name sharing problem.

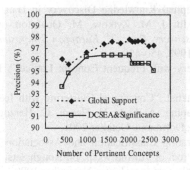

Fig. 8. Precision comparison between DCSEA & Global support-based method on real-world data

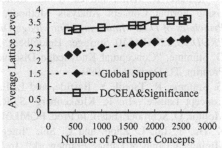

Fig. 9. Different average lattice levels with respect to the number of pertinent concepts on real-world data

4 Conclusion and Future Work

In this paper, we present a novel efficient formal concept mining algorithm, called DCSEA. Attempting repeatedly to self-adjust the normal distribution curve to be as close as the symmetry curve, DCSEA automatically identifies all the pertinent concepts by deleting and masking non-pertinent concepts. In order to evaluate concepts, the notion of significance threshold has been defined. Compared to the support threshold, the proposed significance threshold is easily understandable in statistic sense. Performance evaluations between DCSEA and the global support-based lattice construction method have been conducted using two types of dataset: synthetic data and real-world data. Experimental results show that pertinent concept lattice generated with our method can represent pertinent information effectively and compactly. DCSEA efficiently retrieves all the specific pertinent concepts with high concept retrieval performance from large databases.

Based on Apriori algorithm, the proposed DCSEA pertinent concept lattice construction algorithm works in a bottom-up manner. In order to improve the computational construction time, we plan to propose a top-down version of DCSEA in our future works. Further, inference statistics could be also be applied for very large size of real-world databases, with parallel computation.

References

1. C. Carpineto and G. Roma, "A lattice conceptual clustering system and its application to browsing retrieval", Machine Learning, 1996, 24:95-122.
2. C. Carpineto and G. Roma, "GALOIS: An order-theoretic approach to conceptual clustering", In Proc. ICML 1993, Amherst, July 1993, 33-40.
3. G. Mineau, G. and R.Godin, "Automatic structuring of knowledge based by conceptual clustering", IEEE Transactions on Knowledge and Data Engineering, 1995, 7(5):824-829.
4. G. Stumme, R. Taouil, Y. Bastide, N. Pasquier, and L. Lakhal, "Intelligent structuring and reducing of association rules with formal concept analysis", In Proc. KI 2001, LNAI, Springer, Heidelberg, 2001.
5. G. Stumme, R. Wille, and U. Wille, "Conceptual Knowledge Discovery in Databases Using Formal Concept Analysis Method", In: J. M. Zytkow, M. Quafofou (eds.): *Principles of Data Mining and Knowledge Discovery. Proc. 2ⁿᵈ European Symposium on PKDD'98*, LNAI 1510, Springer, Heidelberg, 1998, 450-458.
6. G. Stumme, "Conceptual Knowledge Discovery with Frequent Concept Lattices", FB4-Preprint, TU Darmstadt, 1999.
7. G. Stumme, R. Taouil, Y. Bastide, and L. Lakhal, "Conceptual Clustering with Iceberg Concept Lattice", In: R. Klinkenberg, S. Raping, A. Flek, N. Henze, C. Herzog, R. Molitor, O. Schroder (Eds.): In Proc. FGML01, Dortmund, October 2001.
8. K. Hu, Y. Lu, L. Zhou, and C. Shi, "Integrating Classification and Association Rule Mining: A Concept Lattice Framework", In Proc. of 7th intl.wksp. on Rough sets, Data mining and Granular computing (RSFDGrcC'99), Springer, Japan, November 1999.
9. K. Waiyamai and L. Lakhal, "Knowledge Discovery from Very Large Databases Using Frequent Concept Lattices", In Proc. ECML 2000, 2000, 437-445.
10. M. Kim and P. Compton, "Formal Concept Analysis for Domain-Specific Document Retrieval Systems", In Proc. of 13th Australian Joint Conference on Artificial Intelligence (AI01), Springer-Verlag, Berlin, Adelaid Australia, December 2001.
11. M. Sahami, "Learning Classification Rules Using Lattices", In Nada Lavrac and Stefan Wrobel (eds.): Machine Learning: ECML-95, Heraclion, Crete, Greece, 1995, 343-346.
12. N. Pasquier, Y. Bastide, R. Taouil, and L. Lakhal, "Efficient mining of association rules using closed itemset lattices", Journal of Information Systems, 1999, 24(1):25-46.

13. N. Rattanakronkul, T. Wattarujeekrit, and K. Waiyamai, "Predicting Protein Structural Class from Closed Protein Sequences", In Proc. PAKDD03, 2003, 136-147.
14. R. Wille, "Concept lattices and conceptual knowledge systems", Computers & Mathematics with Applications, 1992, 23(6-9):493-515.
15. www.itl.nist.gov.
16. G. Stumme, "Using Ontologies and Formal Concept Analysis for Organizing Business Knowledge", In: J. Becker, R. Knackstedt (Eds.): Wissensmanagement mit Referenzmodellen - Konzepte für die Anwendungssystem- und Organisationsgestaltung, Physica, Heidelberg, 2002, 163-174.
17. Y. Kachai and K. Waiyamai, "Representing Large Concept Hierarchies Using Lattice Data Structure", In Proc. PAKDD 2001, 2001, 186-197.
18. Mohammed Javeed Zaki, Ching-Jiu Hsiao, "CHARM: An Efficient Algorithm for Closed Itemset Mining", In Proc. SDM02, 2002.

A Robust Approach to Content-Based Musical Genre Classification and Retrieval Using Multi-feature Clustering

Kyu-Sik Park, Sang-Heon Oh, Won-Jung Yoon, and Kang-Kue Lee

Dankook University,
Division of Information and Computer Science,
San 8, Hannam-Dong, Yongsan-Ku, Seoul Korea 140-714
{kspark, taru74, helloril, lk_sun}@dankook.ac.kr

Abstract. In this paper, we propose a new robust content-based musical genre classification and retrieval algorithm using multi-feature clustering (MFC) method. In contrast to previous works, this paper focuses on two practical issues of the system dependency problem on different input query patterns (or portions) and input query lengths which causes serious uncertainty of the system performance. In order to solve these problems, a new approach called multi-feature clustering (MFC) based on k-means clustering is proposed. To verify the performance of the proposed method, several excerpts with variable duration were extracted from every other position in a queried music file. Effectiveness of the system with MFC and without MFC is compared in terms of the classification and retrieval accuracy. It is demonstrated that the use of MFC significantly improves the system stability of musical genre classification and retrieval performance with higher accuracy rate.

1   Introduction

The amount of available music data in multimedia databases is in rapid increase, and personal and public collections of digital music have become increasingly common over the recent years. As the amount of data increase, user requires more efficient ways of classification, retrieval and browsing for given digital music contents. However, the music data are currently labeled by a file name and a few indexing words manually, which is time-consuming, expensive and difficult. Handling music data using their content becomes a crucial technique for music genre classification and retrieval.

Musical genre classification and retrieval based on music content has been a growing area of research in the last few years. Example applications include search and select music from music digital library (MDL), entertainment industry, virtual reality, and several others on web application.

Most of content-based classification and retrieval of music sound has three common stages of a pattern recognition problem: *feature extraction*, *classification* based on the selected feature, and *retrieval* based on the similarity measure. Depending on

M. J. Maher (Ed.): ASIAN 2004, LNCS 3321, pp. 212–222, 2004

different combinations of these methods, several strategies are employed in these studies. Wold *et al* [1] developed a system called "Muscle Fish". There various perceptual features such as loudness, brightness, pitch, timbre are used to present a sound. A Mahalanobis distance measure and the nearest neighbor (NN) rule are used to classify and retrieve the query sound in the database. However, there work is limited to the sound classes of musical instruments, sound effects, and environmental sounds. Foote [2] chose to use 12 Mel Frequency cepstral coefficients (MFCC) as the audio features. Histograms of sounds are compared and the classification and retrieval is done by using the NN rule. As a first direct attempt of acoustical features to model musical signals, Tzanetakis and Perry [3] combined standard timbral features with representations of rhythm and pitch content, and they achieved classification performance in the rage of 60% for ten musical genres. The classification accuracy based on just rhythm and pitch content was quite poor such as 23%~28%. In Ref. [4], Li performed extensive comparative study on the selection of features between Daubechies wavelet coefficient and the ones used in [3], and they conclude that the timbral feature is more suitable than rhythmic or pitch content for musical genre classification. Jiang et al. [5] proposed a new spectral contrast feature and it performed near 82% for only small set of class music data. Burred et al., [6] suggested hierarchical classification approach and genre dependent feature sets. In 13 musical genres, they can achieve 57.8% classification accuracy. Another interesting approach by Guo and Li [7] was proposed to use support vector machines (SVM) with binary tree recognition. In their work, new metric called distance-from boundary (DFB) with SVM is used to measure music pattern similarity between the classes. Other good works regarding feature extraction, classification and retrieval for music and general audio information are described in [8-12]. On the other hand, in Ref. [10, 13], they pointed out important observation such that a major improvement could be mainly result from a better description of the music signals, i.e., better music features rather than new efficient classification and retrieval schemes.

Although many combinations of music features and classifiers have been evaluated in these works, little attention has been paid to the following two practical issues: First, the classification and retrieval results corresponding to different input query patterns (or portions) within the same music file may be much different. Second, the system performance may also seriously influenced by the input query length. These facts clearly calls for a new robust feature extraction method that can characterize a given full-length music signal.

In this paper, we take a new approach that provides a solution to above problems. As already pointed out, a system dependency to the different input query portions and query lengths within the same music file may cause remarkable uncertainty of the system performance. A new robust feature extraction method called multi-feature clustering (MFC) is proposed to overcome these problems based on k-means clustering algorithm. Basic idea is to extract features over the full-length music signal and then cluster these features in a small number of disjoint subsets. At this point, a SFS (Sequential Forward Selection) feature selection method is applied to reduce the feature dimensionality and to enhance the classification and retrieval accuracy. In this way, the new scheme allows feature set to characterize whole intervals of music file while maintaining a moderate size of database.

This paper is organized as follows. Section 2 describes methods for feature extraction. A SFS feature selection procedure and multi-feature clustering (MFC) method is introduced in section 3. In section 4, extensive experimental results on classification and retrieval are demonstrated with the proposed MFC method. Finally, conclusions are given in section 5.

2   Music Feature Extraction

Before classification and retrieval, the music signals are normalized to have zero mean and unit variance in order to avoid numerical problems caused by small variances of the feature values. At the sampling rate of 22000 Hz, the music signals are divided into 23ms frames with 25% overlapped hamming window at the two adjacent frames. Two types of features are computed from each frame: One is the timbral features such as spectral centroid, spectral rolloff, spectral flux and zero crossing rates. The other is coefficient domain features such as Mel-frequency cepstral coefficients (MFCC) and linear predictive coefficients (LPC). The means and standard deviations of these six original features are computed over each frame for each music file to form a total of 54-dimensional feature vector. Table 1 summarizes the feature set used in this paper. These features are well-known in the literature and only the short description of definition is given in the table.

Table 1. Musical feature definitions

Feature	Definition
Spectral centroid	It is defined as the center of gravity of STFT magnitude spectrum. The centroid is a measure of spectral brightness.
Spectral rolloff	It is defined as the frequency below when 85% of the magnitude distribution is concentrated. It measures spectral shape.
Spectral flux	It is the squared difference between the magnitudes of successive spectral distribution. It is a measure of local spectral change.
ZCR	It is the rate of time-domain zero-crossings. It measures the noisiness of the signal.
MFCC	MFCC is the most widely used feature in speech recognition. It captures short-term perceptual features of human hearing system. Thirteen coefficients are used.
LPC	LPC are a short-time measure of the speech signal with describes the signal as the output of all-pole filter. Ten coefficients are used.

3   SFS Feature Selection and Multi-feature Clustering (MFC)

3.1   SFS Feature Selection Procedure

Not all the 54-dimensional features in the previous section are used for musical genre classification and retrieval purpose. Some features are highly correlated among themselves and some feature dimension reduction can be achieved using the feature redundancy. In order to reduce the computational burden and so speed up the search process, while maintaining a system performance, an efficient feature dimension reduction and selection method is desired. In Ref. [6, 13], a sequential forward selection (SFS) method is used to meet these needs. In this paper, we adopt the same SFS method for feature selection to reduce dimensionality of the features and to enhance the classification accuracy. Firstly, the best single feature is selected and then one feature is added at a time which in combination with the previously selected features to maximize the classification accuracy [13]. This process continues until all 54 dimensional features are selected. After completing the process, we pick up the best feature lines that maximize the classification accuracy. We note that SFS method described here allows not only choosing the best features to form the features vector for music classification and retrieval, but also it helps to keep moderate size of database in multi-feature clustering (MFC) algorithm in next subsection. Further details about this procedure are demonstrated in section 4.

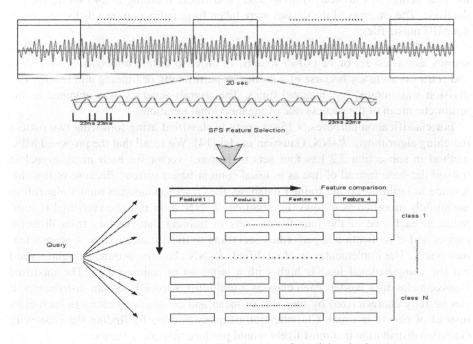

Fig. 1. Multi-feature clustering procedure for training database

3.2   Multi-feature Clustering (MFC)

As pointed out earlier, the classification and retrieval results corresponding to different input query patterns (or portions) and query lengths within the same music file or same class may be much different. It may cause serious uncertainty of the system performance. In order to overcome these problems, a new robust feature extraction method called multi-feature clustering (MFC) is implemented based on k-means clustering algorithm. Basic idea is to extract features over the full-length music signal in a step of 20 sec large window using the SFS method and then cluster these features in four disjoint subsets using k-means clustering. This allows feature set to characterize whole intervals of music signal while maintaining a moderate size of trained database. The system then compares the query pattern to each one of the four feature sets of music file in trained database, and it classifies and retrieves the queried music. Fig. 1 outlines the proposed MFC method with a SFS feature selection method.

4   Experiments on Musical Genre Classification and Retrieval

4.1   Experimental Setup

The proposed algorithm has been implemented and used to classify and retrieve music data from a database of 240 music files. 60 music samples were collected for each of the four genres in Classical, Hiphop, Jazz, and Rock, resulting in 240 music files in database. The excerpts of the dataset were taken from radio, compact disks, and internet MP3 music files.

The 240 music files are partitioned randomly into a training set of 168 (70%) sounds and a test set of 72 (30%) sounds. In order to ensure unbiased classification and retrieval accuracy because of a particular partitioning of training and testing, this division was iterated one hundred times. The overall accuracy was obtained as the arithmetic mean of the success rate of the individual iterations.

For classification purposes, a 15sec query is classified using following two pattern matching algorithms; k-NN, Gaussian model [14]. We recall that the proposed MFC method in subsection 3.2 has four sets of features vector for each music signal in trained database instead of one as in usual content-based system. Because of this difference in building up the trained database, the k-NN, Gaussian model algorithms are slightly modified as follows. In modified k-NN, a query to be classified is compared to each one of the four feature sets in training data vectors from different classes and classification is performed according to the distance to the k nearest feature points. The implementation of modified k-NN classifier is quite straightforward but the computational load is high with a large set of training data. The modified Gaussian classifier models each class as a multidimensional Gaussian distribution. It can be fully characterized by four sets of mean and covariance vectors in each class instead of one as in usual. Classification of query is done by finding the class with Gaussian distribution that most likely would produce this query vector.

For retrieval purposes, a music query is compared to each one of the four feature sets of music file in trained database according to simple Euclidean distance measure.

The retrieval accuracy is then measured base on top 10 ranked list rules in query responses.

4.2 Results and Analysis

Five sets of experiment on classification and retrieval have been conducted in this paper.

- Exp. 1: Performance verification of SFS feature selection method
- Exp. 2: Classifying test music using MFC method with different query patterns
- Exp. 3: Classifying test music using MFC method with different query lengths
- Exp. 4: Retrieval test using MFC method with different query patterns
- Exp. 5: Retrieval test using MFC method with different query lengths.

A   Experimental Results on Classification

Fig. 2 shows average classification accuracy using SFS method described in section 3.1. As seen from the figure with SFS method, simple k (3)-NN shows rapid convergence speed with higher classification accuracy while Gaussian model needs more features to converge. The reason for this is that the SFS method here is not considering any statistical property of the music samples and thus any statistical classification method such as Gaussian model will not properly working with the SFS method. From the figure 1 with k (3)-NN, we can see that the classification performance increases with the increase of features up to 10 with near 90% of accuracy. And it remains constant up to 10 ~13 features. After 13 features, it even makes the system performance worse. Therefore, we can select only first 10 features to represent each music signals and it will be used all through out the experiments in this paper. We note that, in Ref. [13], they end up with 20 features which are twice of our feature dimension.

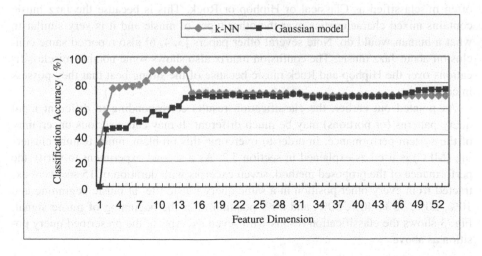

Fig. 2. Classification accuracy of k -NN and Gaussian model using SFS (Exp. 1)

Table 2 shows detailed SFS performance in musical genre classification in a   form of a confusion matrix. As a comparison purpose, the classification results using 54 dimensional feature vector is included in the table. The numbers of correct classification results with SFS lie in the diagonal of the confusion matrix. The numbers shown in parenthesis represent statistics with all 54 dimensional features.

Table 2. Genre confusion matrix with SFS feature selection

	Classical	Hiphop	Jazz	Rock	Classification accuracy
Classical	57 (51)	0(0)	1(5)	2(2)	95% (85%)
Hiphop	0(0)	55(40)	1(6)	4(14)	92%(67%)
Jazz	3(11)	3(12)	49(27)	2(7)	82%(45%)
Rock	1(1)	5(15)	2(8)	52(33)	86%(55%)
Average Classification Accuracy					89%(64%)

From the table 2, we see much improvement of classification performance using only 10 dimensional features derived from SFS feature selection method. It actually can achieves more than 20% improvement even with fewer number of feature set. The SFS method works fairly well over the genre of Classical, Hiphop, and Rock while the average number of correct classification is little lower in Jazz genre. Jazz music is often misclassified as Classical or Hiphop or Rock. This is because the Jazz music contains mixed characteristics over the other types of music and it is very similar to what a human would do. Note several other papers [3, 4, 6] also reported same conclusion about Jazz music. The confusion matrix also shows some notable misclassifications over the Hiphop and Rock music because of the strong beat that they possess in common.

As pointed out earlier, the classification results corresponding to different input query patterns (or portions) may be much different. It may cause serious uncertainty of the system performance. In order to overcome this problem, multi-feature clustering (MFC) is used as explained in section 3.2. As a second experiment to verify the performance of the proposed method, seven excerpts with duration of 15 sec were extracted from every other position in a same query music file- at music beginning and 10%, 20%, 30%, 40%, 50%, and 80% position after the beginning of music signal. Fig. 3 shows the classification results with seven excerpts at the prescribed query position as above.

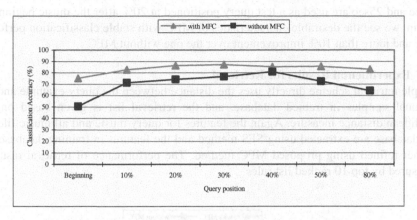

Fig. 3. Classification results with different query portions (Exp. 2)

As we expected, the classification results without MFC greatly depends on the query positions and it's performance is getting worse as query portion towards to two extreme cases of beginning and ending position of the music signal. This is no wonder because, in general, the musical characteristics are not rich enough at those extreme intervals of music signal. On the other hand, we can find quite stable classification performance with MFC method and it even yields higher accuracy rate in the range of 75% ~ 85%. Even at the two extreme cases of beginning and ending position, the system with MFC can achieves classification accuracy as high as 75% which is more than 20% improvement over the system without MFC. This is a consequence of good MFC property which helps the system to build robust musical feature set over the full-length music signal for trained database set.

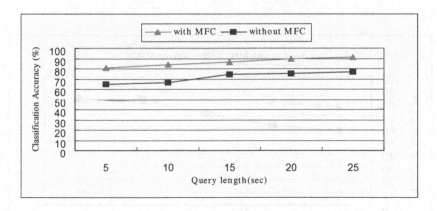

Fig. 4. Classification results with different query lengths (Exp. 3)

As a third experimental outcome, fig. 4 shows an importance of the query length to the overall system performance. Five excerpts with duration of 5sec, 10sec, 15sec,

20sec and 25sec are used as a test query positioned at 20% after the music beginning. Again, we see the desirable characteristics of MFC with stable classification performance and more than 10% improvement over the one without MFC

B   Experimental Results for Retrieval

Simple retrieval scheme directly uses the distance between the query example and individual samples in trained database, and the retrieval list is given based on the Euclidean distance measure. Again the features for query music and all music files in the database are extracted using SFS method and the features in training database are further refined using proposed MFC method. The performance of retrieval result is measured by top 10 ranked list rules.

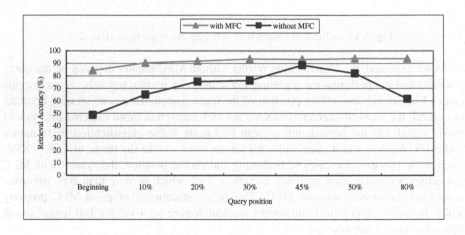

Fig. 5. Retrieval results with different input query portions (Exp. 4)

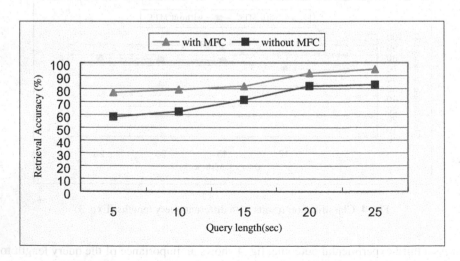

Fig. 6. Retrieval results with different input query lengths (Exp. 5)

Fig. 5 shows retrieval results with different input query patterns while figure 5 displays results for the case of different input query lengths. Same testing conditions regarding query positions and lengths are used as in previous classification test. As seen from figure 4, the retrieval results without MFC greatly depends on the query positions and it's performance is widely fluctuating in the range of 48% ~ 91% as the query position changed. In contrast, we again confirm the superior characteristics of the proposed MFC which guarantees stable retrieval performance in 84%~94% across the whole query positions with even higher accuracy. We can also draw the same conclusion about the results corresponding to different query lengths as shown in figure 6. These results are the further testimonies of robust nature in proposed MFC method with better classification and retrieval performances.

5 Conclusion

In this paper, we propose a new robust content-based music genre classification and retrieval algorithm using multi-feature clustering (MFC) method to overcome the unstable system performance problems due to the different input query patterns and query lengths. For the implementation of the proposed algorithm, sequential forward selection (SFS) is first applied to the 54 dimensional feature set to reduce the feature dimension in the order of one-fifth and then multi-feature clustering (MFC) is adopted to build a robust feature vectors in training database. The system then compares the query pattern to each one of the four feature sets of music file in trained database, and it classifies and retrieves queried music information. To verify the performance of the proposed MFC method, several excerpts with variable duration were extracted from every other position in a same queried music file and the effectiveness of the system with MFC and without MFC is compared in terms of the classification and retrieval accuracy. From the extensive experimental comparisons on classification and retrieval, we show the superiority of MFC method in terms of the performance stability and accuracy. Future work will involve the development of new features, further analysis of classification and retrieval system for practical implementation, and the incorporation of more music classes.

References

1. E. Wold, T. Blum, D. Keislar, and J. Wheaton, "Content-based classification, search, and retrieval of audio," *IEEE Multimedia*, vol.3, no. 2, 1996.
2. J. Foote, "Content-based retrieval of music and audio," in *Proc. SPIE Multimedia Storage Archiving Systems II*, vol. 3229, C.C.J. Kuo *et al.*, Eds., 1997, pp. 138-147.
3. G. Tzanetakis and P. Cook, "Musical genre classification of audio signals," *IEEE Trans. on Speech and Audio Processing*, vol. 10, no. 5, pp. 293-302, July 2002.
4. T. Li, M. Ogihara and Q. Li, "A comparative study on content-based music genre classification," in *Proc. of the 26th annual internal ACM SIGIR*, pp. 282-289, ACM Press, July 2003.
5. D. N. Jiang, L. Lu and H. J. Zhang, "Music type classification by spectra contrast features," in *Proc. ICME2002*, 2002, pp. 113-116.

222   K.-S. Park et al.

6. J. J. Burred and A. Lerch, "A hierarchical approach to automatic musical genre classification," in *Proc. DAFx03*, 2003, pp. 308-311.
7. G. Guo and S. Z. Li, "Content-based audio classification and retrieval by support vector machine," *IEEE Trans. on neural networks*, vol. 14, no. 1, pp. 209-215, Jan. 2003.
8. J. Foote *et al*, "An overview of audio information retrieval," *ACM-Springer Multimedia Systems*, vol. 7, no. 1, pp. 2-11, Jan. 1999.
9. Y. Wang, Z. Liu and J. Huang, "Multimedia content analysis: using both audio and visual clues," *IEEE Signal Proc. Mag.*, Nov. 20000
10. S. Blackburn, "Content based retrieval and navigation of music, 1999, Mini-thesis, University of Southampton.
11. S. Z. Li, "Content-based classification and retrieval audio using the nearest feature line method," *IEEE Trans. on Speech Audio Processing*, vol. 8, pp.619-625, Sept. 2000.
12. T. Zhang and C. Kuo, "Hierarchical system for content-based audio classification and retrieval," *Proceedings of SPIE98*, vol.3527, pp. 398-409, Boston, Nov., 1998
13. M. Liu and C. Wan, "A study on content-based classification retrieval of audio database," *Proc. of the International Database Engineering & Applications Symposium*, pp. 339 - 345. 2001.
14. R. Duda, P. Hart and D. Stork, Pattern Classification, 2nd Ed., Wiley-Interscience Publication, 2001

Registration of 3D Range Images Using Particle Swarm Optimization

Hai V. Phan[1], Margaret Lech[2], and Thuc D. Nguyen[3]

[1] RMIT International University, Vietnam
s3067462@student.rmit.edu.au
[2] School of Electrical and Computer Engineering, RMIT University,
GPO BOX 2476V, Melbourne 3001 Victoria, Australia
margaret.lech@rmit.edu.au
[3] Faculty of Information Technology, College of Natural Science,
HCMC National University, Vietnam
nhbthuy@hcm.vnn.vn

Abstract. Improvements to the application of a relatively new meta-heuristic strategy, called Particle Swarm Optimization (PSO), to the registration of 3D range images are presented. The proposed improvements include: implementation of the search intensification procedure, the von-Neumann particle neighborhood topology and the differential evolution operator. The 3D data registration results show significant improvements compared to the currently used PSO-based registration techniques, as well as techniques involving other meta-heuristic optimization strategies, such as genetic algorithms and adaptive simulated annealing.

1 Introduction

1.1 Registration of Range Images

Range images represent raw 3D data obtained from scanners. The so-called 2½D images are three-dimensional data sets representing single views of 3D objects. Range image registration is the process of aligning 2½D images to minimize the mismatch among their overlapping regions. Its applications include reverse engineering, robotic navigation, medical imaging and manufacturing quality control.

Registration of 2½D range images is a process of finding the alignment transformation that maximizes the matching between overlapping views. The alignment transformation vector **T** is usually represented by six transformation parameters: T_1, T_2, T_3, T_4, T_5 and T_6. The first three parameters T_1, T_2, and T_3 represent translation along the x-axis, y-axis and z-axis, respectively. Parameters T_4, T_5 and T_6 are rotation angles around the x-axis, y-axis and z-axis, respectively.

Due to range image registration's broad range of applications, a considerable amount of research has been performed in this field during the last decade. The most popular method for range image registration is the iterative closest point (ICP) algo-

M. J. Maher (Ed.): ASIAN 2004, LNCS 3321, pp. 223–235, 2004.

rithm introduced by Besl and Chen [1, 2]. Many variants of the ICP algorithm have been developed; these have been summarized by Rusinkiewicz and Levoy [3]. However, without a reasonably good initial alignment, the ICP algorithm tends to converge to local minima, which often do not represent acceptable solutions [4]. Two major approaches have been proposed to overcome this drawback. The first approach is to find a good initial alignment before executing the ICP algorithm. The second approach is to develop a search technique that can escape local minima and navigate the search trajectory towards the global minimum in which case, the algorithm could start from an arbitrary alignment. Some of the proposed techniques have successfully applied stochastic global optimization methods such as very fast simulated re-annealing [5], genetic algorithms [6, 7], tabu search and particle swarm optimization (PSO) [8]. This work investigates application of the PSO algorithm to the range data registration problem.

1.2 Particle Swarm Optimization

The original PSO algorithm introduced by Kennedy and Eberhart in 1995 [9], simulates the social behavior of a school of fish or a flock of birds, called the swarm. The individual swam members are called particles. Particles benefit from experiences and discoveries of others when searching for food. Each particle remembers its own best position called the individual best, as well as the best position found by the whole swarm, and called the global best. Note that in the context of 3D data registration, the term particle position denotes a particular transformation vector $\mathbf{T} = [T_1, T_2, T_3, T_4, T_5, T_6]$. The work by Shi [10] and Clerc and Kennedy [11] improved the original PSO algorithm by the addition of the inertia weight and constriction value parameters. The basic PSO based registration algorithm can be described in the following steps:

1. Initialization: Choose the number N_p of search individuals (population size). For each individual i, set an arbitrary initial transformation vector $\mathbf{T}_{i0}$ and an initial search velocity vector $\mathbf{v}_{i0}$ (where $\mathbf{T}_{i0}, \mathbf{v}_{i0} \in R^n$, and i = 1, 2,...,$N_p$) and calculate the individual cost function values $f(\mathbf{T}_{i0})$. For each individual, assign the individual best transformation vector $\mathbf{T}_i^*$ to $\mathbf{T}_{i0}$. Assign the global best transformation vector $\mathbf{T}_S$ to the transformation vector of the individual with the smallest cost.

2. Iterations: For each iteration j (j=1,2,....N_{iter})

   2a. For each individual i (i=1,2,...,N_p), calculate a new search velocity vector $\mathbf{v}_{ij}$ using:

$$\mathbf{v}_{ij} = K\left[\gamma \mathbf{v}_{ij-1} + \alpha_{ij}\mathbf{d}(\mathbf{T}_i^*, \mathbf{T}_{ij-1}) + \beta_{ij}\mathbf{d}(\mathbf{T}_S, \mathbf{T}_{ij-1}) \right] \qquad (1)$$

where, $\mathbf{v}_{ij} \in R^n$, $\mathbf{d}(.)$ denotes the difference vector, K is a constant constriction factor, γ is a constant inertia weight and α_{ij} and β_{ij} are random numbers such that α_{ij} is uniformly distributed between 0 and a constant value c_1 and β_{ij} is uniformly distributed between 0 and a constant value c_2. The values of α_{ij} and β_{ij} direct, in a random way, the move towards an individual or the global best solution. The inertia weight γ, is a monotonically decreasing function of the iteration number j;

   2b. For each individual i, update the individual transformation vector to $\mathbf{T}_{ij}$ using:

$$\mathbf{T}_{ij} = \mathbf{T}_{ij-1} + \mathbf{v}_{ij} \qquad (2)$$

2c. For each individual i, calculate the individual cost function value $f(T_{ij})$;

2d. If the individual cost $f(T_{ij})$ is less than the individual best cost $f(T_i^*)$, assign the individual best transformation vector T_i^* to T_{ij};

2e. If $f(T_{ij})$ is less than the global best cost $f(T_S)$, assign the global best transformation vector T_S to the individual transformation vector T_{ij};

2f. Termination verification: If the convergence criterion, based on the current value of T_S, is reached then terminate the algorithm, otherwise assign j to j+1 and go to step 2 starting the next iteration.

2 New PSO-Based Registration Technique

2.1 Search Intensification

In order to improve the PSO search efficiency, search intensification is performed using the search-domain contraction technique. Initially, the algorithm is allowed to perform a very diverse search exploring a broad range of possible solutions. It is assumed that after a certain number of iterations, the PSO algorithm finds a promising region; this is a region, which has a high probability of containing the global minimum. The promising region is determined as a region surrounding the current best solution. When the promising region is localized, the search-domain is contracted reducing the set of feasible solutions to a range closely surrounding the current best solution. The contractions are performed at each iteration, reducing recursively the search-domain and increasing the convergence rate of the algorithm. The idea of contracting the search-domain has been successfully used in other global optimization techniques such as genetic algorithms [6, 12] and the tabu search [13]. Here, the search-domain contraction method is applied to the particle swarm optimization.

Let $T^{upper}, T^{lower} \in R^n$ denote vectors defining the upper limit and the lower limit of the transformation vector T respectively. The upper and the lower limits of the transformation vector coordinates indexed with j (j=1,2,...,n), are updated with the reduction rate R_j using:

$$T_{ij}^{upper} = T_{Sj} + \left(T_{(i-1)j}^{upper} - T_{Sj}\right)/\left(R_j\right)^k \qquad (3)$$

$$T_{ij}^{lower} = T_{Sj} - \left(T_{Sj} - T_{(i-1)j}^{lower}\right)/\left(R_j\right)^k \qquad (4)$$

The index k represents the number of already performed contractions, i, is the current iteration number and T_S is the global best transformation vector. In general, the reduction rate R_j can have different values for each coordinate j. In this work, two different values were used; one for the coordinates representing translation parameters and the other for coordinates representing the rotation parameters. The lower and upper limits of the transformation vector coordinates do not have to be reduced at each iteration. A series of iterations during which the boundaries are constant is called the intensification phase. The length of each intensification phase can be kept constant or vary with the increasing number of iterations.

2.2 Topology of the Particle Swarm Optimization

Since the PSO method simulates the information sharing behavior of a group of individuals, the way in which the individuals interact with each other represents an important factor affecting the algorithm's performance. The particle interaction issue
has been addressed in some of the recent studies of the PSO population structures and
their dynamics [14-16]. The population structure can be described using an undirected graph (see Figure 1) in which the black dots denote positions of particles
mapped on a vertical plane and the black lines denote the links between particles. The
direction of a particle movement is determined by its personal best position as well as
the best positions of its neighbors. A particle belongs to a neighborhood of another
particle only if there is a direct link between these two particles. In other words, a
particle interacts only with its neighbors. Let $M=\left\{M_{ij}\right\}_{\substack{i=1,...,N_p \\ j=1,...,N_p}}$ be a square $N_p \times N_p$
matrix representing an arbitrary structure-graph of a population containing N_p parti-
cles. If there is a direct link between particle i and particle j then the matrix element
M_{ij} is equal to 1, otherwise, M_{ij} is equal to 0.

The velocity vector updating formula of Eq.1, can be generalized for the use with
an arbitrary population structure in the following way:

$$\mathbf{v}_{ij} = K\left[\gamma \mathbf{v}_{ij-1} + \alpha_{ij}\mathbf{d}(\mathbf{T}_i^*, \mathbf{T}_{ij-1}) + \beta_{ij}\mathbf{d}(\mathbf{T}_k^*, \mathbf{T}_{ij-1})\right] \tag{5}$$

Where, the index k, refers to the population member that belongs to the neighborhood of an individual with index i (i=1,...,N_p) and such that the personal best cost of
an individual k has the lowest value amongst all neighbors of an individual i. This can
be expressed as follows:

$$k \in \left\{i, M_{i,k} = 1\right\} \quad \text{and} \quad f(\mathbf{T}_k^*) = \min_{r \in \{i, M_{i,r}=1\}} (f(\mathbf{T}_r^*)) \tag{6}$$

The function f(**T**) in Eq.6 is the objective function and $\mathbf{T}_r^*$ represents the personal
best position vector of an individual r.

As described in [14-16], the different population structures of PSO include the
GBEST model, the LBEST model, the pyramid model, the star model, the von-
Neumann model, the sub-population model and the stochastic model. The experimental results of [17] indicate that the von-Neumann structure outperforms, in most cases,
the other structures. However, all the earlier studies of the PSO-based registration
techniques [8] use the traditional GBEST model.

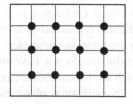 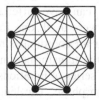

Fig. 1. The von-Neumann model (left) and the GBEST model (right)

As illustrated in Figure 1, in the von-Neumann model [17] each particle has four neighbors. The immediate upper, lower, left and right positions of the neighboring particles form a lattice structure when mapped on a perpendicular plane. In the traditional GBEST model also illustrated in Figure 1, all particles are linked together.

The PSO algorithm using the GBEST model takes into account all population members each time a movement decision is made. This approach may generate a bias towards certain directions represented by 'stronger' population members and causing a premature convergence to a local minimum.

In contrast, the lattice structure of the von-Neumann model does not take into account all population members each time a movement decision is made. Therefore, the PSO search based on the von-Neumann structure may be allowed to explore the solution space more thoroughly.

As indicated in Eq.5, the change of the velocity vector $\mathbf{v}_i$ for an individual i is determined by its personal best position $\mathbf{T}_i^*$ and the best personal position $\mathbf{T}_r^*$ within its neighborhood. In Eq.1, on the other hand, the change of the velocity vector $\mathbf{v}_i$ for an individual i is determined by its personal best $\mathbf{T}_i^*$ and the global best $\mathbf{T}_S$. Thus, Eq.1 can be viewed as special case of Eq.5 assuming the traditional GBEST model.

2.3 Differential Evolution Operator

Zhang [18] introduced the differential evolution operator to the PSO method by dividing the iterations into two categories: swarming iterations and evolutionary iterations. The swarming iterations update the position vectors of the particles using the original formulas of Eq.1 and Eq.2. During the evolutionary iterations, the position vectors of the particles are updated using the differential evolution operator. This means that for each individual i ($i=1,...,N_p$) with the corresponding position vector $\mathbf{T}_i=(T_{i,1}, T_{i,2}, ..., T_{i,n})$ of size n, a coordinate number c is chosen as a random integer ranging from 1 to n. The position vector $\mathbf{T}_i$ is then changed by adjusting only the coordinate number c in the following way:

$$\mathbf{T}_i=[T_{i1},...,T_{i(c-1)},T_{Sc}+\varepsilon,T_{i(c+1)},...,T_{in}] \tag{7}$$

The position vector is changed by adding the differential factor ε to the c-th coordinate of the global best position vector $\mathbf{T}_s$. The differential factor ε is calculated as follows:

$$\varepsilon = [(T_{r_1 c}^* - T_{r_2 c}^*)+(T_{r_3 c}^* - T_{r_4 c}^*)]/2 \tag{8}$$

The indexes r_1 through to r_4 are calculated as random integers ranging from 1 to n and $\mathbf{T}_i^*$ is the personal best position vector of an individual i ($i=1,...,N_p$).

During the even iterations, the algorithm can perform the swarming search and during the odd iterations, the evolutionary search can be pursued. The changes introduced to the position vectors $\mathbf{T}_i$ during the swarming iterations are relatively large compared to the small perturbations introduced during the evolutionary iterations. Such a combination of a global (macroscopic) movements and local (microscopic) explorations allows the algorithm to make wide-range movements in a search for a global solution and at the same time to explore in a fine way local areas of the solution space. This could be particularly beneficial in cases when the global optimum is located at the bottom of a very long narrow valley [22].

The evolutionary change of the position vector T_i given by Eq.7 can be generalized for use with an arbitrary population structure including the von-Neumann model as follows:

$$T_i = [T_{i1}, ..., T_{i(c-1)}, T_{kc}^* + \varepsilon, T_{i(c+1)}, ..., T_{in}] \qquad (9)$$

where, the differential factor ε is calculated according to Eq.8 and the index k of T_{kc}^* is given by Eq.6.

2.4 Dynamic Correspondence Scheme

The registration process of two images I_1 and I_2 includes an iterative evaluation of the objective function. This leads to a repetitive re-calculation of a distance measure between image I_1 image I_2'. Since the image I_2' results from the transformation T applied to the original image I_2, it changes from iteration to iteration. Therefore, the distance between image I_1 and image I_2 has to be iteratively re-calculated.

As described in [19], the three most popular ways of defining the dynamic correspondences between images are: the closest point (or point to point) technique, the point to plane technique, and the camera reverse calibration (or point to projection) technique.

In the proposed PSO registration method, the dynamic correspondence scheme based on the closest point technique was applied. The distance between image I_1 and image I_2' was determined by first, defining a set of N_s points x_k (k=1,...,N_s) that belong to the subset S of I_1 and then, at each iteration, finding a set of the corresponding closest points y_k (k=1,...,N_s) that belong to image I_2'.

It has been indicated in [19] that the closest point technique offers the best trade off between the speed and the accuracy of the algorithm. The camera reverse calibration represents the fastest method, however the accuracy is not always assured. The point to plane technique gives the best accuracy, but at the cost of a very slow speed.

2.5 Objective Function

The objective function used in the presented application of the PSO algorithm, was defined as the median squared error of the alignment between image I_1 and image I_2' measured over a selected set S of N_s points in image I_1. Thus, for a given transformation vector T, the objective function was given as:

$$f(T) = \underset{1 \leq k \leq N_s}{\text{median}} \left(D(x_k, y_k)^2 \right) \qquad (10)$$

where, D(.) denotes the Euclidean distance between two spatial position vectors. Points x_k belong to the subset S of I_1, N_s is the number of points in S and y_k are the closest points of image I_2' to the corresponding points x_k of image I_1. Image I_2' represents the original image I_2 transformed using the transformation vector T. The closest points y_k are usually selected under the condition of $D(x_k, y_k)$ being less than a given threshold value.

As suggested by Masuda and Yokoya [20], the minimization of the median value of the alignment error helps to obtain an even error distribution between the alignment points, and thus to avoid a situation when the quality of matching varies significantly across the alignment area. This is especially beneficial in cases where there are large overleaps between images.

The least median square error works well with images which have large overleaps (more than 50%) and it is considered to be one of the most robust objective function measures [6, 21].

3 Experiments and Results

3.1 Test Algorithms

The 3D data registration tests were conducted using four different registration techniques, including the proposed modified PSO method. The registration techniques used in the experiment are referred to using symbols listed in Table1.

As described in Section 2, the proposed registration algorithm included the implementation of the search intensification procedure, an objective function based on the median squared Euclidean error, the von-Neumann particle neighborhood topology and the differential evolution operator.

The new algorithm was tested against three other, recently introduced, registration algorithms.

The first tested algorithm was the PSO technique of Wachowiak's et al [8], which in [8] is referred to as the PSO8 method with constriction coefficient and relaxed convergence criteria. The speed of convergence was achieved by the application of a local gradient-based search within the promising region determined by the PSO algorithm.

Table 1. 3D data registration algorithms

Symbol	Description
PS1	PSO with constriction coefficient and relaxed convergence criteria.
PS2	Proposed PSO-based technique
DGA	Dynamic genetic algorithm
VFSA	Very fast simulated annealing

The second algorithm used for comparison was the dynamic genetic algorithm proposed by Chow et al [6], which introduced a new fitness function and a genetic operator called the adaptive mutation.

The third comparison algorithm was Ingber's very fast simulated re-annealing [23] applied to the range data registration by Blais and Levine [5]. This algorithm uses individual cooling schedules for each of the optimization parameters. The control parameters for the cooling schedules are re-adjusted according to the individual sensitivities of the parameters. The individual sensitivities are determined based on how each of the parameters affects the cost function.

3.2 Experimental Data

The registration data was obtained from a commercial high-precision optical range scanner (Metricor). The measurement accuracy was 1 in 2000 points of the field view. All visible outliers were removed prior to the registration, using dedicated software incorporated into the measurement system. The initial alignment between the views obtained from the scanner contained only a very small registration error, which could be easily corrected using a gradient-based local optimization method. Therefore, the initial registration positions of images were used as the reference (or 'correct registration') positions to produce more severe random misalignments, which were then used to test the performance of the registration algorithms.

The data were obtained using two different objects; the mannequin's head and the frog's figure. Eight views of each object were taken. The views were separated by a 45^0 angle and each view contained between 7688 and 12112 3D points for the frog data, and between 14004 and 19161 3D points for the head data. Figures 2 and 3 illustrate the original range data sets of the head and frog, respectively. For each object, eight pairs of registration images were generated combining images 1 and 2, images 2 and 3, images 3 and 4, images 4 and 5, images 5 and 6, images 6 and 7 and images 7 and 1.

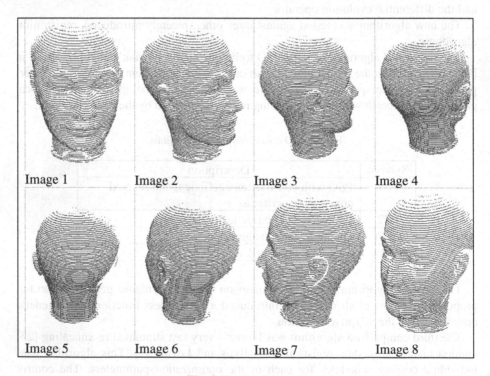

Fig. 2. The head data set

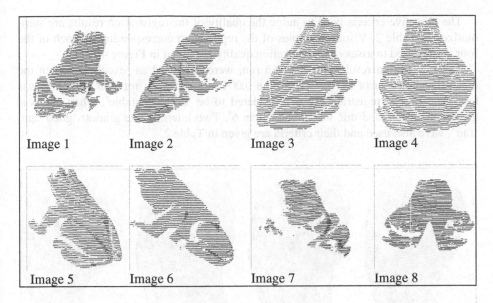

Fig. 3. The frog data set

3.3 Testing Procedure

The sixteen starting points for the registration test were generated by randomly perturbing each of the sixteen pairs of range images. The first view of each pair was kept intact, whereas the second view was randomly rotated and translated along the x, y and z-axes. The random rotations were kept within the range of $\pm 30^0$ and the random translations were within the range of ± 5. All algorithms were tested using the same sixteen starting points.

For each algorithm, the control parameters were set to achieve a convergence after a single run containing about 5000 objective function evaluations. Each algorithm was run 20 times for each of the sixteen data pairs.

The population size for the PS2 algorithm was 50. The reduction rate R in Eq.3 and Eq.4 was constant for all iterations, and the value of R=1.5 was used to update the range of the rotation parameters, and R=4 was used to update the range of the translation parameters. The constriction factor K, of Eq.1, was set to K=0.7298 and the inertial weight γ, of Eq.1 was set to $\gamma =1$ at the beginning of a run, and then linearly reduced to reach $\gamma =0.4$ at the end of a run. The constant parameters c_1 and c_2 limiting the range of the random numbers α and β in Eq.1, had a value of 2.5.

3.4 Algorithm Performance Measures

The performance of the registration algorithms was evaluated using two parameters: mT and mR, where, mT denotes the mean value for three translation directions: along x-axis, along y-axis and along z-axis for a single run, and mR is the mean value for three rotations: about x-axis, about y-axis and about z-axis for a single run.

The objective criteria used to judge the quality of the registration results are summarized in Table 2. Visual examples of the registration corresponding to each of the four grades used to assess the registration quality are given in Figure 4.

The registration results, for a given run, were classified as 'excellent' when the mean translation error mT was less than 0.001 and the mean rotation error mR was less than 2^0. The registration was considered to be 'not acceptable', when mT was greater than 0.02 and mR was greater than 6^0. Two intermediate grades: 'good' and 'fair', were also used and their criteria are given in Table 2.

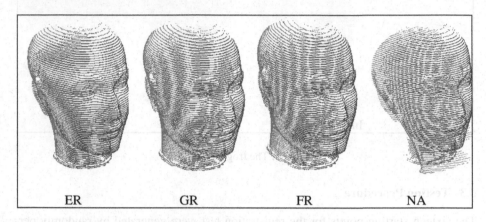

ER   GR   FR   NA

Fig. 4. Visual examples of the registration quality

Table 2. Registration quality measures

Symbol	Description
ER	Excellent registration; $0.0 <= mT < 0.01$ and $0^\circ <= mR < 2^\circ$.
GR	Good registration; $(0.01 <= mT < 0.015$ and $0^\circ <= mR < 4^\circ)$ or $(0.00 <= mT < 0.015$ and $2^\circ <= mR < 4^\circ)$
FR	Fair registration; $(0.015 <= mT < 0.020$ and $4^\circ <= mR < 6^\circ)$ or $(0.015 <= mT < 0.020$ and $0^\circ <= mR < 6^\circ)$ or $(0.000 <= mT < 0.020$ and $4^\circ <= mR < 6^\circ)$.
NA	Not acceptable registration; $mT >= 0.02$ or $mR >= 6^\circ$.

3.5 Registration Results and Discussion

The registration results achieved by the four tested algorithms: PS1, PS2, VFSA and DGA, are summarized in Tables 3 and 4. Table 3 shows the results for the head data set, and Table 4 shows the results for the frog data set. The numbers given in Tables 3 and 4 represent percentages of 160 runs (20 runs for each of the eight data sets), converging to a particular category of the registration quality: 'excellent', 'good', 'fair' or 'not acceptable'.

Table 3. Test result for the head data set

	ER	GR	FR	NA
PS1	4.4%	13.1%	28.8%	53.8%
PS2	74.4%	18.1%	7.5%	0.0%
VFSA	13.8%	13.1%	24.4%	48.8%
DGA	28.8%	33.1%	24.4%	13.8%

Table 4. Test result for the frog data set

	ER	GR	FR	NA
PS1	3.1%	16.3%	49.4%	31.3%
PS2	76.9%	20.6%	2.5%	0.0%
VFSA	13.8%	16.9%	32.5%	36.9%
DGA	35.6%	37.5%	17.5%	9.4%

As indicated in Tables 3 and 4, the proposed registration algorithm PS2, clearly gives the largest percentage of 'excellent' registrations; 74.4% for the head data set and 76.9% for the frog data set compared to all other algorithms. The weakest performance was given by the PS1 algorithm (53.8% of 'not acceptable' solutions for the head-data and 31.3% of 'not acceptable' solutions for the frog data').

It also clear that for the PS2 algorithm, there were no runs converging to 'not acceptable' solutions, whereas all other algorithms showed between 9% to 50% of runs converging to 'non acceptable' solutions. The results also show that the PS2 algorithm gives about 92% to 97% of all solutions within the 'excellent' and 'good' categories, which is in contrast to the other algorithms giving only 20% to 70% solutions within these two categories.

4 Conclusions

A new registration technique based on particle swarm optimization for range images has been presented. Experiments have shown that the proposed technique outperforms three other registration methods recently described in the literature.

The improvement of search efficiency was achieved through the application of a search intensification procedure, the introduction of the differential evolution operator and the von-Neumann particle neighborhood topology. The use of the median squared error objective function helped to eliminate an uneven distribution of the alignment error across the registration region, and thus to improve the quality of registration.

It would be too soon to conclude whether the proposed technique is generally better than the other tested techniques, due to the fact that the author's implementations of these techniques may not be optimized, and the presented experiments were limited to two objects of similar complexity. However, it can be said that the proposed method gives very promising results and represents a strong competition to the existing range data registration methods.

Acknowledgements

The authors thank Kim Ng and Kemal Ajay from the Department of Electrical and Computer Systems Engineering, Monash University, Australia for promptly providing us with data from Metricor, a commercial 3D shape measurement system developed by the Monash University team including K. C. Ng, K. Ajay, N. Smith and M. Lech.

References

[1]   P. J. Besl and H. D. McKay, "A Method for Registration of 3-D Shapes," IEEE Transactions on Pattern Analysis and Machine Intelligence, vol. 14, pp. 239 -256, 1992.

[2]   Y. Chen and G. Medioni, "Object Modelling by Registration of Multiple Range Images," Image and Vision Computing, vol. 10, pp. 145-155, 1992.

[3]   S. Rusinkiewicz and M. Levoy, "Efficient Variants of the ICP Algorithm," Proceedings of Third International Conference on 3-D Digital Imaging and Modeling, pp. 145-152, 2001.

[4]   M. Rodrigues, "Special Issue on Registration and Fusion of Range Images," Computer Vision and Image Understanding, vol. 87, pp. 1-7, 2002.

[5]   G. Blais and M. D. Levine, "Registering Multiview Range Data to Create 3D Computer Objects," IEEE Transactions on Pattern Analysis and Machine Intelligence, vol. 17, pp. 820 -824, 1995.

[6]   C. K. Chow, H. T. Tsui, and T. Lee, "Surface Registration Using a Dynamic Genetic Algorithm," Pattern Recognition, vol. 37, pp. 105 – 117, 2004.

[7]   C. Robertson and R. B. Fisher, "Parallel Evolutionary Registration of Range Data," Computer Vision and Image Understanding, vol. 87, pp. 39-50, 2002.

[8]   M. P. Wachowiak, R. Smolikova, Y. Zheng and J. M. Zurada, "An Approach to Multimodal Biomedical Image Registration Utilizing Particle Swarm Optimization", IEEE Transactions on Evolutionary Computation, vol. 8, no. 3, June 2004.

[9]   J. Kennedy and R. Eberhart, "Particle Swarm Optimization," Proceedings of IEEE International Conference on Neural Networks, vol. 4, pp. 1942 -1948, 1995.

[10] Y. Shi and R. Eberhart, "A Modified Particle Swarm Optimizer," Evolutionary Computation Proceedings of IEEE World Congress on Computational Intelligence, pp. 69-73, 1998.

[11] M. Clerc and J. Kennedy, "The Particle Swarm - Explosion, Stability, and Convergence in a Multidimensional Complex Space," IEEE Transactions on Evolutionary Computation, vol. 6, pp. 58-73, 2002.

[12] R. Chelouah and P. Siarry, "A Continuous Genetic Algorithm Designed for the Global Optimization of Multimodal Functions," Journal of Heuristics, vol. 6, pp. 191-213, 2000.

[13] R. Chelouah and P. Siarry, "Tabu Search Applied to Global Optimization," European Journal of Operational Research, vol. 123, pp. 256-270, 2000.

[14] J. Kennedy and R. Mendes, "Neighborhood Topologies in Fully-Informed and Best-of-Neighborhood Particle Swarms," Proceedings of the 2003 IEEE International Workshop on Soft Computing in Industrial Applications, pp. 45-50, 2003.

[15] J. Kennedy, "Bare Bones Particle Swarms," Proceedings of the 2003 IEEE Swarm Intelligence Symposium, pp. 80-87, 2003.

[16] J. Kennedy, "Small Worlds and Mega-minds: Effects of Neighborhood Topology on Particle Swarm Performance," Proceedings of the 1999 Congress on Evolutionary Computation, vol. 3, pp. 1931-1938, 1999.

[17] J. Kennedy and R. Mendes, "Population Structure and Particle Swarm Performance," Proceedings of the 2002 Congress on Evolutionary Computation, vol. 2, pp. 1671-1676, 2002.

[18] W.-J. Zhang and X.-F. Xie, "DEPSO: Hybrid Particle Swarm with Differential Evolution Operator," Proceedings of IEEE International Conference on Systems, Man and Cybernetics, pp. 3816-3821, 2003.

[19] M. Greenspan, G. Godin, and J. Talbot, "Acceleration of Binning Nearest Neighbor Methods," Proceedings of Vision Interface 2000, Montreal, Quebec, 2000.

[20] T. Masuda, K. Sakaue, and N. Yokoya, "Registration and Integration of Multiple Range Images for 3-D Model Construction," Proceedings of the 13th International Conference on Pattern Recognition, vol. 1, pp. 879-883, 1996.

[21] J. Luck, C. Little, and W. Hoff, "Registration of Range Data Using a Hybrid Simulated Annealing and Iterative Closest Point Algorithm," Proceedings of IEEE International Conference on Robotics and Automation, vol. 4, pp. 3739-3744, 2000.

[22] W.H. Press, S.A. Teukolsky and B.P. Flannery, Numerical Recipes in C, Second Edition, Cambridge University Press.

[23] L. Ingber, "Very Fast Simulated Reannealing (VFSR)", Mathematical and Computer Modelling, vol. 12, no. 8, pp. 967-973, 1989.

Zero-Clairvoyant Scheduling with Inter-period Constraints

K. Subramani

LCSEE,
West Virginia University,
Morgantown, WV
ksmani@csee.wvu.edu

Abstract. This paper introduces two new mathematical modeling paradigms called Periodic Linear Programming and Periodic Quantified Linear Programming. The former is an extension of traditional Linear Programming, whereas the latter extends Quantified Linear Programming. We use these tools to capture the specifications of real-time embedded systems, which are characterized by uncertainty, complex timing constraints and periodicity. The strength of the modeling techniques lies in the ease with which these specifications can be represented and analyzed.

1 Introduction

Real-time scheduling problems are characterized by the presence of complex timing constraints between tasks and the existence of execution time variability. It is important to note that constraints such as relative timing constraints cannot be represented by precedence graphs (which are necessarily acyclic.) A traditional scheduling model, such as the one described in [5] fails to account for either of the above characteristics. A third issue in real-time scheduling problems is the issue of clairvoyance, which specifies when the execution time of a job is known and can be used in the computation of schedules. Towards this end, the E-T-C (Execution Time Constraints) scheduling framework was proposed in [7]. This model addresses the issues of specifying constraints, execution time variability and clairvoyance in real-time scheduling problems in a very broad and flexible manner. The E-T-C scheduling model is built on extensions of Linear Programming and Quantified Linear Programming [8]; these mathematical programming models are single-shot in nature. Consequently, the model itself is suitable only for expressing constraint relationships between jobs in the same period (*intra-period*). However, applications abound, wherein relationships exist between successive invocations of job-sets (*inter-period*).

Consider the operation of a typical traffic control system, such as the one discussed in [1] (See Figure 1).

Each intersection in the traffic system possesses a local controller which carries out local control functions in addition to communicating with a central controller.

M.J. Maher (Ed.): ASIAN 2004, LNCS 3321, pp. 236–247, 2004.
© Springer-Verlag Berlin Heidelberg 2004

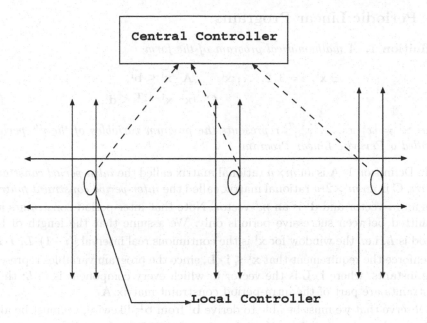

Fig. 1. A Traffic Controller

In any scheduling window the local controller displays red, green and yellow lights in strict sequential order. The repetition of the sequential order in each period represents a *cyclic schedule*. The duration for which a particular light is displayed depends upon the traffic pattern, as monitored by detectors at the intersection. For instance, the green light for a particular direction stays on longer when the traffic is heavy than when it is light. The other lights are controlled in similar fashion. The detectors issue signals when they recognize the presence of an automobile. These signals are used by the local controller to modify the parameters of the cyclic schedule. Some of the information collected from the sensors is passed on to the central controller, which then uses it for improving the traffic flows taking into account the current conditions.

An inter-period constraint would place restrictions between the signal lengths in the same period, whereas an intra-period constraint would place timing restrictions between signal lengths of adjacent periods.

In such systems, the positioning of jobs within a scheduling window is affected by the positioning of jobs in the previous scheduling window, and in turn affects the positioning of jobs in the succeeding scheduling window, thereby necessitating the use of *inter-period constraints*. Observe that even the task of constraint specification, in such systems is a non-trivial task and requires the generalization of familiar mathematical programming models. Accordingly, we focus our efforts on developing a model that permits the specification of such inter-period constraints and introduce the Periodic Linear Programming structure, which is a generalization of Linear Programming. On account of space constraints, we omit additional motivating domains as well as references to related work in the literature.

2   Periodic Linear Programs

Definition 1. *A mathematical program of the form*

$$\exists\, \vec{x^i},\ i = 1, 2, \ldots, \infty \qquad \mathbf{A} \cdot \vec{x^i} \le \vec{b^i},$$
$$\mathbf{C} \cdot [\vec{x^i}\ \vec{x^{i+1}}]^{\mathbf{T}} \le \vec{d^i}, \tag{1}$$

where $\vec{x^i} = [x_1^i, x_2^i, \ldots, x_n^i]^T$ *represents the program variables of the ith period, is called a Periodic Linear Program.*

In Definition 1, $\mathbf{A}$ is an $m \times n$ rational matrix called the *intra-period constraint matrix*, $\mathbf{C}$ is an $m' \times 2 \cdot n$ rational matrix called the *inter-period constraint matrix*, $\vec{b^i}$ is an m-vector and $\vec{d^i}$ is an m'-vector. Note that inter-period constraints are permitted between successive periods only. We assume that the length of the period is L, i.e., the window for $\vec{x^i}$ is the continuous real interval $[(i-1) \cdot L,\ i \cdot L]$. We enforce the requirement that $\vec{x^i} \le \mathbf{i} \cdot \vec{L}$, since the program variables represent time instants, where $\mathbf{i} \cdot \vec{L}$ is the vector in which every component is $i \cdot L$; these constraints are part of the intra-period constraint matrix $\mathbf{A}$.

Observe that we must be able to derive $\vec{b^i}$ from $\vec{b^1}$; likewise, we must be able to derive the vectors $\vec{d^i}$ fom $\vec{d^1}$. Otherwise, the PLP does not have a compact description and the size of the problem is undefined. We only consider PLPs in which $b_j^i = b_j^1 + u_j \cdot (i-1) \cdot L$, $u_j \in \mathcal{Z}$ and $d_j^i = d_j^1 + v_j \cdot (i-1) \cdot L$, $v_j \in \mathcal{Z}$. Further, for each constraint l_i in $\mathbf{A} \cdot \vec{x^i} \le \vec{b^i}$, the u_i value is fixed and part of the input; the same holds for each constraint l_j of the system $\mathbf{C} \cdot [\vec{x^i}\ \vec{x^{i+1}}]^{\mathbf{T}} \le \vec{d^i}$ and the variable v_j. Thus, we see that a PLP has a compact description, in that it is completely described by $\{\mathbf{A}, \mathbf{b^1}, \mathbf{C}, \mathbf{d^1}\}$ and the variable set $M = \{u_1, u_2, \ldots, u_m, v_1, v_2, \ldots, v_{m'}\}$. Each element of M is called a constraint interpreter or a constraint multiplier, for reasons that will become apparent later. In other words, we have a finite and *implicit* description of a (possibly) infinite state system.

For instance, Systems (2) and (3) together represent a PLP.

$$(\mathbf{A}) \begin{bmatrix} 1 & -1 \\ -1 & 1 \\ 1 & 0 \end{bmatrix} \cdot \begin{bmatrix} x_1^i \\ x_2^i \end{bmatrix} \le \begin{bmatrix} -2 \\ 5 \\ 15 + (i-1).15 \end{bmatrix} (\mathbf{\vec{b^i}}) \tag{2}$$

$$(\mathbf{C}) \begin{bmatrix} 1 & 0 & -1 & 0 \end{bmatrix} \cdot \begin{bmatrix} 5 \end{bmatrix} (\mathbf{\vec{d^i}}) \tag{3}$$

In the first two constraints of the intra-period constraint matrix, the constraint multiplier is 0, whereas in the third constraint, the constraint multiplier is 1.

Definition 2. *An Order k restriction of a PLP as described in System (1), is the simple Linear Program that results by restricting the constraint system to k successive periods, starting from the first period.*

Definition 3. *An Order k solution of a PLP is the Linear Programming solution of the Order k restriction.*

Note that an Order k restriction has $k \times n$ variables in all.

Definition 4. *The solution of a PLP is the solution to the Order ∞ restriction; a PLP is said to be feasible, if it has an Order ∞ solution.*

The PLP-decidability problem is as follows: *Given a PLP specification, is it feasible?*

Observe that the above decision problem may not have a succinct solution, in that the Order ∞ restriction may be feasible, but require the specification of infinitely many different program variables for the description of the solution. We now focus on a class of solutions that can be succinctly described.

Definition 5. *The solution of the Order ∞ restriction of a PLP is said to be cyclic, with period k, if the solution vector for the jth period can be computed using Algorithm 2.1.*

Algorithm 2.1: Cyclic Solutions to a PLP

Function Cyclic-Solution $(\mathbf{A}, \vec{\mathbf{b}^i}, \mathbf{C}, \vec{\mathbf{d}^i}, j)$
1: **if** $(j \leq k)$ **then**
2:    Form the Order j restriction of the PLP.
3:    Solve the restriction as a Linear Program over $n \times j$ variables.
4:    Return $\vec{\mathbf{x}^j}$.
5: **else**
6:    Let $\vec{z} = $ Cyclic-Solution $(\mathbf{A}, \vec{\mathbf{b}^i}, \mathbf{C}, \vec{\mathbf{d}^i}, j - k)$.
7:    Set $\vec{\mathbf{x}^j} = \vec{z} + k \cdot \vec{L}$.
8: **end if**

In order to reduce the cumbersomeness of the notation, we shall use $\vec{a} + c_1$ to mean that every component of the vector $\vec{a}$ is increased by c_1.

2.1   Restrictions to PLPs

As defined above, PLPs are extremely general structures and at this point, it is not clear they are even decidable, much less in polynomial time. We now place the following restrictions on the PLPs that we shall be analyzing:

(i)   The vectors $\vec{\mathbf{b}^i}$ and $\vec{\mathbf{d}^i}$ are integral for all $i = 1, 2, \ldots,$.
(ii) Every constraint in $\mathbf{A}$ is either a (strict) difference constraint or an absolute constraint. Note that a constraint of the form $x_i - x_j \leq c_1$ is called a difference constraint [3]. A constraint of the form $x_i \leq c_1$ or $x_i \geq c_2$ is called an absolute constraint.
(iii) The constraints in $\mathbf{C}$ are difference constraints only.
(iv) The constraint multiplier for a difference constraint is 0, whereas the constraint multiplier for a constraint of the form $x_i \leq c_1$ is $+1$ and the constraint multiplier for a constraint of the form $x_i \geq c_2$ is -1.

Restricting the PLP structure as discussed above, results in the following properties.

(a) The vector $\vec{\mathbf{d^i}}$ stays the same for all periods i, i.e., we can set $\vec{\mathbf{d^i}} = \vec{\mathbf{d}}$.

(b) If the jth constraint of the intra-period constraint matrix is a difference constraint, then b_j^i is the same for all periods i. If the jth constraint of the intra-period constraint matrix is an absolute constraint, of the form $x_k \leq ()$, then $b_j^i = b_j^1 + (i-1) \cdot L$ and if it is a constraint of the form $x_k \geq ()$, then $b_j^i = b_j^1 - (i-1) \cdot L$.

(c) The intra-period constraint system for the ith period can be derived from the intra-period constraint system for the first period by shifting the origin to the point $((i-1) \cdot L, (i-1) \cdot L, \ldots, (i-1) \cdot L)$. Hence, the Order 1 restriction PLP, viz., $U_1 : \mathbf{A} \cdot \vec{\mathbf{x^1}} \leq \vec{\mathbf{b^1}}$ is feasible if and only if the intra-period constraint system for the ith period, viz., $U_i : \mathbf{A} \cdot \vec{\mathbf{x^i}} \leq \vec{\mathbf{b^i}}$ is feasible, for $i = 2, 3, \ldots, \infty$. Further, if $\vec{\mathbf{z_1}}$ is a solution to U_1, then $\vec{\mathbf{z_i}} = \vec{\mathbf{z_1}} + (i-1) \cdot L$ is a solution to U_i, and vice versa. This property is termed as *Solution Preservation through Shift of Origin* (SPSO).

Accordingly, we can refer to an arbitrary PLP as:

$$\exists \vec{\mathbf{x^i}}, \ i = 1, 2, \ldots, \infty \quad \mathbf{A} \cdot \vec{\mathbf{x^i}} \leq \vec{\mathbf{b^i}},$$
$$\mathbf{C} \cdot [\mathbf{x^i} \ \mathbf{x^{i+1}}]^{\mathbf{T}} \leq \vec{\mathbf{d}} \tag{4}$$

From this point onwards, when we refer to a PLP, we mean a Periodic Linear Program with the above restrictions, i.e., System (4).

3   Periodic Quantified Linear Programs

Linear Programming models do not permit the specification of uncertainty, even in single-shot situations. One technique to permit the specification of uncertainty in linear programs is through the use of Quantified Linear Programs.

Definition 6. *A linear program in which some of the program variables are universally quantified, over a specified domain is called a Quantified Linear Program (QLP).*

Typically, x variables are existentially quantified and y variables are universally quantified. It is also customary to separate the coefficients of the existentially quantified and universally quantified variables. Accordingly, we write a QLP as:

$$\mathbf{Q}(\vec{\mathbf{x}}, \ \vec{\mathbf{y}}) \ \mathbf{G} \cdot \vec{\mathbf{x}} + \mathbf{H} \cdot \vec{\mathbf{y}} \leq \vec{\mathbf{b}} \tag{5}$$

where $\mathbf{Q}(\vec{\mathbf{x}}, \ \vec{\mathbf{y}})$ represents the quantifier specification of the QLP. As is the case, with Linear Programming models, Quantified Linear Programming models are restricted in that they can only capture single-shot situations. A QLP is completely described by its quantifier string, which specifies the ordering on the variables, and the constraint system. A detailed introduction to Quantified Linear Programming and methodologies to decide QLPs is available in [8].

Definition 7. *A Periodic Quantified Linear Program (PQLP) is a Periodic Linear Program in which some of the program variables are universally quantified, over a specified domain.*

Just as PLPs are implicit descriptions of an infinite Linear Program, we intend that a PQLP is an implicit description of an infinite QLP. There is one issue though that must be addressed, viz., the description of the quantifier string of a PQLP. Definition 7 does not explicitly account for the length of the quantifier specification. However, if a PQLP is to be succinctly described, there must be compact description for its infinite quantifier string, i.e., the description must incorporate some form of periodicity. Observe that a PLP could be specified as in System (4) and the interpretation of the constraint system is straightforward, since all the quantifiers are existential. In the case of a PQLP though, the presence of both existential and quantifiers in the quantifier specification, requires a careful explanation of the semantics involved. We shall postpone the discussion of this issue till Section 5, where a PQLP is used to model a real-world problem and the quantifier string will be interpreted in a manner which is consistent with the demands of the problem. For the present, a PQLP is denoted by:

$$\mathbf{Q_i}(\vec{\mathbf{x}^i}, \vec{\mathbf{y}^i})$$
$$\mathbf{G} \cdot \vec{\mathbf{x}^i} + \mathbf{H} \cdot \vec{\mathbf{y}^i} \le \vec{\mathbf{b}^i}$$
$$\mathbf{I} \cdot \vec{\mathbf{x}^i} + \mathbf{J} \cdot \vec{\mathbf{y}^i} \le \vec{\mathbf{d}}$$
$$i = 1, 2, \ldots, \infty \tag{6}$$

where,

(i) The constraint matrices $\mathbf{G}$ and $\mathbf{I}$ are restricted in exactly the same way as the constraint matrices of a PLP,

(ii) The application of the quantifier specification $\mathbf{Q_i}(\vec{\mathbf{x}^i}, \vec{\mathbf{y}^i})$ will be decided by the problem being modeled.

4   PLP Decidability

In this section, we show that the class of PLPs, that have been restricted as discussed in Section 2 can be decided in polynomial time. Before we proceed with our analysis, we introduce the concept of *projection in polyhedral spaces*, discussed in [2, 6]. Given a polyhedral system in Euclidean space $\Re^d$, we can project this system onto a lower dimensional space $\Re^{d'}$, $d' < d$, while preserving the set of solutions to the original system [6]. One of the more common techniques for achieving this projection is the Fourier-Motzkin (FM) elimination method, which is based on variable elimination.

Let us focus on the PLP described by System (7).

$$\mathbf{A} \cdot \vec{\mathbf{x}^i} \quad \le \quad \vec{\mathbf{b}^i},$$
$$\mathbf{C} \cdot [\vec{\mathbf{x}^i} \ \vec{\mathbf{x}^{i+1}}]^\mathbf{T} \quad \le \quad \vec{\mathbf{d}}$$
$$i = 1, 2, \ldots, \infty \tag{7}$$

Observe that System (7) is in fact a progression of linear programs, with additional constraints being added in every period. Let $S_1 = \mathbf{A} \cdot \vec{\mathbf{x}^1} \leq \vec{\mathbf{b}^1}$ denote the Order 1 restriction to System (7). Now consider the Order 2 restriction of the PLP, i.e.,

$$\mathbf{A} \cdot \vec{\mathbf{x}^1} \leq \vec{\mathbf{b}^1}, \quad \mathbf{A} \cdot \vec{\mathbf{x}^2} \leq \vec{\mathbf{b}^2}, \quad \mathbf{C} \cdot [\vec{\mathbf{x}^1} \ \vec{\mathbf{x}^2}]^\mathbf{T} \leq \vec{\mathbf{d}}. \tag{8}$$

When System (8) is projected onto the space, spanned by $\vec{\mathbf{x}^1}$, by eliminating the variables in $\vec{\mathbf{x}^2}$, we get a polyhedral set $S_2 = \mathbf{A}' \cdot \vec{\mathbf{x}^1} \leq \vec{\mathbf{b}'^1}$. Observe that any solution to System (7) must belong to S_2. Accordingly, we can *reformulate* System (7) as:

$$
\begin{aligned}
\mathbf{A}' \cdot \vec{\mathbf{x}^i} &\leq \vec{\mathbf{b}'^i}, \\
\mathbf{C} \cdot [\vec{\mathbf{x}^i} \ \vec{\mathbf{x}^{i+1}}]^\mathbf{T} &\leq \vec{\mathbf{d}} \\
i &= 1, 2, \ldots, \infty
\end{aligned}
\tag{9}
$$

It is clear that the original PLP (described by System (7)) is feasible if and only if the reformulated PLP (described by System (9)) is. From the properties of Fourier-Motzkin elimination, we know that projecting out a variable in a system of difference and absolute constraint with an integral right-hand side, results in a system with difference and absolute constraints with an integral right-hand side. In other words, the class of restricted PLPs is closed under FM elimination and the intra-period constraint system $\mathbf{A}' \cdot \vec{\mathbf{x}^i} \leq \vec{\mathbf{b}'^1}$ also satisfies the restrictions, discussed in Section 2. Each reformulation of the intra-period constraint matrix, results in a (possibly) different polyhedral set; let S_i denote the polyhedral system that results from projecting the Order i restriction of System (7) onto $\vec{\mathbf{x}^1}$ space.

Lemma 1. $S_1 \supseteq S_2 \supseteq S_3 \ldots$

Proof: Observe that each S_i, $i = 1, 2, \ldots, \infty$ is a polytope formed by adding constraints to the polyhedral set S_{i-1}, further restricting the feasible space. The claim follows.   □

We need to consider the following two possibilities only:

1. The sequence of sets S_i converges to some polyhedron in $\vec{\mathbf{x}^1}$ space.
2. $S_k = \phi$ for some k, i.e., the reformulation process results in an infeasible set.

Corollary 1 follows immediately from Lemma 1.

Corollary 1. *If $S_k \neq S_{k+1}$, then $S_k \supset S_{k+1}$.*

Lemma 2. *If $S_k = S_{k+1}$, for some k, then $S_j = S_k, \forall j \geq k$.*

Proof: The inter-period constraint system $\mathbf{C} \cdot [\mathbf{x^i} \; \mathbf{x^{\vec{i+1}}}]^{\mathbf{T}} \le \vec{\mathbf{d}}$ can be thought of as a *reformulation operator* ∇, which is applied to the intra-period constraint system (input set), in that at each stage, its application results in a (possibly) new intra-period constraint matrix (output set), i.e., $\nabla(S_i) = S_{i+1}$. Since $S_k = S_{k+1}$, it means that $\nabla(S_k) = S_{k+1} = S_k$. Now observe that the reformulation operator is independent of the periods involved; thus reapplying ∇ to S_k will result in the same set S_k. In other words, S_k is a fixed-point of the reformulation operator [4]. □

Lemma 3. *The reformulation operator ∇, described in Lemma 2 can be applied at most $O(n^3 \cdot L)$ times to the intra-period constraint system, at which time, either we would have reached a fixed-point or the reformulated system becomes infeasible.*

Proof: Will be provided in full paper. □

Theorem 1. *A feasible PLP must have a cyclic solution with period 1.*

Proof: Will be provided in full paper. □

Observe that the existence of a period 1 cyclic solution trivially implies that the PLP is feasible. Hence, it follows that,

Theorem 2. *A PLP is feasible if and only if it has a cyclic solution with period 1.*

The principal consequence of the above analysis is that we can set

$$\mathbf{x^{\vec{i+1}}} = \mathbf{x^{\vec{i}}} + L \tag{10}$$

in System (4), while preserving its solution space. Thus the inter-period constraints between window $[(i-1) \cdot L, i \cdot L]$ and $[i \cdot L, (i+1) \cdot L]$ can be merged using Equation (10) and System (7) can be expressed in terms of the $\mathbf{x^{\vec{i}}}$ only. This linear system is denoted by

$$\mathbf{M} \cdot \mathbf{x^{\vec{i}}} \le \vec{\mathbf{f^i}}$$

$$i = 1, 2, \ldots, \infty \tag{11}$$

The polyhedral system (11) is called the `Periodic Polytope` corresponding to the PLP (1); this system is composed of n variables and $m + m'$ constraints. We set $i = 1$ and solve the resultant system; due to the nature of the constraints involved, this system can be solved in $O((m + m') \cdot n)$ steps, using a variation of the Bellman-Ford algorithm [3].

5   Zero-Clairvoyant Scheduling with Inter-Period Constraints

In this section, we describe the modeling of a practical real-time scheduling problem using Periodic Quantified Linear programs.

5.1   Job Model

Assume an infinite time axis that is divided into intervals of length L, starting at time $t = 0$. These intervals are called *scheduling windows*, i.e., the first scheduling window is $[0, L]$, the second scheduling window is $[L, 2 \cdot L]$ and in general the ith scheduling window is $[(i - 1) \cdot L, i \cdot L]$. The scheduling windows are also referred to as periods. We are given a set of ordered non-preemptive jobs $\{J_1, J_2, \dots J_n\}$, which have instances in each scheduling window. The set $\Gamma^1 = \{J_1^1, J_2^1, \dots, J_n^1\}$ corresponds to the instance of the job set in the first scheduling window; the sets $\Gamma^i, i = 2, \dots, \infty$ are defined similarly. The instances of a job within a scheduling window are called tasks. Associated with each job J_i, is its execution time $e_i \in [l_i, u_i]$. During execution, J_i can take anywhere from l_i to u_i to complete; we denote this by $e_i \in [l_i, u_i]$. We let $\vec{\mathbf{e^i}} = [e_1^i, e_2^i, \dots, e_n^i]^T$ denote the execution times of the tasks in the ith window and $\vec{\mathbf{s^i}} = [s_1^i, s_2^i, \dots, s_n^i]^T$ denote their start times. Let $\mathbf{E}$ denote the axis-parallel hyper-rectangle (aph) $\Pi_{i=1}^n [l_i, u_i]$.

5.2   Constraint Model

The constraints on the system are described in terms of scheduling window i:

(a) Intra-period constraints - These constraints are difference constraints between the start or finish time of a task in a period, and the start or finish time of another task *in the same period*. Intra-period constraints also include absolute constraints, i.e., a constraint of the form $s_1^i \geq 7 + (i - 1) \cdot L$. Thus, the intra-period constraints can be described by System (12).

$$\mathbf{G} \cdot \vec{\mathbf{s^i}} + \mathbf{H} \cdot \vec{\mathbf{e^i}} \leq \vec{\mathbf{b^i}} \tag{12}$$

(b) Inter-period constraints - These constraints are difference constraints between the start or finish time of a task in a period, and the start or finish time of a task in the adjacent period.

Thus, the inter-period constraints can be described by System (13).

$$\mathbf{I} \cdot [\vec{\mathbf{s^i}} \ \vec{\mathbf{s^{i+1}}}]^{\mathbf{T}} + \mathbf{J} \cdot [\vec{\mathbf{e^i}} \ \mathbf{e^{i+1}}]^{\mathbf{T}} \leq \vec{\mathbf{d}} \tag{13}$$

Observe that System (12) and System (13) together constitute a PLP, as described in Section 2.1, in that the constraints are restricted to be of the form required by a PLP.

In System (12), $\mathbf{G}$ and $\mathbf{H}$ are $m \times n$ matrices, $\vec{\mathbf{b^i}}$ is an integral m-vector; likewise in System (13), $\mathbf{I}$ and $\mathbf{J}$ are $m' \cdot n$ matrices and $\vec{\mathbf{d}}$ in an integral m'-vector.

5.3   Query Model

In order to completely characterize a real-time scheduling problem, we need to specify the type of schedulability query involved.

In the presence of inter-period constraints, we desire a start-time vector which does not know the execution time of a job, *even after it has completed executing*.

Thus, we are interested in the following schedulability query:

$$\exists \vec{s^1}, \vec{s^2} \ldots \forall \vec{e^1}, \vec{e^2}, \ldots \quad [(12),(13)] \tag{14}$$

In System (14), the start-time vector $\vec{s^i}$ for the ith window does not depend on the execution time vector of any scheduling window.

System (14) can be succinctly described in the following form:

$$\exists \vec{s^i} \quad \forall \vec{e^i} \in \mathbf{E} \quad i = 1, 2, \ldots, \infty$$
$$\mathbf{G} \cdot \vec{s^i} + \mathbf{H} \cdot \vec{e^i} \leq \vec{b^i}$$
$$\mathbf{I} \cdot [\vec{s^i} \ \vec{s^{i+1}}]^{\mathbf{T}} + \mathbf{J} \cdot [\vec{e^i} \ \vec{e^{i+1}}]^{\mathbf{T}} \leq \vec{d} \tag{15}$$

Observe that System (15) is in fact, a Periodic Quantified Linear Program (PQLP). It is to be understood that each execution time vector $\vec{e^i}$ belongs to its own execution time domain $\mathbf{E}$, i.e., the execution time vectors of different scheduling windows are independent of each other. The PQLP, as specified, can be interpreted in a variety of ways; for the purposes of this paper, we attach the semantics of System (14) to System (15). This issue was first raised in Section 3 and has now been adequately addressed.

Definition 8. *The Zero-Clairvoyant scheduling problem, with inter-period constraints (ZCIPC), is concerned with deciding whether System (15) is true.*

System (15) is also referred to as the schedulability query or the schedulability predicate.

6   The Scheduling Algorithm

In this section, we shall focus on developing a polynomial time algorithm for the ZCIPC problem.

As was the case with PLPs, we define the Order k restriction of the PQLP specified by System (15).

Definition 9. *An Order k restriction of a PQLP as described in System (1), is the simple Quantified Linear Program that results by restricting the constraint system to k successive periods, starting from the first period.*

Accordingly, the Order 1 restriction of the PQLP is

$$\exists \vec{s^1} \forall \vec{e^1} \in \mathbf{E}$$
$$\mathbf{G} \cdot \vec{s^1} + \mathbf{H} \cdot \mathbf{e^1} \leq \vec{b^1}$$

i.e., only the intra-period constraints of the first window are in the constraint set.

Likewise, the Order k restriction is given by:

$$\exists \vec{s^1}, \vec{s^2} \ldots \vec{s^k} \; \forall \vec{e^1} \; \vec{e^2}, \ldots, \vec{e^k} \in \mathbf{E}$$
$$\mathbf{G} \cdot \vec{s^i} + \mathbf{H} \cdot \vec{e^i} \leq \vec{b^i}$$
$$\mathbf{I} \cdot [\vec{s^i} \; \vec{s^{i+1}}]^{\mathbf{T}} + \mathbf{J} \cdot [\vec{e^i} \; \vec{e^{i+1}}]^{\mathbf{T}} \leq \vec{d}$$
$$i = 1, 2, \ldots, k-1$$
$$\mathbf{G} \cdot \vec{s^k} + \mathbf{H} \cdot \vec{e^k} \leq \vec{b^k}$$

If P_s^k is used to define the Order k restriction of the PQLP, then the ZCIPC problem can be thought of as deciding the truth value of P_{stat}^∞.

6.1   Constraint Analysis

Note that for each k, P_s^k is a fixed-dimensional Quantified Linear Program and hence can be decided using the quantifier elimination procedure discussed in [8]. The intra-period constraint set, $\mathbf{I} \cdot [\vec{s^i} \; \vec{s^{i+1}}]^{\mathbf{T}} + \mathbf{J} \cdot [\vec{e^i} \; \vec{e^{i+1}}]^{\mathbf{T}} \leq \vec{d}$ which serves as the reformulation operator ∇ is independent of the period involved. Once again, the reformulation operator has to be applied at most $O(n^3 \cdot L)$ times, on account of the restricted structure of the constraint set. It is not hard to see that the entire discussion in Section 4, which demonstrated the existence of fixed-point solutions for PLPs can be applied to the case of the specialized PQLPs , describing the Zero-Clairvoyant Scheduling problem with Inter-Period constraints. The existence of a fixed-point solution implies the existence of a cyclic solution with period 1. We therefore have,

Theorem 3. *System (15) is feasible, if and only if it has a cyclic solution with period 1.*

Proof: Follows from the discussions in Section 4 and this section.   □

Theorem 3 permits us to set $\vec{s^{i+1}} = \vec{s^i} + L$ in the schedulability specification, System (15). By combining all the constraints, we get the following PQLP:

$$\exists \vec{s^i} \; \forall \vec{e^i} \in \mathbf{E} \quad i = 1, 2, \ldots, \infty$$
$$\mathbf{M} \cdot \vec{s^i} + \mathbf{N} \cdot [\vec{e^i} \; \vec{e^{i+1}}]^{\mathbf{T}} \leq \vec{f^i} \qquad (16)$$

Setting $i = 1$, we get the system

$$\exists \vec{s^1} \; \forall \vec{e^1} \; \vec{e^2} \in \mathbf{E}$$
$$\mathbf{M} \cdot \vec{s^1} + \mathbf{N} \cdot [\vec{e^1} \; \vec{e^2}]^{\mathbf{T}} \leq \vec{f^1} \qquad (17)$$

Now, observe that System (17) is a standard QLP. We can therefore apply the quantifier elimination procedure of [8] to obtain a solution, say $\vec{z^1}$. Then the solution vector for the ith scheduling window can be computed as: $\vec{z^i} = \vec{z^1} + (i-1) \cdot L$.

References

1. P. Ancilloti, G. Buttazo, M. Di Natale, and A.K. Mok. Tracs: A flexible real-time environment for traffic control systems. In *Proceedings IEEE Workshop on Real-Time Applications*, pages 50–53, May 1993.
2. V. Chandru and M.R. Rao. Linear programming. In *Algorithms and Theory of Computation Handbook, CRC Press, 1999*. CRC Press, Boca Raton, Florida, 1999.
3. T. H. Cormen, C. E. Leiserson, and R. L. Rivest. *Introduction to Algorithms*. MIT Press and McGraw-Hill Book Company, Boston, Massachusetts, 2nd edition, 1992.
4. Vasile I. Istratescu. *Introduction to Linear Operator theory*. Marcel Dekker Inc., New York, 1981.
5. M. Pinedo. *Scheduling: theory, algorithms, and systems*. Prentice-Hall, Englewood Cliffs, 1995.
6. Alexander Schrijver. *Theory of Linear and Integer Programming*. John Wiley and Sons, New York, 1987.
7. K. Subramani. A specification framework for real-time scheduling. In W.I. Grosky and F. Plasil, editors, *Proceedings of the 29th Annual Conference on Current Trends in Theory and Practice of Informatics (SOFSEM)*, volume 2540 of *Lecture Notes in Computer Science*, pages 195–207. Springer-Verlag, November 2002.
8. K. Subramani. An analysis of quantified linear programs. In C.S. Calude, et. al., editor, *Proceedings of the 4th International Conference on Discrete Mathematics and Theoretical Computer Science (DMTCS)*, volume 2731 of *Lecture Notes in Computer Science*, pages 265–277. Springer-Verlag, July 2003.

A Novel Texture Synthesis Based Algorithm for Object Removal in Photographs

Feng Tang, Yiting Ying, Jin Wang, and Qunsheng Peng

State Key Lab of CAD & CG, Zhejiang University,
310027 Hangzhou, P. R. China
{tang, ytying, jwang, peng}@cad.zju.edu.cn
http://www.cad.zju.edu.cn

Abstract. Natural images and photographs sometimes may contain stains or undesired objects covering significant portions of the images. Inpainting is a method to fill in such portions using the information from the remaining area of the image. In this paper, we propose a novel photograph editing framework that utilizes texture synthesis techniques. Major contributions of our algorithm are: 1) a constraint-based candidate patch searching method which limits the searching within neighboring region with similar texture; 2) a metric of Coherence Confidence for selecting the best fit candidate preventing error accumulation and propagation; 3) integration of graphcut optimization to make the seam visually invisible. Experiments show that our system can efficiently handle different cases especially large regions in complex background.

1   Introduction

In practice, we often meet with such problems as old photos spoiled by ink or old paintings full of scratches after long time reservation. Besides, photos may contain undesired large objects to be removed, for example, a passing-by person may drop into the view when a photo is taking. Although these two types of problems originate from different scenarios, they are commonly known as inpainting. How to robustly and efficiently solve both problems remains a challenge.

Several works have addressed these problems. For the first type of problems, which contains scratches or small missing regions, there are diffusion based and filter based methods. Bertalmio [1]-[2] filled in the tiny spoiled regions by propagating information from the outside of the masked region along level lines (isophotes). Reference [3] proposes Total Variational (TV) inpainting algorithm that invokes Euler-Lagrange equation and applies anisotropic diffusion [5] inside the inpainting domain based on the contrast of the isophotes. Oliveira [6] filtered the input image using a Gaussian convolution kernel to remove the undesired flaws. Yet such algorithms account for only small local regions. Artifacts of blurs may occur when they are applied to large region of inpainting.

For the second type of problems, a kind of synthesis based methods has been proposed. Liang et al. [7] composed the texture of a large region by selecting

M. J. Maher (Ed.): ASIAN 2004, LNCS 3321, pp. 248–258, 2004.
© Springer-Verlag Berlin Heidelberg 2004

rectangle candidate patches recursively from the input texture. Efros and Freeman [8], Vivek Kwatra et al [9] searched for irregularly-shaped patches with optimizations, using dynamic programming and graph-cut respectively. Criminisi [10] employed texture synthesis to remove large objects in photographs and obtained impressive results. Such synthesis based methods consist of two major steps: searching and pasting. Searching is the process of finding best matched patch in the source region based on texture similarity. Pasting is the process of attaching the selected patch to the target region at the desired position. Nevertheless, most current synthesis based algorithms suffer from two major limitations. 1) Searching the source patch globally will easily lead to error match, and such error will accumulate and propagate to other areas quickly, even make the results completely unacceptable. 2) The global search is computationally expensive, which limits the overall fill-in speed. In this paper we propose a photograph editing framework which can robustly remove various kinds of undesired details from photographs. By adopting local constraints for searching and estimating the coherence confidence for the candidate patch, our synthesis-based object removal algorithm achieves better results and performs more robustly and faster than previous methods.

The rest of the paper is organized as follows: section 2 explains our synthesis method for large object removal. Section 3 shows the results of the proposed method and section 4 draws the conclusion.

2   Texture Synthesis Based Image Editing for Object Removal

2.1   Example Based Object Removal

In [10], Criminisi et al proposed an exemplar based inpainting method, which fills in the target region with patches from the source region possessing similar texture. The candidate patches are selected from the whole image with special priority to those along the isophotes (lines of equal gray value) so as to preserve the linear structure during the filling-in. This process is quite similar to patch matching in texture synthesis and the fill-in priority is inspired by the partial differential equations method of physical heat flow.

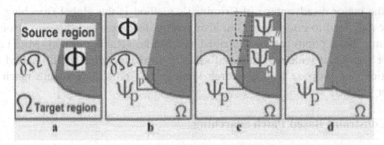

Fig. 1. Example based object removal procedure

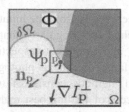

Fig. 2. Priority computation

As shown in Fig.1, suppose the target region is denoted as Ω and its contour as $\delta\Omega$, Φ is the source region. p is a point on the contour. Let ψ_p denote the current patch to fill. Although there exist many matched candidates in Φ indicated as ψ_q', ψ_q'' (Fig.1 (c)), the selection is made based on the priority of the candidate patches. The priority of each contour point is calculated as follows:

$$P(p) = C(p) * D(p) \tag{1}$$

where $C(p)$ is the confidence term that indicates the reliability of the current patch, and $D(p)$ is the data term that gives special priority to isophotes as demonstrated in Fig. 2. $\nabla I_{\bar{p}}^{\perp}$ is the isophote and n_p is the normal at point p. α is the normalization factor.

$$C(p) = \frac{\sum_{q \in \Psi_p \cap \Omega} C(q)}{|\Psi_p|} \tag{2}$$

$$D(p) = \frac{|\nabla I_{\bar{p}}^{\perp} \cdot n_p|}{\alpha} \tag{3}$$

2.2   Our Algorithm

To achieve robust and fast filling results, we propose a novel texture synthesis method called coherence-based local searching (CBLS) for region filling. As we can see from above, the major disadvantage of Criminisi's method is the global searching, which not only may lead to error match but also greatly decreases the system performance. We assume that a photo can be described as a Markov Random Field (MRF) and the neighbor regions provide sufficient information to decide what to fill. Based on this assumption, we propose a local search strategy to find the candidate patch in the neighbor regions.

2.3   Constraint Based Patch Searching

Fig. 3 shows an example of removing some undesired figures from a photo by employing Criminisi's approach. The red region is the target to fill in, the current patch under processing is indicated by a yellow rectangle and the selected candidate is shown in green, both are scaled for clear view in Fig. 3 (b) and 3 (c). Unfortunately, this is an error match. After the error patch is copied to the target region, the consequent patch matching will use this error information to search for the next candidate in the global domain. As shown in Fig. 3 (d), 3 (e) and 3 (f), 3 (g), the initial error will accumulate and propagate to a large region.

To make a correct selection from the candidate pool, we introduce a reference image called segment map which corresponds to a coarse segmentation of the original image. As an initialization step, the whole image is segmented into several separate regions according to the texture similarity. We adopt Meanshift operator here to achieve this segmentation, for technical details please refer to [11]. Suppose the whole image is denoted by I, it is divided into N regions with each region as T_i, then we have:

Fig. 3. Error propagation. Red region is the target to fill in. the yellow rectangles and green rectangles in (b), (c), (d), (e), (f), (g) indicate respectively the current patch and the selected candidate patch under processing and patches at successive steps

$$I = \bigcup_{i=1}^{N} T_i \quad \text{and} \quad T_i \cap T_j = \varnothing \quad (1 \le i,j \le N). \tag{4}$$

During the synthesis, the patch matching is restricted within the relevant regions which the current patch overlaps with (an example is shown in Fig. 4):

1) *If the current patch is completely within one segment T_i, searching is limited within T_i, this eliminates the type of error in Fig. 3.*

2) *If the current patch intersects k segments $T_i \sim T_{i+k}$, searching is limited in the combined region of all the k segments $T_i \cup T_{i+1} \cup ... \cup T_{i+k}$.*

Although a local search is performed, we may still find many candidate patches. We adopt the concept of coherence confidence to select the best fit from candidate pool. The key idea of our approach is that the texture should be spacially coherent, we then choose the one from the candidates such that it has the greatest coherence confidence (measured in formula (7)) with the neighbor patches already filled.

Fig. 4. Original image and corresponding segment map. To remove the elephant, our algorithm only searches in the neighborhood regions 2, 4, 5 and 6

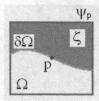

Fig. 5. Structure of a partially synthesized patch. Ψ_p is the patch under processing centered at point p. ζ is the already synthesized region and Ω is the unsynthesized region

2.4   Coherence Confidence

Our definition of the coherence confidence is based on the nearest pixels whose positions on the source region (where it comes from) are known, using neighbor synthesized pixels and make them vote. This is quite similar to that of k-nearest neighbor classifier. As indicated in Fig. 5, suppose the current patch is centered at p and it overlaps with already filled region. We find k candidate patches $\Psi_{p_1}, \Psi_{p_2}, ..., \Psi_{p_k}$. Let point q_i be a point in ζ and it comes from point c_{q_i}, p_i be a point in $\Psi_p\text{-}\zeta$ and it is a corresponding point in current patch Ψ_{p_i} at c_{p_i}. We define the mean distance $MD\zeta$ of region ζ as follows:

$$MD_\zeta = \sum_{i=0}^{N} |\ \mathbf{q}_i\ \mathbf{c}_q| / \ N \tag{5}$$

where N is the number of points in ζ, and $|*|$ is the Euclidian distance.

The mean distance of region $\Psi_p\text{-}\zeta$ can be defined as:

$$MD_{\Psi_p} - \zeta = \sum_{i=0}^{M} \| \mathbf{p}_i, \mathbf{c}_{p_i} \| / M \qquad (6)$$

where M is the number of points in Ψ_p-ζ.

The *coherence confidence* $CC_{\Psi_{P_i}}$ of patch Ψ_{pi} is defined as:

$$CC_{\Psi_{P_i}} = \frac{1}{| MD_\zeta - MD_{\Psi_p} - \zeta |} \qquad (7)$$

From the candidate pool, we select the patch $\Psi_{\hat{q}}$ with the highest *coherence confidence*.

$$\Psi_{\hat{q}} = \max_{\Psi_q} \arg CC_{\Psi_q} \qquad (8)$$

Using the above criteria, we can determine the best match in the candidate pool.

2.5 Graphcut Optimization

Direct pasting of selected candidate patch to the current patch may still lead to great visual discontinuity. We implement a "Graphcut" optimization to find the best cut in the overlap region, such that the newly added patch can seamlessly join the already synthesized regions. Details of "Graphcut" can be found at [12]. The algorithm is briefly described as follows:

    1.   Construct a flow graph as Fig. 6

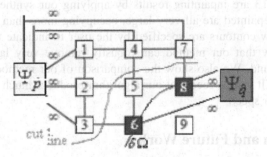

Fig. 6. Flow graph

    The patch is deemed as a start knot in the graph and the candidate patch as the end knot. The flows of points $p \in \partial\Psi_{\hat{p}} \cap \Omega$ on the boundary of the current patch are set to be infinite to the start knot; the already synthesized pixels $p \in \partial\Psi_{\hat{p}} \cap \Phi$ are connected to the end knot and the corresponding flows are set to be infinite. The points in the current patch $\Psi_{\hat{p}} \cap \Phi$ are then connected to its corresponding patch, and the flows are calculated as follows:

$$M(s,t,\Psi_{\hat{p}},\Psi_{\hat{q}}) = \left\| \Psi_{\hat{p}}(s) - \Psi_{\hat{q}}(s) \right\| + \left\| \Psi_{\hat{p}}(t) - \Psi_{\hat{q}}(t) \right\| \tag{9}$$

where $\Psi_{\hat{p}}, \Psi_{\hat{q}}$ are the current patch to fill and the candidate patch respectively, and s, t are corresponding points. $\left\| \Psi_{\hat{p}}(s) - \Psi_{\hat{q}}(s) \right\|$ is the L-2 distance.

$$\left\| \Psi_{\hat{p}}(s) - \Psi_{\hat{q}}(s) \right\| = \sum \sqrt{(R(s_{\hat{p}}) - R(s_{\hat{q}}))^2 + (G(s_{\hat{p}}) - G(s_{\hat{q}}))^2 + (B(s_{\hat{p}}) - B(s_{\hat{q}}))^2} \tag{10}$$

2. Perform the min-cut on the flow graph

3. The points in the starting knot (1, 2, 3, 4, 7 in Fig. 6) keep their original value in $\Psi_{\hat{p}}$; the points in the end knot (5, 6, 8 in Fig. 6) are selected from the candidate patch $\Psi_{\hat{p}}$. Other points are also selected from $\Psi_{\hat{p}}$.

3   Experiment Results

We have implemented the proposed algorithm on a home PC with Duron 1.2GHz CPU, 512M RAM using Visual C++. We provide two methods for the user to select the region for editing. One is an eraser tool with which users can replace the details in the specified region with uniform color. The other is a contour tool with which users can outline the contour of the region to fill.

Fig. 7 – Fig. 13 are inpainting results by applying our synthesis-based method. The areas to be repainted are all very large, occupying more than 10%f the entire image. The yellow contours are specified by the user to indicate the target regions. Experiments show that our method can robustly remove very large objects under complex background. We also show the comparison of our method with Criminisi's method in Fig. 9. It can be seen that our algorithm produces much better results, and attains much faster speed.

4   Conclusion and Future Work

We have proposed a novel synthesis-based inpainting algorithm, by taking into consideration of constraints during synthesis. Our method can effectively and efficiently handle large complex regions. Compared to previous methods, our algorithm has the following three characteristics: 1) A constraint-based candidate patch selection method is suggested which limits the searching within neighboring regions of similar texture; 2) the desired candidate is selected based on coherence confidence preventing the error accumulation and propagation; 3) The selected patch is integrated into the original image with Graphcut optimization to make the seam visually invisible.

Experiments show that our approach provides a powerful photo editing tool for undesired object removal. In the future, we will try to find automatic method that can infer the structure information in the unknown region to guide the texture fill-in. Another possible direction is to find a better texture difference metric that can capture both structure and intensity difference, for example, Garbor based feature representation. In addition, we would extend our framework to video inpainting.

Acknowledgements

This work is partially supported by Natural Science Foundation of China under grant 60033010 and National 973 program (No. 2002CB312101).

Fig. 7. Removal of scratches on Lincon's portrait." Left is the spoiled images and right is our inpainting result

Fig. 8. Removal of the elephant from the photo. Left is the original image, middle is the segment map and right is our result

Fig. 9. Removal of large objects in photos. (a) and (d) are the original images; (b) and (e) are Criminisi's results. As can be seen, errors propagation is serious and the results are not satisfactory. (c) and (f) are our results. The contours in yellow are specified by the user

Fig. 10. Removal of a large object in photo Park"

Fig. 11. Removal of a human in the photo Girl"

Fig. 12. Removal of the islands from the scene

Fig. 13. Removal of ocean wave

References

1. Bertalmio, M, Sapiro, G., Caselles, V., Ballester, C. Image Inpainting. SIGGRAPH 2000, pages 417-424.
2. Bertalmio M. Bertozzi A. L Sapiro. G. Navier-Stokes, Fluid Dynamics, and Image and Video Inpainting Computer Vision and Pattern Recognition (CVPR01) - Volume 1 December 08 - 14, 2001 Kauai, Hawaii
3. Chan, T., Shen, J. Mathematical Models for Local Deterministic Inpaintings. UCLA CAM TR 00-11, March 2000.
4. Chan, T., Shen, J. Non-Texture Inpainting by Curvature-Driven Diffusions (CCD). UCLA CAM TR 00-35, Sept. 2000.
5. Perona, P. Malik, J. Scale-space and edge detection using anisotropic diffusion. IEEE-PAMI 12, pp. 629-639, 1990.
6. Oliveira M., Bowen B., McKenna R., and Chang Y-S., Fast Digital Image Inpainting, in Proceedings of the Visualization, Imaging, and Image Processing IASTED Conference, Marbella, Spain, 261-266, Sept. 2001.
7. Liang, L., Liu, C., X Y., Gguo, B., and Shum, H.-Y. 2001. Real-time texture synthesis using patch-based sampling. ACM Trans. Graphics 20, 3, 127–150.
8. Efros, A., Freeman, W. 2001. Image quilting for texture synthesis and transfer. In SIGGRAPH'01, 341–346.
9. Vivek Kwatra, Arno Schdl, Irfan Essa, Greg Turk and Aaron Bobick Graphcut Textures: Image and Video Synthesis Using Graph Proc. ACM Transactions on Graphics, SIGGRAPH 2003

10. Criminisi A., Perez P. and Toyama K. Object Removal by Exemplar-Based Inpainting. CVPR, Madison, Wisconsin, June, 2003.
11. Comaniciu D., Meer P.: Mean Shift: A Robust Approach toward Feature Space Analysis, IEEE Trans. Pattern Analysis Machine Intell., Vol. 24, No. 5, 603-619, 2002
12. Vivek Kwatra, Arno Schöl, Irfan Essa, Greg Turk and Aaron Bobick Graphcut Textures: Image and Video Synthesis Using Graphcut Proc. ACM Transactions on Graphics, SIGGRAPH 2003.

Highly Efficient and Effective Techniques for
Thai Syllable Speech Recognition

S. Tangwongsan, P. Po-Aramsri, and R. Phoophuangpairoj

Department of Computer Science, Mahidol University,
Bangkok, 10400 Thailand
ccstw@mahidol.ac.th
g4536827@student.mahidol.ac.th

Abstract. This paper presents a Thai syllable speech recognition system with the capability to achieve high accuracy of Thai syllable speech and Thai tone recognition. The recognition accuracy of 97.84%is achieved for Thai syllable speech recognition using the Continuous Density Hidden Markov Model (CDHMM). To provide a faster response, a beam pruning technique is applied, in which the result shows that by using this technique with an appropriate beam width, the recognition time can be reduced by more than 4 times. As Thai is tonal language, tone recognition is crucial for distinguishing meanings of Thai syllables. To obtain high rates of tone recognition in the Thai language, the CDHMM and a mixed acoustic feature method are employed. The tone recognition rates of 97.88%97.36%98.81%90.67%and 100.0% are achieved for mid, low, falling, high and rising tones, respectively.

1 Introduction

Thai speech recognition has been actively studied for more than a decade. Many techniques commonly used in speech recognition of foreign languages, such as English and Chinese, are applied to develop Thai speech recognition systems. Thai speech recognition systems have been developed using various techniques, for example, the Hidden Markov Model (HMM), the Neural Network (NN), and the Segmental Probability Model (SPM). The HMM and the NN are among two most widely used techniques. The recognition accuracy of 72% is obtained when applying the HMM with the Linear Prediction Coefficients (LPC) [1], while the accuracy of 88%is achieved when employing the NN with the Mel Frequency Cepstral Coefficients (MFCC) [2]. The SPM is another efficient technique that aims to reduce recognition time and alleviates human-aided segmentation of training data [3]. By using the SPM, Thai syllables are equally divided into 3 parts and these parts are considered to be Thai initial consonants, Thai vowels, and Thai final consonants, respectively. The accuracy of 88.18%is obt ained when using the SPM with Gaussian mixtures [4]. Concerning Thai tone recognition, the pitch contour is widely used in this respect. Different approaches are used to recognize Thai tones, for example, the linear regression and the HMM. Thai tone recognition using linear regression to model pitch contours can obtain the accuracy of 72.02%69. 71%97.08%97.69%

M. J. Maher (Ed.): ASIAN 2004, LNCS 3321, pp. 259–270, 2004.

and 96.48%for mid, low, falling, high and ri sing tones, respectively [4]. The HMM is one of the approaches providing high rates of recognition. The overall recognition result is 90%when tone models are trained with 200 syllables and tested with another 200 syllables [5]. Although researchers can achieve high accuracy in recognizing Thai syllables and tones, further improvement is still needed in terms of accuracy and speed if one would like to implement the Thai speech recognition system for commercial uses. Efficient techniques should be adapted and created circumspectly to capture the specific characteristics of the Thai language.

Thai syllable structure normally consists of three parts: the initial consonant, vowel and final consonant as C(C)V(C). Besides consonants and vowels, another most important characteristic of the Thai language is tone. In the Thai language, there are five different tones as follows: mid, low, falling, high, and rising. All the five tones play a significant role in determining Thai syllables. A base syllable with a different tone always means different things. To distinguish tones, it is practical to use pitch frequency. Fig. 1 shows the average pitch contours of five different Thai tones when syllables are spoken in isolation by a speaker. Syllables with different sounds but the same tone normally have the same characteristic of pitch transformation or pitch contour. Consequently, if we obtain a pitch contour, we can identify the tone of a syllable. Recognizing a syllable can be considered as a combination of two steps: recognizing a base syllable first and then recognizing the tone of that syllable. The Thai spoken language is a language, in which many words are monosyllabic words. Therefore, Thai speech recognition based on a syllable unit is an important basis for further development of more reliable and complex systems.

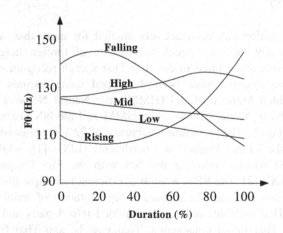

Fig. 1. Pitch contour of five Thai tones

In this work, we employ the CDHMM with mixed acoustic features, combined with a beam pruning method in order to obtain high accuracy of Thai speech recognition with a short response time.

2   CDHMM with Mixed Acoustic Features

The Continuous Density HMM is an efficient way to model speech utterances. It is a parametric model that is particularly suitable for describing speech events. The CDHMM has two stochastic processes which enable the modeling of not only acoustic phenomena, but also acoustic changes corresponding to time. The CDHMM applied in this work is the left to right model consisting of state transition probabilities and Gaussian mixtures as shown in Fig. 2.

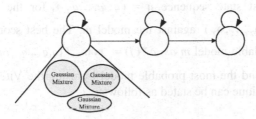

Fig. 2. Continuous Density Hidden Markov Model

The CDHMM is trained by using mixed acoustic features, which is different from the traditional speech recognition system. To improve recognition accuracy, a mixed acoustic feature method is used by including the tone acoustic component (pitch frequency) in feature vectors of syllables. Fig. 3 reveals the mixed acoustic feature method.

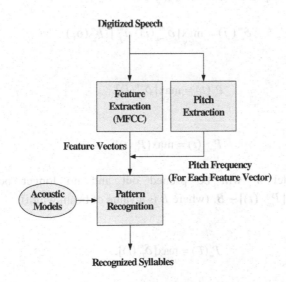

Fig. 3. Mixed acoustic feature method

3 Beam Pruning Technique

To reduce recognition time, a technique called beam pruning is applied to the Viterbi algorithm [6]. Beam search is a widely used search technique for speech recognition systems [7][8]. It is a breadth-first style. However, beam search only expands nodes that are likely to succeed at each level. Only these nodes are kept in the beam, and the rest are pruned [9]. In this work, all syllable models, whose logarithmic probabilities fall below that of the most probable model more than a beam width for each frame of speech utterances, are pruned out. The Viterbi algorithm is used to find the single best state sequence, $\mathbf{q} = (q_1 q_2 \ldots q_T)$, for the given observation sequence $\mathbf{O} = (\mathbf{o}_1, \mathbf{o}_2, \ldots, \mathbf{o}_T)$ against the model m. The best score $\delta_t^m(j)$ is defined when given the syllable model m as $\delta_t^m(j) = \max\limits_{q_1, q_2, \ldots, q_{t-1}} P[q_1 q_2 \ldots q_{t-1}, q_t = j, \mathbf{o}_1 \mathbf{o}_2 \ldots \mathbf{o}_t \mid m]$.

The procedure to find the most probable model by using the Viterbi algorithm with beam pruning technique can be stated as follows:

1. Initialization:

$$\delta_1^m(j) = \pi_j^m \cdot b_j^m(\mathbf{o}_1), \quad 1 \le j \le N \tag{1}$$

where π_j^m is the initial state distribution probability of the state j in the model m and $b_j^m(\mathbf{o}_i)$ is the observation probability of $\mathbf{o}_i$ in the state j and the model m.

2. Induction:
   For each time step $2 \le t \le T$ and $1 \le j \le N$

$$\delta_t^m(j) = \max\limits_{1 \le i \le N}[\delta_{t-1}^m(i) \cdot a_{ij}^m] \cdot b_j^m(\mathbf{o}_t). \tag{2}$$

$$P_m(t) = \max\limits_{1 \le i \le N}[\delta_t^m(i)]. \tag{3}$$

$$P_{max}(t) = \max\limits_{m}[P_m(t)]. \tag{4}$$

where the model m will be pruned out and no longer be computed if $\log[P_m(t)] < \log[P_{max}(t)] - B$. (where B is a value of beam width)

3. Termination:

$$P_m(T) = \max\limits_{1 \le i \le N}[\delta_T^m(i)]. \tag{5}$$

$$m^* = \arg\max\limits_{m}[P_m(T)]. \tag{6}$$

where $P_m(T)$ is a likelihood probability of an observation sequence **O** given a model m (which is not pruned) or $P(\mathbf{O}|m)$. The recognized answer is the model m^*, which gives the highest likelihood score.

4 Tone Recognition Using a Pitch Contour

Tone recognition is essential for Thai speech recognition since a base syllable with a different tone always has different meanings. Many researchers used pitch contours to identify tones [10][11]. Different Thai syllables with the same tone usually have a similar pitch contour characteristic. For this reason, Thai tones can be identified by using the pitch contours as shown in Fig. 1.

In this work, an efficient approach for Thai tone recognition using a weighted HMM is proposed. Although the HMM is one of the most efficient statistical models used to model the shapes of pitch contours, it treats all parts of a pitch contour in a similar manner. Actually, different parts of pitch contours contain different amounts of tone information used in classification. Modeling Thai tones with the HMM and applying a weight on the HMM appropriately is an efficient method for Thai tone recognition.

5 Thai Tone Recognition Using a Mixed Acoustic Feature

The recognition of Thai tones using the mixed acoustic feature is an innovative and efficient method that provides very high recognition rates for Thai tone recognition. Although pitch frequencies or pitch contours can represent Thai tones quite satisfactorily, it may not be only a feature describing Thai tones. To improve the result of Thai tone recognition, we present another way that can recognize Thai tones accurately. Pitch frequency is included in the MFCC and the mixed acoustic feature is used as the tone feature. After obtaining mixed acoustic features, the CDHMM is used to model all 5 Thai tones and tones of Thai syllables are recognized by using the forward procedure.

6 Experimental Results

The experimental results are divided into four parts. In the first part, Thai syllable speech recognition based on the CDHMM is investigated. In the second part, the performance improvement of the HMM recognizer by applying the beam pruning technique is evaluated. In the last two parts, Thai tone recognition using a pitch contour with the weighted HMM and Thai tone recognition using the mixed acoustic feature are developed and evaluated.

As for the acoustic features, the 1st and 2nd experiments as well as the first part of the 4th experiment use the same front-end processing and feature extraction modules. However, the second part of the 4th experiment uses the same acoustic feature but excludes pitch frequency. The acoustic feature is composed of the 12-order MFCC with energy and pitch frequency as well as their 1st and 2nd-order derivatives. In the 3rd experiment only pitch frequencies and their 1st-order derivatives is used as the acoustic feature. The speech database is recorded in a typical room environment. The details of the speech database are shown as follows:

- The vocabulary consists of 551 monosyllabic words.
- Each word is recorded for 5-6 utterances (Except the 3rd experiment, the first four utterances are used for training and the remaining utterances are used for testing).
- Totally, the speech database has 2,759 utterances or approximately 56 minutes of Wav files.
- All speech utterances are recorded in the 16-bit PCM format at 11,025 samples per second.

The system is developed in C/C++. The recognition time is measured on an Intel based PC with P4 1.7 GHz and 256 MB under Microsoft Windows 2000.

6.1 Thai Syllable Speech Recognition Based on CDHMM

In this part, Thai syllable speech recognition using the CDHMM is evaluated in terms of recognition accuracy, and recognition time.

6.1.1 Recognition Accuracy
In this experiment, we study the CDHMM in terms of recognition accuracy rates. All syllables are modeled with the 3-emitting-state CDHMM while the number of Gaussian mixtures is varied according to the experiments.

Table 1. Recognition rates of Thai syllable speech recognition using the CDHMM

Number of Mixtures	Accuracy (%)
4	97.84
8	97.84
16	95.14

The highest recognition accuracy is 97.84%when   the numbers of Gaussian mixtures are 4 and 8. The results in Table 1 show that the CDHMM is a very efficient method for Thai syllable speech recognition.

6.1.2 Recognition Time
Average recognition time is calculated by measuring the total time for recognizing all speech data, and then averaging it with the total number of recognized syllables of the speech data. In addition, the effect of the number of Gaussian mixtures on recognition time is also studied. The results are shown in Table 2.

Table 2. Average recognition time (in second)

Number of Mixtures	Time (Sec)
4	0.744
8	1.283
16	2.326

According to the results, when using the CDHMM, the greatest accuracy of 97.84% is achieved in an average of 0.744 seconds.

6.2   Beam Pruning Technique with the CDHMM

In this experiment, we applied the beam pruning technique to the probability (likelihood) computation by using the Viterbi algorithm. To study effects of using the beam pruning technique with the CDHMM, the beam widths are varied from 40 to 60, 80, 100, 200, 400, 600 and 800, respectively. The results are measured in terms of recognition rates and time. The average recognition time is shown in Table 3.

Table 3. Average recognition time using the Viterbi algorithm with beam pruning technique

Number of Mixtures	Beam Width								
	No beam	40	60	80	100	200	400	600	800
4	0.696	0.054	0.060	0.073	0.073	0.109	0.170	0.219	0.264
8	1.241	0.064	0.069	0.075	0.081	0.116	0.183	0.251	0.316
16	2.315	0.091	0.099	0.108	0.118	0.166	0.266	0.365	0.460

The recognition time decreases from 0.696 seconds to 0.170 seconds when using the Viterbi algorithm and the beam width equals 400 as well as the number of mixtures is 4. The result shows that the recognition time can be reduced by 4.1 times without additional errors. The recognition rates of the testing data set using the Viterbi algorithm with the beam pruning technique are shown in Table 4.

The results show that when the beam width is 800, the recognition rates remain at the highest level, however, a lower beam width usually results in a lower recognition accuracy rate.

Table 4. Recognition rates using the Viterbi algorithm with beam pruning on testing data (in %

Number of Mixtures	Beam Width								
	No beam	40	60	80	100	200	400	600	800
4	97.84	49.91	68.65	80.90	86.49	95.86	97.84	97.84	97.84
8	97.84	30.99	45.41	57.30	64.50	89.91	96.76	97.48	97.84
16	95.14	23.78	33.87	43.42	51.53	80.90	92.43	94.77	95.14

6.3 Tone Recognition Using Pitch Frequency and a Weighted HMM

In this experiment, the pitch frequencies and their 1st-order derivatives obtained from the last 60%of a pitch contour, excluding the last frame, are used as the tone features. The last frame is not used due to the fact that the endpoint detection program may obtain noise occurring around the syllable boundary. All tones are modeled with the 6-state CDHMM and each state is modeled by 5 Gaussian mixtures. The training set comprises 198 mid-tone, 141 low-tone, 133 falling-tone, 112 high-tone, and 152 rising-tone syllables. However, due to a characteristic of the Thai language, the number of syllables that have each tone is different. Thus, the testing set comprises 594 mid-tone, 423 low-tone, 399 falling-tone, 103 high-tone, and 228 rising-tone syllables. The tone recognition rates are 77.61%77.54%91.73%88.35%and 96.05%for mid, low, falling, high and risi ng tones, respectively. The results are shown in Table 5.

Table 5. The results of Thai tone recognition using pitch frequency on the testing data

Result \ Origin	Mid	Low	Falling	High	Rising
Mid	**77.61%**	19.87%	2.02%	0.50%	0.00%
Low	20.09%	**77.54%**	2.36%	0.00%	0.00%
Falling	4.76%	3.51%	**91.73%**	0.00%	0.00%
High	8.74%	0.97%	0.00%	**88.35%**	1.94%
Rising	0.88%	0.88%	0.00%	2.19%	**96.05%**

Since the position from 60%to 85%f pitch contours is an essential part, in tone recognition, the probabilities from the probability density function at this position are weighted by 3. After applying the weight, the tone recognition rates are increased to 78.45%78.49%95.49%90.29%and 97.37%for mi   d, low, falling, high and rising tones, respectively. The details are shown in Table 6.

Table 6. The results of Thai tone recognition using pitch frequency and the weighted HMM on the testing data.

Result / Origin	Mid	Low	Falling	High	Rising
Mid	**78.45%**	19.87%	1.35%	0.34%	0.00%
Low	19.39%	**78.49%**	2.13%	0.00%	0.00%
Falling	1.25%	3.26%	**95.49%**	0.00%	0.00%
High	4.85%	0.97%	0.00%	**90.29%**	3.88%
Rising	0.44%	1.32%	0.00%	0.88%	**97.37%**

6.4   Thai Tone Recognition Using MFCC and the Pitch Frequency

In this part, we show the performance of the Hidden Markov Model for Thai tone recognition. Each tone is modeled by using the CDHMM having 5 to 9 states (the first and last states are not emitting states), and each emitting state has 4, 8, 16, 32, 64 and 128 Gaussian mixtures. The acoustic feature is the 12-order MFCCs with energy and the pitch frequency. In addition, its 1st and 2nd-order derivatives are combined as components of the acoustic feature.

The experiment is arranged into two parts. In the first part, the pitch frequency is included in the acoustic feature while the second part, pitch frequency is excluded.

6.4.1   Pitch Frequency Included

The pitch frequency is included in the acoustic feature in this experiment. The number of states and Gaussian mixtures are varied to study the effects of these two parameters on Thai tone recognition rates on the testing data set. The results are shown in Table 7.

Table 7. Thai tone recognition rates on the testing data (pitch frequency included, in %

Number of Gaussian Mixtures	Number of HMM States				
	5	6	7	8	9
4	81.70	84.45	83.91	84.45	87.20
8	84.72	87.73	89.41	90.15	91.76
16	90.48	91.35	92.56	93.97	93.90
32	93.10	94.57	94.91	95.24	95.51
64	94.57	95.78	96.65	96.85	97.32
128	96.05	97.18	**97.92**	97.59	97.72

The highest accuracy of Thai tone recognition on the testing data set is 97.92% with the 7-state CDHMM and 128 Gaussian mixtures per state. The details of Thai tone recognition results when using the 7-state 128-Gaussian mixture CDHMM are shown in Table 8. The Thai tone recognition rates are 97.88%,97.36%,98.81% 90.67%and 100.0%for th   e mid, low, falling, high and rising tones, respectively. The results show that the high tone has the lowest accuracy of 90.67%with 5.33%false recognition of the high tone to the mid tone, and 4%f the high tone to the rising tone.

Table 8. Thai tone recognition results on the testing data (pitch frequency included)

Origin \ Result	Mid	Low	Falling	High	Rising
Mid	**97.88%**	2.12%	0.00%	0.00%	0.00%
Low	1.32%	**97.36%**	0.00%	0.00%	1.32%
Falling	0.89%	0.30%	**98.81%**	0.00%	0.00%
High	5.33%	0.00%	0.00%	**90.67%**	4.00%
Rising	0.00%	0.00%	0.00%	0.00%	**100.00%**

6.4.2   Pitch Frequency Excluded

No pitch frequency is included in this part of experiment. The recognition rates of Thai tone recognition on the testing data set are shown in Table 9. The highest recognition rate of the Thai tone recognition accuracy on the testing data set dropped to 95.51%with the 7-state CDHMM with 128 Gaussian mixtures per state.

Table 9. Thai tone recognition rates on the testing data (pitch frequency excluded)

Number of Gaussian Mixtures	Number of HMM States				
	5	6	7	8	9
4	66.69	70.58	73.39	76.81	78.08
8	76.68	82.37	84.32	85.32	87.94
16	84.52	88.61	90.42	92.02	92.36
32	88.74	92.49	93.23	93.70	94.17
64	93.50	93.36	94.71	94.44	94.97
128	93.83	94.57	**95.51**	95.38	95.31

Table 10 shows the results of the 7-state 128-Gaussian mixture CDHMM Thai tone recognizer yielding the highest recognition rates on the testing data set.

Table 10. Thai tone recognition results on the testing data (pitch frequency excluded)

Result / Origin	Mid	Low	Falling	High	Rising
Mid	**96.47%**	2.82%	0.18%	0.00%	0.53%
Low	4.62%	**93.40%**	0.00%	0.00%	1.98%
Falling	2.08%	0.00%	**97.62%**	0.30%	0.00%
High	13.33%	0.00%	.33%	**78.67%**	6.67%
Rising	0.95%	0.47%	0.00%	0.00%	**98.58%**

The tone recognition rates of 96.47%, 93.40%, 97.62%, 78.67% and 98.58% are achieved for the mid, low, falling, high and rising tones, respectively. The results show that the high tone obtains the lowest tone recognition accuracy (78.67%). The highest error is incurred by recognizing the high tone as the mid tone (13.33%) and the high tone as the rising tone (6.67%).

7   Conclusions

In this paper, we examine and purpose a number of efficient methods for Thai sylla-ble recognition and Thai tone recognition. We create a reliable Thai syllable speech recognition system and purpose the use of the weighted HMM and acoustic mixed feature to improve the accuracy of Thai tone recognition. The following are some valuable points on the results from the experiments.

1. The CDHMM can provide a very high accuracy for both Thai syllable speech recognition and tone recognition.
2. Based on the beam pruning technique, when a small beam width applied, the recognition accuracy is much lower than that of without a beam. In addition, the recognition rate usually increases when the beam width is expanded. The recog-nition accuracy becomes highest under a particular configuration when the beam width is large enough. According to the experiment, using the beam prun-ing technique with the CDHMM, the recognition time can be reduced by more than 4 times without any additional errors.
3. Pitch frequency is known as a significant feature in tone recognition (including Thai tone recognition). Tone recognition using the pitch frequency and its de-rivative with the weighted HMM can provide quite high recognition rates. Addi-tionally, pitch contours are rather independent from vowels, and pitch contours of syllables that have the same tone are usually similar.
4. MFCC with pitch frequency is another acoustic feature that is powerful for Thai tone recognition. By combining pitch frequency with the MFCC for Thai tone recognition, the recognition rates can be improved. The highest recognition rates of Thai tone recognition using the CDHMM are 97.88%, 97.36%, 98.81%, 90.67% and 100.0% for   mid, low, falling, high and rising tones, respectively.

References

1. Weerawat, T.: Speaker Dependent Voice Recognition of Thai Language, Master Thesis, Electrical Engineering, King Mongkut's Institute of Technology Ladkrabang, 1998
2. Suwancheewasiri, C.: Thai Speech Recognition for Speaker-dependent 500-word Vocabulary Based on Phonemic Distinctive Features of Isolated Syllables and Neural Network, Proceedings of the Fifth National Computer Science and Engineering Conference, pp. 59- 69, 2001
3. Lyu, R.Y., et al.: Isolated Mandarin Base-syllable Recognition Based-upon the Segmental Probability Model, IEEE Trans on Speech and Audio Processing, Vol. 6, No. 3, pp. 293-299, May 1998
4. Thanasanurak, W.: Thai Syllable Speech Recognition by Segmental Probability Model, Master's Thesis, Department of Computer Science, Mahidol University, 2001
5. Tungthangthum, A.: Tone Recognition for Thai, IEEE Asia-Pacific Conference on Circuits
6. Systems, pp. 157-160, 1998
7. Rabiner, L.R., and Juang B.H.: Fundamentals of Speech Recognition, Prentice-Hall, Inc., 1993
8. Ortmanns, S., Eiden, A., Ney, H., and Coenen, N.: Look-ahead Techniques for Fast Beam Search, ICASSP, Vol. 3, pp. 1783-1786, April 1997
9. Yong, Q., Fu-Yuan, M., Chang-Li, L., and Ding-Hua, G.: Chinese Speech Recognition System with Very Large Vocabulary, International Conference on Signal Processing, Vol. 1, pp. 817-820, October, 1996
10. Xedong, H., Alex, A., and Hsiao-wuen, H.: Spoken Language Processing, Prentice-Hall, Inc., 2001
11. Lee, T., Chan, P.C., Chan, L.W., Cheng, Y.H., and Mak, B.: Tone Recognition of Isolated Cantonese Syllables, IEEE Trans on Speech and Audio Processing, Vol. 3, No. 3, pp. 204-209, May 1995
12. X, Y.: Effects of Tone and Focus on the Formation and Alignment of f_0 Contours, Journal of Phonetics, pp. 55-105, 1999

Robot Visual Servoing Based on Total Jacobian

Qingjie Zhao [1], Zengqi Sun [2], and Hongbin Deng [1]

[1] Dept of Computer Science & Engineering, Beijing Institute of Technology,
Beijing, P. R. China
zhaoqj@bit.edu.cn
http://www.bit.edu.cn
[2] Dept of Computer Science & Technology, Tsinghua University,
Beijing, P. R. China
szq-dcs@mail.tsinghua.edu.cn
http://www.tsinghua.edu.cn

Abstract. In robot visual servoing, only the conception of image Jacobian is traditionally utilized, which indicates that the image feature varies only with the robot motion. But for a moving object, the image feature can be affected both by the robot motion and by the object motion. In this paper total Jacobian and object Jacobian are defined. The total Jacobian matrix, consisting of image Jacobian and object Jacobian, is estimated using exponentially weighted least square algorithm. No calibration is required. The control scheme is image Jacobian pseudo-inverse appending an item related to the object motion. Results for both two and six degree-of-freedom systems demonstrate the success of the algorithm.

1 Introduction

Vision based robot control system can overcome many difficulties of uncertain models and unknown environments which limit the application of current robots. Visual servoing either can automatically position a robot at a desired location with respect to an object, or can accurately follow an object as it moves through an unknown trajectory.

Visual servoing systems can be classified into position-based control and image-based control. In image-based robot visual servoing systems, inverse or pseudo-inverse Jacobian controllers are usually used. Hager[1] positions a robot manipulator to a static target using stereo vision. The error signal defined in feature space is actuated on by an inverse Jacobian controller. Castano[2] implements an inverse image Jacobian where the first two rows of the Jacobian correspond to an image-based approach and the third row is position-based acting on depth information provided by the robot joint encoders. Espiau[3] incorporates the interaction matrix or Jacobian into what he denotes as a task function approach. The control algorithm is a type of redundant control where the homogeneous solution is included to optimize additional criterion or tasks. Reasoning that constantly updating the Jacobian is too computationally intensive, they use the Jacobian at the desired target position.

M. J. Maher (Ed.): ASIAN 2004, LNCS 3321, pp. 271–285, 2004.

Most existing control schemes need off-line calibrations to determine robot kinematics, camera models, and coordinate relations between them. If system structure and parameters are a priori known, the Jacobian can be derived analytically and a numerical Jacobian at each sample point can be calculated, but this is difficult for a high DOF (Degree-Of-Freedom) system. The parameters can also be identified in an off-line process. But such systems are not robust to the disturbances, change of parameters, and unknown environments. Some systems require partial analytic modeling of the system [4-5], or using only image location visual measurements under a weak perspective assumption [6].

Several groups [7-11] have shown how the Jacobian matrix itself can be estimated online from the measurements of robot and image motion. Hosoda[7] uses an adaptive control law in which the Jacobian matrix is estimated by exponentially weighted recursive least square update method. Jagersand[8] takes a trust region control scheme in which the Jacobian is updated using Broyden's method. Zhao[11] utilizes a large residual optimization algorithm in vision-based robot positioning, in which the Jacobian matrix is estimated using Broyden's update and the inverse of Hessian matrix is estimated using DFP method. The above instances have focused primarily on static or slow targets.

Piepmeier[9] demonstrates a dynamic quasi-Newton method to control a robot tracking a moving target, where a stationary camera is used in the workspace. When using a stationary camera, the observed image features include two parts: the first is that of the robot end-effector and the second is that of the target. The motion in the image about the target can be easily estimated from the feature change of the target. But for an eye-in-hand case, the observed image feature changes of the target may result from camera (or manipulator) motion, or from target motion, or both of them. It is difficult to decide how much is aroused by the target motion if no additive restrictions used. Asada [10] utilizes two cameras on the robot effecter to track an unknown moving object, which needs three additional stationary marks to predict the motion of the virtually stationary target. The three marks should be always in the field of vision. This introduces an additional computational load and limits the region of object moving.

In this paper, we propose a robot visual servoing method based on Total Jacobian. The Total Jacobian and the object Jacobian are defined. The Total Jacobian matrix, consisting of image Jacobian and object Jacobian, is estimated using exponentially weighted least square algorithm. The control law is image Jacobian pseudo-inverse appending the item related to the object motion. The method does not need a priori knowledge of system parameters or any off-line calibrations. Special marks are neither needed.

This paper is organized as follows. Section 2 discusses the definitions of Jacobian, and Section 3 describes how to estimate total Jacobian. The control scheme is proposed in Section 4. Section 5 describes simulations on a two DOF system and a six DOF system. Finally Section 6 gives some conclusions.

2 Definitions of Jacobian

There are two basic approaches [12] to robot visual control: Imaged-Based Visual Servoing, in which an error signal measured directly in the image is mapped to actuator commands; and Position-Based Visual Servoing, in which actuator commands are computed with respect to the 3D workspace. In image-based visual servoing, an image Jacobian is usually used.

2.1 Traditional Image Jacobian

Traditional image Jacobian relates differential changes in joint angles (or robot's positions) to differential changes in image features.

For an eye-in-hand system, if the object in the workspace is motionless, the changes of image features are only related to the position of robot or camera, which can be expressed as

$$y = f(\theta) \ . \tag{1}$$

Suppose $\theta = [\theta_1, \theta_2, \cdots, \theta_n]^T$ denotes the coordinate of robot position in robot joint space, $s = [s_1, s_2, \cdots, s_p]^T$ is that of robot end-effector in Cartesian space, and $y = [y_1, y_2, \cdots, y_m]^T$ is image feature vector. While $\dot{s}$ is the velocity of robot end-effector, including translational velocity and angular velocity, $\dot{\theta}$ is the velocity in robot joint space, and $\dot{y}$ is the velocity in image feature space.

The relation between $\dot{\theta}$ and $\dot{s}$ is

$$\dot{s} = J_1 \cdot \dot{\theta} \tag{2}$$

$$J_1 = \frac{\partial s}{\partial \theta} = \begin{bmatrix} \dfrac{\partial s_1}{\partial \theta_1} & \dfrac{\partial s_1}{\partial \theta_2} & \cdots & \dfrac{\partial s_1}{\partial \theta_n} \\ \cdots & \cdots & \cdots & \cdots \\ \dfrac{\partial s_p}{\partial \theta_1} & \dfrac{\partial s_p}{\partial \theta_2} & & \dfrac{\partial s_p}{\partial \theta_n} \end{bmatrix} .$$

J_1 is called robot Jacobian [13]. Similarly, the relation between $\dot{s}$ and $\dot{y}$ will be

$$\dot{y} = J_2 \cdot \dot{s} \tag{3}$$

$$J_2 = \frac{\partial y}{\partial s} = \begin{bmatrix} \dfrac{\partial y_1}{\partial s_1} & \dfrac{\partial y_1}{\partial s_2} & \cdots & \dfrac{\partial y_1}{\partial s_p} \\ \cdots & \cdots & \cdots & \cdots \\ \dfrac{\partial y_m}{\partial s_1} & \dfrac{\partial y_m}{\partial s_2} & & \dfrac{\partial y_m}{\partial s_p} \end{bmatrix} .$$

J_2 is called Local Jacobian [14], or image Jacobian. Thus we can get

$$\dot{y} = J_2 \cdot \dot{s} = J_2 \cdot J_1 \cdot \dot{\theta} .$$

Let $J_\theta = J_2 \cdot J_1$, then

$$\dot{y} = J_\theta \cdot \dot{\theta} . \tag{4}$$

J_θ is nominated as image Jacobian, feature Jacobian, feature sensitivity matrix, interaction matrix [3], Visual-Motor Jacobian [8], et al. In this paper we name J_θ as image Jacobian.

2.2 Total Jacobian

In the part 2.1, an object is supposed motionless. For a moving target, the control signals based on the image Jacobian forenamed will be unsuitable, because the image Jacobian only relates the change of image feature to robot's motion, not considering target's motion.

Suppose r is the pose of the target, and $y_d \in \Re^m$ is the desired image feature vector. For a moving target y is a function of both θ and r, that is

$$y = y(\theta, r) . \tag{5}$$

Define $J_\theta = \partial y / \partial \theta$, $J_r = \partial y / \partial r$, then

$$\dot{y} = J_\theta \dot{\theta} + J_r \dot{r} . \tag{6}$$

If the motion of the target is continuous but the velocity is unknown, eventually y is the function of both robot joint θ and the time t.

$$y = y(\theta, t) \tag{7}$$

Expanding with Taylor series about (θ, t) and dropping the higher order terms yields

$$\Delta y = J_\theta(\theta, t)\Delta\theta + J_t(\theta, t)\Delta t \tag{8}$$

$$J_\theta = \partial y / \partial \theta, \ J_t = \partial y / \partial t$$

$$\Delta y = y(k) - y(k-1)$$

$$\Delta\theta = \theta(k) - \theta(k-1)$$

Δt is time interval, J_θ is the traditional image Jacobian, and J_t is called object Jacobian.

In (8) the first term is the change of image feature with camera motion, and the second is that with target motion. Formula (8) can be rewritten as

$$\Delta y = \begin{bmatrix} J_\theta & J_t \end{bmatrix} \cdot \begin{bmatrix} \Delta\theta \\ \Delta t \end{bmatrix} . \tag{9}$$

We define $J = \begin{bmatrix} J_\theta & J_t \end{bmatrix}$ as the total Jacobian. Suppose $\Delta x = \begin{bmatrix} \Delta\theta & \Delta t \end{bmatrix}^T$, then (9) becomes as

$$\Delta y = J \cdot \Delta x . \tag{10}$$

3   Estimation of Total Jacobian

The total Jacobian matrix $J = \begin{bmatrix} J_\theta & J_t \end{bmatrix}$, containing the image Jacobian and the object Jacobian, is estimated using exponentially weighted least square algorithm. J_θ and J_t are all related to the robot joint angles and the time, which is different from [9], where J_θ is only related to the robot joint angles, and J_t ($\partial y^*(t)/\partial t$ in [9]) is only a function of the time .

Because the kinematics and the motion of the object are unknown, we utilize exponentially weighted least square algorithm to estimate the total Jacobian J .

3.1   Exponentially Weighted Least Square

For a MISO (multi-input and single-output) system, suppose the equation of measurement is

$$y = x^T b + e , \tag{11}$$

where $x^T = \begin{pmatrix} x_1 & x_2 & \cdots & x_n \end{pmatrix}$, b is a vector of unknown parameters, whose estimation denotes $\hat{b}$, and e is measurement noise.

The exponentially weighted error equation is

$$\begin{pmatrix} \rho^{(k-1)/2} e(1) \\ \rho^{(k-2)/2} e(2) \\ \vdots \\ e(k) \end{pmatrix} = \begin{pmatrix} \rho^{(k-1)/2} y(1) \\ \rho^{(k-2)/2} y(2) \\ \vdots \\ y(k) \end{pmatrix} - \begin{pmatrix} \rho^{(k-1)/2} x^T(1) \\ \rho^{(k-2)/2} x^T(2) \\ \vdots \\ x^T(k) \end{pmatrix} \cdot \hat{b}$$

or

$$e_k = Y_k - X_k \hat{b} . \tag{12}$$

The objective function is defined as

$$E_k = \sum_{i=1}^{k} \rho^{k-i} \|e(i)\|^2 = (Y_k - X_k \hat{b})^T (Y_k - X_k \hat{b}) \cdot \tag{13}$$

$\rho \, (\rho \in [0,1])$ is forgetting factor. Newer data are more important if with a bigger value of ρ. In order to make E_k least,

$$\hat{b}(k) = (X_k^T X_k)^{-1} X_k^T Y_k \cdot \tag{14}$$

$\hat{b}(k+1)$ can be educed based on the $(k+1)^{th}$ measurements and $\hat{b}(k)$, that is, recursive least square algorithm. For $(k+1)$ measurements, the exponentially weighted error is

$$\begin{pmatrix} \sqrt{\rho} e_k \\ e(k+1) \end{pmatrix} = \begin{pmatrix} \sqrt{\rho} Y_k \\ y(k+1) \end{pmatrix} - \begin{pmatrix} \sqrt{\rho} X_k \\ x^T(k+1) \end{pmatrix} \cdot \hat{b}$$

or

$$\hat{b}(k) = (X_k^T X_k)^{-1} X_k^T Y_k \cdot \tag{15}$$

The objective function is

$$E_{k+1} = \sum_{i=1}^{k+1} \|e(i)\|^2 = (Y_{k+1} - X_{k+1}\hat{b})^T (Y_{k+1} - X_{k+1}\hat{b}) \cdot \tag{16}$$

Suppose

$$P(k) = (X_k^T X_k)^{-1}, \tag{17}$$

then

$$P(k+1) = (X_{k+1}^T X_{k+1})^{-1}$$

$$= \left[\left(\sqrt{\rho} X_k^T \quad x(k+1) \right) \begin{pmatrix} \sqrt{\rho} X_k \\ x^T(k+1) \end{pmatrix} \right]^{-1}$$

$$= \left[\rho X_k^T X_k + x(k+1) \cdot x^T(k+1) \right]^{-1}$$

$$= \frac{1}{\rho} \left[P^{-1}(k) + x(k+1) \cdot \rho^{-1} \cdot x^T(k+1) \right]^{-1}. \tag{18}$$

Using

$$(A + BCD)^{-1} = A^{-1} - A^{-1}B(C^{-1} + DA^{-1}B)DA^{-1}$$

$$A = P^{-1}(k), B = x(k+1)$$

$$C = \rho^{-1}I, D = x^T(k+1),$$

then (18) becomes

$$P(k+1) = \frac{1}{\rho}\{P(k) - P(k)x(k+1) \tag{19}$$

$$\cdot \left[\rho + x^T(k+1)P(k)x(k+1)\right]^{-1}x^T(k+1)P(k)\}$$

Define

$$K(k+1) = P(k)x(k+1)\left[\rho + x^T(k+1)P(k)x(k+1)\right] \tag{20}$$

Replacing (20) into (19), we can get

$$P(k+1) = \frac{1}{\rho}\{P(k) - K(k+1)x^T(k+1)P(k)\} \tag{21}$$

Combining(19), (20), (15), and utilizing

$$\hat{b}(k) = P(k)X_k^T Y_k,$$

we have

$$\hat{b}(k+1) = (X_{k+1}^T X_{k+1})^{-1}X_{k+1}^T Y_{k+1}$$

$$= P(k+1)\left[\sqrt{\rho}X_k^T \quad x(k+1)\right] \cdot \begin{bmatrix} \sqrt{\rho}Y_k \\ y(k+1) \end{bmatrix}$$

$$= P(k+1)\left[\rho X_k^T Y_k + x(k+1)y(k+1)\right]$$

$$= \frac{1}{\rho}\{P(k) - P(k)x(k+1)\left[\rho + x^T(k+1)P(k)x(k+1)\right]^{-1}$$

$$x^T(k+1)P(k)\} \cdot \left[\rho X_k^T Y_k + x(k+1)y(k+1)\right]$$

$$\hat{b}(k+1) = \hat{b}(k) + K(k+1)\left[y(k+1) - x^T(k+1)\hat{b}(k)\right] \tag{22}$$

Rewriting (22) (20) (21) yields

$$\hat{b}(k+1) = \hat{b}(k) + K(k+1)\left[y(k+1) - x^T(k+1)\hat{b}(k)\right]$$

$$K(k+1) = P(k)x(k+1)\left[\rho + x^T(k+1)P(k)x(k+1)\right]^{-1}$$

$$P(k+1) = \frac{1}{\rho}\{P(k) - K(k+1)x^T(k+1)P(k)\}.$$

3.2 Estimation of Total Jacobian

The visual servoing system is a MIMO (multi-input and multi-output) system. The measurement equation is

$$\Delta y = J\Delta x + e \tag{23}$$

$$\Delta y = (\Delta y_1 \quad \cdots \quad \Delta y_m)^T$$

$$\Delta x = (\Delta\theta_1 \quad \cdots \quad \Delta\theta_n \quad \Delta t)^T$$

e is a white noise vector, and

$$J = \begin{pmatrix} J_{11} & \cdots & J_{1n} & J_{1(n+1)} \\ \cdots & \cdots & \cdots & \cdots \\ J_{m1} & \cdots & J_{mn} & J_{m(n+1)} \end{pmatrix}, \text{ whose estimation is } \hat{J}.$$

Suppose

$$\hat{b}_j = \begin{pmatrix} \hat{J}_{j1} & \cdots & \hat{J}_{jn} & \hat{J}_{j(n+1)} \end{pmatrix}^T, \; j = 1, \cdots, m,$$

then

$$\Delta y_j = \hat{b}_j^T \Delta x + e_j = (\Delta x)^T \hat{b}_j + e_j \tag{24}$$

The error based on the kth measurements is

$$\begin{pmatrix} \rho^{(k-1)/2} e_j(1) \\ \rho^{(k-2)/2} e_j(2) \\ \vdots \\ e_j(k) \end{pmatrix} = \begin{pmatrix} \rho^{(k-1)/2} \Delta y_j(1) \\ \rho^{(k-2)/2} \Delta y_j(2) \\ \vdots \\ \Delta y_j(k) \end{pmatrix} - \begin{pmatrix} \rho^{(k-1)/2} (\Delta x(1))^T \\ \rho^{(k-2)/2} (\Delta x(2))^T \\ \vdots \\ (\Delta x(k))^T \end{pmatrix} \cdot \hat{b}_j, \text{ or}$$

$$e_{jk} = Y_{jk} - X_k \hat{b}_j \tag{25}$$

The objective function takes

$$E_k = \sum_{j=1}^m e_{jk}^T e_{jk} = \sum_{j=1}^m \left(Y_{jk} - X_k \hat{b}_j \right)^T \left(Y_{jk} - X_k \hat{b}_j \right) \tag{26}$$

From $\partial E_k / \partial \hat{b}_j = 0$, $\hat{b}_j$ can be exported.

We express the estimation of J as

$$\hat{J} = \begin{pmatrix} \hat{b}_1^T & \cdots & \hat{b}_m^T \end{pmatrix}^T \tag{27}$$

Referring the former results, the estimating algorithm is shown as following

$$\hat{b}_j(k+1) = \hat{b}_j(k) + K(k+1)\left[\Delta y_j(k+1) - (\Delta x(k+1))^T \hat{b}_j(k) \right] \tag{28}$$

$$K(k+1) = P(k) \cdot \Delta x(k+1)\left[\rho + (\Delta x(k+1))^T P(k) \Delta x(k+1) \right]^{-1} \tag{29}$$

$$P(k+1) = \frac{1}{\rho}\left\{ P(k) - K(k+1)[\Delta x(k+1)]^T P(k) \right\} \tag{30}$$

From (28) we can get

$$\hat{\boldsymbol{b}}_j^T(k+1) = \hat{\boldsymbol{b}}_j^T(k) + \left[\Delta y_j(k+1) - \hat{\boldsymbol{b}}_j^T(k) \cdot \Delta x(k+1)\right] \cdot \boldsymbol{K}^T(k+1)$$

For all $\hat{\boldsymbol{b}}_j$ ($j = 1, \cdots, m$), the matrixes $\boldsymbol{P}$ and $\boldsymbol{K}$ are the same. Considering (27), yields

$$\hat{\boldsymbol{J}}(k+1) = \hat{\boldsymbol{J}}(k) + \left[\Delta y(k+1) - \hat{\boldsymbol{J}}(k) \cdot \Delta x(k+1)\right] \cdot \boldsymbol{K}^T(k+1) \tag{31}$$

Then the Total Jacobian can be computed from (31), (29), and (30).

4   Control Scheme

Using the estimated Jacobian matrixes, $\hat{\boldsymbol{J}}_\theta$ and $\hat{\boldsymbol{J}}_t$, we construct a visual servoing scheme which makes image features converge to the desired values.

The error in image feature space is defined as $\boldsymbol{e}_y = \boldsymbol{y}_d - \boldsymbol{y}$.

We propose a visual servoing scheme

$$\dot{\boldsymbol{\theta}}_c = \hat{\boldsymbol{J}}_\theta^+ \boldsymbol{e}_y - \hat{\boldsymbol{J}}_\theta^+ \hat{\boldsymbol{J}}_t \tag{32}$$

$$\hat{\boldsymbol{J}}_\theta^+ = (\hat{\boldsymbol{J}}_\theta^T \hat{\boldsymbol{J}}_\theta)^{-1} \hat{\boldsymbol{J}}_\theta^T .$$

A Lyapunov function is defined as $V = \dfrac{1}{2} \boldsymbol{e}_y^T \boldsymbol{e}_y$, then

$$\dot{V} = \boldsymbol{e}_y^T \dot{\boldsymbol{e}}_y = \boldsymbol{e}_y^T(-\boldsymbol{J}_\theta \dot{\boldsymbol{\theta}} - \boldsymbol{J}_t) = \boldsymbol{e}_y^T(-\boldsymbol{J}_\theta \hat{\boldsymbol{J}}_\theta^+ \boldsymbol{e}_y + \boldsymbol{J}_\theta \hat{\boldsymbol{J}}_\theta^+ \hat{\boldsymbol{J}}_t - \boldsymbol{J}_t).$$

If the values of $\hat{\boldsymbol{J}}_\theta$ and $\hat{\boldsymbol{J}}_t$ are equal to that of $\boldsymbol{J}_\theta$ and $\boldsymbol{J}_t$, we can get $\dot{V} = -\boldsymbol{e}_y^T \boldsymbol{e}_y \le 0$. The second term of the control ruler is compensatory for the motion of the object. If the object is non-moving, the term is naught, and the task of vision control is positioning. Otherwise, if the object is moving, the task of vision control is tracking. If the sample interval is Δt, the discrete expression is

$$\Delta \boldsymbol{\theta}_c(k) = \lambda \hat{\boldsymbol{J}}_\theta^+(k) \boldsymbol{e}_y(k) - \hat{\boldsymbol{J}}_\theta^+(k) \hat{\boldsymbol{J}}_t(k) \Delta t, \ \lambda > 0 \tag{33}$$

5   Simulations

5.1   2 DOF System

Firstly, simulation of tracking is executed on a 2DOF eye-in-hand system. An object is moving on the worktable, and the moving trajectory of the object is

$$\begin{cases} x = 0.5 + 0.3\cos(0.1t) \\ y = 0.5 + 0.2\sin(0.1t) \end{cases}$$

Sample period is 50ms, $\lambda = 0.3$, $\rho = 0.95$. ± 0.5 pixel uniform random noise is added to the image feature.

$$\hat{J}(0) = \begin{pmatrix} -204.02 & 24.98 & 0 \\ 198.50 & 161.28 & 0 \end{pmatrix}.$$

Fig. 1, Fig. 2 and Fig. 3 show the results of Simulation. Fig.1 displays the trajectories of the robot end-effector tracking the moving target, where the dashed ellipse is the trajectory of moving target, thick real line is that of end-effector with the total Jacobian estimated, and the thin real line is that of end-effector with only image Jacobian estimated. The changes of the joint angles are demonstrated in Fig. 2, and the errors of image features are shown in Fig. 3, where $e_y = [e_{y1} \quad e_{y2}]^T$. Similar as in Fig. 1, thick lines are trajectories with total Jacobian estimated, and thin lines are trajectories with only image Jacobian estimated.

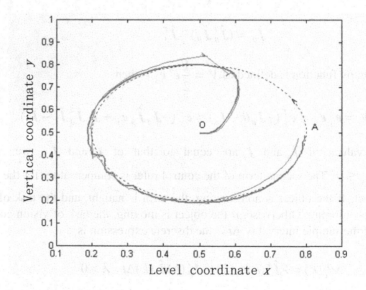

Fig. 1. Trajectories of robot end-effector

We can find that the control results with total Jacobian estimated are much better than that with only image Jacobian estimated.

5.2   6 DOF System

A 6DOF eye-in-hand testbed is constructed using the same robot model as MOTOMAN UP6 (shown in Fig. 4.), with six revolute joints $\theta_1, \cdots, \theta_6$. The focus of

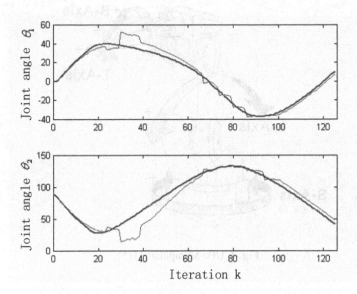

Fig. 2. Trajectories of joint angles

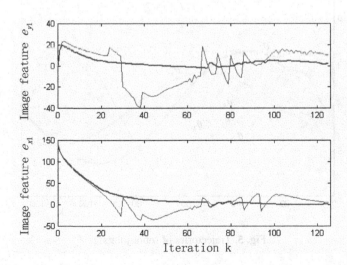

Fig. 3. Trajectories of image features

camera lens is $12mm$. The image resolution is 640×480, and the coordinates of the principal point are (320 , 240) 。

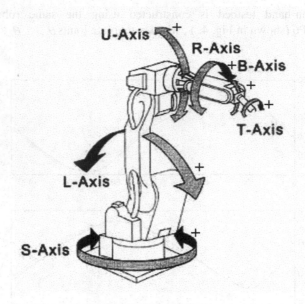

Fig. 4. UP6 Manipulator [15]

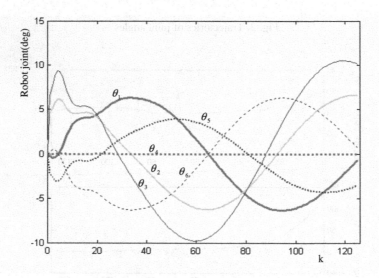

Fig. 5. Trajectories of robot joints

The object is an equilateral triangle with each side $20mm$, moving on a table plane. The image features include the coordinates of three angle points, which form a feature vector with 8 elements, $\mathbf{y} = (y_1 \quad \cdots \quad y_8)^T$. The desired image feature vector is

$$\mathbf{y}_d = (320 \quad 240 \quad 320 \quad 230 \quad 308 \quad 250 \quad 332 \quad 250)^T ,$$

that is, the center of the triangle is hoped in the center of an image. Suppose the trajectory of the object is an ellipse

$$\begin{cases} x_O = 0.06\cos(t) \\ y_O = 0.08\sin(t) \\ z_O = 0 \end{cases}$$

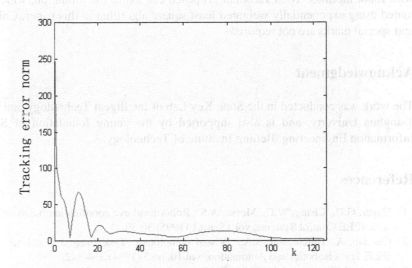

Fig. 6. Error norm of image features

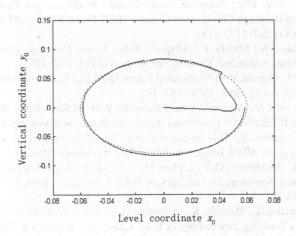

Fig. 7. Trajectories of robot end-effector

Control the robot end-effector to track the moving object so that the image features convergent to the desired values. Fig.5, Fig.6 and Fig.7 are the results of tracking, where $\lambda = 0.4$, $\rho = 0.95$, and ± 0.5 pixel uniform random noise is added to the image feature.

6 Conclusion

In robot visual servoing, image features can be influenced both by robot motion and by object motion. Traditional image Jacobian implies that image features vary only with robot motions. Total Jacobian proposed can avoid the limitation, which is estimated using exponentially weighted least square algorithm in this paper. Calibrations and special marks are not required.

Acknowledgment

The work was conducted in the State Key Lab of Intelligent Technology and Systems, Tsinghua Univerty, and is also supported by the young foundation of School of Information Engineering, Beijing Institute of Technology.

References

1. Hager, G.D., Chang, W.C., Morse, A.S.: Robot hand-eye coordination based on stereo vision. IEEE Control Systems, vol.15, no.1 (1995) 30-39.
2. Castano, A., Hutchinson, S.A.: Visual compliance: Task-directed visual servo control. IEEE Trans Robotics and Automation, vol.10, no.3 (1994) 334-342.
3. Espiau, B., Chaumette, F., Rives, P.: A new approach to visual servoing in robotics. IEEE Trans Robotics and Automation, vol.8, no.3 (1992) 313-326.
4. Feddema, J.T., Lee, C.S.: Adaptive Image Feature Prediction and Control for Visual Tracking with a Hand-eye Coordinated Camera. IEEE Trans. Systems, Man and Cybernetics, vol.20, no.5 (1990) 1172-1183.
5. Papanikolopoulos, N., Khosla, P.: Adaptive Robot Visual Tracking: Theory and Experiments. IEEE Trans Automatic Control, vol.38, no.3 (1993) 429–445.
6. Hollinghurst, N., Cipolla, R.: Uncalibrated Stereo Hand Eye Coordi-nation. Image and Vision Computing, vol.12, no.3 (1994) 187–192.
7. Hosoda, K., Asada, M.: Versatile Visual Servoing Without Knowledge of True Jacobian. Proceedings of IEEE/RSJ/GI Conference Intelligent Robots and Systems, (1994) 186-193.
8. Jägersand, M., Nelson, R.: On-line Estimation of Visual-Motor Models using Active Vision. Proceedings of ARPA Image Un-derstanding Workshop, (1996).
9. Piepmeier, J.A., McMurray, G.V., Lipkin, H.: A Dynamic Quasi-Newton Method for Uncalibrated Visual Servoing. Proceedings of IEEE Conference Robotics & Automation, (1999) 1595–1600.
10. Asada, M., Tanaka, T., Hosoda, K.: Adaptive Binocular Visual Servoing Independently Moving Target Tracking. Proceedings of IEEE Conference Robotics & Automation, (2000) 2076–2081.
11. Zhao Qingjie, Sun Zengqi: Large Residual Optimization Algorithm for Uncalibrated Visual Servoing. Proceedings of International Conference on Computational Intelligence, Robotics and Autonomous Systems, (2001) 7–11.

12. Corke P., Hutchinson, S.: Real-Time Vision, Tracking and Control. Proceedings of IEEE Conference Robotics & Automation, (2000) 2267–2272.
13. Sun Zengqi, Yan Junwei, Qian Zonghua: Robot Intelligent Control, (1995). (in Chinese)
14. Weiss, L.E., Sanderson, A.C., Neuman, C.P.: Dynamic sensor-based control of robots with visual feedback. IEEE Journal Robotics Automation, vol. RA-3, no.5 (1987) 404–417.
15. UP6 Operator's Manual.

Online Stochastic and Robust Optimization

Russell Bent and Pascal Van Hentenryck

Brown University, Box 1910, Providence, RI 02912

Abstract. This paper considers online stochastic optimization problems where uncertainties are characterized by a distribution that can be sampled and where time constraints severely limit the number of offline optimizations which can be performed at decision time and/or in between decisions. It reviews our recent progress in this area, proposes some new algorithms, and reports some new experimental results on two problems of fundamentally different nature: packet scheduling and multiple vehicle routing (MVR). In particular, the paper generalizes our earlier generic online algorithm with precomputation, least-commitment, service guarantees, and multiple decisions, all which are present in the MVR applications. Robustness results are also presented for multiple vehicle routing.

1 Introduction

Online scheduling and routing problems arise naturally in many application areas and have received increasing attention in recent years. Contrary to offline optimization, the data is not available a priori in online optimization. Rather it is incrementally revealed during algorithm execution. In many online optimization problems, the data is a set of requests (e.g., packets in network scheduling or customers in vehicle routing) which are revealed over time and the algorithm must decide which request to process next.

This research considers an online stochastic optimization framework which assumes the distribution of future requests, or an approximation thereof, is a black-box available for sampling. This is typically the case in many applications, where either historical data and predictive models are available. The framework assumes that the uncertainty does not depend on decisions, an assumption which holds in a great variety of applications and has significant computational advantages. Indeed, there is no need to explore trees of scenarios and/or sequences of decisions. In addition, this research focus primarily on online stochastic optimization under time constraints, which assumes that the time to make a decision is severely constrained, so that only a few offline optimizations can be performed at decision time and/or in between decisions. Online problems of this kind arise in many applications, including vehicle routing, taxi dispaching, packet scheduling, and online deliveries.

The paper reviews our recent progress in that area, proposes some new algorithms, generalizes existing ones to accommodate a variety of significant functionalities, and reports some new experimental results. All results are presented in a unified framework, abstracting the contributions spread accross multiple papers and crystallizing the intuition beyond the algorithmic design decisions.

The starting point is the generic online algorithm, initially proposed in [4], which can be instantiated to a variety of oblivious and stochastic approaches. When no time

M.J. Maher (Ed.): ASIAN 2004, LNCS 3321, pp. 286–300, 2004.

constraints are present, the generic algorithm naturally leads to the "traditional" expectation algorithm and to a novel hedging algorithm that provides an online counterpart to robust optimization [10]. When time constraints are present, the critical issues faced by the online algorithms is how to use their time wisely, since only a few scenarios can be optimized within the time constraints. The generic algorithm can then be instantiated to produce consensus [4] and regret [5], two algorithms which approximate expectation.

The generic online algorithm can be elegantly generalized to accommodate many features that are critical in practical applications. In particular, the paper shows how it can incorporate precomputation (to make immediate decisions), least-commitment (to avoid suboptimal decisions), service guarantees (to serve all accepted requests), and agregate decisions (to serve several requests simultaneously).

Various instantiations of the generic online algorithm are evaluated experimentally on two fundamentally different applications: packet scheduling and multiple vehicle routing. These two applications represent two extremes in the landscape of online stochastic optimization. Packet scheduling is of interest because of its simplicity: its offline problem is polynomial and the number of possible actions at each time step is small. As a consequence, it is possible to study how consensus and regret approximate expectation and hedging, as well as how all these algorithms behave under severe and less severe time constraints. Multiple vehicle routing is of interest because of its complexity: its offline problem is NP-hard and it features many of the modeling complexities of practical applications.

The rest of this paper is organized as follows. Sections 2 and 3 present the online stochastic framework and the generic online algorithm. Section 4 presents stochastic algorithms for loose time constraints and Section 5 shows how these algorithms can be approximated by consensus and regret under strict time constraints. Section 6 compares the algorithms on packet scheduling under various time constraints. Section 7 generalizes the online algorithm to incorporate precomputation, service guarantees, least-commitment and pointwise consensus/regret. Finally, Section 8 presents experimental results of the generalized algorithm to a complex multiple vehicle routing applications.

2   The Online Stochastic Framework

The Offline Problem. The framework assumes a discrete model of time. The offline problem considers a time horizon $H = [\underline{H}, \overline{H}]$ and a number of requests R. Each request r is associated with a weight $w(r)$ which represents the gain if the request is served. A solution to the offline problem serves a request r at each time $t \in H$ and can be viewed as a function $H \to R$. Solutions must satisfy the problem-specific constraints which are left unspecified in the framework. The goal is to find a feasible solution σ maximizing $\sum_{t \in H} w(\sigma(t))$. In the online version, the requests are not available initially and become progressively available at each time step.

The Online Problem. The online algorithms have at their disposal a procedure to solve, or approximate, the offline problem. They have also access to the distribution of future requests. The distribution is seen as a black-box and is available for sampling. In practice, it may not be practical to sample the distribution for the entire time horizon and hence the sizes of the samples is an implementation parameter.

ONLINEOPTIMIZATION(H)

```
1   R ← ∅;
2   w ← 0;
3   for t ∈ H
4   do R ← AVAILABLEREQUESTS(R, t) ∪ NEWREQUESTS(t);
5       r ← CHOOSEREQUEST(R, t);
6       SERVEREQUEST(r, t);
7       w ← w + w(r);
8       R ← R \ {r};
```

Fig. 1. The Generic Online Algorithm

Time Constraints. Practical applications often include severe time constraints on the decision time and/or on the time between decisions. To model this requirements, the algorithms may only use the offline procedure $\mathcal{O}$ times at each time step.

Properties of the Framework. The framework is general enough to model a variety of practical applications, yet it has some fundamental computational advantages compared to other models. *The key observation is that, in many practical applications, the uncertainty does not depend on the decisions.* There is no need to explore sequences of decisions and/or trees of scenarios: the distribution can be sampled to provide scenarios of the future without considering the decisions. As a consequence, the framework provides significant computational advantages over more general models such as multi-stage stochastic programming [6] and partially observable Markov decision processes (POMDPs) [9]. The simplicity of the framework also allows us to prove some nice theoretical properties of the algorithms. These will be described in a forthcoming paper.

3   The Generic Online Algorithm

The algorithms in this paper share the same online optimization schema depicted in Figure 1. They differ only in the way they implement function CHOOSEREQUEST. The online optimization schema simply considers the set of available and new requests at each time step and chooses a request r which is then served and removed from the set of available requests. Function AVAILABLEREQUEST(R, t) returns the set of requests available for service at time t and function SERVEREQUEST(r, t) simply serves r at time t ($\sigma(t) \leftarrow r$). To implement function CHOOSEREQUEST, the algorithms have at their disposal two black-boxes:

1. a function OPTIMALSOLUTION(R, t, Δ) that, given a set R of requests, a time t, and a number Δ, returns an optimal solution for R over $[t, t + \Delta]$;
2. a function GETSAMPLE($[t_s, t_e]$) that returns a set of requests over the interval $[t_s, t_e]$ by sampling the arrival distribution.

To illustrate the framework, we specify two oblivious algorithms as instantiations of the generic algorithm. These algorithms will serve as a basis for comparison.

Greedy (G): This algorithm serves the available request with highest weight. It can be specified formally as

CHOOSEREQUEST-G(R, t)
1   $A \leftarrow$ READY(R, t);
2   **return** $argmax(r \in A)\ w(r)$;

Local Optimal (LO): This algorithm chooses the next request to serve at time t by finding the optimal solution for the available requests at t. It can be specified as

CHOOSEREQUEST-LO(R, t)
1   $\sigma \leftarrow$ OPTIMALSOLUTION(R, t);
2   **return** $\sigma(t)$;

4   Online Stochastic Algorithms Without Time Constraints

This section reviews two algorithms exploiting stochastic information. The first algorithm optimizes expectation, while the second one hedges against the worst-case scenario. These algorithms are appropriate when time constraints are loose (i.e., when $\mathcal{O}$ is large enough to produce high-quality results).

Expectation (E): Algorithm E chooses the action maximizing expectation at each time step. Informally speaking, the method generates future requests by sampling and evaluates each available request against that sample. A simple implementation can be specified as follows:

CHOOSEREQUEST-E(R, t)
1   $A \leftarrow$ READY(R, t);
2   **for** $r \in A$
3   **do** $f(r) \leftarrow 0$;
4   **for** $i \leftarrow 1 \dots \mathcal{O}/|A|$
5   **do** $S \leftarrow R\ \cup$ GETSAMPLE$([t + 1, t + \Delta])$;
6   **for** $r \in A$
7   **do** $f(r) \leftarrow f(r) + (w(r) + w(\text{OPTIMALSOLUTION}(S \setminus \{r\}, t + 1)))$;
8   **return** $argmax(r \in A)\ f(r)$;

Line 1 computes the requests which can be served at time t. Lines 2-3 initialize the evaluation function $f(j)$ for each request r. The algorithm then generates a number of samples for future requests (line 4). For each such sample, it computes the set R of all available and sampled requests at time t (line 5). The algorithm then considers each available request r successively (line 6), it implicitly schedules r at time t, and applies the optimal offline algorithm using $S \setminus \{r\}$ and the time horizon. The evaluation of request r is updated in line 7 by incrementing it with its weight and the score of the corresponding optimal offline solution. All scenarios are evaluated for all available requests and the algorithm then returns the request $r \in A$ with the highest evaluation. Observe Line 4 of Algorithm E which distributes the available offline optimizations across all available requests.

Hedging (H): Algorithm H is an online adaptation of robust optimization, whose key idea is to hedge against the worst-case scenario. In other words, the goal is to find, at each time step, a solution whose deviation with respect to the optimal solution is minimal over all scenarios.

Definition 1 (Deviation). *Let R be the set of requests at time t and $r \in R$. The deviation of r wrt R and t, denoted by* DEVIATION(r, R, t), *is defined as*

$$| \ w(\text{OPTIMALSOLUTION}(R, t)) - (w(r) + w(\text{OPTIMALSOLUTION}(R \setminus \{r\}, t + 1))) \ | \ .$$

The differences between algorithms E and H are Line 7 which computes the maximum deviation for the action and Line 8 which selects the action with minimum deviation.

CHOOSEREQUEST-H(R, t)
1    $A \leftarrow$ READY(R, t);
2   **for** $r \in A$
3   **do** $f(r) \leftarrow 0$;
4   **for** $i \leftarrow 1 \ldots \mathcal{O}/|A|$.
5   **do** $S \leftarrow R \cup$ GETSAMPLE$([t + 1, t + \Delta])$;
6   **for** $r \in A$
7   **do** $f(r) \leftarrow max(f(r), \text{DEVIATION}(r, R, t))$;
8   **return** $argmin(r \in A) \ f(r)$;

Observe that algorithm E can be obtained from algorithm H by replacing line 7 with

$$f(r) \leftarrow f(r) + \text{DEVIATION}(r, R, t);$$

since E can be viewed as minimizing the average deviation.

5   Online Stochastic Algorithms Under Time Constraints

This section studies online optimization under time constraints, i.e., when the number of optimizations at each time step t is small. As mentioned earlier, algorithms E and H distribute the available optimizations $\mathcal{O}$ across all requests (line 4). When $\mathcal{O}$ is small (due to the time constraints), each request is only evaluated with respect to a small number of samples and the algorithms do not yield much information. This is precisely why online vehicle routing algorithms [2] cannot use E and H, since the number of requests is very large (about 50 to 100), the time between decisions is relatively short, and optimization is computationally demanding. The section shows how algorithm E can be approximated and presents two approximation algorithms, consensus and regret. The regret algorithm can also be adaped to approximate algorithm H.

Consensus (C): The consensus algorithm C was introduced in [4] as an abstraction of the sampling method used in online vehicle routing [2]. Its key idea is to solve each scenario once and thus to examine $\mathcal{O}$ scenarios instead of $\mathcal{O}/|A|$. More precisely, instead of evaluating each possible request at time t with respect to each sample, algorithm C executes the offline algorithm on the available and sampled requests once per sample. The request scheduled at time t in optimal solution σ is credited $w(\sigma)$ and all other requests receive no credit. Algorithm C can be specified as follows:

CHOOSEREQUEST-C(R, t)

```
1  for r ∈ R
2      do f(r) ← 0;
3  for i ← 1 ... O
4      do S ← R ∪ GETSAMPLE([t + 1, t + Δ]);
5         σ ← OPTIMALSOLUTION(S, t);
6         f(σ(t)) ← f(σ(t)) + w(σ);
7  return argmax(r ∈ R) f(r);
```

Observe line 5 which calls the offline algorithm with all available and sampled requests and a time horizon starting at t and line 6 which increments the number of times request $\sigma(t)$ is scheduled first. Line 7 simply returns the request with the largest score. The main appeal of Algorithm C is its ability to avoid partitioning the available samples between the requests, which is a significant advantage when the number of samples is small and/or when the number of requests is large. Its main limitation is its *elitism*. Only the best request is given some credit for a given sample, while other requests are simply ignored. It ignores the fact that several requests may be essentially similar with respect to a given sample. Moreover, it does not recognize that a request may never be the best for any sample, but may still be extremely robust overall. The regret algorithm shows how to gather that kind of information from the sample solutions without solving additional optimization problems.[1]

Regret (R): The key insight in Algorithm R is the recognition that, in many applications, it is possible to estimate the deviation of a request r at time t quickly. In other words, once the optimal solution σ of a scenario is computed, it is easy to compute the deviation of all the requests, thus approximating E with one optimization. This intuition can be formalized using the concept of *regret*.

Definition 2 (Regret). *A regret is a function that, given a request r, a set R ($r \subset R$), a time t, and an optimal solution $\sigma = $ OPTIMALSOLUTION(R, t), over-approximates the deviation of r wrt R and t, i.e.,*

$$\text{REGRET}(r, R, t, \sigma) \geq \text{DEVIATION}(r, R, t).$$

Moreover, there exists two functions f_o and f_r such that

- OPTIMALSOLUTION(R, t) *runs in time* $O(f_o(R, \Delta))$;
- REGRET(r, R, t, σ) *runs in time* $O(f_r(R, \Delta))$;
- $|A|f_r(R, \Delta)$ *is* $O(f_o(R, \Delta))$.

Intuitively, the complexity requirement states that the computation of the $|A|$ regrets does not take more time than the optimization. Regrets typically exist in practical applications. In an online facility location problem, the regret of opening a facility f can be

---

[1] The consensus algorithms behaves very well on many vehicle routing applications because, on these applications, the objective function is first to serve as many customers as possible. As a consequence, at a time step t, the difference between the optimal solution and a non-optimal solution is rarely greater than 1. It is over time that significant differences between the algorithms accumulate.

estimated by evaluating the cost of closing the selected facility $\sigma(t)$ and opening f. In vehicle routing, the regret of serving a customer c next can evaluated by swapping c with the first customer on the vehicle serving c. In packet scheduling, the regret of serving a packet p can be estimated by swapping and/or serving a constant number of packets. In all cases, the cost of computing the regret is small compared to the cost of the offline optimization and satisfy the above requirements. Note that there is an interesting connection to local search, since computing the regret may be viewed as evaluating the cost of a local move for the application at hand. We are now ready to present the regret algorithm R:

CHOOSEREQUEST-R(R, t)
1   for $r \in R$
2   do $f(r) \leftarrow 0$;
3   for $i \leftarrow 1 \ldots \mathcal{O}$
4   do $S \leftarrow R \cup$ GETSAMPLE$([t + 1, t + \Delta])$;
5           $\sigma \leftarrow$ OPTIMALSOLUTION(S, t);
6           $f(\sigma(t)) \leftarrow f(\sigma(t)) + w(\sigma)$;
7   for $r \in$ READY$(R, t) \setminus \{\sigma(t)\}$
8   do $f(r) \leftarrow f(r) + (w(\sigma) -$ REGRET$(\sigma, r, R, t))$;
9   return $argmax(r \in R) \; f(r)$;

Its basic organization follows algorithm C. However, instead of assigning some credit only to the request selected at time t for a given sample s, algorithm R (lines 7-8) uses the regrets to compute, for each available request r, an approximation of the best solution of s serving r at time t, i.e., $w(\sigma) -$ REGRET(σ, r, R, t). Hence every available request is given an evaluation for every sample at time t for the cost of a single offline optimization (asymptotically). Observe that algorithm R performs $\mathcal{O}$ offline optimizations at time t and that it is easy to adapt algorithm R to approximate algorithm H.

6   Packet Scheduling

This section reports experimental results on the online packet scheduling problem studied in [8]. This networking application is of interest experimentally since (1) the number of requests to consider at each time t is small and (2) the offline algorithm can be solved in polynomial time. As a result, it is possible to evaluate all the algorithms experimentally, contrary to vehicle routing applications where this is not practical. The packet scheduling is also interesting as it features a complex arrival distribution for the packets based on Markov Models (MMs).

The Offline Problem. We are given a set *Jobs* of jobs partitioned into a set of classes C. Each job j is chararacterized by its weight $w(j)$, its arrival date $a(j)$, and its class $c(j)$. Jobs in the same class have the same weight (but different arrival times). We are also given a schedule horizon $H = [\underline{H}, \overline{H}]$ during which jobs must be scheduled. Each job j requires a single time unit to process and must be scheduled in its time window $[a(j), a(j) + d]$, where d is the same constant for all jobs (i.e., d represents the time a job remains available to schedule). In addition, no two jobs can be scheduled at the same time and jobs that cannot be served in their time windows are dropped. The goal is to find a schedule of maximal weight, i.e., a schedule which maximizes the sum of

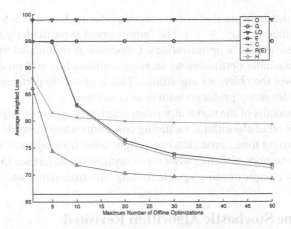

Fig. 2. The Regret Algorithm on Packet Scheduling

the weights of all scheduled jobs. This is equivalent to minimizing weighted packet loss. More formally, assume, for simplicity and without loss of generality, that there is a job scheduled at each time step of the schedule horizon. Under this assumption, a schedule is a function $\sigma : H \rightarrow Jobs$ which assigns a job to each time in the schedule horizon. A schedule σ is feasible if

$$\forall t_1, t_2 \in H : t_1 \neq t_2 \rightarrow \sigma(t_1) \neq \sigma(t_2)$$
$$\forall t \in H : a(\sigma(t)) \leq t \leq a(\sigma(t)) + d.$$

The weight of a schedule σ, denoted by $w(\sigma)$, is given by $w(\sigma) = \sum_{t \in H} w(\sigma(t))$. The goal is to find a feasible schedule σ maximizing $w(\sigma)$. This offline problem can be solved in quadratic time $O(|J||H|)$.

The Online Problem. The experimental evaluation is based on the problems of [8, 4], where all the details can be found. In these problems, the arrival distributions are specified by independent MMs, one for each job class. The results are given for the reference 7-class problems and for an online schedule consisting of 200,000 time steps. Because it is unpractical to sample the future for so many steps, the algorithms use a sampling horizon of 50, which seems to be an excellent compromise between time and quality. The regret function is given in [5] and consists of swapping a constant number of packets in the optimal schedule.

Experimental Results. Figure 2 depicts the average packet loss as a function of the number of available optimizations $\mathcal{O}$ for the various algorithms on a variety of 7-class problems. It also gives the optimal, a posteriori, packet loss (O). The results indicate the value of stochastic information as algorithms E and H significantly outperform the oblivious algoritms G and LO and bridge much of the gap between these algorithms and the optimal solution. Note that LO is worse than LO, illustrating the (frequent) pathological behavior of over-optimizing. Hedging slightly outperforms expectation, although the results are probably not statistically significant.

The results also indicate that consensus outperforms E and H whenever few optimizations are available (e.g., ≤ 15). The improvement is particularly significant when there are very few available optimizations. Consensus is dominated by E and H when the number of available optimizations increases, although it still produces significant improvements over the oblivious algorithms. This is of course pertinent, since E and H are not practical for many problems with time constraints.

Finally, the benefits of the regret algorithm are clearly apparent. Algorithm R indeed dominates all the other algorithms, including consensus when there are very few offline optimizations (strong time constraints) and expectation/hedging even when there are a reasonably large number of them, (weak time constraints). Reference [5] also shows this to be the case for complex online vehicle routing with time windows.

7   The Online Stochastic Algorithm Revisited

This section considers four important generalizations to the framework: precomputation, service guarantees, least-commitment, and multiple decisions.

Precomputation. Some applications are characterized by very short decision times, either because of problem requirements or to produce solutions of higher quality. These applications however allow for some limited number of optimizations in between decisions. For instance, online vehicle routing and deliveries are applications exhibiting these features. The generic online algorithm can generalized to provide these functionalities. The key idea is to maintain a set of scenario solutions during execution. At decision time, these solutions can then be used to choose an appropriate request to serve. The set of solutions can then be updated to remove solutions that are incompatible with the selected decisions and to include newly generated solutions.

Figure 3 depicts the generalized online algorithm and shows how to instantiate it with consensus. The set of solutions Σ is initialized in Line 2. The request is selected in line 5 by function CHOOSEREQUEST which now receives Σ as input as well. Lines 9 and 10 remove the infeasible solutions and generates new ones. The function GENERATESOLUTIONS is also depicted in Figure 3. It is essentially the core of the CHOOSEREQUEST implementation in algorithms C and R with the logic to make decisions abtracted away. The decision code is what is left in the instantiations of function CHOOSEREQUEST. The figure also gives the implementation of CHOOSEREQUEST for algorithm C to illustrate the instantiations.

Service Guarantees. Many applications require service guarantees. The algorithm may decide to accept or reject a new request but, whenever a request is accepted, the request must be served. The online algorithm can be enhanced to include service guarantees. It suffices to introduce a new function to accept/request new requests and to keep only those solutions which can accommodate the requests. Of course, to accept a request, at least one solution must be able to serve it in addition to the current requests. The new online generic algorithm with service guarantees is depicted in Figure 4. The changes are in lines 4-6. Function ACCEPTREQUESTS (line 4) selects the new requests to serve using the existing solutions Σ and function REMOVEINFEASIBLESOLUTIONS removes those solutions which cannot accommodate the new requests.

ONLINEOPTIMIZATION(H, R)
1    $w \leftarrow 0$;
2    $\Sigma \leftarrow$ GENERATESOLUTIONS($R, 0$);
3   **for** $t \in H$
4   **do** $R \leftarrow$ AVAILABLEREQUESTS(R, t) $\cup$ NEWREQUESTS(R, t);
5        $r \leftarrow$ CHOOSEREQUEST(R, t, Σ);
6   SERVEREQUEST(r, t);
7        $w \leftarrow w + w(r)$;
8        $R \leftarrow R \setminus \{r\}$;
9        $\Sigma \leftarrow \{\sigma \in \Sigma \mid \sigma(t) = r\}$;
10       $\Sigma \leftarrow \Sigma \cup$ GENERATESOLUTIONS(R, t);

GENERATESOLUTION(R, t)
1    $\Sigma \leftarrow \emptyset$;
2   **repeat**
3        $S \leftarrow R \cup$ GETSAMPLE($[t + 1, t + \Delta]$);
4        $\sigma \leftarrow$ OPTIMALSOLUTION(S, t);
5        $\Sigma \leftarrow \Sigma \cup \{\sigma\}$;
6   **until** until time $t + 1$
7   **return** Σ;

CHOOSEREQUEST-C(R, t, Σ)
1   **for** $r \in R$
2   **do** $f(r) \leftarrow 0$;
3   **for** $\sigma \in \Sigma$
4   **do** $f(r) \leftarrow f(r) + w(\sigma)$;
5   **return** $argmax(r \in R)\ f(r)$;

Fig. 3. The Generic Online Algorithm with Precomputation

Least-Commitment. In the packet scheduling application, it is always suboptimal not to serve a packet at each time step. However, in many online applications, it may be advisable not to serve a specific request, since this may reduce further choices and/or make this algorithm less adaptive. The ability to avoid or to delay a decision is critical in some vehicle routing applications, as shown later in the paper. It is easy to extend the framework presented so far to accommodate this feature. At every step, the algorithm may select a request $\perp$ which has no effect and no profit/cost. It suffices to use CHOOSEREQUEST($R \cup \{\perp\}, t, \Sigma$) in line 5 of the algorithm.

Multiple Decisions and Pointwise Consensus. Many practical applications have the ability to serve several requests at the same time, since resources (e.g., machines or vehicles) are often available in multiple units. The online algorithm naturally generalizes to multiples decisions. Assume that a solution σ at time t returns a tuple $\sigma(t) = (r_1, \ldots, r_n) = (\sigma_1(t), \ldots, \sigma_n(t))$. It suffices to replace r in the online algorithm by a tuple $(r_1, \ldots, r_n)$ to obtain a generic algorithm over tuples of decisions. However, it is important to reconsider how to choose requests in this new context. A straighforward generalization of consensus would give

ONLINEOPTIMIZATION(H, R)

```
1   w ← 0;
2   Σ ← GENERATESOLUTIONS(R, 0);
3   for t ∈ H
4   do N ← ACCEPTREQUESTS(R, t, Σ);
5       Σ ← REMOVEINFEASIBLESOLUTIONS(R, t, N, Σ);
6       R ← AVAILABLEREQUESTS(R, t) ∪ N;
7       r ← CHOOSEREQUEST(R, t, Σ);
8       SERVEREQUEST(r, t);
9       w ← w + w(r);
10      R ← R \ {r};
11      Σ ← {σ ∈ Σ | σ(t) = r};
12      Σ ← Σ ∪ GENERATESOLUTIONS(R, t);
```

Fig. 4. The Generic Online Algorithm with Precomputation and Service Guarantees

CHOOSEREQUEST-C(R, t)

```
1   for e ∈ R^n
2   do f(e) ← 0;
3   for i ← 1 ... O
4   do S ← R ∪ GETSAMPLE([t + 1, t + Δ]);
5       σ ← OPTIMALSOLUTION(S, t);
6       f(σ(t)) ← f(σ(t)) + w(σ);
7   return argmax(e ∈ R^n) f(e);
```

Unfortunately, this generalized implementation of consensus is not particularly effective, especially when there are many requests and few scenarios. Indeed, the information about decisions is now distributed over tuples of requests instead of over individual requests and consensus does not capture the desirability of serving particular requests. This limitation can be remedied by evaluating the decisions independently accross all scenarios and by selecting the best coupling available among the solutions. This *pointwise consensus* can be formalized as follows:

CHOOSEREQUEST-PC(R, t)

```
1   for r ∈ R, i ∈ 1..n
2   do f_i(r) ← 0;
3   for i ← 1 ... O
4   do S ← R ∪ GETSAMPLE([t + 1, t + Δ]);
5       σ ← OPTIMALSOLUTION(S, t);
6       for i ∈ 1..n
7       do f_i(σ_i(t)) ← f_i(σ_i(t)) + w(σ);
8   σ* = argmax(σ ∈ Σ) Σ_{i=1}^{n} f_i(σ_i(t));
9   return σ*(t);
```

Note that pointwise consensus reduces to consensus when $n = 1$ and that pointwise regret could be derived in the same fashion.

8 Vehicle Routing

This section describes the applications of the online generic algorithm with precomputa-tion, service guarantees, and least-commitment to a multiple vehicle routing applications. Contrary to the applications in [2] where the focus is on feasibility, the difficulty here lies in the lexicographic objective function, i.e., serving as many customers as possible and minimizing travel distance. The interesting result is that approximations of expectation perform remarkably in these two "orthogonal" applications.

The Problem. The application is based on the model proposed in [11] where customers are distributed in a 20km×20km region and must be served by vehicles with uniform speed of 40 km/h. Service times for the customers are generated according to a log-normal distribution with parameters (.8777, .6647). With this distribution, the mean service time is 3 min. and the variance is 5 min. The service times were chosen to mimic the service times of long-distance courier mail services [11]. We use n to denote the expected number of customers and T to denote the time horizon (8 hours). Problems are generated with a degree of dynamism (DOD) (i.e, the ratio of known customers over stochastic customers) in the set $\{0\%, 5\%, \ldots, 100\%\}$. For a DOD x, there are $n(1 - x)$ known customers. The remaining customers are generated using an exponential distribution with parameter $\lambda = \frac{nx}{T}$ for their inter-arrival times. It follows from the corresponding Poisson distribution (with parameter λT) that the expected number of unknown customers is nx, the expected number of customers is n, and the expected DOD is x. The results given here assume that 4 vehicles amd 160 customers. Each vehicle can serve at most 50 customers and the vehicle must return to the depot by the time horizon. The customers are generated using 2-D Gaussians centered at two points in the region. (Similar results are obtained under other distributions). The objective function consists in minimizing the number of missed customers and minimizing the travel distance. The experimental results are based on 15 instances and an average of 5 runs on each instances. See Reference [3] for a more comprehensive evaluation.

Setting of the Algorithms. The online generic algorithm is run with the following settings. Initially, 25 different scenarios are created and optimized for 1 minute using large-scale neighborhood search (LNS) [12, 1]. These initial solutions are used to determine the first customer for each vehicle. An additional 25 scenarios are created and optimized for 1 minute with the first customers fixed. It was verified experimently that this second step improves the quality of the final solutions. Subsequent scenarios are optimized for about 10 seconds using LNS. The parameters for LNS are as follows: 30 for the maximum number of customers to remove at one time, 100 attempts at removing c customers without improvement before removing $c + 1$ customers, 15 for the determinism factor of the relatedness function, and 4 discrepancies. A simple insertion heuristic is used to decide whether a new request should be accommodated. The online algorithm uses precomputations to decide whether to accept requests immediately and to avoid delaying the dispatching of vehicles, service guarantees to serve all accepted requests, least-commitment to be able to postpone vehicle departures to accommodate future requests more effectively, and pointwise consensus to gather as much information as possible from the small number of scenarios available in this application.

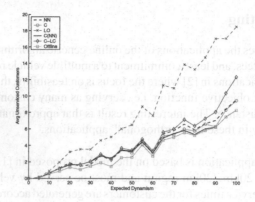

Fig. 5. Results on the Number of Serviced Customers

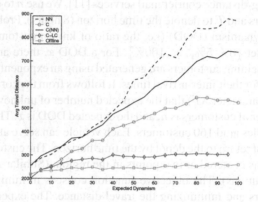

Fig. 6. Results on Travel Distance

Experimental Results. The online generic algorithm is compared with the Nearest Neighbor (NN) heuristic proposed in [11] and generalized to providing guarantees on servicing customers. Whenever a request arrives, the NN algorithm is simulated to determine if it can accommodate the new request. If it cannot, the request is rejected. More generally, the results compare NN and the online algorithm instantiated with local optimization (LO), consensus (C), consensus with least-commitment (C-LC), and consensus using NN instead of LNS (C(NN)) to find solutions to the scenarios. The figures will also give the offline solution found using LNS, which represents the "best" solution the various online algorithm could hope to achieve.

Figure 5 describes the experimental results concerning the number of serviced customers for various degrees of dynamism. The results clearly indicate that the stochastic approaches are superior to LO which is unable to service as many customers. A detailed look at the trace of the decisions performed by LO indicate that it waits too long to deploy some of the vehicles. This is because optimal solutions use as few vehicles as possible to minimize travel distance and LO believes it can use fewer vehicles than necessary until late in the simulation. The remaining approaches service a comparable number of customers. With higher degrees of dynamism, the benefits of using a consensus function for ranking are clear, as it reduces the number of missed customers significantly compared to using travel distance. The online stochastic algorithm do not bring significant benefits in terms of serviced customers compared to NN. C(NN) is generally superior to NN, while C is roughly similar to NN (except for very high degrees of dynamism). Note that C-LC does not perform as well as C for these very high degrees of dynamism: It has a tendency to wait too long, which could be addressed easily by building some slack in C-LC.

Figure 6 depicts the results for the travel distance, which are extremely interesting. No results are given for LO, since it is far from being competitive for customer service. The results indicate that the stochastic instantiations of the online algorithm significantly reduce travel distance compared to NN. The results are particularly impressive for C-LC, whose travel distance is essentially not affected by the degree of dynamism. Observe also that the comparison between C(NN) and the other stochastic approaches tend to indicate that it seems beneficial for these problems to use a more sophisticated optimization algorithms on fewer samples than a weaker method on more samples.

Robustness. It is natural to question how the algorithms behave when the stochastic information is noisy. This situation could arise from faulty historical data, predictions, and/or approximations in machine learning algorithms. Figure 7 shows some results

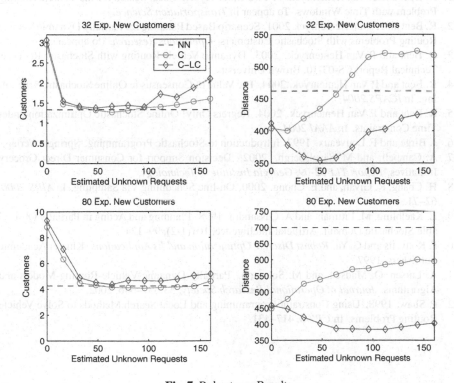

Fig. 7. Robustness Results

when run on the 20% and 50% dynamism instances of M3 (32 and 80 expected new customers respectively). It is interesting to see that, in both cases, it is better to be optimistic when estimating the number of dynamic customers. For example, on 20% dynamism, C-LC is able to service roughly the same number of customers when it expects between 20 and 100 dynamic customers. However, it performs the best in terms of travel distance when it expects 50 dynamic customers, slightly more than the 32 of actual problem sets themselves. In addition, these results show that, even in the presence of significant noise, stochastic approaches are still able to achieve high-quality results.

Acknowledgments

This paper is dedicated to Jean-Louis Lassez for all these long conservations about science, scientific communities, and the joy of research. Happy birthday, Jean-Louis, and thank you for being such an inspiration for so many of us! This research is partially supported by NSF ITR Award DMI-0121495.

References

1. R. Bent and P. Van Hentenryck. A Two-Stage Hybrid Local Search for the Vehicle Routing Problem with Time Windows. To appear in *Transportation Science*.
2. R. Bent and P. Van Hentenryck 2001. Scenario Based Planning for Partially Dynamic Vehicle Routing Problems with Stochastic Customers. *Operations Research*. (to appear).
3. R. Bent and P. Van Hentenryck. 2003. Dynamic Vehicle Routing with Stochastic Requests Technical Report CS-03-10, Brown University.
4. R. Bent and P. Van Hentenryck. 2004. The Value of Consensus in Online Stochastic Scheduling. In *ICAPS 2004*.
5. R. Bent and P. Van Hentenryck. 2004. Regrets Only! Online Stochastic Optimization under Time Constraints. In *AAAI 2004*.
6. J. Birge and F. Louveaux. 1997. Introduction to Stochastic Programming. Springer Verlag.
7. A. Cambell, and M. Savelsbergh. 2002. Decision Support for Consumer Direct Grocery Initiatives. *Report TLI-02-09, Georgia Institute of Technology*.
8. H. Chang, R. Givan, and E. Chong. 2000. On-line Scheduling Via Sampling. In *AIPS'2000*, 62–71.
9. L. Kaelbling, M. Littman, and A. Cassandra. 1998. Planning and Acting in Partially Observable Stochastic Domain. Artificial Intelligence, 101(1-2), 99–124.
10. P. Kouvelis and G. Yu. *Robust Discrete Optimization and Its Applications*. Kluwer Academic Publishers, 1997.
11. A. Larsen, O. Madsen, and M. Solomon. Partially Dynamic Vehicle Routing-Models and Algorithms. *Journal of Operational Research Society*, 53:637–646, 2002.
12. P. Shaw. 1998. Using Constraint Programming and Local Search Methods to Solve Vehicle Routing Problems. In *CP'98*, 417–431.

Optimal Constraint Decomposition for Distributed Databases

Alexander Brodsky[1], Larry Kerschberg[1], and Samuel Varas[2]

[1] Center for Information Systems Integration and Evolution,
Department of Information and Software Engineering,
George Mason University,
Fairfax, VA 22030-4444
[2] Industrial Engineering Department,
University of Chile,
Republica 701 - Santiago - Chile

Abstract. The problem considered is that of decomposing a global integrity constraint in a distributed database into local constraints for every local site, such that the local constraints serve as a conservative approximation, i.e., satisfaction of the local constraints by a database instance guarantees satisfaction of the global constraint. Verifying local rather than global constraints during database updates reduces distributed processing costs and allows most updates, even in the presence of site and network failures. This paper focuses on the problem of deriving the best possible decompositions, both at database design and update processing time. A generic framework is formulated for finding optimal decompositions for a range of design and update-time scenarios. For the case of linear arithmetic constraints, (1) a bounded size parametric formulation of the decomposition optimization problem is introduced which has a possibly smaller search space but is proven to have the same optimum, (2) the decomposition problem is reduced to the problem of resource distribution which simplifies distributed management of constraints, and (3) autonomous optimal decompositions in subsets of local database sites are shown possible and are proven to preserve optimality under the resource bounds constraints.

1   Introduction

Local Verification of Global Integrity Constraints

Often, coordination of components in distributed systems involves constraint-based agreements, which can be viewed as global integrity constraints [9]. These global database integrity constraints are difficult to monitor, update and enforce in distributed environments, so that new, distributed techniques and protocols are desirable.

To reduce the costs of distributed management of global constraints, the idea of local verification of global constraints was introduced and studied (e.g., [2, 1, 4, 6, 12, 14, 16, 5]). The idea is to decompose a global constraint into a set of

M.J. Maher (Ed.): ASIAN 2004, LNCS 3321, pp. 301–319, 2004.

local ones that will serve as a conservative approximation, such that satisfaction of local constraints by a database instance guarantees satisfaction of the global constraint. When a local site i is being updated, if the update satisfies its local constraint C_i, no global constraint checking is necessary. Thus, most of the work can be delegated to local processing, thereby saving communication and other distributed processing costs. The ability to perform updates autonomously is also very important in the presence of site or network failures [17].

While the above-mentioned works have considered many aspects of local verification (see Related Work), the problem of finding optimal constraint decompositions and distributed constraint-management protocols that achieve decomposition optimality along with maximal resource utilization has remained open. This paper focuses on the problem of deriving and managing optimal integrity constraint decompositions, during both database design and update processing.

Contributions

First, we introduce a formal generic framework for finding optimal constraint decompositions for a range of design and update-time scenarios, characterized by the properties of *Safety, Local Consistency, Partial Constraint Preservation*, and *Resource Partition Preservation*, which are defined formally in the paper. The objective function can describe a variety of optimization criteria, such as the probability that an update satisfies its local constraint.

Second, for the case of linear arithmetic constraints, we provide a mathematical programming formulation for finding an optimal safe decomposition. The problem is that the size of parametric description of local constraints in a safe decomposition is, in general, unbounded, which does not allow a direct reduction to mathematical programming. To overcome this problem, we introduce the notion of *Compact Split*, which is a parametric characterization of a subset of safe decompositions, whose size is bounded. We then prove that the optimum of any monotonic objective function among all safe decompositions can always be found in the subspace of *compact splits*.

Third, we introduce a *resource-based* characterization of split decompositions, defined formally in the paper, to simplify the distributed management of constraints. Specifically, every local constraint C_i for site i in a split decomposition D is uniquely associated with a *resource* vector[1] r_i, and the global constraint is associated with the *global resource* vector, for which we prove that: D is a compact split (safe decomposition) if and only if the cumulative *resource* at all sites is bounded by the *global resource*. Furthermore, given a database instance, every site i is also uniquely associated with a *lower resource bound*, for which we prove that: the local database instance at i satisfies its local constraint if and only if its *resource* is bounded from below by its *lower resource bound*. In addition, every site is associated with its *resource upper bound*. The *resource*

---

[1] The dimension of this and other resource vectors equals the number of atomic linear constraints (i.e., linear inequalities over reals) in the global constraint.

and its *bounds* for every site constitute a *resource distribution*, which a protocol can maintain instead of the explicit local constraints and database instance. The key advantage of a resource distribution is its small size of $O(nc)$ as compared with the size $O(nc * nv)$ of a constraint decomposition, where nc and nv are the number of constraints and variables, respectively, in the global constraint. In fact, nv may be as large as the size of a database, for example when the global constraint reflects that the summation of some quantity, one per relational tuple, is bounded by a constant.

Fourth, we study *Concurrent Split Decompositions*. To manage concurrent constraint decompositions, a protocol must be able to (re-) decompose constraints autonomously in a (small) subset θ of sites, when the constraints and database instances outside θ are unknown, and, furthermore, may change.[2] The only imposed limitation is the property of *resource partition* $(B_\theta, B_{\overline{\theta}})$ between sites inside and outside θ, that is, the cumulative resources of sites inside and outside θ must be bounded by their current *resource upper bounds* B_θ and $B_{\overline{\theta}}$, respectively. We show that the decompositions can be done autonomously in θ by proving that, given a database instance for sites in θ, the following are equivalent: (1) there exists a (partial) resource distribution (which we call *permissible*) for sites in θ such that for each site in θ its *resource* is bounded between its *lower* and *upper bounds*, and that the cumulative *resource upper bound* in θ is B_θ, and (2) there exists a (full) compact split (safe decomposition) of the global constraint that satisfies *resource partition* w.r.t. B_θ and local consistency. Furthermore, we show that optimal (full) constraint decompositions adhering to *resource partition* can also be achieved autonomously in θ subject to *resource partition* bounds.

Related Work

The problem of reformulation of global integrity constraints for easier verification, including local verification of global constraints in distributed databases, has drawn much attention (e.g., [16, 17, 15, 14, 2, 1, 4, 6, 12, 5, 7]). More closely associated with our work are the works [2, 1, 17, 13] which deal with local verification of numeric constraints, and the works [10, 11, 8] which consider applications of parametric linear constraint queries and their connection to Fourier's elimination method [3]. While the setting in [10, 11, 8] is very different from ours, we use very similar techniques for parametric characterization of constraints (in our case satisfying safety). However, the work on parametric queries assumes that the number of parameters (i.e., coefficients) is bounded, whereas our main technical difficulty is that *safe* decompositions do not have a bound on a number of parameters. Furthermore, [10, 11, 8] are not concerned with such notions as *resource* characterization, or other issues specific to distributed protocols, including the properties of feasible decompositions, concurrency, decompositions in the presence of partial data, and an optimization framework for safe decompositions.

---

[2] Note that the constraint preservation property is not adequate for this purpose because it assumes that the constraints outside θ are fixed (and known).

Perhaps most closely related to our work is the work of [2, 1], which was the first to consider verification of linear arithmetic constraints in the context of distributed databases. It considers a single atomic linear constraint at the global level. Intuitively, an atomic constraint must be such that it could be decomposed between two sites storing individual variables using some constant *boundaries*[3]. For example, the global constraint $A+B \geq 100$ can be decomposed into two local constraints $A \geq a$ and $B \geq b$, where a and b are constants such that $a+b \geq 100$; or the global constraint $A \leq B$ can be decomposed into $A \leq a$ and $b \leq B$, where $a \leq b$. The focus in [2, 1] is on the "demarcation" distributed protocol which is concerned with efficient (in terms of communication and other costs) negotiation between two sites on synchronizing the change in constant boundaries, in case a local update violates its local constraint. However, the question of how to achieve constraint decompositions was not addressed.

The work [17] uses an idea similar to the demarcation of [2, 1][4], in the context of network partition failures, in order to overcome the problem by trying to perform transactions locally. Similar to [2, 1], a global constraint in the example considered in [17] is a single linear inequality of the form $x_1 + \ldots + x_n \leq c$, which is split among n sites (i.e., a single variable per site) by giving each site a quota of c. However, [17] focuses on the distributed transaction management issue but does not address the problem of how to achieve constraint decompositions, or what constraint types, beyond the form given above, can be handled by its techniques.

The work [13] extends the demarcation protocol of [2, 1] by considering a wider class of constraints: linear, quadratic and polynomial. A global constraint is a single inequality that is decomposed into local constraints involving one variable per site. Since for this case, the property of *safety* corresponds geometrically to containment of a multidimensional rectangle in the shape described by the global constraint (inequality), [13] suggests the use of geometrical techniques for (dynamic) decompositions. However, geometrical techniques (e.g., from computational geometry) are restricted to low dimension (i.e., small overall number of variables) whereas typical distributed databases involve a large number of variables used in the global constraint (e.g., Crisis Management Scenario).

In contrast to [2, 1, 17, 13], our methods for linear arithmetic constraints have none of the above-mentioned restrictions, i.e., we allow atomic linear inequalities of any general form, a global constraint may have any number of atomic constraints, constraints may be partitioned among any number of sites, and each site may have not just one, but any number of variables. Furthermore, ours is the only work that provides a comprehensive framework and a full solution for achieving optimal decompositions for the case of general linear constraints. Moreover, our decompositions can work for different scenarios, i.e., with different assumptions regarding what is known at the time of a decomposition. In fact, among the works on local verification, to the best of our knowledge, only

---

[3] There is no precise formulation of the allowed atomic constraints in [2, 1].
[4] In fact, [17] is earlier.

the work [14] considers a framework for achieving optimal decompositions, but this is done for the first-order (non-numeric) *fragmentation* constraints and in a different setting (i.e., when additional semantic information is known.)

The work [12] dealt with first-order (non-numeric) constraints, in the context of distributed databases, and suggested certain heuristics to select better decompositions. However, the questions of how, and under what conditions, these heuristics relate to optimization criteria such as maximizing the probability of not violating local constraints and the optimality of decomposition were not considered.

The work [15] considered a certain class of first-order (non-numeric) constraints and presented an approach of equivalent reformulation (which is different from our *safety*) in the presence of additional semantic information (not for a distributed environment). They also suggested some heuristics, based on costs of constraint verification and reformulation, but no algorithm or guarantee of optimality in any sense was provided. Finally, [14] applied the techniques of [15] to database design in distributed environments and used an inference engine as a (local constraints) space generator for all (finitely many) alternatives of distributing a global constraint. They also presented heuristics for balancing the cost of constraint distribution and the efficiency of constraint verification.

Organization

The paper is organized as follows. Following the introduction, Section 2 provides a formal framework for selecting optimal decompositions, which is generic for all types of constraints. In Section 3 we review linear arithmetic constraints. Sections 4, 5 and 6 cover compact splits, resource characterization of decompositions, and resource distributions, respectively. Section 7 concludes and identifies some directions for future work.

2   Decomposition Optimization Framework

In this section we define the central notion of *safe decompositions*, and formulate our problem as one of finding the best feasible safe decomposition of a global constraint. The problem formulation in this section, except for the notion of *resources*, is applicable to all types of constraints.

2.1   Safe Decompositions

Definition 1. *A constraint C in variables x is a Boolean function from the domain of x, to the Boolean set, i.e., $C: Domain(x) \rightarrow \{True, False\}$*

By a slight abuse of notation x will denote either a vector or a set of variables.

Definition 2. *A variable partition $\mathbb{P}$ of the set of variables x is defined as $\mathbb{P} = (y_1, \ldots, y_M)$, such that $y_1 \cup y_2 \cup \ldots \cup y_M = x$, and $y_i \cap y_j = \emptyset$ for all i, j $(1 \leq i, j \leq M, i \neq j)$.*

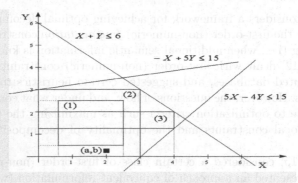

Fig. 1. Safe decompositions of Ω

Definition 3. *Let Ω be a constraint, and $\mathbb{P} = (\boldsymbol{y}_1, \ldots, \boldsymbol{y}_M)$ be a partition of variables. We say that $\mathbb{C} = (C_1, \ldots, C_M)$ is a decomposition of Ω, if in every constraint C_i all free variables are from $\boldsymbol{y}_i$. Sometimes we will use $\mathbb{C}$ to indicate the conjunction $C_1 \wedge .. \wedge C_M$. We say that a decomposition $\mathbb{C} = (C_1, \ldots, C_M)$ is safe if $C_1 \wedge .. \wedge C_M \models \Omega$, i.e., $C_1, \ldots, C_M$ is subsumed by Ω.*

Definition 4. *Let $\boldsymbol{x}^0 = (\boldsymbol{y}_1^0, \ldots, \boldsymbol{y}_M^0)$ be a database instance. We say that $\boldsymbol{x}^0$ satisfies a decomposition $\mathbb{C} = (C_1, \ldots, C_M)$ if $\boldsymbol{y}_i^0$ satisfies C_i for all $1 \leq i \leq M$.*

Example 1. Consider the following set Ω of linear constraints: $X + Y \leq 6$, $-X + 5Y \leq 15$, $5X + 4Y \leq 15$, and both variables X and Y are non-negative. A graphic representation of Ω is given in Figure 1. For the partition $\mathbb{P} = (\{X\}, \{Y\})$ (i.e., two sites with a single variable each), consider three safe decompositions $\mathbb{C}_1, \mathbb{C}_2, and\, \mathbb{C}_3$ as follows:

$$\mathbb{C}_1 = (C_{11}, C_{12})$$
$$= (\{0.5 \leq X \leq 2.5\}, \{0.5 \leq Y \leq 2.5\})$$
$$\mathbb{C}_2 = (C_{21}, C_{22}), and$$
$$= (\{0.0 \leq X \leq 3.0\}, \{0.0 \leq Y \leq 3.0\}), and$$
$$\mathbb{C}_3 = (C_{31}, C_{32})$$
$$= (\{0.0 \leq X \leq 4.0\}, \{1.25 \leq Y \leq 2.0\}).$$

$\mathbb{C}_1$ is a safe decomposition of Ω because every point (X,Y) that satisfies C_{11} and C_{12} will also satisfy Ω. Geometrically, this means that the space (1) defined by $\mathbb{C}_1$ is contained in the space defined by Ω. Similarly, $\mathbb{C}_2$ and $\mathbb{C}_3$ are safe decompositions of Ω. Note that the database instance (a, b) satisfies $\mathbb{C}_2$, but not $\mathbb{C}_1$ and $\mathbb{C}_3$.

Note that rectangle (1) (for $\mathbb{C}_1$) is strictly contained in rectangle (2) (for $\mathbb{C}_2$). Hence, the decomposition $\mathbb{C}_2$ is better than $\mathbb{C}_1$ in the sense that, in $\mathbb{C}_2$ we will have to perform global updates less frequently than in $\mathbb{C}_1$, i.e., less overhead. This notion is defined formally as follows:

Definition 5. *Given a constraint Ω and its two decompositions $\mathbb{C}_1 = (C_{11}, \ldots, C_{1M})$ and $\mathbb{C}_2 = (C_{21}, \ldots, C_{2M})$, we say that $\mathbb{C}_2$ subsumes $\mathbb{C}_1$ (or $\mathbb{C}_1$ is subsumed by $\mathbb{C}_2$) if:*

$$\bigwedge_{i=1}^{M} C_{1i} \models \bigwedge_{i=1}^{M} C_{2i}$$

We will denote this by $\mathbb{C}_1 \models \mathbb{C}_2$. We say that $\mathbb{C}_2$ strictly subsumes $\mathbb{C}_1$ if $\mathbb{C}_1 \models \mathbb{C}_2$, but $\mathbb{C}_2 \not\models \mathbb{C}_1$. Furthermore, we say that a safe decomposition $\mathbb{C}$ is minimally-constrained, if there is no safe decomposition $\mathbb{C}'$ that strictly subsumes $\mathbb{C}$. Finally, we say that $\mathbb{C}_1$ is equivalent to $\mathbb{C}_2$, denoted by $\mathbb{C}_1 \equiv \mathbb{C}_2$, if $\mathbb{C}_1 \models \mathbb{C}_2$ and $\mathbb{C}_2 \models \mathbb{C}_1$.

Note that in Example 1, $\mathbb{C}_2$ and $\mathbb{C}_3$ are minimally-constrained safe decompositions, while $\mathbb{C}_1$ is not.

Proposition 1. *Let $\mathbb{P} = (y_1, \ldots, y_M)$ be a variable partition and $\mathbb{C}$ be a conjunction of constraints $(C_1, \ldots, C_M)$, where C_i $(1 \leq i \leq M)$ is over y_i. Then, $\mathbb{C}$ is satisfiable iff for every i, $1 \leq i \leq M$, C_i is satisfiable.*

Proof. The ONLY-IF part is immediate, while the IF part follows from the fact that $C_1, \ldots, C_M$ do not share variables.

Proposition 2. *Let $\mathbb{C}_1 = (C_{11}, \ldots, C_{1M})$ and $\mathbb{C}_2 = (C_{21}, \ldots, C_{2M})$ be two lists of constraints over $y_1, \ldots, y_M$ respectively, for partition $\mathbb{P} = (y_1, \ldots, y_M)$. Then:*

1. *If $\mathbb{C}_1$ is satisfiable, then $\mathbb{C}_1 \models \mathbb{C}_2$ iff for all i, $1 \leq i \leq M$, $C_{1i} \models C_{2i}$.*
2. *If both $\mathbb{C}_1$ and $\mathbb{C}_2$ are satisfiable, then $\mathbb{C}_1 \equiv \mathbb{C}_2$ iff for all i, $1 \leq i \leq M$, $C_{1i} \equiv C_{2i}$.*

Proof. Part 2 immediately follows from Part 1. In Part 1, the IF direction is obvious, while the ONLY-IF direction is due the fact that the variable partition $\mathbb{P} = (y_1, \ldots, y_M)$ is disjoint as follows: Assume that $\mathbb{C}_1 \models \mathbb{C}_2$, but, by way of contradiction, for some i, $1 \leq i \leq M$, $C_{1i} \not\models C_{2i}$. Then, there exists a_i over y_i that satisfies C_{1i}, but not C_{2i}. Since, $\mathbb{C}_1$ is consistent, each $C_{11}, \ldots, C_{1M}$ must be consistent, and therefore there must exist $b_1, \ldots, b_M$ over $y_1, \ldots, y_M$, that satisfy $C_{11}, \ldots, C_{1M}$, respectively. Then, $b_1, \ldots, b_{i-1}, a_i, b_{i+1}, \ldots, b_M$ satisfies $\mathbb{C}_1$, but not $\mathbb{C}_2$, contradicting the fact that $\mathbb{C}_1 \models \mathbb{C}_2$.

In practical cases, we are only interested in the case when Ω is satisfiable, because otherwise the database must be empty and no update would be allowed. Technically, however, every unsatisfiable (i.e., inconsistent) decomposition will be safe for an unsatisfiable Ω. If Ω is satisfiable, then we have the following:

Proposition 3. *Let Ω be a satisfiable constraint. Then every minimally-constrained safe decomposition $\mathbb{C}$ of Ω is satisfiable.*

Proof. Since Ω is satisfiable, there exists $a = (a_1, \ldots, a_M)$ over x that satisfies Ω. Then, the decomposition $\mathbb{C}_1 = (y_1 = a_1, \ldots, y_M = a_M)$ is always a safe decomposition of Ω. Consider now an arbitrary minimally-constrained safe decomposition $\mathbb{C}$. If, by way of contradiction, $\mathbb{C}$ is not satisfiable, then $\mathbb{C} \models \mathbb{C}_1$ and $\mathbb{C}_1 \nvDash \mathbb{C}$, contradicting the minimality of $\mathbb{C}$.

Clearly, safe or even minimally-constrained safe decompositions are not unique. In our example, both $\mathbb{C}_2$ and $\mathbb{C}_3$ are minimally-constrained, because there is no other safe decomposition that strictly subsumes $\mathbb{C}_2$ or $\mathbb{C}_3$.

Since safe decompositions are not unique, an important question is how to choose a safe decomposition that is optimal according to some meaningful criterion. In our example, the rectangle with the maximal area may be a good choice. In fact, if update points (X,Y) are uniformly distributed over the given space (defined by Ω), then the larger area (volume in the general case) corresponds to greater probability that an update will satisfy local constraints, and thus no global processing will be necessary. We defer the discussion on optimality criteria to Section 2.3.

2.2   Optimization Problem Formulation

We suggest the following general framework for selecting optimal feasible decompositions:

$$
\begin{aligned}
maximize\ f(s) \\
s.t.\ \ s \in \mathbb{S}
\end{aligned}
\tag{1}
$$

where $\mathbb{S}$ is the set of all feasible decompositions, and $f : \mathbb{S} \to \mathbb{R}$ (real numbers) is the objective function discussed in the next subsection.

Definition 6. *Let Ω be a constraint, $\mathbb{C} = (C_1, \ldots, C_M)$ be a decomposition of Ω, $\theta = \{k + 1, \ldots, M\}$ be a subset of sites $\{1, \ldots, M\}$, and $x^0 = (y_1^0, \ldots, y_M^0)$ be a database instance. We consider the following properties of $\mathbb{C}$:*

1. Safety. $\mathbb{C}$ *has this property if it is a safe decomposition of Ω.*
2. Local Consistency. $\mathbb{C}$ *has this property w.r.t. $x^0 = (y_1^0, \ldots, y_M^0)$ if every local instance y_i^0 satisfies its local constraint C_i $(1 \leq i \leq M)$. Clearly, local consistency and safety imply global consistency.*
3. Partial Constraint Preservation. $\mathbb{C} = (C_1, \ldots, C_M)$ *has this property w.r.t. local constraints $C_1', \ldots, C_M'$ outside $\theta = \{k + 1, \ldots, M\}$ if $C_i = C_i'$ for all $1 \leq i \leq M$, i.e., local constraints outside θ are fixed to $(C_1', \ldots, C_k')$, respectively.*
4. Resource Partition B_θ. *The resource partition property is given only for families of constraints in which the resource characterization, defined in Section, is possible (e.g., linear constraints considered in this paper). Namely, the global constraint Ω is associated with the global resource bound b, each local constraint C_i in the decomposition is associated with a resource r_i, and each subset of sites θ is associated with the cumulative resource r_θ. A partition of*

the global resource bound between θ and $\bar{\theta}$ (i.e., all sites except θ) is a pair $(\boldsymbol{B}_\theta, \boldsymbol{B}_{\bar{\theta}})$, such that $\boldsymbol{B}_\theta + \boldsymbol{B}_{\bar{\theta}} = \boldsymbol{b}$, which is identified by $\boldsymbol{B}_\theta$.

We say that a decomposition $\mathbb{C}$ has a resource bound partition property w.r.t. a partition $\boldsymbol{B}_\theta$, if the cumulative resource $\boldsymbol{r}_\theta$ is bounded by $\boldsymbol{B}_\theta$, and $\boldsymbol{r}_{\bar{\theta}}$ is bounded by $\boldsymbol{B}_{\bar{\theta}}$.[5]

Given a set Pr of properties that contains safety and possibly other properties above, the set $\mathbb{S}_{Pr}$ of feasible (safe) decompositions w.r.t. Pr is the set of all decompositions of Ω that satisfy the properties in Pr.

Before we discuss how the decomposition problems can be solved effectively, we consider possible candidates for function f.

2.3 Optimization Criteria

There are many feasible (minimally-constrained) safe decompositions in $\mathbb{S}$, and we would like to formulate a criterion to select the best among them. This criterion should represent the problem characteristics, and the decomposition goals. Possible criteria include:

- Maximize the probability that an update will not violate the existing local constraints (decomposition).
- Minimize overall expected cost of computations during an update.
- Maximize the expected number of updates before the first update that violates local constraints.
- Maximize the expected length of time before an update violates local constraints.

Many other optimization criteria are possible. However, any reasonable criteria should be monotonic, as defined below.

Definition 7. *Let f be a function from the set of safe decompositions of Ω to $\mathbb{R}$. We say that f is monotonic if for every two decompositions $\mathbb{C}_1, \mathbb{C}_2$ of Ω, $\mathbb{C}_1 \models \mathbb{C}_2$ implies $f(\mathbb{C}_1) \leq f(\mathbb{C}_2)$.*

Intuitively, being monotonic for an optimization criterion means that enlarging the space defined by a decomposition can only make it better.

Note, that if f is monotonic, then $f(\mathbb{C}_1) = f(\mathbb{C}_2)$ for any two equivalent decompositions $\mathbb{C}_1$ and $\mathbb{C}_2$.

As we will see in Section, it is often necessary to consider a subspace of all safe decompositions (without loosing an optimal decomposition).

Definition 8. *Let $\mathbb{S}$ be a set of safe decompositions of Ω.[6] A subset $\mathbb{S}'$ of $\mathbb{S}$ will be called a monotonic cover of $\mathbb{S}$ if for every decomposition $\mathbb{C}$ in $\mathbb{S}$ there exists a decomposition $\mathbb{C}'$ in $\mathbb{S}'$, such that $\mathbb{C}'$ subsumes $\mathbb{C}$ (i.e., $\mathbb{C} \models \mathbb{C}'$).*

---

[5] Note that the notion of resource bound partition is more flexible than constraint preservation, and allows one to perform concurrent constraint decompositions.

[6] Not necessarily the set of all safe decompositions of Ω.

The following proposition states that optimal decompositions are not missed when the search space is restricted to a monotonic cover and the optimization function is monotonic.

Proposition 4. *Let $\mathbb{S}$ be a (sub) set of all safe decompositions of Ω, $\mathbb{S}'$ be a monotonic cover of $\mathbb{S}$, and f be a monotonic function from $\mathbb{S}$ to $\mathbb{R}$. Then, the following two optimization problems yield the same maximum.*

1. *Problem 1. max $f(s)$, s.t. $s \in \mathbb{S}$.*
2. *Problem 2. max $f(s)$, s.t. $s \in \mathbb{S}'$.*

Proof. Suppose the maxima of f in Problems 1 and 2 are achieved by $s = \mathbb{C}$ in $\mathbb{S}$ and $s = \mathbb{C}'$ in $\mathbb{S}'$ respectively. Since $\mathbb{S}' \subseteq \mathbb{S}$, $f(\mathbb{C}') \leq f(\mathbb{C})$. Now, since $\mathbb{S}'$ is a monotonic cover of $\mathbb{S}$, there must exist $\mathbb{C}'' \in \mathbb{S}'$, such that $\mathbb{C} \models \mathbb{C}''$. Therefore, because f is monotonic, $f(\mathbb{C}) \leq f(\mathbb{C}'')$. Finally, since the maximum of Problem 2 is achieved at $\mathbb{C}'$, $f(\mathbb{C}') \geq f(\mathbb{C}'') \geq f(\mathbb{C})$. Thus, $f(\mathbb{C}') = f(\mathbb{C})$ which completes the proof.

3   Linear Arithmetic Constraints

Definition 9. *An atomic linear constraint is an inequality of the form $a_1 x_1 + a_2 x_2 + ... + a_n x_n \leq b$, where $a_1, a_2, ..., a_n$, and b are real numbers, and $x_1, x_2, .., x_n$ are variables ranging over the reals.*

Definition 9 defines an atomic linear constraint as a symbolic expression. However, for each instantiation of values into the variables $(x_1, x_2, \ldots, x_n)$, the inequality $a_1 x_1 + a_2 x_2 + \ldots + a_n x_n \leq b$ evaluates to TRUE or FALSE, and thus defines a Boolean function $C : \mathbb{R}^n \to \{TRUE, FALSE\}$.

Definition 10. *A linear system Ω is a conjunction of atomic linear constraints.*

A linear system Ω containing n variables and (a conjunction of) m atomic linear constraints, can be written as follows:

$$
\begin{array}{l}
a_{11}x_1 + a_{12}x_2 + \ldots + a_{1n}x_n \leq b_1 \\
a_{21}x_1 + a_{22}x_2 + \ldots + a_{2n}x_n \leq b_2 \\
\quad \vdots \qquad \vdots \qquad \vdots \qquad \vdots \qquad \vdots \\
a_{m1}x_1 + a_{m2}x_2 + \ldots + a_{mn}x_n \leq b_m
\end{array}
\tag{2}
$$

This system Ω can also be written in matrix notation as the system $A\boldsymbol{x} \leq \boldsymbol{b}$, where A is the matrix

$$
\begin{pmatrix}
a_{11} & a_{12} & \ldots & a_{1n} \\
a_{21} & a_{22} & \ldots & a_{2n} \\
\vdots & \vdots & \vdots & \vdots \\
a_{m1} & a_{m2} & \ldots & a_{mn}
\end{pmatrix}
\tag{3}
$$

and $\boldsymbol{b}$ is the column vector $(b_1\ b_2\ \ldots\ b_m)$, and $\boldsymbol{x}$ is the vector $(x_1\ x_2\ \ldots\ x_n)$.

Parametric Optimization Problem

To solve the optimization problem for linear arithmetic constraints we want to rewrite the problem (1), i.e.,

$$max\ f(s)$$
$$s.t.\ s \in \mathbb{S}$$

where $\mathbb{S}$ is the set of feasible safe decompositions, into the form

$$max\ f(\boldsymbol{w})$$
$$s.t.\ \Phi(\boldsymbol{w})$$

where $\boldsymbol{w}$ is the set of variables describing coefficients (i.e., parameters) of constraints on a decomposition $D(\boldsymbol{w})$, and $\Phi(\boldsymbol{w})$ is a logical condition in terms of $\boldsymbol{w}$ defining the search space

$$\mathbb{S}' = \{D(\boldsymbol{w}) \mid \Phi(\boldsymbol{w})\}$$

such that $\mathbb{S}'$ is a monotonic cover of $\mathbb{S}$.

By Proposition 4, the two problems are equivalent for any user-given monotonic optimization function, but the latter allows the use of known mathematical programming methods to solve it.

4   Split Decompositions

The first problem is that, for a safe decomposition, a constraint C_i at site i may be characterized by an unbounded set of atomic linear constraints. Thus the size of a parametric description (using coefficients of those constraints) is unbounded. To overcome this problem, we reduce the search space to the set of what we call *compact split* decompositions, for which we prove that: (1) there does exist a parametric description of bounded size and (2) the optimum of the objective function among all safe decompositions *can always be found* in the subspace of split decompositions.

Definition 11. *Let $\Omega = A\boldsymbol{x} < \boldsymbol{b}$ be a constraint on $\boldsymbol{x}$, and $\mathbb{P} = (\boldsymbol{y}_1, \ldots, \boldsymbol{y}_M)$ be a variable partition of $\boldsymbol{x}$. A split of Ω, denoted by $D(\boldsymbol{r}_1, \ldots, \boldsymbol{r}_M)$, is a tuple $(A_1\boldsymbol{y}_1 \leq \boldsymbol{r}_1, \ldots, A_M\boldsymbol{y}_M \leq \boldsymbol{r}_M)$ of constraints, where A_i, $1 \leq i \leq M$, is the matrix composed of the columns of A that are associated with $\boldsymbol{y}_i$. We say that a split $D(\boldsymbol{r}_1, \ldots, \boldsymbol{r}_M)$ is safe (respectively, minimally-constrained) if it is a safe (respectively, minimally-constrained) decomposition of Ω. For a subset θ of sites, say $\{k + 1, \ldots, M\}$, a (partial) θ-split of Ω, denoted by $D(\boldsymbol{r}_{k+1}, \ldots, \boldsymbol{r}_M)$, is a tuple $(A_{k+1}\boldsymbol{y}_{k+1} \leq \boldsymbol{r}_{k+1}, \ldots, A_M\boldsymbol{y}_M \leq \boldsymbol{r}_M)$ of constraints.*

Note that the vectors $\boldsymbol{r}_i$ above have the same dimension as the vector $\boldsymbol{b}$, which equals to the number of constraints in Ω. Recall that, by Proposition 3, $D(\boldsymbol{r}_1, \ldots, \boldsymbol{r}_M)$ is satisfiable if and only if for all i, $1 \leq i \leq M$, $A_i\boldsymbol{y}_i \leq \boldsymbol{r}_i$ is satisfiable.

For our classification we introduce the notion of *tight form* for a system $Ax \leq b$,[7] which states, intuitively, that the values of b are tight. This is formalized as follows.

Definition 12. *We say that a constraint $Ax \leq b$ is tight, if there does not exist b', such that $b' \leq b$, $b' \neq b$, and $Ax \leq b$ is equivalent to $Ax \leq b'$. We say that a split $D(r_1, \ldots, r_M)$ is tight if every satisfiable constraint $A_i y_i \leq r_i$ in it $(1 \leq i \leq M)$ is tight.*

Claim 1. *For any satisfiable system $Ax \leq b$ (respectively split) there exists an equivalent system (respectively split) that is tight. Furthermore, every tight constraint $Ax \leq b$ is satisfiable.*

Definition 13. *We say that a split $D(r_1, \ldots, r_M)$ of Ω is compact if*

$$\sum_{i=1}^{M} r_i \leq b$$

Lemma 1 (Split Properties). *Let $\Omega = Ax \leq b$. Then:*

1. *Every compact split is safe.*
2. *If $D(r_1, \ldots, r_M)$ is a tight split, it is compact iff it is safe.*
3. *For every safe decomposition $\mathbb{C}$ of Ω, there exists a minimally-constrained safe split $D(r_1, \ldots, r_M)$ of Ω that subsumes $\mathbb{C}$, i.e., $\mathbb{C} \models D(r_1, \ldots, r_M)$.*
4. *Every minimally-constrained safe decomposition of Ω is equivalent to (1) a minimally-constrained safe split of Ω and to (2) a compact split of Ω.*

Theorem 1. *Let $\Omega = Ax \leq b$ be a satisfiable global constraint, f be a monotonic function from the set of all safe decompositions of Ω to $\mathbb{R}$, $\mathbb{P} = (y_1, \ldots, y_M)$ be a variable partition of x, and $x^0 = (y_1^0, \ldots, y_M^0)$ be an instance of x. Let $\mathbb{S}$ and $\mathbb{SS}$ be the sets of all safe decompositions and all compact splits of Ω, respectively, and let $\mathbb{S}_{x^0}$ and $\mathbb{SS}_{x^0}$ be the sets $\mathbb{S}$ and $\mathbb{SS}$ restricted to decompositions that satisfy x^0. Then,*

1. *$max\ f(s)$ s.t. $s \in \mathbb{S} = max\ f(s)$ s.t. $s \in \mathbb{SS}$*
2. *$max\ f(s)$ s.t. $s \in \mathbb{S}_{x^0} = max\ f(s)$ s.t. $s \in \mathbb{SS}_{x^0}$*

Proof. 1. Using Lemma 1, we can easily derive that $\mathbb{SS}$ is a *monotonic cover* of $\mathbb{S}$. Then, by Proposition 4, both problems yield the same maximum.

2. Similar to Part 1, by Lemma 1, $\mathbb{SS}_{x_0}$ is a monotonic cover of $\mathbb{S}_{x_0}$, and then, by Proposition 4, both problems yield the same maximum. This completes the proof.

Following Theorem 1, from now on we only consider compact splits. Vectors $(r_1, \ldots, r_M)$ can be viewed as resources assigned to sites, because they represent how much of vector b is distributed to each site. The following section presents a parametric resource characterization of splits and a parametric formulation of the optimization problem in terms of resources.

---

[7] $Ax \leq b$ in this definition denotes any system of linear constraints, not just the global constraint.

5   Resource Characterization

This section characterizes (compact) splits in terms of *resources*, which allows (in the next subsection) to reduce *compact split* (safe decompositions) to *resource distributions*. In turn, *resource distributions* can significantly simplify the management of a distributed protocol, because of their small size as compared with the size of a (safe) decomposition. Also, as explained in the next subsection, *resource distributions* support concurrent constraint (re-)decompositions.

More specifically, we formulate the properties of *compactness*, *local consistency (lc)*, *partial constraint preservation (pcp)*, and *resource bound partition (rp)* for splits in terms of resources. Then, the optimization problem is formulated in terms of such a characterization. First, we introduce the concept of *resources* of (compact) splits.

Definition 14 (Resource Parameters). *Let $\Omega = A\boldsymbol{x} \leq \boldsymbol{b}$ be a satisfiable global constraint, $D(\boldsymbol{r}_1, \dots, \boldsymbol{r}_M)$ be a tight compact split of Ω, $\boldsymbol{x}^0 = (\boldsymbol{y}_1^0, \dots, \boldsymbol{y}_M^0)$ be an instance of $\boldsymbol{x}$, and θ be a subset $\{k+1, \dots, M\}$ of sites $\{1, \dots, M\}$. Then, we say that:*

1. *$\boldsymbol{b}$ is the global upper bound of resources in Ω.*
2. *$\boldsymbol{r}_i$ is the resource assigned to site i, $1 \leq i \leq M$.*
3. *$\boldsymbol{r} = \sum_{i=1}^{M} \boldsymbol{r}_i$ is the global resource.*
4. *$\boldsymbol{\delta} = \boldsymbol{b} - \boldsymbol{r}$ is the global passive slack of Ω w.r.t. $D(\boldsymbol{r}_1, \dots, \boldsymbol{r}_M)$.*
5. *$(\boldsymbol{\delta}_1, \dots, \boldsymbol{\delta}_M)$ such that $\boldsymbol{\delta}_i \geq 0$, $1 \leq i \leq M$ and $\sum_{i=1}^{M} \boldsymbol{\delta}_i = \boldsymbol{\delta}$, is a partition of $\boldsymbol{\delta}$. Each $\boldsymbol{\delta}_i$, $1 \leq i \leq M$, is called the passive slack at site i.*
6. *$\boldsymbol{ur}_i = \boldsymbol{r}_i + \boldsymbol{\delta}_i$ is the upper resource bound at site i, $(1 \leq i \leq M)$.*
7. *Given an instance $\boldsymbol{y}_i^0$ at site i, $\boldsymbol{lr}_i = A_i \boldsymbol{y}_i^0$, $1 \leq i \leq M$, is the lower resource bound at site i w.r.t. $\boldsymbol{y}_i^0$.*
8. *Given an instance $\boldsymbol{y}_i^0$ at site i, $\boldsymbol{\Delta}_i = \boldsymbol{r}_i - \boldsymbol{lr}_i$, $1 \leq i \leq M$, is the active slack at site i w.r.t. $\boldsymbol{y}_i^0$.*

Finally, we define the cumulative parameters for θ, namely cumulative resources, resource upper and lower bounds, and passive and active slacks by $\boldsymbol{r}_\theta = \sum_{i \in \theta} \boldsymbol{r}_i$, $\boldsymbol{ur}_\theta = \sum_{i \in \theta} \boldsymbol{ur}_i$, $\boldsymbol{lr}_\theta = \sum_{i \in \theta} \boldsymbol{lr}_i$, $\boldsymbol{\delta}_\theta = \sum_{i \in \theta} \boldsymbol{\delta}_i$, and $\boldsymbol{\Delta}_\theta = \sum_{i \in \theta} \boldsymbol{\Delta}_i$, respectively.

The above resource parameters are shown in Figure 2. In this figure, each resource $\boldsymbol{r}_i$ is bounded between its lower and upper bound ($\boldsymbol{lr}_i$ and $\boldsymbol{ur}_i$), the difference between upper bound ($\boldsymbol{ur}_i$) and the resource is the passive slack $\boldsymbol{\delta}_i$, and the difference between the resource ($\boldsymbol{r}_i$) and its lower bound ($\boldsymbol{lr}_i$) is the active slack ($\boldsymbol{\Delta}_i$).

The following proposition characterizes the split properties of *compactness*, *local consistency (lc)*, *partial constraint preservation (pcp)*, and *resource bound partition (rp)* in terms of the resource parameters.

Proposition 5 (Parametric Feasible Properties). *Let $\Omega = A\boldsymbol{x} \leq \boldsymbol{b}$ be a global satisfiable constraint, $\mathbb{P} = (\boldsymbol{y}_1, \dots, \boldsymbol{y}_M)$ be a variable partition of $\boldsymbol{x}$, $\boldsymbol{x}^0 = (\boldsymbol{y}_1^0, \dots, \boldsymbol{y}_M^0)$ be an instance of $\boldsymbol{x}$, θ be a subset of sites, say $\theta = \{k+1, \dots, M\}$,*

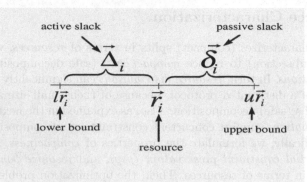

Fig. 2. Resource Representation at site i

$\bar{\theta}$ be the set $\{1, \ldots, k\}$, and $D(r_1^0, \ldots, r_k^0)$ be a (partial) $\bar{\theta}$-split that satisfies $(y_1^0, \ldots, y_k^0)$. Then, for any split $D(r_1, \ldots, r_M)$

1. $D(r_1, \ldots, r_M)$ is compact iff the global resource r is bounded by the global upper bound b, i.e., $r \leq b$. We denote this condition by $\Phi_{compact}(r_1, \ldots, r_M)$.
2. $D(r_1, \ldots, r_M)$ satisfies local consistency w.r.t. x^0 iff the resource r_i assigned to site i is bounded from below by its lower bound lr_i, i.e., for every site i, $1 \leq i \leq M$, $lr_i \leq r_i$. We denote this condition by $\Phi_{lc}(r_1, \ldots, r_M)$.
3. $D(r_1, \ldots, r_M)$ satisfies partial constraint preservation w.r.t. $\bar{\theta}$-split $D(r_1^0, \ldots, r_k^0)$ iff the resources at each site outside θ are fixed, i.e., $r_1 = r_1^0, \ldots, r_k = r_k^0$. We denote this condition by $\Phi_{pcp}(r_1, \ldots, r_M)$.

We will also denote by $\Phi_{rp, B_\theta}(r_1, \ldots, r_M)$ the condition stating that $D(r_1, \ldots, r_M)$ satisfies resource partition w.r.t. a resource bound partition B_θ.

Proof. 1) follows directly from Lemma 1 part 1, 2) follows from the definition of local consistency, i.e., for a given instance y_i^0, it satisfies its local constraint iff $A_i y_i^0 \leq r_i$, and 3) follows directly from partial constraint preservation definition. This completes the proof.

In the following, we denote by $SS_{compact}$, SS_{lc}, SS_{pcp} and SS_{rp} the set of splits satisfying compactness, local consistency w.r.t. x^0, partial constraint preservation w.r.t. a θ-split $D(r_1, \ldots, r_k)$, and resource bound partition B_θ w.r.t. θ, respectively. We will use Pr to denote a subset of the set of properties $\{compact, lc, pcp, rp\}$. Finally, set $\mathbb{S}_{Pr}$ will denote the set of all splits that satisfy the properties Pr, i.e., $SS_{Pr} = \cap_{p \in Pr} SS_p$, and $\Phi_{Pr}(r_1, \ldots r_M)$ will be the conjunction of the corresponding conditions, i.e., $\Phi_{Pr}(r_1, \ldots r_M) = \wedge_{p \in Pr} \Phi_p (r_1, \ldots r_M)$. We can present the optimization problem in terms of resource characterization.

Theorem 2 (Resource Optimization). *Let $\Omega = Ax \leq b$ be a satisfiable global constraint, f be a monotonic function from the set of all safe decompositions to $\mathbb{R}$, $\mathbb{P} = (y_1, \ldots, y_M)$ a variable partition of x, $x^0 = (y_1^0, \ldots, y_M^0)$ be an instance of x, and Pr be the subset of properties $\{compactness, lc, pcp, rp\}$*

that must contain compactness or resource bound partition. Then, solving the optimization problem

$$max\ f(s)$$
$$s.t.\ s \in \mathbb{SS}_{Pr}$$

is equivalent to solving the parametric problem[8]

$$max\ f(D(r_1, \ldots, r_M))$$
$$s.t.\ \Phi_{Pr}(r_1, \ldots, r_M)$$

Proof. The proof follows from Lemma 1, Proposition 5, and the fact that resource bound partition is a stronger property than compactness as follows:

(1) The *compactness* part. By Proposition 5, the set $\mathbb{SS}_{compactness}$ is characterized by condition $\Phi_{compactness}(r_1, \ldots, r_M)$. Then, both problems yield the same maximum.

The other Pr cases follow directly from Proposition 5. This completes the proof.

6 Resource Distributions and Concurrent Splits

The resource characterization of the previous section assumes that information from all sites is used. However, in order to support distributed and autonomous protocols we would like to make constraint decompositions and re-decompositions involving only a (small) subset of sites, say $\theta = \{(k+1), \ldots, M\}$ of sites $\{1, \ldots, M\}$. To do that a formulation of the decomposition problem can only involve or affect information that is stored in sites θ. We capture this idea using the notion of (full or partial) *resource distribution* as follows.

Definition 15. *Let $\Omega = A\boldsymbol{x} \leq \boldsymbol{b}$ be a global constraint, $D(r_1, \ldots, r_M)$ be a compact split of Ω, $(\delta_1, \ldots, \delta_M)$ be a partition of the global passive slack δ, $\boldsymbol{x}^0 = (\boldsymbol{y}_1^0, \ldots, \boldsymbol{y}_M^0)$ be a database instance, θ is a subset, say $\{k+1, \ldots, M\}$, of sites. A resource distribution is a tuple (of triples) $((\boldsymbol{lr}_1, \boldsymbol{r}_1, \boldsymbol{ur}_1), \ldots, (\boldsymbol{lr}_M, \boldsymbol{r}_M, \boldsymbol{ur}_M))$[9]; a θ-resource distribution is $((\boldsymbol{lr}_{k+1}, \boldsymbol{r}_{k+1}, \boldsymbol{ur}_{k+1}), \ldots, (\boldsymbol{lr}_M, \boldsymbol{r}_M, \boldsymbol{ur}_M))$. We say that the resource distribution is permissible if*

$$\sum_{i=1}^{M} \boldsymbol{ur}_i = \boldsymbol{b} \text{ and } \boldsymbol{lr}_i \leq \boldsymbol{r}_i \leq \boldsymbol{ur}_i, \text{for every } 1 \leq i \leq M$$

Given a resource bound partition $\boldsymbol{B}_\theta$, we say that the θ-resource distribution $((\boldsymbol{lr}_{k+1}, \boldsymbol{r}_{k+1}, \boldsymbol{ur}_{k+1}), \ldots, (\boldsymbol{lr}_M, \boldsymbol{r}_M, \boldsymbol{ur}_M))$ is permissible w.r.t. $\boldsymbol{B}_\theta$ if

$$\sum_{i \in \theta} \boldsymbol{ur}_i = \boldsymbol{B}_\theta \text{ and } \boldsymbol{lr}_i \leq \boldsymbol{r}_i \leq \boldsymbol{ur}_i, \text{for every } i \in \theta$$

---

[8] we will sometimes write $f(D(r_1, \ldots, r_M))$ as $f(r_1, \ldots, r_M)$.
[9] Note $(\boldsymbol{lr}_i, \boldsymbol{r}_i, \boldsymbol{ur}_i)$'s are defined in Definition 14.

Note that if θ is the set of all sites, the resource distribution permissibility is equivalent to θ resource distribution permissibility.

The key advantage of a resource distribution is its small size of $O(nc)$ as compared with the size $O(nc * nv)$ of a constraint decomposition, where nc and nv are the number of constraints and variables, respectively, in the global constraint. In fact, nv may be as large as the size of a database, for example when the global constraint reflects that the summation of some quantity, one per relational tuple, is bounded by a constant.

The following proposition motivates the notion of permissible distribution and provides a local criterion for a subset θ of sites to decide whether a feasible resource distribution exists, after database instances have been updated (i.e., lower bounds).

Proposition 6 (θ-Resource Distribution Feasibility). *Let θ be a subset, say $\{k+1, \ldots, M\}$, of sites $\{1, \ldots, M\}$, and $\bar{\theta}$ be its complement, B_θ be a resource bound partition. Then,*

1. *Given a database instance $(y_{k+1}^0, \ldots, y_M^0)$ at sites θ (and thus lower bounds $(lr_{k+1}, \ldots, lr_M)$), the following are equivalent:*
   (a) *There exists a compact split of Ω satisfying resource partition B_θ and local consistency w.r.t. to an instance $(y_{k+1}^0, \ldots, y_M^0)$ at θ.*
   (b) *There exists a θ-permissible resource distribution w.r.t. B_θ (with the above lower bounds).*
   (c) *$lr_\theta = \sum_{i \in \theta} lr_i \leq B_\theta$*
2. *The combination of θ-permissible resource distribution w.r.t. B_θ and $\bar{\theta}$-permissible resource distribution w.r.t. $B_{\bar{\theta}}$ constitutes a permissible resource distribution.*

The optimization functions f are defined in terms of the information of all sites (i.e., $D(r_1, \ldots, r_M)$), whereas we need to work with the information only on a subset θ of sites. To do that, we define the notion of θ-localizer as follows.

Definition 16. *Let Ω be a constraint over x, and $\mathbb{S}$ be the set of all splits of Ω, $f : \mathbb{S} \to \mathbb{R}$ be a function. Let $\theta = \{k+1, \ldots, M\}$ be a subset of sites $\{1, \ldots, M\}$, $\bar{\theta}$ its complement, $\mathbb{S}_\theta$ be the set of all θ-splits. Then, function $f_\theta : \mathbb{S}_\theta \to \mathbb{R}$ is called θ-localizer of f if for any $r_1^0, \ldots, r_k^0$, and for every two splits $D(r_1^0, \ldots, r_k^0, r_{k+1}, \ldots, r_M)$ and $D(r_1^0, \ldots, r_k^0, r_{k+1}', \ldots, r_M')$ in $\mathbb{S}$,*

$$f(r_1^0, \ldots, r_k^0, r_{k+1}, \ldots, r_M) \geq f(r_1^0, \ldots, r_k^0, r_{k+1}', \ldots, r_M') \Leftrightarrow$$
$$f_\theta(r_{k+1}, \ldots, r_M) \geq f_\theta(r_{k+1}', \ldots, r_M')$$

(i.e., f_θ preserves monotonicity for any resource instantiation outside of θ).

We are now ready to formulate a theorem to be used for concurrent (re-) decompositions of the global constraints.

Theorem 3. *Let $\Omega = Ax \leq b$ be a global constraints, θ be a subset, say $\{k+1, \ldots, M\}$, of sites $\{1, \ldots, M\}$ and $\bar{\theta}$ be its complement, $\mathbb{S}$ be the set of* *all compact splits of* Ω, $f : \mathbb{S} \to \mathbb{R}$ *be a monotonic function and* f_θ *be its* θ-*localizer,* $\boldsymbol{B}_\theta$ *be a resource bound partition, and* $D(\boldsymbol{r}_1^0, \ldots, \boldsymbol{r}_k^0)$ *be a (partial)* $\bar{\theta}$-*split for sites outside* θ. *Let* Pr *be a subset of properties that contains* rp *and* $\Phi_{\theta-Pr}$ *be the condition (1)* $\boldsymbol{r}_\theta \leq \boldsymbol{B}_\theta$ *for the case that* Pr *contains just* rp, *and the conditions (2)* $\boldsymbol{r}_\theta \leq \boldsymbol{B}_\theta, \boldsymbol{l}\boldsymbol{r}_i \leq \boldsymbol{r}_i$ *for every* $i \in \theta$, *for the case of* Pr *that contains both* rp *and* lc. *Then,*

1. *For* Pr *being the set of properties* $\{rp, pcp\}$ *or* $\{rp, pcp, lc\}$, *let* $(\boldsymbol{r}'_{k+1}, \ldots, \boldsymbol{r}'_M)$ *be a solution to the problem*

$$max f_\theta(\boldsymbol{r}_{k+1}, \ldots, \boldsymbol{r}_M)$$
$$s.t. \Phi_{\theta-Pr}$$

   Then, $(\boldsymbol{r}_1^0, \ldots, \boldsymbol{r}_k^0, \boldsymbol{r}'_{k+1}, \ldots, \boldsymbol{r}'_M)$ *is a solution to the problem*

$$max f(\boldsymbol{r}_1, \ldots, \boldsymbol{r}_M)$$
$$s.t. \Phi_{Pr}$$

2. *For* Pr *being the set of properties* $\{rp\}$ *or* $\{rp, lc\}$, *let* $(\boldsymbol{r}'_{k+1}, \ldots, \boldsymbol{r}'_M)$ *be a solution to the problem*

$$max f_\theta(\boldsymbol{r}_{k+1}, \ldots, \boldsymbol{r}_M)$$
$$s.t. \Phi_{\theta-Pr}$$

   and $(\boldsymbol{r}'_1, \ldots, \boldsymbol{r}'_k)$ *be a solution to the problem*

$$max f_{\bar{\theta}}(\boldsymbol{r}_{k+1}, \ldots, \boldsymbol{r}_M)$$
$$s.t. \Phi_{\bar{\theta}-Pr}$$

   Then, $(\boldsymbol{r}'_1, \ldots, \boldsymbol{r}'_k, \boldsymbol{r}'_{k+1}, \ldots, \boldsymbol{r}'_M)$ *is a solution to the problem*

$$max f(\boldsymbol{r}_1, \ldots, \boldsymbol{r}_M)$$
$$s.t. \Phi_{Pr}$$

Proof. The proof follows from the fact that f has a θ-localizer, and from Propositions 5 and 6. This completes the proof.

7 Conclusions

This paper has formulated a generic framework for finding optimal constraint decompositions for a range of design and update-time scenarios, and provided a comprehensive solution for the case of general linear constraints, which are widely used in distributed applications such as resource allocation, reservations, financial transactions, and logistics.

Although the main motivation has been distributed integrity constraint maintenance, the technical results seem considerably more general, and applicable to other problems that can be expressed as constraint decompositions. However, more study is necessary to identify other domains where our approach can work.

Extending our decomposition techniques to non-linear numeric constraints, as well as to non-numeric constraint is another area for future study. Technically, many non-numeric constraints can be captured by linear constraints. For example, [2, 1] showed that referential integrity constraints can be expressed using linear constraints. However, it remains to be carefully studied how efficient our decomposition methods would be if combined with the reduction, as compared to possible direct decomposition techniques working directly on non-numeric constraints.

Finally, we have worked on related issues, which are outside the scope of this paper and will be reported elsewhere. They include the development of distributed protocols for constraint verification during database updates and constraint (re-) decompositions, and finding parametric forms of specific objective functions, such as minimizing the probability of local constraints.

Acknowledgements

We would like to thank Daniel Barbará for his very valuable and insightful comments and suggestions for improving this paper. This research was sponsored in part by the Defense Advanced Research Project Agency (DARPA) within the Advanced Logistic Program under contract number N00600-96-D-3202, and by the National Science Foundation (NSF) grants IIS-9734242 and IRI-9409770.

References

1. D. Barbará and H. Garcia-Molina. The Demarcation Protocol: A technique for maintaining constraints in distributed database systems. *VLDB Journal*, 2(3).
2. D. Barbará and H. Garcia-Molina. The demarcation protocol: A technique for maintaining arithmetic constraints in distributed database systems. In *Proc. 3rd International Conference on Extending Data Base Technology, EDBT92*, pages 373–388. Springer-Verlag, 1992.
3. J.B.J. Fourier. *Reported in: Analyse de travaux de l'Academie Royale des Sciences, pendant l'annee 1824, Partie Mathematique, Historyde l'Academie Royale de Sciences de l'Institue de France 7 (1827) xlvii-lv. (Partial English translation in: D.A. Kohler, Translation of a Report by Fourier on his work on Linear Inequalities. Opsearch 10 (1973) 38-42.).* 1824.
4. H. Garcia-Molina. Global consistency constraints considered harmful. In *Proc. First International Workshop on Interoperability in Multidatabase Systems (IMS 91)*, pages 248–250, 1991.
5. S. Grufman, F. Samson, S. Embury, P. Gray, and T. Risch. Distributing semantic constraints between heterogeneous databases. In *13th International Conference on Data Engineering (ICDE'97), (IEEE), Birmingham, England*, 1997.
6. A. Gupta and J. Widom. Local verification of global integrity constraints in distributed databases. In *Proc. ACM-SIGMOD International Conference on Management of Data*, pages 49–58, Washington, D.C., 1993. ACM.

7. N. Huyn. Maintaining global integrity constraints in distributed databases. *CON-STRAINTS, An Internation Journal*, 2(3–4):377–399, 1997.
8. T. Huynh, L. Joskowicz, C. Lassez, and J.L. Lassez. Practical tools for reasoning about linear constraints. *Fundamenta Informaticae, Special issue on Logic and Artificial Intelligence*, 15(4):357–379, 1991.
9. S. Jajodia and L. Kerschberg. *Advanced Transaction Models and Architectures*. Norwall, MA, Kluwer Academic Publishers, first edition, 1997.
10. J-L. Lassez. Querying constraints. In *Proc. 9th ACM SIGACT-SIGMOD-SIGART Symp. on Principles of Database Systems*, 1990.
11. Jean-Louis Lassez and Michael Maher. On Fourier's algorithm for linear arithmetic constraints. *Journal of Automated Reasoning*, 9:373–379, 1992.
12. S. Mazumdar. Optimizing distributed integrity constraints. In *Proc. Third International Symposium on Database Systems for Advanced Applications (DASFAA-93)*, pages 327–334, Taejon, Korea, 1993.
13. S. Mazumdar and Z. Yuan. Localizing global constraints: A geometric approach. In *In Proceedings of the 9th International Conference on Computing and Information. ICCI'98*, 1998.
14. X. Qian. Distributed desing of integrity constraints. In L. Kerschberg, editor, *Proc. Second International Conference on Expert System Database Systems*, pages 417–425, Redwood City, California, 1988. Benjamin Cummings.
15. X. Qian and D. Smith. Constraint reformulation for efficient validation. In *Proc. Thirteenth International Conference on Very Large Databases*, pages 622–632, 1987.
16. E. Simon and P. Valduriez. Design and implementation of an extendible integrity subsystem. In *Proc. Nineteenth Hawaii International Conference on System Sciences*, pages 622–632, 1986.
17. N. Soparkar and A. Silberschatz. Data-value partitioning and virtual messages. In ACM, editor, *Proc. 9th ACM SIGACT-SIGMOD-SIGART Simposium on Principles of Database Systems*, Nashville, Tennessee, 1990.

Adaptive Random Testing

T.Y. Chen[1,*], H. Leung[2], and I.K. Mak[3]

[1] School of Information Technology, Swinburne University of Technology,
Hawthorn, Victoria 3122, Australia
tchen@it.swin.edu.au
[2] Department of Computer Science, New Mexico State University,
Las Cruces, NM 88003, U.S.A.
hleung@nmsu.edu
[3] School of Professional and Continuing Education,
The University of Hong Kong, Pokfulam Road, Hong Kong
keith.mak@hkuspace.hku.hk

Abstract. In this paper, we introduce an enhanced form of random testing called *Adaptive Random Testing*. Adaptive random testing seeks to distribute test cases more evenly within the input space. It is based on the intuition that for non-point types of failure patterns, an even spread of test cases is more likely to detect failures using fewer test cases than ordinary random testing. Experiments are performed using published programs. Results show that adaptive random testing does outperform ordinary random testing significantly (by up to as much as 50%) for the set of programs under study. These results are very encouraging, providing evidences that our intuition is likely to be useful in improving the effectiveness of random testing.

1 Introduction

There are basically two approaches towards the selection of test cases, namely the white box and the black box approach. Among the black box techniques, random selection of test cases is generally regarded as not only a simple but also an intuitively appealing technique (e.g. see White [1]). In random testing, test cases may be randomly chosen based on a uniform distribution or according to the operational profile. As pointed out by Hamlet [2], the main merits of random testing include the availability of efficient algorithms to generate its test cases, and its ability to infer reliability and statistical estimates.

In all random testing studies, only the rate of failure-causing inputs (hereafter referred to as the failure rates) is used in the measurement of effectiveness. For example, the expected number of failures detected and the probability of detecting at least one failure are all defined as functions of the failure rates. However, in a recent study by Chan et al. [3], it has been found that the performance of a partition testing strategy depends not only on the failure rate, but also on the

---

* Corresponding author.

M.J. Maher (Ed.): ASIAN 2004, LNCS 3321, pp. 320–329, 2004.

geometric pattern of the failure-causing inputs. This has prompted us to investigate whether the performance of random testing can be improved by taking the patterns of failure-causing inputs into consideration. We have developed a new type of random testing, namely adaptive random testing. Our studies show that the effectiveness of random testing can be significantly improved without incurring significant overheads. We observe that adaptive random testing outperforms ordinary random testing in general.

2   Preliminaries

In this study, we assume that the random selection of test cases is based on a uniform distribution and without replacement. As a reminder, most of the analytical studies of random testing assume that selections are with replacement [4]. The assumption of selection with replacement has long been criticised by practitioners, because in reality, test cases should not be repeated. It has been used mainly because it has a simpler mathematical model and hence of easier analysis. Thus, the model of our study reflects the reality more closely.

We basically follow the notation used by Chen and Yu [4]. Elements of an input domain are known as failure-causing inputs, if they produce incorrect outputs. For an input domain D, we use d, m and n to denote the size, number of failure-causing inputs and number of test cases, respectively. The sampling rate σ and failure rate θ are defined as $\frac{n}{d}$ and $\frac{m}{d}$, respectively.

In previous studies of random testing, the two most popular effectiveness metrics are: the probability of detecting at least one failure (referred to as the P-measure) and the expected number of failures detected (referred to as the E-measure). There have been some criticisms on these two metrics despite their popularity. The main criticism of using the E-measure is that higher E-measures do not necessarily imply more faults or more distinct failures; and the main disadvantage of using the P-measure is that there is no distinction between cases of detecting different number of failures. Although these two measures are not ideal, they have been used extensively in the literature.

In this paper, rather than using the P-measure and the E-measure as the effectiveness metrics as done in previous studies, we propose another effectiveness metric. We use the expected number of test cases required to detect the first failure (referred to as the F-measure), as the effectiveness metric. For random selection of test cases with replacement, the F-measure, denoted by F, is equal to $\frac{1}{\theta}$, or equivalently $\frac{d}{m}$. Intuitively speaking, the F-measure reflects the effectiveness of a testing strategy more naturally and directly, as the lower the F-measure the more effective the testing strategy because fewer test cases are required to reveal the first failure. In practice, when a failure is detected, testing is normally stopped and debugging starts. The testing phase would normally be resumed only after fixing of the fault. Hence, the F-measure is not only more intuitively appealing but also more realistic from a practical perspective.

In a recent study, Chan et al. [3] have observed that the performance of some partition testing strategies varies with the patterns of failure-causing inputs (hereafter referred to as the failure patterns). They have classified the patterns of failure-causing inputs into three categories: point, strip and block patterns. To illustrate this, let us assume that the input domain has two dimensions. Figures 1-3 show the point, strip and block patterns, respectively. The outer boundaries represent the borders of the input domain and the filled regions denote the failure-causing inputs, that is the failure patterns, in each of the figures.

Fig. 1. Point Pattern   **Fig. 2.** Strip Pattern   **Fig. 3.** Block Pattern

The main characteristic of the point pattern is that either the failure-causing inputs are stand alone points, or they form regions of a very small size which are scattered over the whole domain. For the strip pattern, the failure-causing inputs form the shape of a narrow strip. A typical example of this failure pattern is White and Cohen's [5] domain errors. For the block pattern, the main characteristic is that the failure-causing inputs are concentrated in either one or a few regions.

3   Adaptive Random Testing

With ordinary random testing, the chances of hitting the failure patterns, that is selecting failure-causing inputs as test cases, depends solely on the magnitude of the failure rate. However, a closer inspection shows that for non-point patterns which include both the strip and block patterns, the failure detection capability can be significantly improved by slightly modifying the ordinary random testing technique. Let us use an example to illustrate the intuition behind our modified random testing.

Consider an input domain D. Suppose D has a "regular" geometry, that is, we assume that it is easy to generate inputs randomly from D. Let there be a fault in the program. For example, let the input in D be consisting of values x and y where $0 \leq x, y \leq 10$. Suppose the fault lies in a conditional expression

$$x + y > 3$$

at a program statement while the correct expression should have been

$$x + y > 4$$

Specifically, the domain D is a square $\{(x,y) \mid 0 \leq x,y \leq 10\}$. The fault corresponds to a failure region $\{(x,y) \mid 3 < x + y \leq 4\}$, which is a "strip" of width 1 across the square domain D.

We consider applying random testing to the program. Suppose we generate one test case $(2.2, 2.2)$. Since $2.2+2.2$ is larger than 4, the fault is not revealed. Let the next test case generated be $(2.1, 2.1)$. Should we know in advance that the error region is a strip of width 1, we could argue that the choice of the second test case is too "humble". The test set should better be more spaced out such that two neighbouring test cases are kept apart by a distance of length at least 1.

Thus, we propose to modify random testing as follows. To generate a new test case, we need to make sure that the new test case should not be too close to any of the previously generated ones. One way to achieve this is to generate a number of random test cases and then choose the "best" one among them. That is, we try to distribute the selected test cases as spaced out as possible.

To summarize, we conjecture that *test cases should be as evenly spread over the entire input domain as possible* in order to achieve a small F-measure. This forms the basis of our new method of random testing, namely *Adaptive Random Testing*.

An implementation of *Adaptive Random Testing* is as follows: Adaptive random testing makes use of two sets of test cases, namely the executed set and the candidate set which are disjoint. The executed set is the set of distinct test cases that have been executed but without revealing any failure; while the candidate set is a set of test cases that are randomly selected without replacement. The executed set is initially empty and the first test case is randomly chosen from the input domain. The executed set is then incrementally updated with the selected element from the candidate set until a failure is revealed. From the candidate set, an element that is *farthest away* from all executed test cases, is selected as the next test case.

Obviously, there are various approaches to implement the intuition of "*farthest away*". In this paper the criterion for implementing this intuition is defined as follows. Let $T = \{t_1, t_2, \cdots, t_n\}$ be the executed set and $C = \{c_1, c_2, \cdots, c_k\}$ be the candidate set such that $C \cap T = \emptyset$. The criterion is to choose the element c_h such that for all $j \in \{1, 2, \cdots, k\}$,

$$\min_{i=1}^{n} dist(c_h, t_i) \geq \min_{i=1}^{n} dist(c_j, t_i)$$

where $dist$ is defined as the Euclidean distance.

In other words, in an m-dimensional input domain, for inputs $a = (a_1, a_2, \cdots, a_m)$ and $b = (b_1, b_2, \cdots, b_m)$, $dist(a, b) = \sqrt{\sum_{i=1}^{n} (a_i - b_i)^2}$. The rationale of this criterion is to evenly spread the test case through maximising the minimum distance between the next test case and the already executed test cases.

It should be noted that there are also various ways to construct the candidate set giving rise to various versions of adaptive random testing. It will be quite difficult, if not impossible, to carry out an analytical study for the performance

Table 1. Program name, dimension (D), input domain, seeded error types, and total number of errors for each of the error-seeded programs. The error types are: arithmetic operator replacement (AOR); relational operator replacement (ROR); scalar variable replacement (SVR) and constant replacement (CR)

Prog Name	D	Input Domain From	To	AOR	ROR	SVR	CR	Total Errors	Failure Rate
airy	1	(-5000.0)	(5000.0)				1	4	0.000716
bessj	2	$(2.0, -1000.0)$	$(300.0, 15000.0)$	2	1		1	4	0.001298
bessj0	1	(-300000.0)	(300000.0)	2	1	1	1	5	0.001373
cel	4	$(0.001, 0.001, 0.001, 0.001)$	$(1.0, 300.0, 10000.0, 1000.0)$	1	1		1	3	0.000332
el2	4	$(0.0, 0.0, 0.0, 0.0)$	$(250.0, 250.0, 250.0, 250.0)$	1	3	2	3	9	0.000690
erfcc	1	(-30000.0)	(30000.0)	1	1	1	1	4	0.000574
gammq	2	$(0.0, 0.0)$	$(1700.0, 40.0)$		3		1	4	0.000830
golden	3	$(-100.0, -100.0, -100.0)$	$(60.0, 60.0, 60.0)$		3	1	1	5	0.000550
plgndr	3	$(10.0, 0.0, 0.0)$	$(500.0, 11.0, 1.0)$	1	2		2	5	0.000368
probks	1	(-50000.0)	(50000.0)	1	1	1	1	4	0.000387
sncndn	2	$(-5000.0, -5000.0)$	$(5000.0, 5000.0)$			4	1	5	0.001623
tanh	1	(-500.0)	(500.0)	1	1	1	1	4	0.001817

of the adaptive random testing in comparison with the ordinary random testing with respect to the F-measure. Therefore, we conduct an empirical study in this paper.

4   An Empirical Study on Adaptive Random Testing

In this section, an empirical investigation was conducted to compare the performance between adaptive random testing and ordinary random testing, using the F-measure as the effectiveness metric, which is defined as the expected number of test cases required to detect the first failure. In this paper, F_a and F_r are used to denote the F-measures for the adaptive random testing and ordinary random testing, respectively. Unless otherwise specified, ordinary random testing would be abbreviated as random testing or RT.

In this empirical study, we use a set of 12 error-seeded programs. They are all published programs ([6, 7]), which are written in Fortran, Pascal or C with program sizes ranging from 30 to 200 statements. All of them involve numerical computations and have been converted into C++ programs. Table 1 lists details of the failure rate, type and number of seeded errors for each program.

The candidate set is of constant size and a new candidate set is constructed each time a test case is selected. Let us refer it as the *Fixed Size Candidate Set Version of the Adaptive Random Testing* (abbreviated as the FSCS). Algorithm 1 describes how to generate a candidate set and select a test case for this version of adaptive random testing.

Firstly, we conducted a preliminary study to see how the size of the candidate set would affect the performance of the adaptive random testing for our sample of 12 programs. The range of the size of the candidate set, k, was set to vary from 2 to 10 with an increment of 2 and then from 10 to 100 with an increment of 10. We have observed that in general, the larger the k, the smaller the number of test cases required to detect the first failure. Furthermore, for $k \geq 10$, there not much difference in the number of test cases required to detect the first failure in our sample of programs. Hence, we set $k = 10$ in our experiment.

Algorithm 1:
```
/*
selected_set := { test data already selected };
candidate_set := {};
total_number_of_candidates := 10;
*/
function Select_The_Best_Test_Data(selected_set, candidate_set,
                   total_number_of_candidates);
    best_distance := -1.0;
    for i := 1 to total_number_of_candidates do
            candidate := randomly generate one test data from the program
                         input domain, the test data cannot be in
                         candidate_set nor in selected_set;
            candidate_set := candidate_set + { candidate };
            min_candidate_distance := Max_Integer;
            foreach j in selected_set do
                    min_candidate_distance := Minimum(min_candidate_distance,
                            Euclidean_Distance(j, candidate));
            end_foreach
            if (best_distance < min_candidate_distance) then
                    best_data := candidate;
                    best_distance := min_candidate_distance;
            end_if
    end_for
    return best_data;
end_function
```

For each program, we applied both FSCS and RT with the same first randomly selected test case. We obtained a pair of numbers (u^a, u^r), where u^a and u^r were the numbers of test cases required to detect the first failure, using FSCS (Algorithm 2) and RT, respectively. We called such a pair of numbers a sample. Obviously, (u^a, u^r) depends on the first test case. Hence, the process was repeated with various randomly selected inputs as the first test case. The central limit theorem [8] was used to determine the size of the sample set $S = \{(u_1^a, u_1^r), (u_2^a, u_2^r), \cdots\}$, that is, to determine when the process could be stopped so that we have enough samples in S to provide reliable statistic estimates.

Algorithm 2:
initial_test_data := *randomly generate a test data from the input domain*;
selected_set := { initial_test_data };
counter := 1;
total_number_of_candidates := 10;
use initial_test_data *to test the program*;
if (*program output is incorrect*) **then**
        reveal_failure := **true**;
else
        reveal_failure := **false**;
end_if
while (**not** reveal_failure) **do**
        candidate_set := {};
        test_data := Select_The_Best_Test_Data(selected_set, candidate_set,
                        total_number_of_candidates);
        use test_data *to test the program*;
        if (*program output is incorrect*) **then**
                reveal_failure := **true**;
        else
                selected_set := selected_set + { test_data };
                counter := counter + 1;
        end_if
end_while
output counter;

Suppose we want to estimate the mean of the number of test cases required to reveal the first failure, that is, the F-measure, for FSCS with an accuracy of $\pm r\%$ and a confidence level of $(1 - \alpha) \times 100\%$, where $1 - \alpha$ is the confidence coefficient. According to the central limit theorem, the size of S required to achieve this goal should be at least as

$$|S| = \left(\frac{100 \cdot z \cdot \sigma^a}{r \cdot \mu^a} \right)^2 \tag{1}$$

where z is the normal variate of the desired confidence level, μ^a is the population mean and σ^a is the population standard deviation. Similarly, for RT,

$$|S| = \left(\frac{100 \cdot z \cdot \sigma^r}{r \cdot \mu^r} \right)^2 \tag{2}$$

where μ^r and σ^r are the population mean and standard deviation, respectively. To ensure the size of S satisfying both FSCS and RT, we take the maximum value of equations (1) and (2), that is

$$|S| = max \left[\left(\frac{100 \cdot z \cdot \sigma^a}{r \cdot \mu^a} \right)^2, \left(\frac{100 \cdot z \cdot \sigma^r}{r \cdot \mu^r} \right)^2 \right] \tag{3}$$

Obviously, the larger the sample size $|S|$, the higher the associated confidence z, or the smaller the accuracy range r. However, larger samples mean more effort and resources. In our experiment, the confidence level was set to 95% and r was set to 5%. In other words, we collect samples until the sample mean is accurate within 5% of its value at 95% confidence. From the statistical tables, we know that for 95% confidence, $z = 1.96$. Moreover, since μ^a, μ^r, σ^a and σ^r are unknown, their estimators $\bar{u}^a$, $\bar{u}^r$, s^a and s^r were used instead, respectively, where $\bar{u}^a = \frac{\sum_{i=1}^{n} u_i^a}{n}$, $\bar{u}^r = \frac{\sum_{i=1}^{n} u_i^r}{n}$, s^a is the standard deviation of $\{u_1^a, u_2^a, \cdots, u_n^a\}$, s^r is the standard deviation of $\{u_1^r, u_2^r, \cdots, u_n^r\}$ and n is the current size of S. Thus, equation (3) becomes

$$|S| = max \left[\left(\frac{100 \cdot 1.96 \cdot s^a}{5 \cdot \bar{u}^a} \right)^2, \left(\frac{100 \cdot 1.96 \cdot s^r}{5 \cdot \bar{u}^r} \right)^2 \right] \qquad (4)$$

The above equation was used to decide when the process of collecting (u_i^a, u_i^r) could be stopped.

As shown in Table 2, the sizes of S vary with programs, but less than 3000. In this experiment, we have chosen 3000 as the sample size for all programs to calculate the means and standard deviations for both FSCS and RT.

Table 2. Mean Comparison Summary

Pg ID	k	$|S|$	F_a	95% CI of F_a	F_r	95% CI of F_r
AIRY	10	1506	799.29	(779.43, 819.15)	1381.44	(1332.51, 1430.37)
BESSJ	10	1598	466.66	(451.95, 481.38)	802.17	(722.90, 831.44)
BESSJ0	10	1567	423.93	(412.55, 435.30)	733.96	(707.44, 760.48)
CEL	10	1550	1607.80	(1552.49, 1663.11)	3065.25	(2955.10, 3175.40)
EL2	10	1650	686.48	(661.78, 711.17)	1430.76	(1377.71, 1483.81)
ERFCC	10	1543	1004.68	(980.34, 1029.02)	1803.62	(1738.94, 1868.30)
GAMMQ	10	1557	1081.43	(1044.35, 1118.51)	1220.28	(1176.32, 1264.24)
GOLDEN	10	1569	1829.74	(1765.68, 1893.80)	1860.81	(1793.52, 1928.10)
PLGNDR	10	1438	1806.94	(1754.47, 1859.41)	2741.66	(2646.75, 2836.57)
PROBKS	10	1480	1442.50	(1407.24, 1477.76)	2634.86	(2542.32, 2727.40)
SNCNDN	10	1617	628.68	(606.16, 651.20)	636.19	(612.83, 659.54)
TANH	10	1469	306.86	(299.21, 314.51)	557.96	(538.43, 577.48)

where

Pg ID	Program ID
k	Size of the candidate set
$|S|$	Size of the sample set
F_a	Mean of $\{u_1^a, u_2^a, \cdots, u_{|S|}^a\}$
95% CI of F_a	95% Confidence Interval of FSCS mean
F_r	Mean of $\{u_1^r, u_2^r, \cdots, u_{|S|}^r\}$
95% CI of F_r	95% Confidence Interval of RT mean

Table 2 lists the sample means within an accuracy of 5% and a confidence level of 95%, for all 12 programs. Also included are their 95% confidence interval (CI).

Table 2 shows that in all 12 programs the F_a's are smaller than the corresponding F_r's, that is, on the average, FSCS required fewer test cases than RT to reveal the first failure. Furthermore, in 10 programs (AIRY, BESSJ, BESSJ0, CEL, EL2, ERFCC, GAMMQ, PLGNDR, PROBKS and TANH), the corresponding FSCS's and RT's 95% confidence intervals do not overlap each other. This implies that F_a is smaller than F_r with a probability of $(0.95)^2$ at least. For the remaining 2 programs (GOLDEN and SNCNDN), the 95% confidence intervals of F_a and F_r overlap each other.

To calculate the performance improvement, we use the following formula:

$$\frac{F_r - F_a}{F_r} \times 100 \tag{5}$$

Table 3 shows that the performance improvement for FSCS over RT ranges from the best end of 52.02% to the worst end of 1.18%. Nine programs (AIRY, BESSJ, BESSJ0, CEL, EL2, ERFCC, PLGNDR, PROBKS and TANH) show a very significant improvement; one program (GAMMQ) shows a moderate improvement; two programs (GOLDEN and SNCNDN) show small improvement.

In summary, FSCS has a considerably smaller F-measure than RT. The experimental results show that FSCS has a good chance at outperforming RT by a very significant margin, which can be as high as 50%. On the other hand, FSCS needs more resources to select the next test case from the candidate set than RT does. Let n be the value of the F-measure and k be the size of the candidate set in FSCS. The selection overheads is of the order kn^2.

5 Conclusion

Random testing is simple in concept and is easy to be implemented. Besides, it can infer reliability and statistical estimates. Hence, it has been used by many testers. Since random testing does not make use of any information to generate test cases, it may not be a powerful testing method and its performance is solely dependent on the magnitude of failure rates.

Table 3. Improvement of the F-measure of FSCS over RT

Pg ID	Improvement in %
AIRY	42.14
BESSJ	41.83
BESSJ0	42.24
CEL	47.55
EL2	52.02
ERFCC	44.30
GAMMQ	11.38
GOLDEN	1.67
PLGNDR	34.09
PROBKS	45.25
SNCNDN	1.18
TANH	45.00

A recent study has shown that failure patterns may be classified as point, strip or block failure patterns. Intuitively speaking, when the failure pattern is not a point pattern, more evenly spread test cases have a better chance of hitting the failure patterns. Based on this intuition, we propose a modified version of random testing called *adaptive random testing*. An empirical analysis of 12 published programs has shown that adaptive random testing outperforms random testing significantly for most of the cases.

Our experimental results have been very encouraging, providing evidences that our intuition of spreading test cases more evenly within the input space is potentially very useful. Nevertheless, there are a number of issues of adaptive random testing that need to be considered, such as various criteria of evenly spreading of test cases, ways of defining the candidate sets. We anticipate that analysis of these issues would further improve the effectiveness of adaptive random testing. In fact, we have already obtained some very interesting results [9, 10].

Acknowledgment

We would like to thank F. T. Chan, T. H. Tse and Z. Q. Zhou for their invaluable discussions. We are also grateful to Dave Towey, who has converted the programs used in the former experiment into C++ languages and rerun the experiment.

T. Y. Chen is particularly indebted to Jean-Louis Lassez for showing how to be a model mentor, how to identify the incompleteness of an apparently complete solution, and how to find simple solutions.

References

1. White, L.J.: Software testing and verification. Advances in Computers **26** (1987) 335–391
2. Hamlet, R.: Random testing. Encyclopedia of Software Engineering. Edited by Marciniak, J. Wiley (1994)
3. Chan, F.T., Chen, T.Y., Mak, I.K., Yu, Y.T.: Proportional sampling strategy: guidelines for software testing practitioners. Information and Software Technology **38** (1996) 775–782
4. Chen, T.Y., Yu, Y.T.: On the relationship between partition and random testing. IEEE Transactions on Software Engineering **20** (1994) 977–980
5. White, L.J., Cohen, E.I.: A domain strategy for computer program testing. IEEE Transactions on Software Engineering **6** (1980) 247–257
6. Association for Computing Machinery: Collected Algorithms from ACM, Vol. I, II, III. Association for Computing Machinery (1980)
7. Press, W.H., Flannery, B.P., Teukolsky, S.A., Vetterling, W.T.: Numerical Recipes. Cambridge University Press (1986)
8. Freund, J.E.: Modern Elementary Statistics. Fifth edn. Prentice–Hall (1979)
9. Chen, T.Y., Kuo, F.C., Merkel, R.G., Ng, S.P.: Mirror adaptive random testing. Information and Software Technology (Accepted for publication)
10. Chen, T.Y., Eddy, G., Merkel, R., Wong, P.K.: Adaptive random testing through dynamic partitioning. In: Proceedings of the 4th International Conference on Quality Software (QSIC 04), IEEE Computer Society Press (2004)

Minimal Unsatisfiable Sets: Classification and Bounds

Sudeshna Dasgupta[1] and Vijay Chandru[2]

[1] Department of Computer Science, Mount Carmel College, Bangalore, India
sudeshna_dasgupta@hotmail.com, sudeshna@cs.umd.edu
[2] Indian Institute of Science & Strand Genomics, Bangalore, India
chandru@alum.mit.edu

For Jean-Louis Lassez on the occasion of his 5th cycle birthday.

Abstract. Proving the unsatisfiability of propositional Boolean formulas has applications in a wide range of fields. Minimal Unsatisfiable Sets (MUS) are signatures of the property of unsatisfiability in formulas and our understanding of these signatures can be very helpful in answering various algorithmic and structural questions relating to unsatisfiability. In this paper, we explore some combinatorial properties of MUS and use them to devise a classification scheme for MUS. We also derive bounds on the sizes of MUS in Horn, 2-SAT and 3-SAT formulas.

Keywords: Boolean formulas, propositional logic, satisfiability, Minimal Unsatisfiable Sets.

1 Introduction

Proving the unsatisfiability of propositional Boolean formulas has applications in a wide range of fields. Minimal Unsatisfiable Sets (MUS) are signatures of the property of unsatisfiability in formulas and our understanding of these signatures can be very helpful in answering various algorithmic and structural questions relating to unsatisfiability. Several authors have done a lot of work on Minimal Unsatisfiable Sets (MUS) of clauses [6, 7, 9–11]. Algorithms to search for MUS within a clause-set have been formulated by Renato Bruni and Antonio Sassano[7]. Polynomial time recognition of MUS with fixed clause-variable difference have been developed by Herbert Fleischner, Oliver Kullmann and Stefan Szeider[6]. Relationship between graphs and MUS has been investigated by Stefan Szeider[11]. Generalised MUS and relationship between linear algebra and combinatorics of clause sets have been studied by Oliver Kullman[10]. A framework for enumeration of MUS has been developed by Hans Kleine Büning and Xishun[9]. MUS also have a close link with matroid theory[5] and algorithms for logical inference [1–3]. This paper extends these investigations by presenting new structural properties of MUS, and utilising these properties to obtain a new classification scheme of MUS and bounds on the size of Horn, 2-SAT and 3-SAT MUS.

M.J. Maher (Ed.): ASIAN 2004, LNCS 3321, pp. 330–342, 2004.

This paper has 6 sections. Section 2 introduces some well known and some new definitions. Section 3 investigates basic properties of an MUS. In section 4 we classify literals and hence derive a classification of MUS. In section 5 we use the structural properties of MUS to derive bounds of Horn, 2-SAT and 3-SAT MUS and in the final section we present conclusions and future directions.

2   Preliminaries

In this section we repeat well-known definitions for a propositional CNF (conjunctive normal form) formula and introduce some new definitions.

Definition 1 (Literal). *A literal can be defined as a basic piece of information. A literal will be indicated by P, Q etc.*

Definition 2 (Atom). *An atom is a non-negated or negated literal, denoted respectively by a positive or negative sign preceeding a literal. An atom with a positive sign is called a positive atom and an atom with a negative sign is called a negative atom. An atom will be indicated by A, B or $+P, -P, +Q, -Q$ etc. If A is an atom then its negation will be denoted by $\neg A$. Negation of a positive atom is a negative atom and vice versa.*

Definition 3 (Conflicting Atoms). *If 2 atoms say A and B belong to the same literal but have opposing signs then they conflict or form a conflicting-pair indicated by $(A, B) \in Conf$. Thus $(A, \neg A) \in Conf$, $(+P, -P) \in Conf$ but $(+P, +P) \notin Conf$. A conflicting-pair of atoms cannot be assigned the same truth value, true or false, simultaneously.*

Definition 4 (Clause). *A clause is a finite set of distinct and non-conflicting atoms that are disjuncted from each other. A clause will be indicated by $C, C1, C2$ etc.*

Definition 5 (Conjunctive Formula). *A conjunctive formula is a finite set of distinct clauses that are conjuncted with each other. A conjunctive formula will be indicated by F, F' etc.*

Note 1. Let us note the following:
- Henceforth we shall just use formula to mean a conjunctive formula. The disjunctions between atoms of a clause, and conjunctions between clauses of a formula will be assumed. Thus a clause will be just a finite set of atoms and a formula will be just a finite set of clauses.
- The usual set operations between clauses as sets of atoms and between formulas as sets of clauses apply.

Definition 6 ((n,m) Formula). *A formula with n clauses and m literals will be called a (n, m) formula.*

Definition 7 (Satisfier). *A satisfier of a clause is an atom belonging to that clause.*

Definition 8 (Satcom). *A satcom, or satisfier combination of a formula F with n clauses, is a vector of n atoms, such that it contains exactly one satisfier of each clause of F. A satcom will be indicated by W,W1 etc.*

Though a satcom is a vector, we shall, whenever required, treat it as a set of atoms and apply set operations to it.

Definition 9 (Conflicting Satcom). *A satcom of a formula is said to be conflicting if it contains at least one pair of conflicting atoms. If a satcom is not conflicting then it is non-conflicting.*

Definition 10 (Satcomset). *A set of satcoms, where each satcom contains the same number of satisfiers, is a satcomset. A satcomset will be indicated by Z,Z' etc.*

Definition 11 (Conflicting Satcomset). *If every satcom of a satcomset is conflicting then it is a conflicting satcomset else it is a non-conflicting satcomset.*

Definition 12 (ALLSATCOMS(F)). *Given a formula F, then the satcomset containing all possible satcoms of F, is denoted by ALLSATCOMS(F).*

Definition 13 (SUBSATCOM). *Given a formula F, a satcomset Z such that $Z \subseteq ALLSATCOMS(F)$, and formula $F' \subseteq F$, we define SUBSATCOM(Z,F') as a satcomset that retains from each satcom of Z, only the satisfiers of F'.*
Thus $SUBSATCOM(Z,F') \subseteq ALLSATCOMS(F')$.
Similarly if satcom $W \in ALLSATCOMS(F)$, then SUBSATCOM(W,F') is a satcom obtained by retaining from W, only the satisfiers of F'.
Thus $SUBSATCOM(W,F') \in ALLSATCOMS(F')$.

Definition 14 (Conflicting Formula). *A formula is said to be conflicting if ALLSATCOMS(F) is a conflicting satcomset, else it is a non-conflicting formula.*

Note 2. Intuitively a conflicting formula represents an unsatisfiable formula and a non-conflicting formula represents a satisfiable formula.

Lemma 1. *We state the obvious facts:*

— Subset of a non-conflicting formula is non-conflicting.
— Superset of a conflicting formula is conflicting.

Proof: *The proof follows directly from a similar result for satcoms.* □

Definition 15 (Essential Satcom wrt a Clause). *Let F be a conflicting formula and clause $C \in F$, then satcom $W \in ALLSATCOMS(F)$ is called an essential satcom of F wrt clause C iff SUBSATCOM(W,F-{C}) is non-conflicting. W is called a non-essential satcom of F wrt clause C iff SUBSATCOM(W,F-{C}) is conflicting.*

Lemma 2. *Given a conflicting formula F, a clause $C \in F$, if there exists an essential satcom of F wrt clause C, then every conflicting subset of F must contain C.*

Proof: *This follows from the fact that SUBSATCOM(W,F-{C}) is non-conflicting, where W is an essential satcom of F wrt clause C, and lemma 1.* □

Lemma 3. *If F is a conflicting formula, clause $C \in F$, atom $A \in C$, and W is an essential satcom of F wrt clause C, then $\neg A \in W$.*

Proof: *Let us assume the contradiction that $\neg A \notin W$. Let W1 be a satcom obtained by attaching atom A of C to W, then W1 is a non-conflicting satcom of F. But this is a contradiction, since F is an MUS. Thus our assumption is false.* □

Definition 16 (Essential Satcom wrt an Atom). *Let F be a formula, clause $C \in F$, atom $A \in C$, then satcom $W \in ALLSATCOMS(F)$ is called an essential satcom of F wrt atom $A \in C$ if both the following hold:*

1. *Atom A of clause C is a part of every conflicting-pair of W.*
2. *Atom A appears exactly once in W.*

Lemma 4. *Let F be a conflicting formula, clause $C \in F$ then the following hold:*

– *If satcom W is an essential satcom of F wrt clause C, and the satisfier of clause C in W is atom A then W is an essential satcom of F wrt atom A of clause C.*
– *If satcom W is an essential satcom of F wrt atom A of clause C, then W is an essential satcom of F wrt clause C.*

Proof: *The proof follows from the definitions of essential satcom of F wrt a clause and an atom.* □

Definition 17 (MUS or Minimal Unsatisfiable Set). *Given a formula F then it is called an MUS, if it satisfies the following conditions:*

– *F is a conflicting formula.*
– *If $F' \subset F$ then F' is non-conflicting.*

Example 1. Let us look at some examples of an MUS

1. F = { {+P}, {-P, +Q}, {-P, -Q} } is an MUS since it satisfies both the conditions of an MUS.
2. F = { {+P}, {-P}, {-Q} } is not an MUS, since second condition fails, { {+P}, {-P} } is conflicting.
3. F = { {+P}, {-P, +Q}, {+P, -Q} } is not an MUS since first condition fails, (+P, +Q, +P) is a non-conflicting satcom.

Note 3. The only MUS with 1 literal say P, is { { +P },{ -P } }.
An (n,m) MUS refers to an MUS with n clauses and m literals.

Lemma 5. *Given a conflicting formula F, then there exists a formula F',*
$F' \subseteq F$, such that F' is an MUS.

Proof: *Since F has a finite number of clauses, it is easy to prove the lemma.*□

Definition 18 (Matrix Representation of a Formula). *Given a formula F*
with n clauses. Let us number them say C_1, C_2, ..., C_n. Let F have m distinct
literals. Let us number them say P_1, P_2, ..., P_m. Thus F has an equivalent n by m
matrix representation, rows corresponding to clauses and columns corresponding
to literals, where $F(C_i, P_j)$ can be either $\emptyset$, $+P_j$ or $-P_j$ and $1 \leq i \leq n$, $1 \leq j$
$\leq m$. Note that since a clause consists of distinct and non-conflicting atoms,
thus such a representation accounts for all the atoms appearing in a formula.
We shall use the matrix representation of a formula as its convenient equivalent
representation.

Definition 19 (Matrix Operations). *If P is a literal of formula F, where F*
has no empty clause i.e. a clause containing no atoms, we define the following
matrix operations on the matrix representation of F:

- *The formula obtained by deleting the column corresponding to literal P is*
  denoted by $\Delta_P(F)$. All empty clauses of $\Delta_P(F)$ are dropped.
- *The formula obtained by deleting the rows where atom $+P$ appears is denoted*
  by $\Delta^{P+}(F)$.
- *The formula obtained by deleting the rows where atom $-P$ appears is denoted*
  by $\Delta^{P-}(F)$.
- *The formula obtained by retaining only the rows where atom $+P$ appears is*
  denoted by $\Sigma^{P+}(F)$.
- *The formula obtained by retaining only the rows where atom $-P$ appears is*
  denoted by $\Sigma^{P-}(F)$.
- *If F^* is a formula such that $F^* \subseteq \Delta_P(F)$, then we define $\zeta_P(F, F^*)$ as the*
  formula obtained by adding the atoms of the deleted P literal to F^*.
  Thus $\zeta_P(F, F^*) \subseteq F$ and $\Delta_P(\zeta_P(F, F^*)) = F^*$.

3   Basic Properties

Definition 20 (Literal-Set). *Given formula F, then*
literal-set(F) = { P | +P or -P appears in some clause of F }. Similarly we have
literal-set(W) = { P | +P or -P is a satisfier appearing in W } if W is a satcom,
literal-set(C) = { P | Either +P $\in$ C or -P $\in$ C } if C is a clause.

Proposition 1. *In an MUS F, there cannot exist formula $F^* \subset F$, such that*
literal-set(F^*) $\cap$ literal-set(F - F^*) = $\emptyset$.

Proof: *Let F be an MUS. Assume the contradiction that there exists formula*
$F^* \subset F$, such that literal-set(F^*) $\cap$ literal-set(F - F^*) = $\emptyset$. Since F^* is a proper
subset of an MUS, it has a non-conflicting satcom say W1. Similarly (F - F^*)
must have non-conflicting satcom say W2. Let W3 be obtained by attaching W1

to W2. W3 is then a non-conflicting satcom of F by our assumption. But this is a contradiction since F is an MUS. Thus our assumption is false.   □

Proposition 2. *Given an MUS F, then the following hold:*

1. *For every atom $A \in C$, where clause $C \in F$, there exists a satcom $W \in ALLSATCOMS(F)$ such that W is an essential satcom of F wrt atom A of clause C.*
2. *For every clause $C \in F$, there exists a satcom $W \in ALLSATCOMS(F)$ such that W is an essential satcom of F wrt clause C.*
3. *For every atom $A \in C$, where clause $C \in F$, $\exists$ atom $A' \in C'$ where clause $C' \in F$ and $C \neq C'$, such that $(A,A') \in Conf$.*
4. *If $P \in literal\text{-}set(F)$, then $\exists$ clauses $C1, C2 \in F$ such that $+P \in C1$ and $-P \in C2$.*
5. *There exists no clauses $C1, C2 \in F$, such that $C1 \subseteq C2$.*

Proof:

1. Let atom $A \in C$, where clause $C \in F$. Assume the contradiction that every satcom W of F is such that either W has a conflicting-pair without atom A of clause C, or A is repeated in W. Let Z be a satcomset defined as $Z = \{ W \mid W \in ALLSATCOMS(F)$ and in W the satisfier of clause C is atom A $\}$. Z is a conflicting satcomset, since F is an MUS. Let satcomset $Z' = SUBSATCOM(Z, F - \{C\})$. By our assumption, Z' is a conflicting satcomset. Also $Z' = ALLSATCOMS(F - \{C\})$. But F is an MUS, thus we have reached a contradiction. Thus our assumption is false.
2. Assume the contradiction that there exists a clause C such that there is no essential satcom of F wrt clause C. Thus if $W \in SUBSATCOM(ALLSATCOMS(F), F - \{C\})$, then W is conflicting. So $(F - \{C\})$ is conflicting. Since F is an MUS, we have reached a contradiction. Thus our assumption is false.
3. Assume the contradiction that there exists an atom A in clause C, and there exists no atom A' in clause C', such that $(A,A') \in Conf$ and $C \neq C'$. Let Z be a satcomset defined as $Z = \{ W \mid W \in ALLSATCOMS(F)$ and in W the satisfier of clause C is atom A $\}$. Z is a conflicting satcomset, since F is an MUS. Let $Z' = SUBSATCOM(Z, F - \{C\})$. Since there exists no A' such that $(A,A') \in Conf$, thus Z' is a conflicting satcomset. Also $Z' = ALLSATCOMS(F - \{C\})$. But F is an MUS, thus we have reached a contradiction. Thus our assumption is false.
4. The proof follows from the definition of literal-set(F) and the proof of part 3.
5. Assume the contradiction that there exist clauses $C1, C2 \in F$, such that $C1 \subseteq C2$. Let Z be a satcomset defined as $Z = \{ W \mid W \in ALLSATCOMS(F)$ and in W the satisfiers of clauses C1 and C2 are equal $\}$. Z is a conflicting satcomset since F is an MUS. Let $Z' = SUBSATCOM(Z, F - \{C2\})$. Z' remains a conflicting satcomset. Also since $C1 \subseteq C2$ so $Z' = ALLSATCOMS(F - \{C2\})$. But F is an MUS, thus we have reached a contradiction. Thus our assumption is false.   □

Definition 21 (Unit Clause). *Clause C of a formula is called unit with respect to literal P, if clause C has exactly one non-empty entry i.e. $F(C,P) \neq \emptyset$. If a clause is not unit with respect to literal P, then it is called non-unit with respect to P.*

Definition 22 (Unit Literal). *Literal P of a formula F is called unit if F has a unit clause with respect to literal P. If literal P is not unit then it is non-unit.*

We make the following trivial observation:

Lemma 6. *If F is a (n,m) MUS with a non-unit literal P, then $m \geq 2$.*   □

Theorem 1. *If F is a (n,m) MUS with unit literal P and $m \geq 2$, then $\Delta_P(F)$ is a (n-1,m-1) MUS.*

Proof: *Let the unit clause of F be C. Let C contain +P. Let us prove the 2 conditions of an MUS for $\Delta_P(F)$ as follows:*

1. *By part 5 of proposition 2, F cannot have +P in any clause other than C. F cannot have a unit clause with -P, since the 2 unit clauses would be a conflicting subset of F, which contradicts the fact that F is an MUS. Thus $\Delta_P(F)$ is a (n-1,m-1) formula. Let Z be a satcomset defined as*
   $Z = \{ W \mid W \in ALLSATCOMS(F)$ *and W contains no -P* $\}$. *Such a choice is possible since F has no unit clause containing -P. Z is a conflicting satcomset, since F is an MUS. Let $Z' = SUBSATCOM(Z, F - \{ C \})$. Since no satcom of Z has a -P, thus Z' is a conflicting satcomset. Also $Z' = ALLSATCOMS(\Delta_P(F))$. Thus condition 1 of an MUS is fulfilled.*
2. *Assume the contradiction that there exists $F^* \subset \Delta_P(F)$ such that F^* is a conflicting formula. Now $(\zeta_P(F, F^*) \cup \{ C \}) \subset F$.*
   Let W be a satcom of $(\zeta_P(F, F^*) \cup \{ C \})$. The following cases arise:
   - *If W contains a -P, then W is conflicting since W contains the +P of unit clause C.*
   - *If W contains no -P, then W is conflicting since $SUBSATCOM(W, F^*)$ is conflicting.*
   Thus $(\zeta_P(F, F^*) \cup \{ C \})$ is a conflicting formula. But this is a contradiction since F is an MUS. Thus our assumption is false.

Thus it is proved that $\Delta_P(F)$ is a (n-1,m-1) MUS if C contains +P. Similarly if C contains -P, we can prove that $\Delta_P(F)$ is a (n-1,m-1) MUS.   □

Theorem 2. *If F is a (n,m) MUS with non-unit literal P then the following hold:*

1. *$\Delta_P(\Delta^{P-}(F))$ is a conflicting formula*
2. *$\Delta_P(\Delta^{P+}(F))$ is a conflicting formula*
3. *There exists a (n',m') MUS defined as $F^{(P,0)}$, $m' \leq (m-1)$, such that $\Delta_P(\Sigma^{P+}(F)) \subseteq F^{(P,0)} \subseteq \Delta_P(\Delta^{P-}(F))$*
4. *There exists a (n',m') MUS defined as $F^{(P,1)}$, $m' \leq (m-1)$, such that $\Delta_P(\Sigma^{P-}(F)) \subseteq F^{(P,1)} \subseteq \Delta_P(\Delta^{P+}(F))$*

Proof:

1. Let Z be a satcomset defined as $Z = \{\ W \mid W \in ALLSATCOMS(F)$ and W contains -P of clauses of $\Sigma^{P-}(F)$ and W contains no +$P\ \}$. Such a choice is possible since P is a non-unit literal. Z is a conflicting satcomset, since F is an MUS. Let $Z' = SUBSATCOM(Z, \Delta^{P-}(F))$. By the way Z is defined, Z' must be a conflicting satcomset. Also $Z' = ALLSATCOMS(\Delta_P(\Delta^{P-}(F)))$, so $\Delta_P(\Delta^{P-}(F))$ is conflicting.

2. The proof is similar to the above proof.

3. By lemma 5, there exists a (n',m') MUS, say $F^{(P,0)}$, $m' \leq (m-1)$, such that $F^{(P,0)} \subseteq \Delta_P(\Delta^{P-}(F))$, since $\Delta_P(\Delta^{P-}(F))$ is conflicting. Let us now show that $\Delta_P(\Sigma^{P+}(F)) \subseteq F^{(P,0)}$. Assume the contradiction that there exists a clause $C \in \Sigma^{P+}(F)$, such that $\Delta_P(\{\ C\ \}) \notin F^{(P,0)}$. By part 1 of proposition 2, let W be an essential satcom of +P of C. Let satcom $W1 = SUBSATCOM(W, F^{(P,0)})$. Since $\Delta_P(\{\ C\ \}) \notin F^{(P,0)}$, thus $W1$ is a non-conflicting satcom of $F^{(P,0)}$. But this is a contradiction since $F^{(P,0)}$ is an MUS. Thus our assumption is false.

4. The proof is similar to the above proof. $\qquad\square$

Note 4. Intuitively $F^{(P,0)}$ is the MUS contained in formula F when literal P is set to false. Similarly $F^{(P,1)}$ is the MUS contained in formula F when literal P is set to true.

Note 5. The contents of theorem 1 and 2 of our paper is similar to the splitting theorem which has been used by several authors[9, 11], however here these results have been independently derived and extended.

Lemma 7. *Let P be a non-unit literal of formula F, let $F0$ be a $F^{(P,0)}$, $F1$ be a $F^{(P,1)}$ such that $F0 \cap F1 \neq \emptyset$, if $C \in \zeta_P(F,F0 \cap F1)$ then $F(C,P) = \emptyset$.*

Proof: *This follows from part 3 and 4 of theorem 2.* $\qquad\square$

The following result is crucial for deriving bounds of special formulas.

Proposition 3. *In an MUS F, if literal P is non-unit and F0 is a $F^{(P,0)}$ and F1 is a $F^{(P,1)}$ then $\zeta_P(F,F0) \cup \zeta_P(F,F1) = F$*

Proof: *Assume the contradiction that there exists a clause C of F such that $\zeta_P(F,F0) \cup \zeta_P(F,F1) = F - \{C\}$. Let W be a satcom of $(\zeta_P(F,F0) \cup \zeta_P(F,F1))$. ¿From part 3 and 4 of theorem 2, C has neither +P nor -P. The following cases arise:*

- *If W contains both +P and -P, then W is conflicting.*
- *If W contains no -P then it must contain a satcom of $F1$, and so W is conflicting.*
- *If W contains no +P then it must contain a satcom of $F0$, and so W is conflicting.*

Thus W is conflicting. But this is a contradiction since $(F - \{C\})$ is a proper subset of F and F is an MUS. Thus our assumption is false. $\qquad\square$

4   Classification

In this section we classify literals of an MUS. We then use this classification to categorize MUS into three exhaustive and mutually exclusive groups.

Definition 23 (Omnipresent Literal). *Literal P of formula F is called omnipresent if $F(C,P) \neq \emptyset$ for every clause $C \in F$. If literal P is not omnipresent then it is non-omnipresent.*

Definition 24 (Lonesign Literals and Clauses). *Literal P of formula F is called lonesign if either atom +P appears exactly once in F or atom -P appears exactly once in F. The clause corresponding to the unique appearance of +P or -P is called the lonesign clause with respect to P. Literal P of formula F is called non-lonesign if both +P and -P appear more than once in F. A clause C is called non-lonesign wrt P if $F(C,P) \neq \emptyset$, and the atom $F(C,P)$ appears more than once in the formula F.*

Proposition 4. *In an MUS F, if literal P is omnipresent and lonesign then P must also be a unit literal.*

Proof: *Let F be an MUS and literal P be omnipresent and lonesign. Let clause C be the corresponding lonesign clause with respect to P. Let C contain +P.*

By part 5 of proposition 2, if $C' \in (F - \{C\})$, then $-P \in C'$.

Assume the contradiction that P is not a unit literal. Thus C is not a unit clause. Clause C has at least one more non-empty atom say A. Let us construct a satcom W of F by choosing A from clause C and -P from each clause of $(F - \{C\})$. Then W is non-conflicting which is a contradiction since F is an MUS. Thus the assumption is false. Similarly if clause C contains -P, it can be shown that P is a unit literal. $\square$

Definition 25 (COM(F,P,F0,F1)). *If literal P of MUS F is non-unit, and F0 is a $F^{(P,0)}$ and F1 is a $F^{(P,1)}$, then COM(F,P,F0,F1) is defined as $F0 \cap F1$.*

Definition 26 (Rowsplit,Non-rowsplit Literal). *Given an MUS F, let literal P be non-unit, then :*

- *P is called rowsplit if $\exists$ F0 and F1, F0 is a $F^{(P,0)}$ and F1 is a $F^{(P,1)}$, such that $COM(F,P,F0,F1) = \emptyset$.*
- *P is called non-rowsplit if for every F0 and F1, F0 is a $F^{(P,0)}$ and F1 is a $F^{(P,1)}$, $COM(F,P,F0,F1) \neq \emptyset$.*

Definition 27 (Basic-Types). *Unit, rowsplit and non-rowsplit are the basic types of a literal of an MUS. A literal of an MUS must belong to exactly one of these categories. Given an MUS F and a literal P of F, BASICTYPE(P,F) indicates its basic type - unit, rowsplit or non-rowsplit.*

Definition 28 (Columnsplit,Non-columnsplit Literal). *Given an MUS F, let literal P be non-unit, then :*

- P is called *columnsplit* if $\exists$ $F0$ and $F1$, $F0$ is a $F^{(P,0)}$ and $F1$ is a $F^{(P,1)}$, such that *literal-set(F0)* $\cap$ *literal-set(F1)* $= \emptyset$.
- P is called *non-columnsplit* if for every $F0$ and $F1$, $F0$ is a $F^{(P,0)}$ and $F1$ is a $F^{(P,1)}$, *literal-set(F0)* $\cap$ *literal-set(F1)* $\neq \emptyset$.

Proposition 5. *A columnsplit literal of an MUS must be a rowsplit literal as well.*

Proof: Let P be a columnsplit literal of MUS F. Assume the contradiction that P is non-rowsplit. Thus $\exists$ $F0$ and $F1$, $F0$ is a $F^{(P,0)}$ and $F1$ is a $F^{(P,1)}$, such that $COM(F,P,F0,F1) \neq \emptyset$. Let clause $C \in COM(F,P,F0,F1)$. By lemma 7, $+P$ or $-P$ cannot appear in C, thus if $P' \in$ *literal-set(C)*, then $P' \in$ *literal-set(F0)* $\cap$ *literal-set(F1)*. But this is a contradiction since P is columnsplit. Thus our assumption is false. $\square$

Note 6. A rowsplit literal of an MUS however need not be a columnsplit literal as well.

Definition 29 (Type-1,2,3 MUS). *We classify an MUS into the following exhaustive and mutually exclusive categories:*

1. *Type-1 MUS: If an MUS has atleast one unit literal then it is a type-1 MUS.*
2. *Type-2 MUS: If an MUS has no unit literal but has at least one rowsplit literal then it is a type-2 MUS.*
3. *Type-3 MUS: If all the literals of an MUS are non-rowsplit then it is a type 3-MUS.*

5   Structural Properties of Special Formulas

In this section we derive interesting properties of well known special formulas. Using the structural properties of an MUS we derive an exact bound of a Horn MUS and upperbounds for 2-SAT and 3-SAT MUS. First we review some known definitions.

Definition 30 (Horn Formula). *A Horn formula has atmost 1 positive atom in each clause.*

Definition 31 (2-SAT Formula). *A 2-SAT formula has exactly 2 atoms in each clause.*

Definition 32 (3-SAT Formula). *A 3-SAT formula has exactly 3 atoms in each clause.*

We define a pure-unit MUS as follows:

Definition 33 (Pure-Unit (n,m) MUS). *A pure-unit (n,m) MUS is a (n,m) MUS which is recursively defined as follows:*

- *A (2,1) MUS is a pure-unit (2,1) MUS.*
- *A (n,m) MUS F is a pure-unit (n,m) MUS, where $m > 1$, iff F has a unit literal say P such that $\Delta_P(F)$ is a pure-unit (n-1,m-1) MUS.*

It is clear from the above definition that in a pure-unit (n,m) MUS F we have
$n = m + 1$.

Lemma 8. *Every MUS must have at least one clause such that every atom belonging to the clause is positive.*

Proof: *The proof is easy to see since if the above was false then there would exist a non-conflicting satcom consisting of only negative atoms, which is a contradiction since F is an MUS.* □

We derive the well-known result regarding the size of a Horn MUS in the present framework as follows:

Proposition 6. *If F is a Horn (n,m) MUS then F is a pure-unit (n,m) MUS and $n = m + 1$.*

Proof: *By lemma 8, F must have a clause with only positive atoms. But F is Horn, thus F has a unit literal say P. Again $\Delta_P(F)$ is a Horn MUS. Continuing to reason this way, it is clear that F is a pure-unit MUS, thus $n = m + 1$.* □

Proposition 7. *We state the following properties of a pure-unit MUS that have direct proofs:*

1. *If F is a conflicting Horn formula, and $F' \subseteq F$ is an MUS, then F' is a pure-unit MUS.*
2. *Every Horn formula without a unit clause is non-conflicting.*

We obtain a coarse upperbound for 2-SAT MUS as follows:

Theorem 3. *The number of clauses in a 2-SAT MUS with m literals is atmost $O(2 * m)$.*

Proof: *Let F be a 2-SAT MUS with m literals. Let P be any literal of F, then P is a non-unit literal. Let F0 be a $F^{(P,0)}$, and F1 be a $F^{(P,1)}$. Since F is a 2-SAT MUS, by part 4 of proposition 2, both F0 and F1 are type-1 MUS. Let P' be a unit literal of F0. Again by part 4 of proposition 2, theorem 1 and the fact that F is a 2-SAT MUS we conclude that either $\Delta_{P'}(F0)$ is empty or $\Delta_{P'}(F0)$ is also a type-1 MUS. Reasoning this way we conclude that F0 is a pure-unit MUS. Similarly we can prove that F1 is also a pure-unit MUS. By the definition of a pure-unit MUS, $|F0| \leq (m - 1) + 1$, i.e $|F0| \leq m$. Similarly $|F1| \leq m$. Using proposition 3, we get an upperbound on $|F|$, by assuming that P is a row-split literal of F and $F0 \cap F1 = \emptyset$. Thus $|F| \leq O(2 * m)$.* □

Let us define $MUS^a{}_m$, which will be required while proving an upperbound of a 3-SAT MUS.

Definition 34 ($MUS^a{}_m$, $\|MUS^a{}_m\|$). *$MUS^a{}_m$, where a and m are positive integers, is defined as a set of MUS as follows:*
$MUS^a{}_m = \{ F \mid F$ is an MUS with m literals, such that if $C \in F$, then $|C| \leq a$ and $\exists\ C' \in F$ such that $|C'| < a \}$.
$\|MUS^a{}_m\|$ is a positive integer defined as follows:
$\|MUS^a{}_m\| = Max_{F \in MUS^a_m} |F|$.

We obtain a coarse upperbound for 3-SAT MUS as follows:

Theorem 4. *The number of clauses in a 3-SAT MUS with m literals is atmost $O(1.62^{(m-1)})$.*

Proof: *Let F be a 3-SAT MUS with m literals. Let P be any literal of F, then P is a non-unit literal. Let F0 be a $F^{(P,0)}$, and F1 be a $F^{(P,1)}$. Since F is a 3-SAT MUS, by part 4 of proposition 2, F0, F1 $\in MUS^3{}_{m-1}$. Let us try to obtain an upperbound for $\|MUS^3{}_{m-1}\|$. By part 4 of proposition 2, $\exists$ clause $C' \in F0$, such that $|C'| = 2$.*

Let literal-set(C') = { P', P" }. Since F is a 3-SAT MUS, P' and P" must be non-unit literals of F0. Let F00 be a $F0^{(P',0)}$, and F01 be a $F0^{(P',1)}$.

Thus F00, F01 $\in MUS^3{}_{m-2}$. If $+P' \in C'$ then F00 must be a type-1 MUS with unit literal P". Similarly if $-P' \in C'$ then F01 must be a type-1 MUS with unit literal P". Let F" denote the type-1 MUS either F00 or F01, with unit literal P".

Thus $\Delta_{P"}(F") \in MUS^3{}_{m-3}$. Let r_m denote an upperbound for $\|MUS^3{}_m\|$. Using proposition 3 and assuming that P' is a row-split literal of F0, and $F00 \cap F01 = \emptyset$, we get an upperbound on $|F0|$, in the form of the following recurrence relation:

$r_{m-1} = r_{m-2} + r_{m-3} + 1$. The solution of this recurrence has the form $r_m = c_1 \alpha^{(m-1)} + c_2 \hat{\alpha}^{(m-1)} + c_3$, where c_1, c_2 and c_3 are constants[1], $\alpha = (1 + \sqrt{5})/2$ and $\hat{\alpha} = (1 - \sqrt{5})/2$. Thus we get r_{m-1} is $O(1.62^{m-1})$, i.e. $|F0| \leq O(1.62^{m-1})$. Similarly we can derive that $|F1| \leq O(1.62^{m-1})$. Using proposition 3, we get an upperbound on $|F|$, by assuming that P is a row-split literal of F and $F0 \cap F1 = \emptyset$. Thus $|F| \leq O(1.62^{(m-1)})$.$\square$

Note 7. It is possible to generalise the idea of theorem 4 to obtain a bound for a n-SAT MUS.

6   Conclusion

This concludes our investigation of the Minimal Unsatisfiable Sets. We have presented structural properties of these sets, formulated a classification scheme for them and derived bounds on the sizes of such sets in Horn, 2-SAT and 3-SAT formulas. We believe that the work presented in this paper could have the following uses:

- The structural properties of MUS may be utilised to obtain an alternative framework for enumeration of MUS.
- Tighter upperbounds of 2-SAT and 3-SAT MUS and upperbounds of other special formulas can be derived using the structural properties of an MUS. These bounds may affect algorithms for logical inference of these formulas.

---

[1] Utilizing the facts that $\|MUS^3{}_1\| = r_1 = 2$, $\|MUS^3{}_2\| = r_2 = 4$, $\|MUS^3{}_3\| = r_3 = 7$, c_1, c_2 and c_3 can be computed

- This work was motivated by theorem proving in first order using partial instantiation methods [4, 8, 12]. The knowledge of the structure of MUS can be utilised to direct the choice of clauses for instantiation in these methods to improve convergence.
- The connections of the fields of linear algebra and graph theory to special propositional formulas have been studied extensively in recent years [10, 11]. Knowledge about the structural properties of an MUS may help in the understanding of these connections.
- A new theoretical framework involving satcoms, subsatcoms etc. has been presented in this paper, to facilitate the derivation of the structural properties of an MUS. This framework may be used to explore other combinatorial properties of MUS.

References

1. Chandru, V., and Hooker, J. N., *Optimization Methods for Logical Inference*, Wiley, New York, 1999.
2. Chang and Lee. *Symbolic Logic and Mechanical Theorem Proving*, Academic Press, 1987.
3. Hans Kleine Büning and Theodor Lettman, *Propositional Logic: Deduction and Algorithms*, Cambridge University Press, 1999.
4. Jeroslow, R. G., *Logic-Based Decision Support - Mixed Integer Model Formulation*, North-Holland 1989.
5. Truemper, K., *Effective Logic Computation*, Wiley, New York, 1998.
6. Stefan Szeider, Herbert Fleischner and Oliver Kullmann, *Polynomial-Time Recognition of Minimal Unsatisfiable Formulas with Fixed Clause-Variable Difference*, Theoretical Computer Science 289 no. 1, pp. 503-516, 2002. Draft version (updated August 2001).
7. Bruni, R., Sassano, A., *Finding Minimal Unsatisfiable Subformulae in Satisfiability Instances*, in proc. of 6th International Conference on Principles and Practice of Constraint Programming (CP2000), Singapore, Lecture Notes in Computer Science, Vol. 1894, Springer-Verlag, 2000.
8. J. N. Hooker, G. Rago, V. Chandru, and A. Shrivastava, *Partial instantiation methods for inference in first order logic*, Journal of Automated Reasoning 28, pp. 371-396, 2002.
9. Hans Kleine Büening and Zhao Xishun, *Minimal Unsatisfiability:Results and Open Questions*, a technical report available at http://www.uni-paderborn.de/cs/ag-klbue/projects/index.html.
10. Oliver Kullman, *On some Connections between Linear Algebra and the Combinatorics of Clause-sets*, SAT 2003 informal proceedings available at http://www.cs.swan.ac.uk/ csoliver/papers.html.
11. Stefan Szeider, *Acyclic Formulas and Minimal Unsatisfiability*, September 2000, available at http://www.cs.toronto.edu/ szeider/.
12. Sudeshna Dasgupta, *A Declarative Constraint Logic Programming Language based on Partial Instantiation Technique for Inference*, Ph.D. Dissertation, Birla Institute of Technology and Science, Pilani, Rajasthan, India, 2002.

LPOD Answer Sets and Nash Equilibria

Norman Foo[1], Thomas Meyer[1], and Gerhard Brewka[2]

[1] National ICT Australia; and The School of Computer Science and Engineering,
University of New South Wales, Sydney NSW 2052, Australia
[2] Intelligent Systems Department, Computer Science Institute, University of Leipzig,
Augustusplatz 10-11, 04109 Leipzig, Germany

Abstract. Logic programs with ordered disjunctions (LPODs) are natural vehicles for expressing choices that have a preference ordering. They are extensions of the familiar extended logic programs that have answer sets as semantics. In game theory, players usually prefer strategies that yield higher payoffs. Since strategies are choices, LPODs would seem to be a suitable logical formalism for expressing some game-theoretic properties. This paper shows how pure strategy normal form games can be encoded as LPODs in such a way that the answer sets that are mutually most preferred by all players are exactly the Nash equilibria. A similar result has been obtained by researchers using a different, but related, logical formalism, viz., ordered choice logic programs that were used to encode extensive games.

1 Introduction

A variety of computer science areas, particularly artificial intelligence, are now importing concepts and techniques from game theory. In multiagent systems where interactions involve bargaining, negotiation, collaboration, competition, etc., game theory can provide underpinnings for rational choices. On the other hand, agent intentions and preferences can be succinctly and precisely represented as logic programs of one kind or another. The central result of this paper concerns a natural dovetailing of game theory and logic programming, so that optimal agent preferences captured as most preferred answer sets of logic programs with ordered disjunctions (LPODs) coincide with the well-known Nash equilibria for pure strategy games. De Vos and Vermeir [De Vos and Vermeir 99] have obtained a similar result using their ordered-choice logic programs (OCLPs) with a new semantics that is suitable for encoding the extensive form of games. On the other hand, LPODs semantics are an extension of the familiar answer sets and may afford a more congenial formalism for generalizing to the mixed strategies that we do not address here. LPODs also appear to be more suitable for encoding the normal form of games. However, the similarity of the results and the inter-translatability of extensive and normal form games suggests an equivalence between LPODs and OCLPs that should be investigated.

The structure of the paper is as follows. Section 2 is a brief review of the concepts in game theory needed for our exposition. Of necessity this is only a skeletal treatment, and standard works like Luce and Raiffa [Luce and Raiffa 57] or Watson [Watson 02] (from which our notation and examples are adapted) should be consulted for details.

M.J. Maher (Ed.): ASIAN 2004, LNCS 3321, pp. 343–351, 2004.

Section 3 is a recapitulation of extended logic programs (ELPs) with answer set semantics. This is largely a condensation of the original paper by Gelfond and Lifschitz [Gelfond and Lifschitz 91]. (In fact, for our paper we only need the weaker form of ELPs that do not have negative literals.) Then in section 4 LPODs are described, by summarizing the content of the paper by Brewka [Brewka 04]. The main result follows in section 5. On-going and future work is outlined in the concluding section.

2   Game Theory

Game theory has an extensive literature and many profound results. However for the purpose of this paper we need only a small part of it.

We consider finite games of players $1, \ldots, n$ each of whom has a finite set S_i of moves or strategies. A profile s is a tuple of strategies, one from each player; formally $s \in S_1 \times \ldots \times S_n$. For brevity we denote $S_1 \times \ldots \times S_n$ simply by S. A profile can be informally regarded as a "play" or a "round" of the game, with each player choosing a move independently of the choice of the others. Given a profile s, by s_{-i} is meant the tuple $\langle s_1, \ldots, s_{i-1}, s_{i+1}, \ldots, s_n \rangle$, i.e. the strategies or moves of the players except i. Thus s_{-i} is an element of $S_1 \times \ldots \times S_{i-1} \times S_{i+1} \ldots S_n$. Again for brevity we write S_{-i} for $S_1 \times \ldots \times S_{i-1} \times S_{i+1} \ldots S_n$.

Each player i has an associated payoff function $u_i : S \to R$, where R is the set of real numbers. $u_i(s)$ is the payoff to player i as a result of i's choice of strategy s_i while the others have chosen s_{-i}. In the special case when n is 2, these payoff functions are denoted by u_1 and u_2, and their respective strategies are denoted by s_1 and s_2 for a profile s.

A player i may have some belief or guess about the probabilities over the strategic choices of other players. A particular belief is represented by a probability distribution μ_{-i} over S_{-i}, where (by our abbreviation) the latter is the set of all combinations of strategies by the players other than i. We may then write $\mu_{-i} \in \Delta S_{-i}$ to indicate that μ_{-i} is in the collection ΔS_{-i} of probability distributions over S_{-i}. With this belief μ_{-i}, if player i chooses the strategy s_i the expected payoff $u_i(s_i, \mu_{-i})$ for i is therefore $\Sigma_{s_{-i} \in S_{-i}} \mu_{-i}(s_{-i}) u_i(s_i, s_{-i})$.

Player i's strategy s_i is a best response to its belief μ_{-i} (about the others' strategies) if $u_i(s_i, \mu_i) \geq u_i(s_i', \mu_i)$ for every $s_i' \in S_i$, i.e, s_i has the best expected payoff relative to its other strategies. There can be more than one best response to μ_{-i}. In the two-player examples below the probability distributions are all very simple — they place probability 1 on each of the opponent's possible strategy choices in turn, so that it suffices to reason about the best response to opponent strategies one at a time.

A profile s is a pure strategy[1] *Nash equilibrium* if for each player i s_i is a best response to s_{-i}. What this entails is that the choice of s_i for each player is rationalizable with respect to the hypothesis that every player knows that each player can reason completely about the best strategic choices other players will make in any circumstance.

We illustrate this using four standard two-player games as shown in figures 1, 2, 3 and 4 chosen as representatives of the features to be addressed in later sections.

---

[1] A mixed strategy is one in which player i "randomizes" its own strategies according to some probability distribution. We consider such strategies in a later paper.

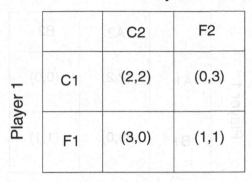

Player 2

	C2	F2
C1	(2,2)	(0,3)
F1	(3,0)	(1,1)

Fig. 1. Prisoners' Dilemma

Player 2

	OP2	MV2
OP1	(2,1)	(0,0)
MV1	(0,0)	(1,2)

Fig. 2. Battle of the Sexes

Perhaps the most familiar of these four is the Prisoners' Dilemma. Its scenario is as follows. Two persons (players 1 and 2 in this abstract setting) are arrested for joint commission of a crime. They are interrogated separately and cannot collaborate. The circumstances are such that if they both cooperate by keeping silent, both will get off. However if player 1 keeps silent but player 2 confesses ("finks"), then player 1 will be severely punished while player 2 gets only a short sentence; and vice-versa. In the remaining case, they both fink, and here they are both jailed for a long time. The payoffs corresponding to these possibilities are shown in the matrix. The rows are indexed by player 1's strategies, and the columns by those of player 2. The matrix entries are pairs in which the first component is the payoff to player 1 and the second is the payoff to player 2 for the corresponding strategy pairs. For instance, the entry for cell $(F1, C2)$ is $(3,0)$, meaning that $u_1(F1, C2) = 3$ and $u_2(F1, C2) = 0$. The striking feature of this game is that the best response of each player is to fink, resulting in a Nash equilibrium $(F1, F2)$ which is a bad outcome compared with the $(C1, C2)$ possibility.

Player 2

	A2	B2
A1	(2,2)	(0,0)
B1	(0,0)	(1,1)

Player 1

Fig. 3. Pareto Coordination

Player 2

	H2	T2
H1	(1,−1)	(−1,1)
T1	(−1,1)	(1,−1)

Player 1

Fig. 4. Matching Pennies

The next game is the Battle of the Sexes where the scenario is that a couple has to decide whether to go to the opera or the movies. One prefers the opera and the other prefers the movies, but they would rather forgo their preference than go alone. Here there are two Nash equilibria, $(OP1, OP2)$ and $(MV1, MV2)$, as can be verified.

In the Pareto Coordination game there are also two Nash equilibria $(A1, A2)$ and $(B1, B2)$, as again can be verified.'

The last example is the Matching Pennies game. It does not have a Nash equilibrium[2]. To recapitulate the well-known argument for this, the best response of player 1 to the strategy $H2$ of player 2 is $H1$, and to $T2$ of player 2 is $T1$. But the best response of player 2 to player 1's strategy $H1$ is $T2$, and to $T1$ of player 1 is $H2$. Thus none of $(H1, H2)$, $(H1, T2)$, $(T1, H2)$ or $(T1, T2)$ have the respective mutual best reponses of players 1 and 2 in any pair. For example, in $(H1, H2)$, $H2$ is not player 2's best response to player 1's $H1$.

---

[2] However, it does have a mixed strategy Nash equilibrium.

3   ELPs — Extended Logic Programs

Extended logic programs (ELPs) were introduced by Gelfond and Lifschitz (op.cit.) to increase the expressive power of logic programs by permitting classical negation to co-exist with negation as failure. They achieved this by a slight tweak of their notion of *stable models* of logic programs that only have negation as failure (but not classical negation). Our main result actually depends on stable models, but in anticipation of future generalization to situations in which agents can reason about other agents, and can promise or commit to do or not to do certain actions, we might as well use the vocabulary of the analogs of stable models in the ELP setting — *answer sets*.

Given an ELP Π and a set S of literals, the *Gelfond-Lifschitz reduct of Π* with respect to S, denoted by Π^S, is a definite logic program obtained from Π by (i) deleting every clause which has *not L* in its body for which $L \in S$, and (ii) dropping all *not L* in the surviving clauses. Intuitively this can be justified by regarding S as a "guess" at a solution (successful queries) of the ELP Π. If the guess S is correct and L is in it then any clause with *not L* in its body cannot be used. In all other clauses, any *not L* is guaranteed to be such that $L \notin S$, so it *not L* might as well be dropped since L is bound to finitely fail. To close the circle of this intuition, we call a guess S an *answer set* if $lfp(\Pi^S) = S$, where $lfp(\Pi^S)$ is the least fixed point [3] of the definite program Π^S.

This simple example of an ELP is taken from Gelfond and Lifschitz (op.cit.).

> $eligible \leftarrow highGPA$
> $eligible \leftarrow minority, fairGPA$
> $\neg eligible \leftarrow \neg fairGPA, \neg highGPa$
> $interview \leftarrow not\ eligible, not\ \neg eligible$

Assume also that the facts about a certain candidate *Anne* are given: $fairGPA$, $\neg highGPA$. It can be verified that the only answer set is $\{fairGPA, \neg highGPA, inteview\}$. If instead the facts had been $minority, fairGPA$, then the only answer set will contain *eligible*.

4   LPODs — Logic Programs with Ordered Disjunctions

Logic Programs with ordered disjunctions were introduced by Brewka and his colleagues (op.cit.) to rank multiple answer sets of ELPs further extended with disjunctive literals in the *heads* of clauses. A clause of this kind may look like

$$A \lor B \lor C \leftarrow D \land E \land F \land not\ G \land not\ H$$

where $A, B, \ldots H$ are literals. The disjunct $A \lor B \lor C$ in the head essentially says that whenever the body is true, at least one of A or B or C is true, and in fact it is the *minimal* models that suffice. However, even with this extended syntax we do not have a way to express a preference for, say, the answer sets that contain A to those that contain B, and those with B to those with C. The desire to express such preferences led Brewka,

---

[3] This can also be defined as the smallest Herbrand model of Π^S.

et al. to introduce a new class of disjunctive ELPs. The syntactic modification to the above clause to express preference of A to B is this:

$$A \times B \times C \leftarrow D \wedge E \wedge F \wedge not\ G \wedge not\ H$$

Informally, this means that (when the body is true), if we can have A we are done, but if not then we will settle for B, and if even B cannot be had, we will be satisfied with C. For details of how this can be formally achieved with a variety of preferences on answer sets of programs that contain such ordered disjunctions we refer the reader to Brewka's exposition (op.cit.). In our present context this will be achieved in a manner that in fact reflects one simple preference.

5   Answer Sets and Nash Equilibria

We will encode the normal form specification of games by LPODs. Player i will "own" a set of clauses, and each clause encodes the player's most preferred responses to a hypothetical strategic choice profile of other players (s_{-i} in the notation above). The ordering of the disjunction is exactly this preference, with the best response strategy s_i being the first disjunct, followed by other strategies of i's with less and less payoffs. There is also a classical disjunction that encodes the deterministic strategic choices of player i, but that is explained in the simpler 2-person context in the next paragraph.

In the specific cases of two person pure strategy games, a typical clause will therefore be of the form $p_1 \times p_2 \ldots \times p_k \leftarrow q$ where p_i are the strategies of player 1 in response to the strategy q of player 2. In the disjunction, p_1 will be 1's best response to q, p_2 will be 1's next best response to q, and so forth. If player 2 has n stratgeies, then there will be n such clauses for player 1. Dually, player 2 will have a set of k clauses encoding 2's responses to each of 1's k strategies. To say that player 1 has to make a move, or equivalently choose a strategy, we write a clause of the form $p_1 \vee p_2 \ldots \vee p_k \leftarrow$, i.e. a classical disjnctive fact, which we call "move clauses" below. Under the minimal model semantics [Brewka 04] that is conventional for classical disjunctions this forces models to be those with exactly one of the disjuncts. There is a dual such fact for player 2.

We now exhibit the clauses that encode the four examples above. Although we separate the clauses of the two players, this is only for convenience — there is only one program for each normal form. The answer sets are ranked according to the ranks of the preferred strategies for each player, so that the individual ranks of each pair of strategies (s_1, s_2) — s_1 being 1's move, and s_2 being 2's move — induces a corresponding rank in the answer sets. The minimal model semantics for the "move clauses" ensure that each (s_1, s_2) pair occurs in one and only one answer set, so this rank is well-defined.

Prisoner's Dilemma

Player 1 clauses.

$$F1 \times C1 \leftarrow C2 \tag{1}$$

$$F1 \times C1 \leftarrow F2 \tag{2}$$

Player 2 clauses

$$F2 \times C2 \leftarrow C1 \tag{3}$$

$$F2 \times C2 \leftarrow F1 \tag{4}$$

Move clauses:

$$F1 \vee C1. \tag{5}$$

$$F2 \vee C2. \tag{6}$$

Answer set rankings (player 1, player 2):
$(C1, C2)$ is $(2, 2)$; $(C1, F2)$ is $(2, 1)$; $(F1, C2)$ is $(1, 2)$; and $(F1, F2)$ is $(1, 1)$.
The only Nash equilibrium is $(F1, F2)$, which has rank $(1, 1)$.

Battle of the Sexes

Player 1 clauses.

$$OP1 \times MV1 \leftarrow OP2 \tag{7}$$

$$MV1 \times OP1 \leftarrow MV2 \tag{8}$$

Player 2 clauses

$$OP2 \times MV2 \leftarrow OP1 \tag{9}$$

$$MV2 \times OP2 \leftarrow MV1 \tag{10}$$

Move clauses:

$$OP1 \vee MV1. \tag{11}$$

$$OP2 \vee MV2. \tag{12}$$

Answer set rankings (player 1, player 2):
$(MV1, MV2)$ is $(1, 1)$; $(MV1, OP2)$ is $(2, 2)$; $(OP1, MV2)$ is $(2, 2)$;
and $(OP1, OP2)$ is $(1, 1)$.
There two Nash equilibria are $(OP1, Op2)$ and $(MV1, MV2)$ both of which have rank $(1, 1)$.

Pareto Coordination

Player 1 clauses.

$$A1 \times B1 \leftarrow A2 \tag{13}$$

$$B1 \times A1 \leftarrow B2 \tag{14}$$

Player 2 clauses

$$A2 \times B2 \leftarrow A1 \tag{15}$$

$$B2 \times A2 \leftarrow B1 \tag{16}$$

Move clauses:

$$A1 \vee B1. \tag{17}$$

$$A2 \vee B2. \tag{18}$$

Answer set rankings (player 1, player 2):
$(A1, A2)$ is $(1, 1)$; $(A1, B2)$ is $(2, 2)$; $(B1, A2)$ is $(2, 2)$;
and $(B1, B2)$ is $(1, 1)$.
There two Nash equilibria are $(A1, A2)$ and $(B1, B2)$ both of which have rank $(1, 1)$.

Matching Pennies

Player 1 clauses:

$$H1 \times T1 \leftarrow H2 \tag{19}$$

$$T1 \times H1 \leftarrow T2 \tag{20}$$

Player 2 clauses

$$T2 \times H2 \leftarrow H1 \tag{21}$$

$$H2 \times T2 \leftarrow T1 \tag{22}$$

Move clauses:

$$H1 \vee T1. \tag{23}$$

$$H2 \vee T2. \tag{24}$$

Answer set rankings (player 1, player 2):
$(H1, H2)$ is $(1, 2)$; $(H1, T2)$ is $(2, 1)$; $(T1, H2)$ is $(2, 1)$; and $(T1, T2)$ is $(1, 2)$.
There is no (pure strategy) Nash equilibrium[4] and there is no answer set with rank $(1, 1)$.

In these examples their known Nash equilibria coincide with answer sets of rank $(1, 1)$, and when there is no Nash equilibrium there are also no answer sets of rank $(1, 1)$. They are instances of the main result of this paper. We say that an answer set is *most preferred* if all of the components of the inidividual strategies are ranked 1, i.e., the n-tuple of player preferences are all 1's in the answer set.

Proposition 1. *The answer sets which are most preferred are exactly the pure strategy Nash equilibria.*

Proof Outline: Each possible strategy profile $(s_1, s_2, \ldots, s_n)$ of the n players is ranked by a tuple $(p_1, p_2, \ldots p_n)$ where p_i is the position of s_i in the disjunct of player i's clause that has in its body the strategies s_{-i} (i.e., those in the strategies of the other players). The profile is a Nash equilibrium if and only if each such disjunct is the most preferred, i.e., is the left-most disjunct, and therefore $p_1 = 1$ for each i.

However, as the Battle of the Sexes and the Pareto Coordination examples show, the LPOD characterization of Nash equilibria does not discriminate between the asymmetry of the equilibria in the Battle of Sexes game and the symmetry of the equilibria in the Pareto Coordination game. In the former, one of the equilibria benefits player 1 more while the other benefits player 2 more. In the latter, no player has an advantage in either equilibrium. There are ways in LPODs to make these distinctions, and they lead to the encoding of the game-theoretic concept of *Pareto efficiency* [Watson 02] that we have not fully investigated.

On the other hand, the rankings provided by LPODs have information about what the players might next prefer if for some reason they cannot all have their most preferred

---

[4] However, this game has a mixed strategy Nash equilibrium.

choices. Further, in the Prisoner's Dilemma example the most Pareto-efficient strategy profile $(C1, C2)$ can be achieved if the players are allowed to promise or commit to cooperate. There are also ways in LPODs to encode this. The alternative to normal forms for specifying games is the extensive form, which is a tree-like representation showing the time-sequential choice of strategies (moves) in a game. LPODs can also be used to encode this by appealing to experience in using them to discover plans.

6   Conclusion

We have shown that LPODs are a suitable formalism to express finite pure strategy games. In particular Nash equilibria correspond to most preferred answer sets. There are a number of loose ends that need to be tied up. Among them are the following. Not all Nash equilibria are equal, and an extension to LPODs can encode distinctions among them. Many games only have mixed strategy Nash equilibria. LPODs do not have the probabilistic structure to encode them, so they will have to be extended with (perhaps) existing notions of probabilistic disjunctive logic programs. Finally, it would be a service to the logic programming as well as the game theory community to investigate what we suspect is an equivalence between LPODs and OCLPs (ordered choice logic programs).

Acknowledgement

This research is supported by the National ICT Australia, which is funded through the Australian Government's *Backing Australia's Ability* initiative, in part through the Australian Research Council.

References

[Brewka 04]   G. Brewka, "Answer Sets and Qualitative Decision Making", Synthese, 2004.

[Gelfond and Lifschitz 91]   M. Gelfond and V. Lifschitz, "Classical negation in logic programs and disjunctive databases", New Generation Computing, pp. 365-385, 1991.

[Luce and Raiffa 57]   R.D. Luce and H. Raiffa, *Games and Decisions: Introduction and Critical Survey*, 1957, reprinted by Dover Publications.

[De Vos and Vermeir 99]   M. De Vos and D. Vermeir, "Choice Logic Programs and Nash Equilibria in Strategic Games" Proceedings of the 13th CSL'99 conference, pp.266-276, Springer LNCS 1683, 1999.

[Watson 02]   J. Watson, *Strategy: An Introduction to Game Theory*, 2002, W.W. Norton.

Graph Theoretic Models for
Reasoning About Time*

Martin Charles Golumbic

Caesarea Rothschild Institute,
Department of Computer Science,
University of Haifa,
Haifa, Israel
golumbic@cs.haifa.ac.il.

1 Introduction

Reasoning about time is a very ancient discipline, perhaps as old as prehistoric man. These ancient humans had discovered how long to roast their hunted meat and how to dry and age the skins of animals. They learned how and when to plant seeds, and were guided by the cycles of the sun, moon and the seasons. Our ancestors knew that day followed night and night followed day, and they had some notion of duration of day and night. This basic temporal knowledge was exploited to develop a sense of planning, taking advantage of observation and experience. For example, they would have observed that deer drink at the river at a certain time of the day, or that fish are easier to catch in the early morning. Early humans could recognize the changing seasons, and adapted their behavior in order to expect and avoid some of the dangers of cold and hunger by preparing in advance.

If we start our solar-earth stopwatch at 100-200 thousand years ago, and follow it through to the beginning of recorded history, we will see the evolution of human understanding of time. Reasoning about time is an integral component of every advance of mankind. The annual flooding of the Nile brought fertile soil to the river banks long before the Egyptian kingdoms existed, but the quantum leap in exploiting and controlling this natural phenomenon did not happen until the great empires had been formed. As Persia, Mesopotamia, Egypt, China and India developed writing and scholarship, the process of human knowledge and reasoning entered a new era. The seasonal planter became the seasonal shipper, and our temporal planner skipped past a thousand generations.

Our current notions of reasoning about time are not very different from the ideas of the "literate elite" among those of the ancient Greeks, Hebrews, Romans, Confucians. With today's reasoning algorithms and techniques, temporal models and logics, and advances in philosophy, cognitive science and psychology, the

---

* This paper was written in honor of Jean-Louis Lassez with whom the author has often engaged in discussion on reasoning about time, and was supported in part by the IRST Trento – CRI Haifa Cooperation Project.

M.J. Maher (Ed.): ASIAN 2004, LNCS 3321, pp. 352–362, 2004.

major difference between us and those 2500 year old parents of ours is the literacy rate.

I recall a wonderful lecture by Isaac Asimov that I heard in the mid-1970's at the New York Academy of Science. Asimov toyed with the audience's sense of accomplishment and superiority. He made the perceptive argument that intellectuals can imagine how wonderful it might be to find ourselves sitting with Socrates and engaged in a lively, stimulating discussion on a deep philosophical problem (about the nature of time). Ha! Think again. The reality of Ancient Greece says that we would all have been slaves, not knowing an alpha from a zeta.

We may be privileged to live at the beginning of the 21st century of the common era when education and opportunity seem to be at an all time high. Yet, to learn how to reason temporally or otherwise, one must work hard, and many fail at it. Driving to work every day, I see the driver of one car suddenly cut in front of another car, making no "temporal calculation" of speed or braking time. I see police detectives reconstructing a "time line" in their attempt to determine how a gunman escaped or a terrorist entered a secure area. I see advertisements in the newspaper for youthful parties with elements which actually may shorten one's lifetime.

But enough of this pessimistic philosophical rambling. Let us move on to the more interesting models and beneficial applications of reasoning about time. The detective really does have to understand what happened when a crime occurs. The facts and temporal "clues" have to be analyzed to reveal new information which can be deduced from it. My favorite example is the Berge Mystery Story which is adapted from a mathematical puzzle attributed to the French graph theorist Claude Berge.

2   Solving the Berge Mystery Story

Some of you who have read my first book, *Algorithmic Graph Theory and Perfect Graphs*, know the Berge mystery story. For those who don't, here it is:

> Six professors had been to the library on the day that the rare tractate was stolen. Each had entered once, stayed for some time and then left. If two were in the library at the same time, then at least one of them saw the other. Detectives questioned the professors and gathered the following testimony: Abe said that he saw Burt and Eddie in the library; Burt said that he saw Abe and Ida; Charlotte claimed to have seen Desmond and Ida; Desmond said that he saw Abe and Ida; Eddie testified to seeing Burt and Charlotte; Ida said that she saw Charlotte and Eddie. One of the professors lied!! Who was it?

This is a temporal problem that you might want to try to solve on your own before reading my solution. [STOP READING; SOLVE PUZZLE OR GIVE UP; CONTINUE].

(a)   (b)

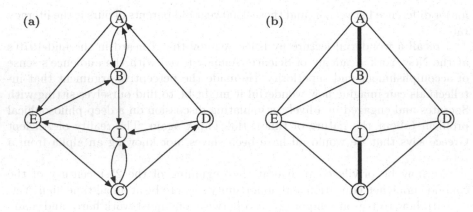

Fig. 1. The testimony graphs

Figure 1(a) shows a diagram representing the testimony written in the story: Abe saw Burt and Eddie, Burt saw Abe and Ida, etc. There is an arrow pointing from X to Y if X "claims" to have seen Y. By reasoning about this directed graph, we will discover some inconsistancies which will lead us to the professor who lied.

Remember that the story said each professor came into the library, was there for an interval of time, during that interval of time he saw some other people. If he saw somebody, that means their intervals intersected. So that provides some data about the intersection, and we can construct an intersection graph G, as in Figure 1(b): In G, there is an edge between two vertices if the time intervals of the corresponding professors intersect. The difficulty is that the testimony may be neither truthful nor complete.

In fact, we know that there is a lie here. Why? Because looking at the intersection graph G, we see a chordless cycle of length 4 (for example, A, B, I, D) which is an impossibility, due to the properties of intervals on a line. Consider the following argument:

If I were to try to construct 4 time intervals I_A, I_B, I_I, I_D for the cycle A, B, I, D, what would happen? I would have to start by drawing two disjoint intervals for I_A and I_I, and then draw the interval I_B which intersects both I_A and I_I, thus, covering the entire gap between them. Now try to add the final interval I_D which must intersect both I_A and I_I, but cannot intersect I_B, but this is impossible. Therefore, we conclude that a chordless 4-cycle is a forbidden and cannot be part of an interval intersection graph.

Now let us use this knowledge about forbidden 4-cycles to solve the Berge Mystery Story. Notice in Figure 1(a), that some pairs have arrows going in both directions, for example, Burt saw Abe and Abe saw Burt, and other pairs are just one way. That gives us some further information. Some of the edges in the intersection graph are more confident edges than others. A bold black edge in Figure 1(b) indicates a double arrow in Figure 1(a), and it is pretty confident because B saw A and A saw B so if at least one of them is telling the truth and the edge really exists. Similarly, for I and C. But all the one-way arrows are possibly true and possibly false. How shall we argue? Well, if we have a 4-cycle, one of those four professors is the liar. I do not know which one, so I will list all the cycles and see who is common. There are three cycles of length 4 without a chord: A, B, I, D and A, D, I, E and A, E, C, D. What can we deduce? We can deduce that the liar is one of these on a 4-cycle. That tells us Burt is not a liar. Why? Burt is one of my candidates in the first cycle, but he is not a candidate in the second, so he is telling the truth. The same goes for Ida; she is not listed in the third cycle, so she is also telling the truth. Charlotte is not in the first cycle, so she is ok. The same for Eddie, so he is ok. Four out of the six professors are now known to be telling the truth. Now it is only down to Abe and Desmond. What were to happen if Abe is the liar? If Abe is the liar, then ABID still remains a cycle because of the testimony of Burt, who is truthful. That is, suppose Abe is the liar, then Burt, Ida and Desmond would be truth tellers and ABID would still be a chordless cycle, which is a contradiction. Therefore, Abe is not the liar. The only professor left is Desmond. Desmond is the liar.

Was Desmond Stupid or Just Ignorant?

If Desmond had studied algorithmic graph theory [3], he would have known that his testimony to the police would not hold up. He could have said that he saw everyone, in which case, no matter what the truthful professors said, the graph would be an interval graph. His (false) interval would have simply spanned the whole day, and all the data would be consistent. Of course, the detectives would probably still not believe him.

3   What Is Temporal Reasoning?

Reasoning about time is essential for applications in artificial intelligence and in many other disciplines. Given certain explicit relationships between a set of events, we would like to have the ability to infer additional relationships which are implicit in those given. For example, the transitivity of "before" and "contains" may allow us to infer information regarding the sequence of events. Such inferences are essential in story understanding, planning and causal reasoning. There are a great number of practical problems in which one is interested in constructing a time line where each particular event or phenomenon corresponds to an interval representing its duration. These include seriation in archeology, behavioral psychology, scheduling, and combinatorics. Other applications arise in non-temporal contexts, for example, in molecular biology, where arrangement of DNA segments along a linear DNA chain involves similar problems.

A system to reason about time should have a knowledge base consisting of temporal and other information, a mechanism for determining consistency of temporal data, and routines for answering queries and discovering or inferring new facts. We will present several formal models to represent temporal knowledge which can be used in reasoning systems. Our interests are in the mathematical and computation properties of these models and the algorithms which can be applied to effectively solve temporal problems.

There are many issues that arise in temporal reasoning. First, there is the *chronology* of a story, where the narrative follows a time line reporting actions in the order in which they happen. Second, the story may provide specific *temporal information* on other actions, like explicit dates and times. Third, the story may give *temporal clues* through the use of words and expressions: just as, before finishing, half hour before, together with, at separate times, etc.

All of these allow one to reconstruct a portion of a temporal scenario, that is, what happened, when or in what order, and by whom. Detectives and puzzle enthusiasts, consultants and strategic planners, chefs and building contractors, politicians and college students all reason about time in this way. Let us look a bit more closely at the process of reasoning about time intervals.

4   Characterizing Interval Graphs

Our solution to the Berge Mystery Story illustrates the use of a mathematical, graph theoretic model to help reason about events represented by time intervals. The model is called an *interval graph* or the *intersection graph of a collection of intervals*. Interval graphs have become quite important because of their many applications. They started off in genetics and in scheduling, and have applications in seriation in archeology, artificial intelligence, mobile radio frequency assignment, computer storage and VLSI design. For those who are interested in reading more in this area, several good books are available and referenced at the end.

Let's look at another example. Suppose we have some lectures that are supposed to be scheduled at a university, meeting at certain hours of the day. Lecture 1 starts at 09:00 in the morning and finishes at 10:15; lecture 2 starts at 10:00 and goes until 12:00 and so forth. We can depict this on the real line by intervals, as in Figure 2. Some of these intervals intersect, for example, lectures 1 and 2 intersect at about 10:07, a time when they are both in session. There is a point, in fact, where four lectures are "active" at the same time. Figure 3(a) shows the interval graph for this example.

Intersecting intervals here indicate a temporal conflict or competition, in which a resource usually cannot be shared. In our example of scheduling university lectures, we cannot put two lectures in the same classroom if they are meeting at the same time, thus, they would need different classes. Problems such as these, where the intervals cannot share the same resource, can be visualized mathematically as a graph coloring problem on the interval graph. Although graph coloring is an NP-complete problem in general, for interval graphs it can be done in linear time, see [3, 7].

Those pairs of intervals that do not intersect are called disjoint. It is not surprising that if you were to consider a graph whose edges correspond to the pairs

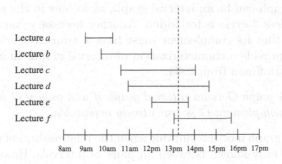

Fig. 2. An interval representation

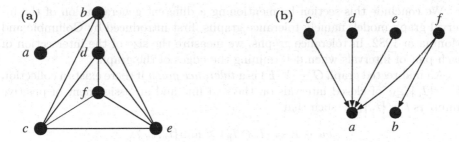

Fig. 3. (a) The interval graph of the interval representation in Figure 2 and (b) its complement

of intervals that are disjoint from one another, you would get the complement[1] of the intersection graph, which we might call the *disjointness graph*. It also happens that since these are intervals on the line, when two intervals are disjoint, one of them is before the other. In this case, we can assign an orientation on the (disjointness) edge to show which interval is earlier and which is later. See Figure 3(b). Mathematically, this orientation is a partial order, and as a graph it is a *transitive orientation*, see [3].

Formally, a graph is an *interval graph* if it is the intersection graph of some collection of intervals on the line. The problem of which graphs could be interval graphs goes back to the Hungarian mathematician Gyorgy Hajös in 1957 and independently to the American biologist, Seymour Benzer in 1959. Hajös posed the question with overlapping time intervals, whereas Benzer was looking at the structure of genetic material and the structure of what we would call genes today, asking whether the sub-elements could be arranged in a linear arrangement. Their original statements of the problem are quoted in [3] page 171.

---

[1] The *complement* $\overline{G}$ of a graph G is obtained by interchanging the edges and the non-edges.

Not every graph can be an interval graph, as we saw in the previous section, since the chordless 4-cycle is forbidden. Another necessary condition for an interval graph is that its complement must have a transitive orientation. These two conditions provide a characterization of interval graphs in a classical result of Gilmore and Hoffman from 1964.

Theorem 1. *A graph G is an interval graph if and only if G has no chordless 4-cycle and its complement $\overline{G}$ is transitively orientable.*

The interval graph model is suitable when the relationships of intersection and disjointness are fully known between all pairs of intervals. However, when only partial information is known, or when other interval relationships are known, then we must move on to a richer model for reasoning about time intervals. This will be the topic of our next section.

We conclude this section by mentioning a different generalization of the interval graph model, namely tolerance graphs, first introduced by Golumbic and Monma in 1982. In tolerance graphs, we measure the size of the intersection of each pair of intervals when determining the edges of the graph.

An undirected graph $G = (V, E)$ is a *tolerance graph* if there exists a collection $\mathcal{I} = \{I_v\}_{v \in V}$ of closed intervals on the real line and an assignment of positive numbers $t = \{t_v\}_{v \in V}$ such that

$$vw \in E \Leftrightarrow |I_v \cap I_w| \geq \min\{t_v, t_w\}.$$

Here $|I_u|$ denotes the length of the interval I_u. The positive number t_v is called the *tolerance* of v, and the pair $\langle \mathcal{I}, t \rangle$ is called an *interval tolerance representation* of G. Notice that interval graphs are just a special case of tolerance graphs, where each tolerance t_v equals some sufficiently small $\epsilon > 0$. A tolerance graph is said to be *bounded* if it has a tolerance representation in which $t_v \leq |I_v|$ for all $v \in V$.

The definition of tolerance graphs was motivated by the idea that a small or "tolerable" amount of overlap between two intervals, may be ignored, and hence not give an edge. Since a tolerance is associated to each interval, we put an edge between a pair of vertices when at least one of them (the one with the smaller tolerance) is "bothered" by the size of the intersection.

We can see this too in the same sort of scheduling problems that we saw in Figure 2. In that example, the chief university officer of classroom scheduling needs 4 rooms to assign to the 6 lectures. But what would happen if she had only 3 rooms available? In that case, would one of the lectures c, d, e or f have to be cancelled? Probably so. However, suppose some of the professors were a bit more tolerant, then an assignment might be possible.

Consider, in our example, if the tolerances (in minutes) were:

$$t_a = 10, t_b = 5, t_c = 65, t_d = 10, t_e = 20, t_f = 60.$$

Then according to the definition, lectures c and f would no longer conflict, since $|I_c \cap I_f| \leq 60 = \min\{t_c, t_f\}$. Notice, however, that lectures e and f remain in conflict, since Professor e is too intolerant to ignore the intersection. The tolerance graph for these values would therefore only erase the edge cf in Figure 3.

Golumbic and Siani gave an $O(qn + n \log n)$ algorithm for coloring a tolerance graph, given the tolerance representation with q vertices having unbounded tolerance. For details and further study of tolerance graphs and related topics, we refer the reader to the new book by Golumbic and Trenk [7].

5   Qualitative Temporal Reasoning[2]

Temporal reasoning problems that make no mention of numbers, clock times, dates, durations, etc. are often referred to as "qualitative"; instead, they use relations such as *before*, *during*, *after* or *not after* between pairs of events. This is often the case in story telling, where we get clues from the natural language text about temporal relationships without specific times stated, as we saw in the Berge Mystery Story.

The *seriation problem* asks for a mapping of the events onto the time line such that all the given relations are satisfied, that is, a consistent scenerio. Unlike the metric temporal constraint problems (MTCPs) where numerical methods, including shortest path algorithms and constraint based methods are useful, here a more combinatorial, graph theoretic approach is needed. This reconstruction of the events on the time line from qualitative data is similar to DNA mapping problems in molecular biology where the true DNA sequence of the gene must be built up using overlap data of clone fragments.

One of the techniques most useful with qualitative problems is the propagation of constraints between pairs of events. We will illustrate this on a simple example taken from [4].

Example 1. Once upon a time there were four bears, Papa bear, Mama bear, Baby bear and Teddy Bear. Each bear sat at the table to eat his morning porridge, but since there were only two chairs (the third chair was broken in a previous story), the bears had to take turns eating. Baby and Teddy always ate together sharing the same chair, and on this day Mama was seated for part of Baby bear's meal when the door opened and in walked their new neighbor, Goldie.

"What a great aroma," she said. "Can I join for a bowl?" Mama replied, "Sure, but you will have to *wait* for a chair!" "Yeah, I know all about chairs," said Goldie. So Goldie sat down when Baby and Teddy got up.

Papa entered the kitchen. Looking at Mama he said, "I wouldn't sit at the same table with that girl." Mama answered, "Then it's good you ate already."

In this story, we assume that each character sat at the table for an interval of time without interruption. Let's call these intervals I_B, I_G, I_M, I_P, I_T, subscripts indicating the first initial of the character's name. Since Baby and Teddy always eat together, we have $I_B = I_T$, so there are only four intervals that we have to place on the time line.

---

[2] This section is based largely on introductory material from the author's chapter [4].

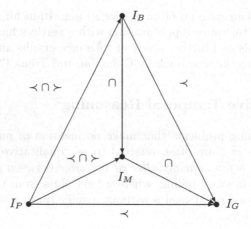

Fig. 4. The constraint graph before propagation

Fig. 5. The multiplication table for A_3 constraint propagation

Here are some questions to ask:

1. Could Papa and Baby both be at the table together?
2. Could Papa and Mama both be at the table together?
3. Could Papa have spent some time at the table with both Baby and Mama?
4. Did anyone sit at the table with Goldie?

Consider the three relations $\prec, \cap, \succ$ between intervals, where $I \prec J$ (equiv. $J \succ I$) denotes that I and J are disjoint intervals with I occurring before J, and $I \cap J$ denotes that I and J intersect. We can represent the interval events of our story as the vertices of an oriented graph with edges labelled by the possible relations between pairs of events. For example, if $I \prec J$ or $I \cap J$ are both possible, and the edge $e = (I, J)$ is oriented from I to J, then e will be labelled $\prec \cap$. The graph for our example appears in Figure 5. This kind of labelled graph can be viewed as a generalization of an interval graph: in that case, labels would be restricted to intersection $\cap$ or disjointness $\prec \succ$.

Our story provides the following facts:

(a) Only two chairs (This is not temporal but resource information.)
(b) $I_B \cap I_M$ Mama and Baby seated when the door opened.
(c) $I_B \prec I_G$ Goldie sat down when Baby got up.
(d) $I_P \prec I_G$ Papa ate before Goldie.
(e) $I_M \cap I_G$ Language clue Papa to Mama, "I wouldn't sit ... with that girl."

Although no relation is stated between I_P and I_B or I_M, we could "reason" that $I_P \prec \cap I_M$, that is, Papa could not eat *after* Mama since that would contradict the last paragraph of the story. Let's formalize that reasoning process using constraint propagation.

The simplest form of propagation is ordinary transitivity, namely if $I \prec J$ and $J \prec K$, then $I \prec K$. But we can define other implications as well between three intervals, for example, $I \prec J$ and $J \cap K$ implies $I \prec \cap K$, which can easily be seen by trying to draw the intervals I, J, K on the time line. This is precisely the rule we used combining facts (d) and (e) to deduce $I_P \prec \cap I_M$. Thus, by propagation, we can delete the relation $\succ$ from the edge from I_P to I_M.

These generalized transitivity rules are given in the multiplication table in Figure 5, and illustrates one of the *interval algebras*, namely the algebra A_3 studied in [6]. This language is limited, as it cannot express the idea that one interval contains another. More complex interval algebras A_6, A_7, and the original interval algebra A_{13} introduced by Allen [1], as well as the ord-Horn algebra [9], allow containment and other primitive relations. For a comprehensive survey of this topic, we refer the reader to [4].

6 Conclusion

In this paper, we have sought to whet the appetite of those interested in seeing how graph theoretic models can be used in reasoning about time. The next step would be to read the author's chapter [4] and follow its citations temporally forward and backward. We have focussed on qualitative methods and models for temporal reasoning. We have not, in this brief paper, treated any of the work on quantitative temporal reasoning [2, 8] nor any of the many logical techniques and models. We encourage the eager reader and researcher to pursue further exploration in this area.

Acknowledgements

I thank the organizers for the invitation to deliver this lecture honoring Jean-Louis Lassez whose contributions to the fields of automated reasoning, constraints and logic programming have influenced the careers of many researchers. I would also like to thank Sara Kaufman and Guy Wolfovitz for assisting me in the preparation of this paper.

References

1. J.F. Allen, Maintaining knowledge about temporal intervals, *Communications of the ACM* **26** (1983), 832–843.
2. R. Dechter, I. Meiri and J. Pearl, Temporal constraint networks, *Artificial Intelligence* **49** (1991), 61–95.

3. M.C. Golumbic, *"Algorithmic Graph Theory and Perfect Graphs"*, Academic Press, New York, 1980. Second edition, *Annals of Discrete Mathematics* **57**, Elsevier, Amsterdam, 2004.

4. M.C. Golumbic, Reasoning about time, in *"Mathematical Aspects of Artificial Intelligence"*, F. Hoffman, ed., *Proc. Symposia in Applied Math.* **55**, American Math. Society, 1998, pp. 19–53.

5. M.C. Golumbic, Algorithmic graph theory and its applications, in *"Graph Theory, Combinatorics and Algorithms: Interdisciplinary Applications"*, M.C. Golumbic and I. Ben-Arroyo Hartman, eds., Kluwer Academic, 2005.

6. M.C. Golumbic and R. Shamir, Complexity and algorithms for reasoning about time: a graph-theoretic approach, *J. Assoc. Comput. Mach.* **40** (1993), 1108–1133.

7. M.C. Golumbic and A.N. Trenk, *"Tolerance Graphs"*, Cambridge University Press, 2004.

8. I. Meiri, Combining qualitative and quantitative constraints in temporal reasoning, *Artificial Intelligence* **87** (1996), 343–385.

9. B. Nebel and H.-J. Bürckert, Reasoning about temporal relations: A maximal tractable subclass of Allen's interval algebra, *J. ACM* **42** (1995), 43–66.

10. P. VanBeek, Reasoning about qualitative information, *Artificial Intelligence* **58** (1992), 297–326.

11. P. VanBeek and R. Cohen, Exact and approximate reasoning about temporal relations, *Computational Intelligence* **6** (1990), 132–144.

12. M. Vilain and H. Kautz, Constraint propagation algorithms for temporal reasoning, *Proc. AAAI-86*, Philadelphia, PA, pp. 377–382, 1986.

13. A.B. Webber, Proof of the interval satisfiability conjecture, *Annals of Mathematics and Artificial Intelligence* **15** (1995), 231–238.

Rule-Based Programming and Proving:
The **ELAN** Experience Outcomes

Claude Kirchner[1] and Hélène Kirchner[2]

[1] LORIA & INRIA, 615 rue du Jardin Botanique 54602 Villers-lès-Nancy, France
[2] LORIA & CNRS, 615 rue du Jardin Botanique 54602 Villers-lès-Nancy, France
First.Last@loria.fr

All mgus are equal, but some mgus are more equal than others
Jean-Louis Lassez, Michael Maher and Kim Marriott [28]

Abstract. Together with the Protheo team in Nancy, we have developed in the last ten years the ELAN rule-based programming language and environment. This paper presents the context and outcomes of this research effort.

1 Introduction

Unification problems and their one-sided version *matching* are at the heart of schematic programming languages. These languages rely on pattern matching where patterns may be as different as first-order term or higher-order graphs or matrices; matching may be syntactic, equational or higher-order, and may identify the pattern and the subject using various forms of substitutions.

Because they offer a very high level of abstraction, many such languages have been designed and implemented to allow for high-level decision-making. We indeed assist currently to a main surge of interest in these languages from different point of views. First, in general purpose languages, matching capabilities are available for instance in ML or Haskell, while Prolog uses unification. Second, production systems and now business rule languages are fully based on pattern matching. They are prominent languages for higher-level decision-making and are central in expert systems and in companies prospective analysis. Third, the raising of XML, in particular as a general term-based syntax for the Web, allows the emergence of rewrite based languages like XSLT as well as for the so-called semantic web, where again languages like RuleML now emerge.

Indeed matching-based and more specifically rule-based languages have been prominent in algebraic specifications since at least twenty years. Several specification languages or programming environments, such as LARCH, OBJ, ASF+SDF [27], Maude [13], ELAN, to cite a few, are using term rewriting as their basic evaluation mechanism. From these experiences, we have inherited a deep knowledge somehow summarized in the CASL language [1].

In all the above mentioned languages and environments, rewriting and matching play an essential role, but few is done to give the user the possibility to control

M.J. Maher (Ed.): ASIAN 2004, LNCS 3321, pp. 363–379, 2004.

the rewrite relation. Moreover, the semantics of some programming languages is imprecise with regard to the way rewritings are applied. One of the main originality of the ELAN project that we are going to detail in this paper, is to have pioneered strategic rewriting, i.e. strategy controlled rewriting.

This paper presents the main outcomes in terms of emerging concepts and lessons of this ten years experience of development of the ELAN language and practice. Several concepts presented in this paper are directly relevant to the context of high-level decision-making.

One of them is the conceptual difference between computation and deduction in a programming environment. Computations are described in the context of rewriting with normalizers, while deductions needs control, expressed by strategies. These two concepts are combined in ELAN to provide strategic rewriting. Another related outcome is the definition of a declarative strategy language that gives to the programmer the capability of precisely defining control.

Understanding and formalizing the concept of strategies in this context led to higher-order functionalities and to the rewriting calculus [11] which provides in particular a smooth integration of first-order rewriting and λ-calculus.

Providing strategic rewriting capability in existing programming language is a further step leading to pervasive rewriting and formal islands that are compiled into the hosting programming language.

This paper is organized as follows. The next section summarizes the main features of the ELAN language in a smooth way, explains and analyzes design choices and mentions missing capabilities. Section 3 enlights the main theoretical concepts that have emerged: strategic rewriting, strategy language, rewriting calculus, rewriting proof terms, pervasive rewriting. Then we share our experience on the practicality of strategic rewriting in Section 4. It illustrates by a few examples related to decision processes, the power of the strategic rewriting approach.

2   Main Features of the Language

The ELAN language is fully described in its user's manual [4] and at url elan.loria.fr. We can summarize its main characteristics by the "equation":

ELAN= computation rules + (deduction rules + strategies)

The syntax of ELAN programs is given by a signature provided by the user and written in mixfix syntax. The semantics of the programs is given by computation and deduction rewrite rules together with strategies to control application of rules. ELAN programs are structured in modules, possibly parameterized and importing other modules.

Programming in ELAN is very easy when just computation is needed and can be quite elaborated when this is combined with powerful deduction and associated strategies. To program in ELAN, one should at least have an intuitive understanding of the two fundamental concepts of computation and deduction. The important difference between them has been formally identified since one century by Henri Poincaré. Both concepts play a central role in today's proof theory as well as in semantics of programming languages. In proof theory, the status of what we search for and what needs to be computed should be identified and treated appropriately, in order to get proofs where only useful (and often difficult) parts are described. This is typical of *Deduction Modulo* as developed in [17]. In semantics because of its close relationship with computation as well as solving, in particular prominent in declarative programming languages.

Computation and deduction steps are defined in ELAN thanks to unlabeled and labeled rules respectively.

2.1   Computation

What is a computation? Here we call computation the normalization by a set of confluent and innermost terminating set of unlabeled rewrite rules.

The simplest kind of *rewrite rule* used in ELAN is an ordered pair of terms denoted $[]\ l \to r$ such that l, r are terms (we denote $\mathcal{T}(\mathcal{F}, \mathcal{X})$ the set of terms build over the signature $\mathcal{F}$ and the set of variables $\mathcal{X}$) and satisfying the usual restriction of their respective set of variables: $Var(r) \subseteq Var(l)$. The empty square brackets $[]$ is important here, as it denotes the fact that this rule has no name.

A typical example of simple computation rules is the definition of addition on Peano numbers build using zero, denoted 0 and successor, denoted *succ*. The computation rules are

$$DEF(+) = \{[]\ x + 0 \to x \qquad []\ x + succ(y) \to succ(x + y)\}$$

They define, in particular in this simple example, a confluent and terminating rewrite system for which normal forms are the computation results in the tradition of algebraic specifications [2].

These rewrite rules define a *congruence* on the set of terms, therefore they are potentially applied everywhere in a term: at the root as well as at any occurrence. The built-in strategy used in ELAN to implement the normalization process by these simple rewrite rules is *left-most inner-most* and the ELAN compiler is able to apply more than 15 millions of such rewrite steps per second. These "simple" rules are already extremely powerful since a single left-linear and even regular rewrite rule is enough to be Turing complete (but of course in this case non-terminating!) [15].

The behavior of these rewrite rules is simple as their application is decided locally, independently of their application context, and could implicitly involve an equality check when they are non-linear. Such rule are at the heart of many algebraic languages like OBJ [21], ASF [34] or LPG [3]. When one wants to add the possibility to specify contextual information, condition are added to simple rules, and they are denoted in ELAN: $[]\ l \to r$ **if** c, where c is a boolean expression.

The next extension of the conditional rewrite rule provided by ELAN, consists in adding the capability to control not only the application context, but also the way computation are done. This is provided through different features. First, ELAN provides the capability to share some results. This is done by generalized conditional rewrite rules of the form $[]\ l \to r$ **where** $p_1 := c_1, \ldots,$ **where** $p_n :=$

c_n whose behavior consists of (1) searching in the term t to be evaluated a subterm at occurrence ω that matches l with a substitution σ (i.e. $\sigma(l) = t_{|\omega}$), then (2) finding a match σ_1 from p_1 to $\sigma(c_1)$ (i.e. $\sigma_1(p_1) = \sigma(c)$),..., then $(n+1)$ find a match σ_n from p_n to $\sigma_{n-1}\ldots\sigma_1\sigma(c_1)$ (i.e. $\sigma_n(p_n) = \sigma_{n-1}\ldots\sigma_1\sigma(c)$) and finally replace $t_{|\omega}$ by $\sigma_n\sigma_{n-1}\ldots\sigma_1\sigma(r)$.

This more technical explanation can be easily understood using the following example. Consider an operation *doubleflat* on lists, that takes a list (for instance $((1.2).(3.4))$) and builds the concatenation of its flattened form with the reverse of its flattened form (in this case $(1.2.3.4.4.3.2.1)$) can be defined by the rule with two matching conditions:

$$[] \; doubleflat(l) \rightarrow append(x, y) \; \textbf{where} \; x := flatten(l)$$
$$\textbf{where} \; y := reverse(x)$$

The interest of this form with respect to the classical rule

$$[] \quad doubleflat(l) \; \rightarrow \; append(flatten(l), reverse(flatten(l)))$$

is indeed to factorise the expression of $flatten(l)$ giving to the programmer the possibility to avoid computing twice the flattened form of l.

Using unlabeled rewrite rules, ELAN provides the programmer with the ability to define *normalizers* i.e. functions that return the unique normal form. This is quite powerful but sometimes surprising for the programmer as well as the end-user: in the above example on Peano arithmetic, the term $succ(0) + succ(0)$ will never be accessible, only the term $succ(succ(0))$ will be visible.

2.2   Strategic Rewriting

As already said, rewrite rules are natural to express computation but also deduction, i.e. rewrite systems that are neither necessarily confluent nor terminating.

While keeping the control over the evaluation of confluent and terminating rewrite rules is not essential (even through it could be useful for describing an efficient way to reach the normal form), it is mandatory for either non-terminating or non-confluent systems.

To provide the control on the execution order of a rewrite rule system, we use in ELAN a non empty rule label. These labels have indeed two purposes: first they signal that the rule is a deduction rule, second they provide a name that will be useful when describing elaborated strategies. A simple labeled rule is therefore of the form $[\ell] \; l \; \rightarrow \; r$ where ℓ is a non-empty label, l and r are terms which variables satisfy as usual $Var(r) \subseteq Var(l)$. Typical simple examples of such labeled rewrite rules are (assuming x, y to be variables and a a constant) $[\texttt{id}] \; x \; \rightarrow \; x$ or $[\texttt{constant-a}] \; x \; \rightarrow \; a$ or $[\texttt{sum}] \; x + succ(y) \rightarrow succ(x + y)$ or $[\texttt{proj1}] \; x + y \; \rightarrow \; x$ or $[\texttt{proj2}] \; x + y \; \rightarrow \; y$. Notice that an equational interpretation of such rules is in general useless.

To make a clear cut between the normalizers —that implicitly embed a traversal strategy— and the deduction rules, the application of a labeled rewrite rule is performed *only* at the top most occurrence of the term on which it is applied and this application consumes the rewrite rule. This is totally similar to a function application in functional programming. For example, the application of id to any term t returns in one step t, and the application of constant-a to t terminates and returns in one step a. But, the application of sum fails on $0 + (0 + succ(0))$, since the redex is not at the top occurrence.

At the current state of our description, we can define a *primal* strategy as a labeled rewrite rule. Applying such a strategy $\mathcal{S}$ on a term t is denoted $\mathcal{S}(t)$. When l matches t (i.e. $\exists \sigma, s.t. \; \sigma(l) = t$), the application result is $\sigma(r)$ and we say that the strategy succeeds. When l does not match t, we say that the strategy fails and the application result is the empty set $\emptyset$. Indeed we will see below that in general the application of a strategy to a term, when it terminates, is a finite (ordered) multiset of terms (i.e. a flat list).

Two strategies can be concatenated by the symbol ";", i.e. the second strategy is applied on all results of the first one. $\mathcal{S}_1 ; \mathcal{S}_2$ denotes the sequential composition of the two strategies. It fails if either $\mathcal{S}_1$ fails or $\mathcal{S}_2$ fails on all results of $\mathcal{S}_1$ and in this case the application result is $\emptyset$. Otherwise, its results are all results of $\mathcal{S}_1$ on which $\mathcal{S}_2$ is successfully applied.

The next natural strategy combinator used in ELAN is dk. It takes a list of strategies and $dk(\mathcal{S}_1, \ldots, \mathcal{S}_n)$ applies all strategies and for each of them returns all its results. Its application result is the union of all the application results of the individual strategies $\mathcal{S}_i$. It may be empty, in which case we say that the strategy fails.

Together with the non-determinism capability provided by the dk operator, the analog of a cut operation is provided par the first_one strategy constructor. $first\_one(\mathcal{S}_1, \ldots, \mathcal{S}_n)$ chooses the first strategy $\mathcal{S}_i$ in the list that does not fail, and returns one of its first results. This strategy returns at most one result or fails if all sub-strategies fail.

Iterators are also provided: for example, $repeat^*(\mathcal{S})$ applies repeatedly the strategy $\mathcal{S}$ until it fails and returns the results of the last unfailing application. This strategy can never fail (zero application of $\mathcal{S}$ is always possible) and may return more than one result.

The full description of strategy combinators available in ELAN can be found in [6, 5]. This simple language allows us to build more elaborated strategies and a rewrite rule is a natural way to give a name to such an expression like in [] simpleStrat $\rightarrow$ dk(id, constant-a). But notice that here, we are not only rewriting terms but strategy expressions.

We are ready to define the general form of an ELAN rewrite rule: A *labeled rewrite rule with general matching conditions* is denoted

$$[\ell] \; l \rightarrow r \; \textbf{where} \; p_1 := \mathcal{S}_1(u_1), \ldots, \textbf{where} \; p_n := \mathcal{S}_n(u_n)$$

and $l, r, p_i, u_i \in \mathcal{T}(\mathcal{F}, \mathcal{X})$, the $\mathcal{S}_i$ are strategy expressions and a variable is used somewhere only if it is "well defined", a natural technical condition detailed in the user's manual. The relation induced on terms by these rules is called *strategic rewriting*.

The application result of such a rule on a term t is defined as follows: (1) match l against t using a substitution σ, then (2) match p_1 against all the results of the application of $\mathcal{S}_1$ on $\sigma(u_1)$. Let σ_1^i be such a match, (3) match p_2 against

$\sigma_1^i \sigma(u_2)$, ... and finally the result consists of the multiset of all the instances of r computed in the **where** part.

When the label of the above rule is empty, the application process is performed everywhere in the term t as for simple unlabeled rules.

From the basic strategy combinators provided by the language, the user can define his own ones, such as in the following strategy expression used in Colette [9] to describe and solve CSP:

```
[] FLAChoicePointSplitLastToFirstAll =>
dk (LocalConsistencyForECandDC);
repeat* (
    dk (first one (SplitDomainSecondMiddle),
        first one (SplitDomainFirstMiddle));
    first one (first one (ExtractConstraintsOnEqualityVar);
                first one (Elimination, id);
                LocalConsistencyForEC
                ,
                first one (ExtractConstraintsOnDomainVar);
                LocalConsistencyForEC
                ,
                id)
        );
first one (GetSolutionCSP)
```

2.3   Summing Up the Main Language Design Choices

In ELAN, computation and deduction are modeled using unlabeled and labeled rules respectively. The control over labeled rules is performed using strategies described by a simple strategy language where iterators and non-determinism operators are available.

The matching process underlying all operations in ELAN has been designed to be first-order, therefore we do not allow higher-order variables in rewrite rules. This is a strong design decision relying on the facts that, if needed, higher-order can be encoded at the first-order level (see for example [16]) and second, that first-order interpretation and compilation techniques are much better understood than for higher-order (where indeed at order five, matching decidability is still an open problem for beta-eta conversion [29]).

Since the strategy language allows non-determinism, the implementation choices have been between returning explicit multisets of results or enumerating these results using backtracking. Even though extensions of the language support the "set-of" capability for returning multisets of results, the basic evaluation mechanism relies on backtracking.

2.4   Added and Missing Capabilities

The design of a programming language is a subtil blend of conceptual choices about its main characteristics with several additional non essential but useful capabilities, altogether carefully and smartly implemented. We give here our view of the goodies and missing of the current ELAN language.

Goodies. Indeed, ELAN comes with many goodies that make it more attractive and usable as a practical programming language.

Pre-processing. Since higher-order matching (and in particular second-order) is not directly available, a pre-processor is quite convenient to overcome this limitation. A typical example is for syntactic unification, where the decomposition rule which indeed depends on the signature, could be expressed as follows:

```
FOR EACH SS:pair[identifier,int]; F:identifier; N:int
SUCH THAT SS:=(listExtract) elem(Fss)
       AND  F:=()first(SS) AND N:=()second(SS) :{
   rules for unifPb
      s_1,...,s_N:term; t_1,...,t_N:term;
   local
   [decompose] P ^ F(s_1,...,s_N)=F(t_1,...,t_N) => P { ^ s_I=t_I }_I=1...N
   end
end  }
```

If the signature contains two symbols, g of arity 1 and f of arity 2, then the pre-processor will generate the two rules:

```
[decompose] P ^ g(s1)=g(t1)       => P ^ s1 = t1 end
[decompose] P ^ f(s1,s2)=f(t1,t2) => P ^ s1 = t1 ^ s2 = t2 end
```

Earley Parsing. In the tradition of several algebraic programming languages like OBJ, ELAN provides the user with the capability to define its own mixfix syntax in a very liberal way. It is then analyzed using Earley's algorithm. This is quite convenient to adapt the syntax to the user's description of the problem, at the price of a cubic parsing process, in the worst case.

Modularity and Parameterization. A serious programming language cannot come without modularization capabilities. ELAN provides the possibility to define local and non-local operators and rules, to import modules and to defined parameterized modules.

Rewriting Modulo AC. An important improvement in the rewriting agility is the capability to take into account, at run time, properties of operators like associativity and commutativity. This frees the programmer of tediously describing all the possible variations in rule application when one of its symbol is associative and commutative (AC for short) but AC matching is not available. The code reduces significantly at the price of a matching algorithm running in exponential time in the worst case. A clever compiler have been designed to keep this drawback as much under control as possible [26].

Missing. A language cannot contain every fashionable features. We have made some design choices but some goodies could have been retrospectively usefully introduced. This section sums up our current views.

Enriched Signatures. The current system is based on many-sorted signatures. The agility of the language could have been enhanced by using order-sorted or membership constraints [33]. But the implementation becomes much more intricated, in particular since the sort of a term may in this case change at run time. Secondly, the interaction of the matching theories and the extended sort capabilities can be quite subtle, even for experienced programmers. Therefore we did not provide this feature as it is indeed done in Maude [13].

Traversals. ELAN provides elaborated strategies as we have seen before, but term traversals are not provided. This is due to the fact that the concept emerged after the main design of the language was achieved [35]. Traversals are useful and completely in line with the language design. Even if some early versions of the language prototyped them, they are not provided in the current distribution and could have been a useful extension.

Matching Theories. The debate on which matching theories are useful and could be usefully implemented has been (and is still!) long-standing in the ELAN team. On one side, one can say that the more the better, on the other hand, the more theory the most complex the implementation becomes. Moreover, when combining several theories, completeness of matching becomes a real difficulty, in particular on the complexity side. Therefore ELAN provides only syntactic and AC matching and their combination. Adding associativity is natural and useful: it is indeed the first theory available in TOM (see Section 3.4).

Deduction modulo relies on the ability to embed potentially complex theories in the modulo part, and therefore in the matching process. It would be useful to give the possibility to match modulo user-defined theories: a capability far beyond the currently available knowledge and rewrite technology if we want to keep the system efficient.

Deep Inference. The design decision to have unlabeled rules applied everywhere (acting as normalizers) and labeled ones to be applied only on top of term (acting as deduction rules), comes from the usual view of deduction rule applied at the top level of formulas as in the calculus of sequent or in natural deduction. Recent works on deep inference [8] show all the interest to have inference rules also applied inside terms or formulas. ELAN is not designed this way, but this is certainly a challenging capability to have deep inference available, for example via a clever combination of labeled rule with traversals.

3   Emerging Concepts

The general setup, design and implementation of the system leaded the ELAN team to elaborate or refine several main ideas and results. We review here the main emerging concepts.

3.1   Semantics of Strategic Rewriting

One of the main originality and useful feature of ELAN is the ability to define and efficiently execute rewriting strategies.

The concern of giving to the programmer some control on the normalization process is already present in OBJ with the so-called local strategies. Further extension of this idea is also related to the control of concurrent evaluation [22]. On the proof side, strategies are called tactics or plans and are mandatory in proof assistants.

What ELAN brings first here is that term rewriting is used for both computation (and therefore normalization) and deduction. This means that rewriting is used to model both equality and transition. As a consequence strategies are not only a (useful) addition but a mandatory one.

However in order to understand the semantics of strategic rewriting, one needs to understand the concept of strategy. It is indeed very simple and natural to define a *strategy as a set of proof terms of a rewrite theory* as proposed in [25, 36] using the rewriting logic [30] framework.

3.2   Strategy Language

Clearly, an arbitrary set of proof terms may be very elaborated or irregular from the computational point of view, it could be in particular non-recursive. This is why languages describing special subclasses of strategies are needed. The ELAN strategy constructors described in Section 2.2 contribute to define such a subclass.

However this is not expressive enough to allow recursive and parameterized strategies. This is why the more general notion of defined strategies has been introduced [5, 6]. Their definition is given by a strategy operator with a rank, and a set of labeled rewrite rules. The example of a map functor on lists illustrates this.

Let map be of rank map : $(\langle s \mapsto s \rangle)\ \langle list[s] \mapsto list[s] \rangle$ where $\langle s \mapsto s \rangle$ and $\langle list[s] \mapsto list[s] \rangle$ are strategy sorts. The argument of map must be a strategy S that applies to a term of sort s and returns results of sort s. The strategy map(S) applies to a term of sort $list[s]$ and returns results of sort $list[s]$. It is defined by the rewrite rule:

$$[\ ]\quad \mathsf{map}(S) \rightarrow \mathsf{first}(\mathrm{nil}, S \cdot \mathsf{map}(S))\quad (1)$$

where S is a variable of sort $\langle s \mapsto s \rangle$. The right-hand side of this definition means that whenever the strategy map(S) is applied to a term t, either t is nil, or the strategy S is applied to the head of t (i.e. t should be a non-empty list) and map(S) is further applied to the tail of t.

Allowing the description of a strategy by a set of rewrite rules, as above, strongly increases the expressive power of ELAN: strategies may be recursive, parameterized and typed as well. But this also leads to understand a strategy as a function, or later on as a functional. Also adding an explicit application operator leads to the formalization of higher-order objects. This gave rise to the rewriting calculus which provides both a language to express strategies and an operational semantics for the constructions of ELAN.

3.3   The Rewriting Calculus

The rewriting calculus, whose initial design is detailed in [11], is also called ρ-calculus. It is a natural but prominent outcome of the ELAN design, implementation and usage. As we have seen in the previous sections, rewriting strategies are central in ELAN and this leaded to the simple idea that the simplest strategy is indeed just a rule (i.e. $l \to r$) and that applying this strategy on a term t is just (explicitly) applying this rule, i.e. $(l \to r\ t)$. Pushing this remark further leaded to the first main idea of the calculus, that is to provide a uniform combination of term rewriting and lambda-calculus. This is in particular fully adapted to the description of computation and deduction transitions.

The second main idea of the calculus is to make all basic ingredients of rewriting explicit objects, in particular the notions of rule *formation*, *application* and *result*. Terms, rules, rule application and therefore rule application strategies are all treated as first class objects. To make this more explicit, the rule constructor "$\to$" uses a different arrow symbol in all this Section. For example, using a syntax close those of lambda-calculus, application of the rule $2 \to s(s(0))$, to a term, e.g. the constant 2, is explicitly represented as the object $(2 \to s(s(0))\ 2)$ which evaluates to $s(s(0))$.

The third important concept in the ρ-calculus is that rules are fired modulo some theory e.g. associativity and commutativity. For example, provided the commutativity of $+$, the ρ-term $(x + 0 \to x\ 0 + 1)$ reduces to 1. The fact that results are explicit allows us to give a precise meaning to the reduction of the ρ-term $(x + y \to x\ a + b)$ as the structure $a \mathbin{\text{\textsf{I}}} b$ that may be informal understood as a set of results.

Since the beginning, we wanted to integrate explicit substitutions [10] but the link with matching constraints has been done later [12] and a combined version is presented in [23]. Indeed to represent explicitly substitution is useful not only from the foundational point of view, but also for representing proof term, as we will see later.

As usual, for a calculus with binder, we work modulo the α-*conversion* and adopt Barendregt's *hygiene-convention i.e.* free and bound variables have different names. The syntax is the following:

$$
\begin{array}{lll}
\mathcal{P} & ::= & \mathcal{T} \qquad\qquad\qquad\qquad\qquad\qquad\qquad\text{Patterns} \\
\mathcal{T} & ::= & \mathcal{X} \mid \mathcal{K} \mid \mathcal{P} \to \mathcal{T} \mid \mathcal{T}\ \mathcal{T} \mid [\mathcal{P} \ll \mathcal{T}]\mathcal{T} \mid \mathcal{T} \mathbin{\text{\textsf{I}}} \mathcal{T} \quad \text{Terms}
\end{array}
$$

1. $A \to B$ denotes a *rule abstraction* with pattern A and body B; the free variables of A are bound in B.
2. $(A\ B)$ denotes the *application* of A to B.
3. $[P \ll A]B$ denotes a *delayed matching constraint* with pattern P, body B and argument A; the free variables of P are bound in B but not in A.

To obtain good properties for the calculus (*e.g.* confluence) the form of patterns has to be restricted to particular classes of ρ-terms. The operator $[\_ \ll \_]\_$ can be decorated by (matching) theory $\mathbb{T}$ and become $[\_\ll_\mathbb{T}\_]\_$ if this is not implicit when working in a given context. The set of solutions of the matching constraint $[P\ll_\mathbb{T}B]A$ is denoted $\mathcal{S}ol(P\mathbin{\twoheadleftarrow}_\mathbb{T}B)$.

The *small step semantics* of the ρ-calculus is given by the following rules:

$[\rho] \quad (P \rightarrow A \ B) \quad \rightarrow_\rho \quad [P \ll_T B]A$

$[\sigma] \quad [P \ll_T B]A \quad \rightarrow_\sigma \quad A\theta_1, \ldots, A\theta_n \quad$ with $\{\theta_1, \ldots, \theta_n\} = \mathcal{S}ol(P \ll_T B)$

$[\delta] \quad (A \mathsf{I} B \ C) \quad \rightarrow_\delta \quad (A \ C) \mathsf{I} (B \ C)$

For example, the β-redex $(\lambda x.t \ u)$ is nothing else than the ρ-redex $(x \rightarrow t \ u)$ (i.e., the application of the rewrite rule $x \rightarrow t$ to the term u) which reduces to $[x \ll u]t$ and then to $\{x/u\}t$ (i.e., the application of the higher-order substitution $\{x/u\}$ to the term t).

The small-step semantics can then be customized, via the rule $[\sigma]$, in order to consider non-unitary and even infinitary theories [12].

These ρ-calculus principles, that can also be understood as a kind of constrained rewriting, emerged from the ELAN design: they soon attracted a lot of attention and have been studied for themselves (see the latest developments at url `rho.loria.fr`). For example, a version of the rewriting calculus with explicit substitutions has been used to represent rewriting derivations modulo AC [32] for the Coq proof assistant.

3.4   TOM and Formal Islands

Since the beginning of the ELAN project, we have been strongly concerned with the feasability of strategic rewriting as a practical programming paradigm. Therefore, the development of efficient compilation concepts and techniques took an important place in the language support design. The results presented in [26] leaded to a quite efficient implementation and thus demonstrated the practicality of the paradigm.

But even if ELAN is a nice language cleverly and efficiently implemented, it requires an existing application to be totaly rewritten in ELAN to benefit from its capabilities. Strategic rewriting is therefore available but hardly usable in the large.

This is the main concern that leaded to the emergence of the idea of *formal island*, a general way to make formal methods, and in particular matching and rewriting available in virtualy any existing environment. TOM [31] is an implementation of this idea. In its Java instance, TOM provides matching and rewriting primitives that are added to the Java language. These specific instructions are then compiled to the host language (e.g. Java), using similar techniques as those used for compiling ELAN. The good things are that one can then use the normal forms provided by rewriting to get conciseness and expressiveness in Java programs, but moreover one can prove that these sets of rewrite rules have useful properties like termination or confluence. Once the programmer has used rewriting to specify functionalities and to prove properties, the compiled dissolves this formal island in the existing code just by compilation. The use of rewriting and TOM therefore induces no dependence: once compiled, a TOM program contains no more trace of the rewriting and matching statements that were used to build it.

TOM and its Eclipse environment are available at url tom.loria.fr: they provide an efficient way to integrate rewriting in Java as well as to perform easy and formaly safe XML rewriting.

4   Applications to High-Level Decision Making

How does this research contribute to high-level decision making? The interested reader may consult the ELAN web page to get an exhaustive idea of developped applications, but we choose here to select four of them to illustrate how ELAN and strategic programming can be used to model different kinds of processes, namely solving, computing and proving.

4.1   Constraint Solving

One of the first concern of the ELAN project was to model constraint solving, from unification problems to complex Constraint Satisfaction Problems (CSP). Research has been very active on CSP since the seventies and often in relationship with traditional Operational Research techniques and Constraint Logic Programming, in particular with the seminal works on Prolog [14] and CLP [24]. Our concern on this topics was to express in a simple and clear way the underlying concepts used to solve CSP, formalized as a deduction process. The ELAN language is especially well-suited, thanks to the explicit definition of deduction rules and control. Actions are associated with rewrite rules and control with strategies that establish the order of applications of deductions. Expressing the algorithms developed for solving CSP as rewrite rules driven by strategies leads to a better understanding, easy combination and potentially to improvements and proof of correctness. Colette [9] is a CSP solving environment implemented in ELAN to validate this approach. Various searching techniques for clever exploration of the solutions space, problem reduction techniques that transform a CSP into an equivalent problem by reducing the values that the variables can take, as well as various forms of consistency algorithms are described by ELAN strategies.

4.2   Chemical Computations

Rule-based systems and strategies have been used for modelling a complex problem of chemical kinetic: the automated generation of reaction mechanisms. The generation of detailed kinetic mechanisms for the combustion of a mixture of organic compounds in a large temperature field requires to consider several hundred chemical species and several thousands of elementary reactions. An automated procedure is the only convenient and rigorous way to write such large mechanisms. Flexibility is often absent or limited to menu systems, whereas the actual use of these systems, during validation of generated mechanisms by chemists, as well as during their final use for conception of industrial chemical processes, requires modifications, activations or deactivations of involved rules according to new experimental data, reactor conditions, or chemist expertise. The purpose of an automated generator of detailed kinetic mechanisms is to take as input one or more hydrocarbon molecules and the reaction conditions and to give as output the list of elementary reactions applied and the corresponding thermodynamic and kinetic data. The GasEl[7] system has been designed in ELAN for that purpose. The representation of the chemical species uses the notion of molecular graphs, encoded by a term structure called GasEl terms. The chemical reactions are expressed by rewrite rules on molecular graphs, encoded by a set of conditional rewrite rules on GasEl terms. ELAN's strategy language is quite appropriate to express the reactions chaining in the mechanism generator. The required flexibility is provided by the high-level specification of the declarative strategy language, that can reflect the chemist's decisions.

4.3   Proving

Several proof tools have been designed in ELAN, ranging from a predicate logic prover, or a completion procedure, to various model checkers. In the context of rule-based programming, termination is a key property that warrants the existence of a result for every evaluation of a program. CARIBOO is a termination proof tool for rewrite programs, given by sets of rewrite rules. Its foundation is a termination proof method based on an explicit induction mechanism on the termination property. CARIBOO is able to deal with different term traversal strategies, corresponding to call-by-value (innermost strategy [19]), call-by-name (outermost strategy [20]), or more local calls (local strategies on the operators [18]). Such proof tools might be later used in proof environments able to combine them in order to guarantee safety of programs. The interesting point here is that the proof procedure is described in each case by deduction rules and a control specification, which are directly reflected in the ELAN implementation.

4.4   Combining Computation and Deduction

The rewriting calculus is quite useful in combining rewriting-based automated theorem proving and user-guided proof development, with the strong constraint of safe cooperation of both. We addressed this problem in practice in combining the Coq proof assistant and the ELAN rewriting based system.

The approach followed for equational proofs relies on a normalization tactic in associative and commutative theories written in ELAN. It generates a proof term in the rewriting calculus, which is then translated into a proof term written in the calculus of constructions syntax that can finally be checked by Coq to get the proof of the normalization process [32]. The advantages of this approach are to take benefit from the efficient (conditional AC) rewriting performed by the ELAN compiler, and to ease the size reducing transformations of the proof terms before sending them to Coq. For that, the ELAN compiler has been extended by a proof term producer that builds the rewriting proof term, and by a proof term translator that transforms this formal trace of ELAN into the corresponding Coq proof term for checking. In this cooperation scheme, ELAN can be seen as a computing server and Coq proof sessions as its clients.

Actually this work goes beyond the specific use of Coq and ELAN. It raises the general problem of incorporating decision procedures in proof assistants based on type theory, in a reliable and efficient way. Reliability is handled here through the concept of proof term, that contains all information about the proof and is exchanged between the two systems. Built by ELAN during the rewriting proof construction, it is then checked by Coq or by any proof assistant.

4.5   When Is ELAN Not Appropriate?

While rules and strategies are really natural in defining normalizers or when having to model deductive or transition systems, some applications are not easy to program in the current version of ELAN.

In a first place, since ELAN is a *term* rewrite rule language, data structures such as graphs or matrices, are not easy to deal with. When working with such structures, encodings, sometimes clever like in the chemistry application reported above, are necessary to take benefit of the language features.

We did not develop any elaborated numerical computations and types nor fancy input/output and graphical libraries. This means that connecting ELAN programs with the outside world is possible but not easy. Any application heavily relying on such characteristics is not currently appropriate to develop in the language.

In the context of ambitious applications, it is fundamental to interface ELAN or its fundamental concepts of rules and strategies with other programming languages. This is the place where TOM comes into play and is already quite promising.

5   Conclusion

One picture is better than hundred explanations: this popular sentence summarizes the interest to deal with modeling environments making a fundamental use of patterns. The research community in informatics is rich in such languages where Prolog, CLP, and rule-based programming play a central role.

Our researches on these topics for the last ten years has been very fruitful, both in practical terms as well as in fundamental research advances: among them, the demonstration that rewrite-based languages could be as efficient as functional ones, the emergence of the rewriting calculus as a unified environment for strategic rewriting and functional evaluation, and the explicitation of the computation and deduction concepts as a consistent programming paradigm.

Because of the pattern matching capability of the language, the applications developped in the ELAN language have shown that programming an algorithm, solving a constraint or searching for a proof are of the very same essence and, indeed, are in general collaborative tasks for which rule-based programming is very well adapted.

As the programming activity is abstracting more and more to allow for higher-level decision-making, rule-based programming offers a clever and useful paradigm that becomes more and more attractive. We hope that the ELAN experience outcomes will contribute to make it more useful and popular.

Acknowledgements

Our sincere and warmest thanks go first to the past and present members of the Protheo team who contributed to the ELAN project, either as designers, users or contributors to theoretical or practical problems.

Such a ten years project could not fruifully happen without the strong support of our host institutions: INRIA, CNRS and the Universities of Nancy.

Our interactions with thematically related groups have been quite fruitful and we are very pleased to thank the SRI International team leaded by Joe Goguen, José Meseguer and Carolyn Talcott and from which emerged OBJ and Maude, the CWI team leaded by Paul Klint and Mark van den Brand who are developping ASF+SDF and who share with us the successful Aircube project.

Our initial interest for rewriting and its applications was triggered and driven by Jean-Pierre Jouannaud. His long support and interest as well as the interactions with his teams have always been particularly fruitful.

Jean-Louis Lassez has been at the start of this project in 1993 for the first presentation of the concepts and system: it is our great pleasure to dedicate him this paper.

References

1. E. Astesiano, M. Bidoit, H. Kirchner, B. Krieg-Br?ckner, P. D. Mosses, D. Sannella, and A. Tarlecki. CASL: The Common Algebraic Specification Language. *Theoretical Computer Science*, 286(2):153–196, September 2002.
2. J. A. Bergstra and J. V. Tucker. Algebraic specifications of computable and semi-computable data structures. *Theoretical Computer Science*, page 24, 1983.
3. D. Bert, P. Drabik, R. Echahed, O. Declerfayt, B. Demeuse, P.-Y. Schobbens, and F. Wautier. Reference manual of the specification language LPG- version 1.8 on SUN workstations. Internal report, LIFIA-Grenoble (FRANCE), January 1989.
4. P. Borovanský, H. Cirstea, H. Dubois, C. Kirchner, H. Kirchner, P.-E. Moreau, Q.-H. Nguyen, C. Ringeissen, and M. Vittek. ELAN *V 3.6 User Manual*. LORIA, Nancy (France), fifth edition, February 2004.
5. P. Borovansky, C. Kirchner, H. Kirchner, and P.-E. Moreau. ELAN from a rewriting logic point of view. *Theoretical Computer Science*, 2(285):155–185, July 2002.
6. P. Borovanský, C. Kirchner, H. Kirchner, and C. Ringeissen. Rewriting with strategies in ELAN: a functional semantics. *International Journal of Foundations of Computer Science*, 12(1):69–98, February 2001.
7. O. Bournez, G.-M. Côme, V. Conraud, H. Kirchner, and L. Ibănescu. A rule-based approach for automated generation of kinetic chemical mechanisms. In *Proceedings 14th Conference on Rewriting Techniques and Applications, Valencia (Spain)*, Lecture Notes in Computer Science. Springer-Verlag, 2003.
8. K. Brünnler. *Deep Inference and Symmetry in Classical Proofs*. Logos Verlag, Berlin, 2004.
9. C. Castro. Building Constraint Satisfaction Problem Solvers Using Rewrite Rules and Strategies. *Fundamenta Informaticae*, 34:263–293, September 1998.
10. H. Cirstea. *Calcul de réécriture : fondements et applications*. Thèse de Doctorat d'Université, Université Henri Poincaré - Nancy I, 2000.

11. H. Cirstea and C. Kirchner. The rewriting calculus — Part I *and* II. *Logic Journal of the Interest Group in Pure and Applied Logics*, 9(3):427–498, May 2001.

12. H. Cirstea, C. Kirchner, and L. Liquori. Matching Power. In A. Middeldorp, editor, *Rewriting Techniques and Applications*, volume 2051 of *Lecture Notes in Computer Science*, Utrecht, The Netherlands, May 2001. Springer-Verlag.

13. M. Clavel, F. Durán, S. Eker, P. Lincoln, N. Martí-Oliet, J. Meseguer, and J. F. Quesada. Maude: Specification and programming in rewriting logic. *Theoretical Computer Science*, 2(285), 2001.

14. A. Colmerauer. An introduction to Prolog III. *Communications of the ACM*, 33(7):69–90, 1990.

15. M. Dauchet. Simulation of Turing machines by a left-linear rewrite rule. In N. Dershowitz, editor, *Proceedings 3rd Conference on Rewriting Techniques and Applications, Chapel Hill (N.C., USA)*, volume 355 of *Lecture Notes in Computer Science*, pages 109–120, April 1989.

16. G. Dowek, T. Hardin, and C. Kirchner. HOL-$\lambda\sigma$ an intentional first-order expression of higher-order logic. *Mathematical Structures in Computer Science*, 11(1):21–45, 2001.

17. G. Dowek, T. Hardin, and C. Kirchner. Theorem proving modulo. *Journal of Automated Reasoning*, 31(1):33–72, Nov 2003.

18. O. Fissore, I. Gnaedig, and H. Kirchner. Termination of rewriting with local strategies. In M. P. Bonacina and B. Gramlich, editors, *Selected papers of the 4th International Workshop on Strategies in Automated Deduction*, volume 58 of *Electronic Notes in Theoretical Computer Science*. Elsevier Science Publishers B. V. (North-Holland), 2001.

19. O. Fissore, I. Gnaedig, and H. Kirchner. CARIBOO : An induction based proof tool for termination with strategies. In *Proceedings of the Fourth International Conference on Principles and Practice of Declarative Programming*, pages 62–73, Pittsburgh (USA), October 2002. ACM Press.

20. O. Fissore, I. Gnaedig, and H. Kirchner. Outermost ground termination. In *Proceedings of the Fourth International Workshop on Rewriting Logic and Its Applications*, volume 71 of *Electronic Notes in Theoretical Computer Science*, Pisa, Italy, September 2002. Elsevier Science Publishers B. V. (North-Holland).

21. J. A. Goguen. Some design principles and theory for OBJ-0, a language for expressing and executing algebraic specifications of programs. In E. Blum, M. Paul, and S. Takasu, editors, *Mathematical Studies of Information Processing*, volume 75 of *Lecture Notes in Computer Science*, pages 425–473. Kyoto (Japan), 1979. Proceedings of a Workshop held August 1978.

22. J. A. Goguen, C. Kirchner, and J. Meseguer. Concurrent term rewriting as a model of computation. In R. Keller and J. Fasel, editors, *Proceedings of Graph Reduction Workshop*, volume 279 of *Lecture Notes in Computer Science*, pages 53–93, Santa Fe (NM, USA), 1987.

23. Horatiu Cirstea, Germain Faure, and Claude Kirchner. A rho-calculus of explicit constraint application. In *Proceedings of the 5th workshop on rewriting logic and applications*. Electronic Notes in Theoretical Computer Science, 2004.

24. J. Jaffar and J.-L. Lassez. Constraint logic programming. In *Proceedings of the 14th Annual ACM Symposium on Principles Of Programming Languages, Munich (Germany)*, pages 111–119, 1987.

25. C. Kirchner, H. Kirchner, and M. Vittek. Designing Constraint Logic Programming Languages using Computational Systems. In P. Van Hentenryck and V. Saraswat, editors, *Principles and Practice of Constraint Programming. The Newport Papers.*, pages 131–158. MIT press, 1995.

26. H. Kirchner and P.-E. Moreau. Promoting rewriting to a programming language: A compiler for non-deterministic rewrite programs in associative-commutative theories. *Journal of Functional Programming*, 11(2):207–251, 2001.
27. P. Klint. A meta-environment for generating programming environments. *ACM Transactions on Software Engineering and Methodology*, 2:176–201, 1993.
28. J.-L. Lassez, M. J. Maher, and K. Marriot. Unification revisited. In J. Minker, editor, *Foundations of Deductive Databases and Logic Programming*. Morgan-Kaufman, 1988.
29. R. Loader. Higher order β matching is undecidable. *Logic Journal of the IGPL*, 11(1):51–68, 2003.
30. J. Meseguer. Conditional rewriting logic as a unified model of concurrency. *Theoretical Computer Science*, 96:73–155, 1992.
31. P.-E. Moreau, C. Ringeissen, and M. Vittek. A Pattern Matching Compiler for Multiple Target Languages. In G. Hedin, editor, *12th Conference on Compiler Construction, Warsaw (Poland)*, volume 2622 of *LNCS*, pages 61–76. Springer-Verlag, May 2003.
32. Q.-H. Nguyen, C. Kirchner, and H. Kirchner. External rewriting for skeptical proof assistants. *Journal of Automated Reasoning*, 29(3-4):309–336, 2002.
33. G. Smolka, editor. *Special issue on Order-sorted Rewriting*, volume 25 of *Journal of Symbolic Computation*. Academic Press inc., April 1998.
34. A. van Deursen, J. Heering, and P. Klint. *Language Prototyping*. World Scientific, 1996. ISBN 981-02-2732-9.
35. E. Visser. Language independent traversals for program transformation. Technical report, Department of Computer Science, Universiteit Utrecht, Utrecht, The Netherlands, 2000.
36. M. Vittek. *ELAN: Un cadre logique pour le prototypage de langages de programmation avec contraintes*. Thèse de Doctorat d'Université, Université Henri Poincaré - Nancy 1, octobre 1994.

Towards Flexible Graphical Communication Using Adaptive Diagrams

Kim Marriott[1], Bernd Meyer[1], and Peter J. Stuckey[2]

[1] School of Comp. Sci. & Soft. Eng., Monash University, Australia
{marriott, berndm}@mail.csse.monash.edu.au
[2] NICTA Victoria Laboratory,
Department of Computer Science and Software Engineering,
University of Melbourne, 3010 Australia
pjs@cs.mu.oz.au

Abstract. Unlike today where the majority of diagrams are static, life-less objects reflecting their origin in print media, the computer of the near future will provide more flexible visual computer interfaces in which diagrams adapt to their viewing context, support interactive exploration and provide semantics-based retrieval and adaptation. We provide an overview of the Adaptive Diagram Research Project whose aim is to provide a generic computational basis for this new type of diagrams.

1 Introduction

Our computing environment is rapidly changing to a ubiquitous information environment in which we access and interact with a variety of digital media using an incredible variety of devices: wearable computers, hand-held PDAs, tablets with styluses, TV like devices, large wall mounted displays, etc. We believe that diagrams and sketches will play an increasingly important role in this interaction, but that unlike today where the majority of diagrams are static, lifeless objects reflecting their origin in print media, the computer of the near future will provide more flexible visual computer interfaces that use diagrams—often in combination with text or speech—for natural communication.

There are five main differences between the old print media based view of diagrams and this new view based on interactive media and the web:

- *Adaption to Viewing environment:* The first difference is the need for a diagram's layout and appearance to adapt to the viewing environment. The obvious reason for this is to make the best use of the viewing device by taking into account its capabilities and display characteristics such as size and aspect ratio. Secondly, adaptation can take into account user needs and desires, for example by adapting text in the diagram to a different language or by using larger fonts and more readily distinguishable colors for users with vision impairment. For example, if the user specifies a larger font size for the organisation chart shown in Figure 1(a) we would expect the layout to adapt by increasing the size of the boxes so that they continue to frame the text and then to appropriately modify the placement of the boxes, text and connectors.

M.J. Maher (Ed.): ASIAN 2004, LNCS 3321, pp. 380–394, 2004.

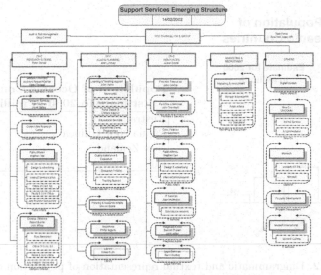

(a) Full layout suitable for large display

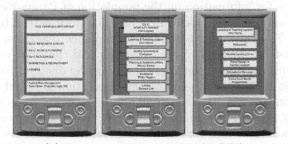

(b) Interactive view suitable for PDA

Fig. 1. Adaptive viewing of organization chart (from [23])

- *Interactive Exploration:* The second key difference is interaction. At the very least one would like the ability to interactively explore large hierarchical diagrams by collapsing or expanding components, thus allowing suppression of detail in areas of little interest so that the other areas can be displayed at a more detailed level. Such *semantic zooming* is particularly important on devices with small display areas, such as PDAs or mobile phones. Consider again the diagram shown in Figure 1(a). This layout is fine for an A4 sized page or laptop but not for a PDA. Figure 1(b) shows how the same diagram can be viewed on a PDA by using semantic zooming to expand and collapse nodes of interest. More ambitiously, we can imagine interactive on-line textbooks that help students to understand and learn by letting them manipulate and experiment via manipulation of diagrams.
- *Access to Semantics:* The ability to access the information contained in diagrams in a structured way is also very important. This is for at least three reasons. The first is semantics based retrieval. For instance, we might wish to

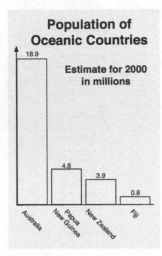

Population of Oceanic Countries
Estimate fo 2000 in millions

Australia	18.9
Papua New Guinea	4.8
New Zealand	3.9
Fiji	0.8

Fig. 2. Diagrammatic and textual representation of population data

query the web for information on the population of Papua New Guinea. This information may well be hidden in a diagram such as the bar chart shown in Figure 2. To enable search engines to find such diagrams, a semantic description must be attached to each diagram and the search engine must be able to use this determine the applicability of the diagram to the query. The second reason is that if the semantic representation is linked with the visual representation, we can use this to adapt the diagram so as to customize it to the user's needs. For example, when the bar chart shown in Figure 2 is retrieved it would be nice for the bar and value associated with Papua New Guinea to be highlighted. Or, as another example, we can use the contextual information that a user is from a particular department to determine which nodes should be collapsed or expanded in the initial view of a large organisation chart. The third reason is that associating a semantic representation with a diagram facilitates access by blind users and users with audio-only views. Given a semantic representation for a bar chart it is simple to generate an equivalent textual representation (Figure 2). Such extreme adaptivity is required if the web is to truly support universal access.

- *Dynamic Generation of Content:* Another important issue is the need to handle dynamic content such as that coming from a database. This requires automatic layout to be performed at display time since the diagram's layout cannot be determined beforehand. For example, the diagrams in Figure 1 and Figure 2 might well be generated automatically from information in a database.

- *Authoring:* The above requirements not only affect how the diagrams behave from the user's perspective but also raise significant new issues for the author. Traditional diagram authoring tools are designed for creating static fixed diagrams. For the full promise of this new kind of dynamic, flexible diagrams to be met, authoring tools which allow the construction of diagrams with the above capabilities by graphic designers, not programmers are required.

This paper provides an overview of the Adaptive Diagram Research Project whose aim is to provide a generic computational basis for this new kind of diagram. Reflecting the differences identified above, our research is built around four themes. The first is to extend web document standards for graphics to better support adaptive layout and viewer interaction. The second is to investigate formalisms and techniques for describing and reasoning with the semantics of diagrams. The third is to develop better techniques for automated static layout of a diagrams from abstract specifications and for dynamic re-layout of a diagram during interaction. The final theme is the development of authoring tools. Each theme will be examined from three complementary perspectives: mathematical theory (algorithm design), software engineering (generic software tools and methodologies) and human-computer interaction (HCI). In the remainder of the paper we discuss each of these in more detail. However as the project is only beginning much of the present paper consists of questions and possible approaches rather than solutions.

Geometric constraint solving provides a common foundation for our approach. Viewing diagram layout (and more generally any sort of document layout) as constrained optimization provides a general framework for understanding and supporting both static and dynamic layout in which geometric constraints capture the structure of the diagram or document and an objective function measures the aesthetic desirability of a particular layout.

This extends earlier suggestions for constraint-based specifications of web document layout, e.g. [34, 10, 6, 7]. The key idea is that geometric constraints allow the author to give a flexible, partial specification of document layout. The author can specify, for example, that a box should be large enough to contain its text or that certain objects should be aligned. The user's preferences and the viewing environment characteristics can be taken into account as additional constraints. Constraints can be preferences as well as strict requirements. The appearance of the document is the result of a *negotiation* between the author, the user and the viewing tool. Thus, although the author does not rigidly dictate the precise appearance of the document, the use of constraints allows the author to maintain a high degree of control over the document's appearance even in unanticipated viewing situations.

The other main foundation for our project is a tripartite view of a diagram consisting of its visual appearance, its semantics, and its abstract syntactic structure which provides a bridge between the visual appearance and the semantics. For example, in the case of an organisation chart the visual appearance is the low-level graphic elements such as dashed lines and text elements making up the diagram, the abstract syntax representation is a labelled tree while the semantics might be a description of the logical relationships between the entities in the organisation chart such as "x reports to y." An important role of an authoring tool is to invisibly support this tripartite view of a diagram by allowing the designer to (largely) work at the level of the visual appearance and for the authoring tool to automatically update the associated abstract syntax tree and semantics.

2 Graphic Standards That Support Interaction and Adaptive Layout

If flexible, interactive, adaptive diagrams are to become a reality, one of the most basic requirements is for web graphics standards to support and encourage such behaviour.

The desire to separate form from content, in part to allow adaptive layout, has been a major concern underlying the design of web document standards. But until recently the focus has been almost exclusively on textual documents. Older raster-based image formats such as JPEG and GIF do not even re-scale satisfactorily. The new resolution independent graphics web format Scalable Vector Graphics (SVG) [31] is a considerable improvement. It supports uniform rescaling of diagrams, specification of text attributes through style sheets, and allows alternate versions of document elements for different media and languages.

SVG provides the standard graphic primitives such as rectangles, circles, paths and standard graphical transformations. It also provides simple animation. The following SVG code specifies some text and a rectangle which frames the text.

```
<text id="t" x="300" y="50" text-anchor="middle">
The rect should contain this text
</text>
<rect x="200" y="45" width="200" height="10"
    stroke="red" stroke-width="2" fill="none"/>
```

Unfortunately, the current SVG standard does not adequately support adaptation of the document's layout to the viewing context such as by adapting the layout to the size of the browser window or to the user's desire for larger fonts. In the above example, if the user changes the font size by, say, using a style sheet, the rectangle will not adapt to the change in text size.

Adding constraint-based layout mechanisms to SVG would provide such adaptive behaviour [2, 23] and we have explored this in an extension of SVG called Constraint SVG (CSVG) [21]. CSVG extends SVG by allowing attribute values to be specified with expressions that are evaluated at display time to determine the actual attribute value. In other words, CSVG extends SVG with one-way constraints.

One-way (or data-flow) constraints are the simplest, most widely used approach to constraint-based layout [33]. They form the basis for a variety of commercial products including spreadsheets, widget layout engines, and the customizable graphic editors Visio and ConceptDraw. A one-way constraint has form $x = f_x(y_1, ..., y_n)$ where the function f_x details how to compute the value of variable x from the variables $y_1, ..., y_n$. Whenever the value of any of the y_i's changes, the value of x is recomputed, ensuring that the constraint remains satisfied. One-way constraints can be used to ensure that, for instance, a box remains large enough to contain its component text or that lines or arrows connecting two objects follow the objects when they are moved.

Thus, using a CSVG example taken from [21], we can rewrite our previous SVG example so that the rectangle's attributes are specified in terms of the

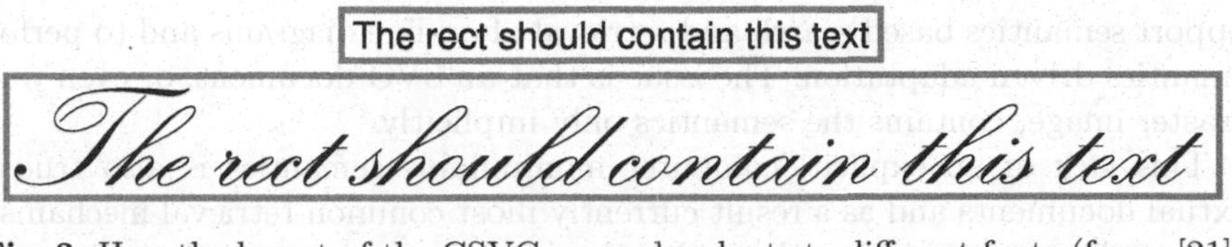

The rect should contain this text

The rect should contain this text

Fig. 3. How the layout of the CSVG example adapts to different fonts (from [21])

text's bounding box, so allowing the rectangle's location and size to adapt to different text sizes as illustrated in Figure 3.

```
<text id="t" x="300" y="50" text-anchor="middle">
   The rect should contain this text
</text>
<rect width="0" height="0" stroke="red" stroke-width="2" fill="none">
   <c:constraint attributeName="x"
                 value="c:x(c:bbox(id('t'))) - 4"/>
   <c:constraint attributeName="y"
                 value="c:y(c:bbox(id('t'))) - 4"/>
   <c:constraint attributeName="width"
                 value="c:width(c:bbox(id('t'))) + 8"/>
   <c:constraint attributeName="height"
                 value="c:height(c:bbox(id('t'))) + 8"/>
</rect>
```

Furthermore, if the location of the text is animated then the rectangle will automatically follow the text or if the text is changed by user interaction then the rectangle will automatically resize.

We have found that adding expression-based evaluation of attributes to SVG adds little implementation effort but allows the document to adapt to different sized browsing windows, to adapt to changes in the size of text elements such as those arising from the use of a larger font or a different language, and to adapt to changes resulting from user interaction and animation [21]. Remarkably all three kinds of adaptation are supported by the same simple mechanism.

Currently the only way to obtain such adaptive layout with SVG is to use script to modify the SVG DOM at display-time. Constraint-based adaptation has two main advantages over script-based adaptation. First, it removes the need for the document author or authoring tool to worry about how to perform efficient incremental updates when document elements are modified since this is the responsibility of the SVG browser. Second, it removes most of the need for SVG authoring tools and document authors to understand script, thus facilitating editing and interchange of SVG documents.

3   Associating Semantics with Diagrams

The second theme of our research is to investigate how to associate an explicit representation of a diagram's semantics with a diagram and how to use this to support semantics based search and retrieval of on-line diagrams and to perform semantics driven adaptation. The issue is that an SVG document, or even worse a raster image, contains the semantics only implicitly.

This lack of an explicit, computer manipulable semantics is also true for textual documents and as a result currently most common retrieval mechanisms on the web use statistical methods that do not support more complex queries and retrieve many irrelevant documents. For this reason there is world-wide interest in the "semantic web" [5] and the use of logic-based languages for specifying document content and their associated ontologies which are designed to support semantics-based information processing and inference.

RDF (Resource Description Framework) forms the core of the semantic web [29]. It allows the specification of objects ("resources") and relations between them and provides a simple semantics for this datamodel. RDF Schema and OWL allow the specification of ontologies. Both provide a vocabulary for describing properties and classes of RDF resources organised into generalization-hierarchies. Of these OWL is the more powerful. It comes in three increasingly powerful variants: OWL Lite, OWL DL which is based on description logic and endeavours to provide maximum expressiveness while retaining decidability, and OWL Full which provides the power of first-order logic but is consequently undecidable.

RDF forms one of the two generic approaches to high-level data representation currently used for the web. XML [37] provides the other approach. It is a generic markup language primarily designed for data transfer between application programs. It allows the specification of customised languages for hierarchical description of data. RDF, RDF Schema, OWL and SVG all have an XML-compliant syntax.

We believe it makes sense to explore both of these approaches for diagram specification since the XML approach is suited to specifying the diagram's abstract syntax while the RDF approach is suited to detailing the diagram semantics. There has been some previous work on XML-based specification of particular kinds of diagram, e.g. XML notations for representing UML [36] and geographic data [12]. As a simple example, we might represent the bar-chart shown in Figure 2 by:

```
<ex:bar-chart orientation="vertical" style="value-annotated">
    <ex:title> Population of Oceanic Countries</ex:title>
    <ex:subtitle>Estimate for 2000 in millions</ex:subtitle>
    <ex:scale bottom="0" top="20" />
    <ex:bars>
      <ex:bar name="Australia" value="18.9"/>
      <ex:bar name="Papua New Guinea" value="4.8"/>
      <ex:bar name="New Zealand" value="3.9"/>
      <ex:bar name="Fiji" value="0.8" />
    </ex:bars>
</ex:bar-chart>
```

Using XML to provide a high-level description of a diagram's abstract syntax is useful but does not directly support reasoning about the diagram. We believe a generic approach to specification of the diagram semantics is also required in which the same generic logic engine can be used to reason about any diagram given some specific inference rules for the class of diagrams it belongs to.

The obvious approach is to use RDF and OWL since this allows us to leverage from the techniques and systems currently being developed to support the semantic web. Such support is important because more complex query answering and retrieval requires the fusion of information from different diagrams and from other contextual information such as surrounding text.

However, for this approach to succeed, three issues need to be addressed. The first is whether or not first-order logic based formalisms and, in particular, description logics are powerful enough to capture the semantics of most diagrammatic notations and does logical inference provide sufficient query answering power?

There is some history of using description logics for representing diagram semantics. For instance, the use of description logics to capture the semantics of sketched queries to a GIS system and visual programming languages [13] and for capturing the semantics of UML diagrams [4]. Thus we believe that description logic and so OWL DL is powerful enough to handle abstract notations, such as graphs and UML notations, which rely only on topological relationships and have a well-defined semantics and syntax. However, in the case of notations in which spatial location and distance between objects is important it is not clear that standard description logic is adequate. For instance, the boundary between the kitchen and living area may be ambiguous in a house plan and in a weather map it is not clear where the boundary between a high-pressure system and a low-pressure system occurs. Even though some approaches have integrated description logic with concrete continuous domains [13], vagueness in a notation can be a severe problem and appears to limit the usefulness of the logic-based approach for representing the semantics of vague notations such as sketches.

The second issue is the need to develop new (good!) user interaction metaphors for search and retrieval that utilise semantic information to highlight the relevant parts of the diagram. More ambitiously, we can imagine a system that merges information from a variety of media and generates a mixture of diagrams and text to provide a summary.

User interaction is even more of an issue when presenting a diagram to a blind user: If one is only interested in the population of Papua New Guinea one does not want to have to listen to an audio description of the entire bar-chart in Figure 2 and the population of all countries in it. On the other hand one might be interested in summary information such as trends which is implicit in the diagram. There has been relatively little work in this field, although see [27].

The final issue is how to associate a diagram with an RDF description of its semantics in terms of a notation specific ontology. The simplest method is for the RDF description to be generated automatically by the authoring tool from the diagram's abstract syntactic structure. The harder (but arguably more useful method) is to search for relevant diagrams and then generate the RDF description from the visual representation of the diagram on the fly.

Computational approaches to understanding of diagrams have been studied for the last 40 years and syntax-based diagram interpretation is now comparatively well understood [24], but it relies on a precise and narrow definition of the notation to be interpreted. In the context of the web this is inadequate and more flexible interpretation techniques are required. To understand the difficulties consider the bar-chart in Figure 2. This has many semantically equivalent representations: the axes could be rotated, values could be associated with elements indirectly by using a scale on the axis rather than by directly annotating the bar with a value, or bars might not be rectangles but instead vertical stacks of stick figure icons. On the other hand, interpretation of diagrams specified in SVG is considerably simpler than for raster-based graphics since SVG's vector-based graphic specifications details immediately the graphic primitives such as lines, text, polygons etc that the diagram is made up of.

4   Automated Layout

Automated layout is at the core of diagram adaptation. It is useful to distinguish between static layout in which a new layout is computed from scratch, and dynamic layout in which the current layout is modified.

Static layout takes an abstract specification of a diagram, for instance an organisation hierarchy, and determines a good layout for the diagram. This is most relevant to generation of diagrams from on-line data but is also required in authoring tools when data is imported from other application programs such as spreadsheets or databases.

Constrained optimization provides a general framework for modelling static layout. Geometric constraints capture the structure of the diagram and an objective function measures the aesthetic desirability of a particular layout. For some diagrammatic notations, such as trees, we can solve the resulting constrained optimisation problem sufficiently rapidly using standard operations research techniques such as linear or quadratic programming. Typically, however, the resulting constrained optimisation problems are impractical to solve using standard generic techniques since they involve complex non-linear and non-convex arithmetic constraints and objectives, such as minimising the number of edge crossings. For this reason, we must usually devise algorithms that are specialised for a particular kind of diagram and drawing convention.

Most research into specialised algorithms for diagram layout has been for graphs and this has largely focussed on the static layout of idealized mathematical graphs (including the important sub-case of trees) composed of nodes and edges [3]. Surprisingly little work has gone into layout of other types of diagrammatic notations.

We plan to explore layout of blob charts. These are an important and common kind of diagram used in business, science and engineering for illustrating processes and sub-processes and relationships between them. Concept maps and state charts are both examples of blob chart like notations. Syntactically, blob charts are a labelled graph notation enriched with hierarchical grouping of nodes indicated by inclusion. Often they contain significant amounts of text.

These seemingly minor extensions make their layout significantly harder than that of normal graphs and to date no general layout procedure for such diagrams is known, despite their practical importance [14]. One possible approach to such more complex layout problems is the use of optimization meta-heuristics. Unfortunately meta-heuristic approaches usually suffer from being slow which makes them unsuitable for interactive applications, so that specialized algorithms are required.

Dynamic layout is required to support interactive exploration of graphs and other notations such as organisation charts and blob-charts in which the diagram is hierarchically organized and sub-parts can be collapsed or expanded to show internal structure. This is naturally modeled as a constrained optimization problem in which the new layout is required to be as "similar" as possible to the old layout so as to preserve the user's mental map [8]. Unfortunately, most research has focused on static layout and there has been surprisingly little work on dynamic layout even in the context of graphs.

We are planning to investigate how to best model dynamic layout as constrained optimization by conducting user studies to better understand which layout properties (e.g. ordering of nodes) it is most important to preserve during re-layout. Then we will develop techniques for solving constrained optimization problems of the form identified. These need to be fast enough to support user interaction. Inspired by the approach taken in [16], we plan to extend dynamic linear approximation to handle the non-linear objective functions arising in force directed graph layout [3].

5   Authoring Tools

The final research theme is the design and development of authoring tools. We believe that the best approach is to use a generic constraint-based graphics editor which can be customized for particular diagrammatic notations by providing notation specific graphic elements and layout templates. Tools such as Microsoft Visio and ConceptDraw are a first step into this direction. They are based on one-way constraints and provide extensible templates for common notations, but we believe that they fall short when it comes to usability, flexible customisability and more complex geometric constraints. Furthermore, they are not intended to generate adaptive diagrams or an associated semantic description.

As we discussed in the introduction we believe that the authoring tool should have tripartite view of a diagram consisting of its visual representation, its abstract syntax and its semantics. As the author changes the visual representation it is the role of the editor to appropriately update the abstract syntax representation and to generate the adaptive visual representation, say in CSVG, and associated semantic description, say in RDF, from this tree.

Creating an editor for a new notation requires a (possibly implicit) specification of the abstract syntax and ontology for that language as well as mappings between abstract syntax and semantics and abstract syntax and visual representation. One possible starting point are grammar-based techniques for specifying abstract syntax for diagrams and automatic generation of customised diagram editors [19].

Requiring the authoring tool to keep a high-level representation of the diagram allows more than generation of RDF. It allows the tool to provide high-level notation specific transformations, such as for instance transforming a bar chart to a pie chart. Furthermore, it can provide default adaptive layout behaviour for a particular notation.

Of course the default adaptive layout behaviour will not be appropriate for all diagrams. A key question is how the author can specify diagram specific adaptive layout behaviour. Clearly they should not have to write textual attribute expressions when, say, specifying one-way constraints in CSVG. Previously, we have suggested two possible approaches that have been used for specifying widget layout in user interfaces: learning from examples and spring-based specification [23]. However, it is unclear whether either of these will be practical.

We would like the authoring tool to behave as if it understands the abstract syntax and semantics of a diagram during editing. Constraint solving plays a crucial role in supporting such natural behaviour since it allows the abstract syntax (and so the semantics) of the diagram to be maintained during manipulation by preserving the geometric relationships between the diagram components. For example, constraints can be used to ensure that if a component in a network diagram is moved, its internal components and external connections move with it. Thus, constraint-solving is the heart of our approach to diagram editing.

Starting with SketchPad [30], constraint solving for graphics editors and other interactive graphical applications has been investigated for four decades [17]. The requirement for very fast re-computation of solutions during direct manipulation has led to the development of specialized constraint solving algorithms.

As previously discussed, one-way constraints are the most widely used approach. They are simple to implement and can be solved extremely quickly. Their main limitation is that constraint solving is directional and cyclic dependencies between variables are not allowed. This means that only fixed, pre-determined modes of interaction can be supported. For instance, a text box will change size if the text is changed, but if the size of the box is changed, the text remains the same size.

The customizable grahics editor Microsoft Visio provides one-way constraint-based tools, such as alignment and distribution for layout, but a preliminary user study [35] suggests that many users find it difficult to understand what will happen if an object is involved in more than one constraint since only one constraint will be enforced and the others ignored. Our study suggests that if so-called multi-way constraints are used this difficulty vanishes. Thus, we believe that multi-way constraints are required in graphic editing.

Unfortunately, there is no single best approach to solving multi-way constraints. One popular approach is to use propagation based solvers, e.g. [28, 32], while linear arithmetic constraint solvers are a more recent approach, e.g.

[11, 1, 22]. Iterative numeric approaches have also been tried, e.g. [26], as well as degrees of freedom analysis for CAD [20].

However, it is fair to say that none of the existing techniques for solving multi-way constraints is powerful enough. For example, consider a constraint-based authoring tool for blob chart diagrams or a document authoring tool that allows constraint-based placement of floating figures in a page. In both cases, we would like to support direct manipulation of arbitrary diagram components in the context of non-overlap, containment, alignment and distribution constraints. Existing approaches to constraint solving for interactive graphical applications cannot handle the kind of (possibly cyclic) multi-directional non-linear constraints arising in these scenarios: we need to develop more powerful constraint solving techniques.

Currently we are exploring a very simple idea: we model more complex geometric constraints, such as non-overlap and containment in non-convex objects, by a dynamically changing conjunction of linear constraints. Consider a complex non-linear "non-overlapping" constraint on two rectangles. At each stage during an interaction, this constraint can be locally approximated by a conjunction of linear constraints (e.g. "left-of" and "above") which can be efficiently solved using linear arithmetic constraint solvers. The solver automatically switches between local approximations at appropriate moments during the interaction. Our initial prototype implementation [25, 18] is very promising, but we need to develop a methodology and theoretical justification for designing such linear approximations.

However, the lack of powerful constraint-solving algorithms is not the only reason limiting the adoption of constraint solving techniques in graphics editors. Probably the most pressing issue is the need for good user interaction techniques and metaphors for constraint-based authoring. A key question is how to provide adequate visualization of constraints without cluttering large diagrams with representations of the constraints such as guidelines etc.

It is natural to imagine that the authoring tool and the browser provide the same constraint solving capabilities. However, we believe that this may not be the best model. Instead in our approach the authoring tool provides a powerful multi-way constraint solver for constructing the diagram and then compiles the resulting constraints into much less powerful constraints such as one-way constraints or even scripting commands that encode a plan for satisfying the author's constraints. The advantage is that the browser and web standards such as CSVG do not need to support more powerful constraints. This is possible as long the browser only allows the user to interact in limited ways such as by changing browser window size or the default font size.

We therefore plan to study "compilation" of linear arithmetic constraints and more complex geometric constraints into one-way constraints to support this model and also to improve the efficiency of constraint solving in general. We have previously investigated the use of a projection-based algorithm for compiling linear arithmetic constraints [15], but there is considerable work to be done.

6   Conclusion

Diagrammatic notations are ubiquitous, but current web and computer-mediated communication technologies have primarily focussed on text, video and static image data. The research program we have sketched is intended to remedy this bias by developing the basis for computational handling of diagrams which supports sophisticated, intelligent user interaction, visualization and retrieval. It will support universal access from different devices and better access to diagrammatic information by people with vision impairment and blindness.

However, it is important to realise that diagrams usually occur as part of a larger (usually textual) document. Thus an important question in the longer term is how to provide better support for adaptation and user interaction for compound documents containing text, diagrams and variety of other media. Adaptation of multimedia documents so as to take into account device capabilities, bandwidth, temporal and simple layout constraints has been studied for some time, see e.g. [9], but sophisticated layout adaptation, semantics based retrieval and generic support for user interaction has not been explored. In particular, adaptive layout is a hard problem since we may need to trade-off the areas assigned to different document components in order to find the best overall layout. We are hopeful that the techniques we are developing for diagrams will provide a suitable basis for more complex, compound documents.

Acknowledgements

It is a pleasure to acknowledge the many useful (sometimes heated but always interesting) discussions with the other members of the Adaptive Diagrams Project: Stan Gunawan, Nathan Hurst, Cameron McCormack, Linda McIver, Peter Moulder, Michael Wybrow and those colleagues who have worked on precursors to the project: Sitt Sen Chok, Tony Jansen, Weiqing He and Laurent Tardif.

References

1. G. Badros, A. Borning, and P.J. Stuckey. The Cassowary linear arithmetic constraint solving algorithm. *ACM Transactions on Computer Human Interaction*, 8(4):267–306, 2001.
2. G. Badros, J. Tirtowidjojo, K. Marriott, B. Meyer, W. Portnoy, and A. Borning. A constraint extension to Scalable Vector Graphics. *ACM Conference on the World Wide Web (WWW10)*, pages 489–498, Hong Kong, May 2001.
3. G. Di Battista, P. Eades, R. Tomassia, and I. Tollis. *Graph Drawing*. Prentice Hall, 1999.
4. D. Berardi, D. Calvanese, and G. De Giacomo. Reasoning on UML class diagrams using description logic based systems. *Proc. of the KI'2001 Workshop on Applications of Description Logics*, CEUR Electronic Workshop Proceedings http://ceur-ws.org/Vol-44/, 2001.
5. T. Berners-Lee, J. Hendler and O. Lassila. The Semantic Web. *Scientific American*, May 2001.

6. A. Borning, R. Lin and K. Marriott. Constraints for the Web. In *Fifth ACM International Multi-Media Conference*, pp. 173–182. Seattle, November 1997.
7. A. Borning, R. Lin and K. Marriott. Constraint-based document layout for the Web. *ACM/Springer Verlag Multimedia Systems Journal* 8(3): 177-189, 2000.
8. P. Eades, W. Lai, K. Misue and K. Sugiyama. Preserving the mental map of a diagram. In *CompuGraphics'91*, Vol. 1, pp. 34–41, 1991.
9. A semantic framework for multimedia document adaptation. J. Euzenat, N. Layada and V. Dias. *International Joint Conference on Artificial Intelligence (IJCAI'2003)*, 2003.
10. W.H. Graf. The constraint-based layout framework LayLab and its applications. In *ACM Workshop on Effective Abstractions in Multimedia* in conjunction with ACM Multimedia, 1995.
11. M. Gleicher and A. Witkin. Drawing with constraints. *The Visual Computer* 11(1), 1994.
12. GML - the Geography Markup Language. http://opengis.net/gml/
13. V. Haarslev, R. Möller, M. Wessel. Visual spatial query languages: A semantics using description logic. In *Diagrammatic Representation and Reasoning*, M. Anderson, B. Meyer, P. Olivier (eds), 387–402.
14. D. Harel and G. Yashchin. An algorithm for blob hierarchy layout. *The Visual Computer*, 18:164–185, 2002.
15. W. Harvey, P.J. Stuckey, and A. Borning. Fourier elimination for compiling constraint hierarchies. *Constraints*, 7:199–219, 2002.
16. W. He and K. Marriott. Constrained graph layout. *Constraints* 3(4): 289-314. 1998.
17. W. Hower, W.H. Graf. A bibliographical survey of constraint-based approaches to CAD, graphics, layout, visualization, and related topics. *Knowledge-Based Systems*, 9:449–464, 1996.
18. N. Hurst, K. Marriott and P. Moulder. Dynamic approximation of complex graphical constraints by linear constraints. *ACM Symposium on User Interface Software and Technology*, 191–200, Paris, Nov. 2002.
19. A. Jansen, K. Marriott, and B. Meyer. CIDER: A component-based toolkit for creating smart diagram environments. *Proc. of the 2003 International Conference on Visual Languages and Computing (VLC 2003)*, Miami, Sep. 2003, pp 353-359.
20. G. Kramer. A geometric constraint engine. *Artificial Intelligence*, 58:327–360, 1992.
21. C. McCormack, K. Marriott and B. Meyer. Adaptive layout using one-way constraints in SVG. *3rd Annual Conference on Scalable Vector Graphics (SVG Open)*. Japan, Sep. 2004.
22. K. Marriott and S.S. Chok. QOCA: A constraint solving toolkit for interactive graphical applications. *Constraints* 7(3/4): 229–254, 2002.
23. K. Marriott, B. Meyer, and L. Tardif. Fast and efficient client-side adaptivity for SVG. *ACM Conference on the World Wide Web (WWW 2002)*, 496–507, Honolulu, May 2002.
24. K. Marriott, B. Meyer and K. Wittenburg. A survey of visual language specification and recognition. In K. Marriott and B. Meyer, editors, *Theory of Visual Languages*. Pages 5–85. Springer-Verlag. 1998.
25. K. Marriott, P. Moulder, P.J. Stuckey, and A. Borning. Solving disjunctive constraints for interactive graphical applications. In T. Walsh, editor, *Proceedings of the Sixth International Conference on Principles and Practices of Constraint Programming*, LNCS, 361–374. Springer-Verlag, 2001.
26. G. Nelson. Juno: A constraint-based graphics system. In *SIGGRAPH '85 Conference Proceedings*, 235–243, ACM Press, 1985.

27. H. Petrie, C. Schlieder, P. Blenkhorn, D. Evans, A. King, A-M. O'Neill, G. Ioannidis, B. Gallagher, D. Crombie, R. Mager, M. Alafaci. TeDUB: A System for Presenting and Exploring Technical Drawings for Blind People. *Computers Helping People with Special Needs, 8th International Conference (ICCHP).* LNCS Springer-Verlag, pages 537-539, 2002.
28. M. Sannella, J. Maloney, B. Freeman-Benson, B., and A. Borning Multi-way versus one-way constraints in user interfaces: Experience with the DeltaBlue algorithm. *Software—Practice and Experience* 23(5), 529–566, 1993.
29. Semantic Web. http://www.w3.org/2001/sw/
30. I.E. Sutherland, Sketchpad: a man-machine graphical communication system. In *Proceedings of the Spring Joint Computer Conference*, pp. 329–346, IFIPS
31. Scalable Vector Graphics (SVG) http://www.w3.org/Graphics/SVG/
32. B. Vander Zanden An incremental algorithm for satisfying hierarchies of multi-way dataflow constraints. *ACM Transactions on Programming Languages and Systems* 18(1), 30–72, 1996.
33. B. Vander Zanden, R. Halterman, B. Myers, R. McDaniel, R. Miller, P. Szekely, D. Giuse, and D. Kosbie. Lessons learned about one-way, dataflow constraints in the Garnet and Amulet graphical toolkits. In *ACM Transactions on Programming Languages and Systems* 23(6), 776–796, 2001.
34. L. Weitzman and K. Wittenburg. Automatic presentation of multimedia documents using relational grammars. In *Proceedings of 2nd ACM Conference on Multimedia*, 443–451, 1994.
35. M. Wybrow, K. Marriott, L. McIver and P. Stuckey. The usefulness of constraints for diagram editing. *Proc. of the 2003 Australasian Computer Human Interaction Conference (OzCHI 2003)*, Brisbane, Nov. 2003, pp 192-201.
36. XML Metadata Interchange (XMI). http://www.omg.org/technology/documents/formal/xmi.htm
37. Extensible Markup Language (XML). http://www.w3.org/XML/

A Framework for Compiler Driven Design Space Exploration for Embedded System Customization[*]

Krishna V. Palem, Lakshmi N. Chakrapani, and Sudhakar Yalamanchili

Center for Research on Embedded Systems and Technology,
Georgia Institute of Technology, Atlanta GA 30308, USA
{palem, nsimhan, sudha}@ece.gatech.edu
http://www.crest.gatech.edu

Abstract. Designing custom solutions has been central to meeting a range of stringent and specialized needs of embedded computing, along such dimensions as physical size, power consumption, and performance that includes real-time behavior. For this trend to continue, we must find ways to overcome the twin hurdles of rising non-recurring engineering (NRE) costs and decreasing time-to-market windows by providing major improvements in designer productivity. This paper presents compiler directed design space exploration as a framework for articulating, formulating, and implementing global optimizations for embedded systems customization, where the design space is spanned by parametric representations of both candidate compiler optimizations and architecture parameters, and the navigation of the design space is driven by quantifiable, machine independent metrics. This paper describes the elements of such a framework and an example of its application.

1   Introduction

Over the last two decades Moore's Law has remained the primary semiconductor market driver for the emergence and proliferation of embedded systems. Chip densities have being doubling and cost halving every 18 months. For this trend to continue, the non-recurring engineering (NRE) costs must be amortized over high unit volumes, thus making customization attractive only in the context of high-volume applications. This generally implies uniform designs for a large set of applications. However, customization is central to a range of existing and emerging embedded applications including high-performance networking, consumer and medical electronics, industrial automation and control, electronic textiles and computer security, to name a few. These applications are characterized by evolving computational requirements in moderate volumes with stringent time-to-market pressures.

In the past, customization meant investing the time and money to design an application specific integrated circuit (ASIC). However, the non-recurring engineering (NRE) cost of ASIC product development has been rapidly escalating with each new generation of device technology. For example, a leading edge embedded system based on an 80

---

[*] This work was supported in part by DARPA PCA program under contract #F33615-03-C-4105 and DARPA DIS program under contract #F33165-99-1-1499.

M.J. Maher (Ed.): ASIAN 2004, LNCS 3321, pp. 395–406, 2004.

million transistor custom chip in 100 nm technology was projected to cost upwards of $80M [13]. Less aggressive chip designs still incur significant NRE costs in the range of tens of millions of dollars with mask costs alone having surpassed $1M. Furthermore, product life cycles and competitive pressures are placing increasing demands on shortening the time-to-market. This in turn has been a significant limiting factor to the growth of the industry that is increasingly demanding customized solutions. If the embedded systems industry is to grow at the projected rates it is essential that the twin hurdles of increasing NRE costs and longer time-to-market be overcome [24].

Thus, to sustain Moore's Law and its corollary through the next decade, a major challenge for embedded systems design is to produce several orders of magnitude improvement in designer productivity to concurrently reduce both NRE costs and time-to-market. The required gains in productivity and cost can only be realized by fundamental shifts in design strategy rather than evolutionary techniques that retain legacy barriers between hardware and software design methodologies. This paper presents an overview of compiler directed design space exploration as a framework for addressing the challenges of embedded systems customization for the next decade.

2   Evolution of Approaches to Customization

Increasing hardware design complexity and design costs have historically driven innovations in the development of higher levels of hardware abstractions and accompanying design tools for translation into efficient hardware designs. The concept of automated, customized hardware design, termed *silicon compilation*, referred to automated approaches for translating high level specifications to hardware implementations that were subsequently subjected to hardware-centric optimizations. A departure from this conventional view articulated a new approach wherein optimizations were applied early in the compilation process to more abstract representations of the hardware [19]. This framework enabled the development of computationally tractable solutions to several geometric optimizations that, at the time, were applied to lower level hardware descriptions. As the progression of Moore's Law continued through the next decade, each new technology generation placed formidable challenges to designer productivity. Silicon compilation consequently evolved into two distinct disciplines: optimizing compilers and electronic design automation (EDA).

Through the 1970's instruction set architectures (ISAs) emerged as a vehicle for focusing the NRE costs of processor hardware development. Through re-use via software customization, the processor hardware NRE cost could be amortized through high volumes achieved by using the processor across many applications. As technology moved through the sub-micron barrier, to sustain the cost and performance benefits of Moore's Law, we saw the emergence of reduced instruction set architectures (RISC) and accompanying optimizing compiler technology in the early 80's. For the next 15 years the demands for instruction level parallel (ILP) processors and automated parallelization drove developments in compiler technology in the areas of program analysis, e.g., array dependence analysis, and program transformations eg. loop transformations. The goal of program analysis and transformation was to optimize the execution of a program on a fixed hardware target that was abstracted in the form of an ISA. Program transformations also evolved to exploit facets of the micro-architecture and memory hierarchy that were exposed to the compiler, such as the cache block size, memory latency and functional unit latencies. However, in all of these cases the target hardware remained fixed. Central to these program analysis and transformation techniques was the evolution of rich graph-based intermediate program representations, or *program graphs*.

Independently, and concurrently, electronic design automation (EDA) continued to innovate in automating the design of increasingly complex generations of chips. The abstractions for automated EDA tool flows evolved from switch-level abstractions of transistors, through the gate-level, to the register transfer level, with much of the current interest focused at the Electronic System Level (ESL), utilizing hardware specifications in the form of languages such as SystemC and System Verilog. The ISA remained the primary interface between program graphs (software) and the hardware. Decisions concerning functionality that should be in hardware, versus that which should be implemented in software were made with respect to subsequently inflexible and rigid partitions between software and hardware implementations. Thus, through the 80's and much of the 90's the disciplines of optimizing compiler technology for microprocessors and EDA technology for the development of ASICs matured into distinct, vertically integrated design flows with little to no cross-fertilization.

In the 90's a few research efforts began investigating design techniques for customization of embedded systems wherein the compiler optimizations were applied in the context of traditional EDA design tasks. The Program-In-Chip-Out (PICO) system at HP Laboratories [27] pioneered an approach that leverages well known program transformations such as software pipelining to synthesize non-programmable hardware accelerators to implement loop nests found in ANSI C programs [28]. The PICO system also synthesizes a companion custom Very Long Instruction Word (VLIW) processor that is selected through a framework for trading area (through the number of functional units) for performance (through compiler scheduling of instructions) [27]. While PICO generates custom ASIC solutions, the Adaptive Explicitly Parallel Instruction Computing (AEPIC) model [22, 17] targets an EPIC processor coupled to a reconfigurable fabric such as a field programmable gate array (FPGA). Custom instructions are identified from an analysis of the program while hardware implementation of these custom instructions are realized in the reconfigurable fabric.

The EPIC ISA is extended with instructions to communicate with the custom fabric. Traditional compiler optimizations for register allocation, instruction scheduling, and latency hiding are extended to manage the reconfigurable fabric. While hardware description languages emerged as the dominant specification vehicle of choice for hardware design, Handel-C emerged as a C-based language for specifying hardware attributes of concurrency and timing [1] for hardware synthesis leading to a C-based design flow for hardware/software partitioning and hardware synthesis. More recent versions of C-based design flows include ImpulseC [2], SilverC [4], and Single-Assignment C [26]. While all of these approaches attempt to leverage the C language and use an HDL as an intermediate representation, the dynamically variable instruction set architecture approach [11] [18] compiled ANSI C programs to FPGA fabrics without a HDL intermediate representation. In the same spirit of using programming language based specification, a more recent effort describes a comprehensive approach to customization for embedded control applications [7]. A high level specification based on an extended finite state machine formalism is analyzed and used for partitioning functionality into hardware and software driven by performance constraints. The target architecture is an embedded processor coupled with a reconfigurable fabric. Another recent approach for the customization of the memory hierarchy combines locality enhancing transformations with exploration of cache design parameters for programs with pointers – ubiquitous to embedded applications (e.g., C) – and aimed at minimizing the physical size and energy consumed by embedded cache memories [21] [20]. The results demonstrated halving of the cache size and energy consumption, while in many cases simultaneously improving the performance by a similar factor. Thus, a powerful algorithm based on the principles of average-case analysis could achieve power and cost (silicon area) improvements corresponding to those achieved by an entire generation of Moore's Law. Several other recent projects have also begun exploring the co-operative use of optimizing compilers and behavioral synthesis compilers recognizing the strengths of each and seeking a principled application of both targeted to FPGAs [30] [15] [14]. These research efforts further underscore the trend towards the cross-fertilization between the disciplines of optimizing compilers and EDA.

The productivity and performance improvements required to sustain Moore's Law through the next decade must come from a major shift in design principles rather than continued evolutionary improvements in separate, vertically integrated design flows for optimizing compiler technology and EDA. Global optimizations that span these two disciplines afford opportunities for major improvement in designer productivity and system performance. Central to the development of such analysis and optimization techniques will be the redefinition of the hardware/software interface, to enable customization techniques that can operate directly on program graph representations, in bridging to hardware abstractions and thus avoiding the performance and productivity hurdles of utilizing a fixed ISA customized for an application. Design space exploration is a framework for articulating, designing, and implementing such global optimizations where the design space is spanned by parametric representations of both candidate compiler optimizations and architecture parameters and the navigation of the design space is driven by quantifiable, machine independent metrics. In a manner reminiscent of the re-definition of the hardware-software interface that took place in innovating RISC architectures, productivity improvements will come from re-defining the interface between, and roles of, optimizing compilers and EDA tools.

3   Compiler-Directed Design Space Exploration

Functionally, the process of customization via design space exploration (DSE) transforms both a program and the target architecture to generate a "best fit". The essential elements of DSE are describe in this section.

3.1   Compiler As a Vehicle For Design Space Exploration

As illustrated in figure 1, traditionally compilers were designed to target a fixed architecture with a rigid hardware-software interface in the form of an ISA. Compiler optimizations were developed with the aim of efficient utilization of fixed architectural resources such as storage (eg. registers) and data-path (eg. functional units). However, since the hardware is fixed at design time, hardware vendors are motivated to produce

hardware cost-performance trade-off studies have been difficult–this is an artifact of the target being fixed–application characterization using compiler analysis and profile-feedback have been instrumental for detecting optimization opportunities.

The implementation of candidate optimizations are guided by hardware models that quantify the impact of the optimizations, for example area and clock speed. Development of such hardware models is accompanied by decision models that guide the application of candidate program optimizations.

Such a framework that combines application characterization with high level hardware models serves as a point of departure for current generation design space exploration tools. Exploration itself can be carried out employing techniques ranging from heuristics to linear programming. Section 5 provides and example drawn from customization of the memory hierarchy and illustrates the use of machine independent metrics to characterize the demands of the application in the context of the memory hierarchy.

4 Attributes of the Design Space

Our framework for design space exploration has three main aspects: the space of target architectures, the space of compiler optimizations and the metrics that characterize each point in such a design space.

4.1 The Space of Target Architectures

The spectrum of target architectures for embedded systems range from fine grained field programmable gate arrays (FPGAs) [31] at one end, through coarse grained "sea of tiles" architectures (for example [33] [25] [12]) to traditional fixed-ISA microprocessors at the other end.

Fine grained FPGA architectures can implement complex logic and exploit fine grained logic-level parallelism but have very poor logic densities, delay, area and power consumption [35] relative to ASICs. Coarse grained "sea of tiles" architectures, though not as flexible as fine grained FPGAs greatly reduce power, area, delay and configuration times. In addition they map well to traditional compiler representations and abstractions of programs and hence serve as efficient compilation and design space exploration targets. Further, compiler optimizations such as modulo scheduling can readily leverage the tiled nature of such architectures for efficient mapping of applications as well as to influence traditional EDA tasks such as placement and routing. More recently "structured" ASIC design flows have emerged as a mechanism to reduce the cost of chip development [35]. Cost reductions are achieved by reducing the degrees of freedom in physical layout and optimization and by using pre-designed cores [3]. Tiled based architectures with their coarse grain structure and limited space of interconnect and placement options (relative to a full custom gate design) are a good fit for structured ASIC design flows. Most recently, a new class of coarse grained configurable architectures referred to as *polymorphic computing architectures* (PCAs) have emerged as a powerful class of architectures for satisfying demanding embedded computing requirements. PCAs represent a new class of computing architectures whose micro-architecture can *morph* to efficiently and effectively meet the requirements of a range of diverse applications thereby also amortizing development cost. Such architectures have the advantages of smaller footprint (by replacing several ASIC cores), lower risk via post deployment hardware configuration, and sustained performance over computationally diverse applications. This class of architectures is particularly amenable to compiler directed DSE. Finally, A-EPIC [22] and more recent efforts in customizable instruction set architectures [6] [5] [8] recognize the need for more flexible microprocessor targets and are amenable to compiler-directed DSE for custom instruction selection or configuration. In particular, powerful program analysis techniques can be brought to bear to "discover" custom instructions. The challenge in discovering suitable custom instructions lay in the complex interactions instructions have with the data-path micro-architecture and the memory hierarchy.

Along this spectrum of target architectures, the application of compiler directed DSE is based on the ability to adequately characterize the hardware target via high level models utilized to navigate the design space. An additional crucial decision is the selection of the set of architectural parameters which form a design space that is self-contained (and hence can be characterized well) and thereby makes the process of design space exploration tractable. For example, target architectures range from conventional microprocessors to fine grained re-configurable targets like FPGAs. A subset of such target architectures which are easy to explore, self-contained as well as meet the customization requirements, will immediately leverage the benefits of compiler directed DSE.

4.2   The Space of Compiler Optimizations

Compiler optimizations can be broadly divided into two classes. Optimizations that influence the design of the memory subsystem and optimizations that influence the design of the data path of the embedded system. In the subsequent sections, we broadly sketch examples of optimizations in each class relevant to embedded systems customization.

Data Locality Optimizations. In general, data locality optimizations can be classified into three classes. Code transformations which change the data object access pattern for a given data layout [9], data transformations that change the data object layout for a given object access pattern [10] and optimizations which manage the memory hierarchy through techniques such as software based prefetching [32]. Among these techniques the last two are especially attractive in the context of memory hierarchy customization, as code transformation needs to be leveraged for co-optimizing the application and the data path of the target architecture. In this regard novel analysis techniques based on metrics like neighborhood affinity probability (NAP) [23] [20], characterize object access patterns in a program region specific manner, enable integrated optimizations that can drive data re-layout in memory as well as drive software based prefetching to reduce the memory hierarchy cost (in terms of area as well as power) while maintaining performance. Novel metrics also serve as an architecture independent measure of required memory hierarchy resources. Optimizations based on NAP not only reduce the cost of the memory hierarchy but also enable numerous trade-offs among various cache parameters.

Optimizations that improve data locality have been shown to improve performance in the context of conventional microprocessors. In addition, recent work [21] has shown how such data locality optimizations not only yield performance improvements in the case of fixed architectures, but can also be used for the design space exploration of memory hierarchies resulting in reduced costs, in terms of area and power, while maintaining the same performance as that of an unoptimized application. Metrics derived from profile-feedback and application analysis techniques introduced in previous studies [21] not only indicate optimization opportunities, but also serve as an indicator of required hardware resources.

Concurrency Enhancing Optimizations. Traditional optimizations have focused on concurrency enhancement in the presence of fixed hardware resources. In the context of application-architecture co-optimization, since architectural resources are no longer fixed, concurrency in the application can be exploited to gain performance by investing in additional architectural resources. This is a natural way to trade off cost for performance. Several compiler optimizations ranging from fine grained ILP (Instruction level parallelism) techniques [16] to coarse grained loop parallelization techniques [34] can be applied to explore the cost (area) vs. performance of a variable number of functional units.

Scheduling Techniques. Apart from data locality enhancing optimizations which influence the design of the memory hierarchy, and concurrency enhancing optimizations which influence the number of functional units, scheduling techniques can be leveraged to influence EDA steps such as placement and routing. For example, compile-time program analysis can be applied to characterize communications between functional resources and hence influence the placement (e.g., co-location) of these resources [27] for increased performance. In addition, program transformations such as software pipelining can be leveraged to perform integrated placement and routing of regular, systolic array-like structures [15].

Collectively, the classes of compiler optimizations described in the preceding can be leveraged to co-optimize the application with the memory hierarchy, and/or the datapath.

5   An Example: Memory Hierarchy Customization

In this section, we illustrate how a compiler can be leveraged for design space exploration to customize the memory hierarchy for a particular application. We also show how novel metrics, characterized through compiler analysis, aid in such an exploration and serve to quantify hardware resource needs. Formally the problem can be stated as follows: Given a cache with block size $\mathcal{B}$, size $\mathcal{S}$ and bandwidth $\mathcal{W}$ and a cost function F : $(\mathcal{B} \times \mathcal{S} \times \mathcal{W}) \to \Re$ and a program $\mathcal{P}$, what are the values of $\mathcal{B}$, $\mathcal{S}$ and $\mathcal{W}$ for the best performance such that the cost is less than a specified value c ?

To explore a design space, the program behavior should be characterized using architecture independent metrics, and the cost of a particular cache design and its impact on the program performance should be evaluated. The architecture independent metrics introduced in [20] completely characterize the intrinsic or virtual behavior of a program and relate the virtual characteristics to the realized or observed behavior when a micro-architectural model is imposed during execution. These metrics – *Turnover factor, Packing factor* and *Demand bandwidth* –are defined below:

Turnover Factor. Intuitively, turnover factor can be defined as the amount of change in the working set of the program $\mathcal{P}$. Consider a data reference trace $\mathcal{T}$ for the program $\mathcal{P}$ as a string of addresses. $\mathcal{T}$ is partitioned into smaller non-overlapping substrings, $s_1, s_2, \cdots, s_n$. The number of unique characters in each substring s_i equals $\mathcal{V}$, the virtual working-set size. In addition, the substrings do not overlap, and therefore $\mathcal{T} = s_1|s_2|\cdots|s_n$. We now define the cost of a transition between two substrings s_i and s_{i-1} as the turnover factor.

$$\Gamma(s_{i>0}) = \mathcal{V} - |\hat{s}_i \cap \hat{s}_{i-1}|$$

where $\hat{s}_i$ is the set formed from s_i and $\hat{s}_0$ is the null set.

Packing Factor. Packing factor can be loosely defined as the ratio of the total data in the cache to the useful data in the cache. Formally, given a mapping function $\mathcal{M}$ (derived from $\mathcal{B}$ and $\mathcal{S}$) which maps addresses in the trace $\mathcal{T}$ to blocks, let $\mathcal{R}_i$ be the number of blocks needed to map every address $k \in \hat{s}_i$. Now the packing factor can be defined as

$$\Phi(s_i) = \frac{\mathcal{R}_i \times \mathcal{B}}{\mathcal{V}}$$

This measures the efficiency with which useful data is packed into the cache.

Demand Bandwidth. The demand bandwidth of an application can be defined as

$$\mathcal{D}(s_{i>0}) = \frac{\Gamma(s_i) \times \Phi(s_i)}{|s_i|}$$

Abstractly, it captures the rate at which data should be delivered to the caches to meet the program requirements. If the architectural bandwidth does not match the demand bandwidth, the program incurs stalls.

In light of these novel metrics, it is clear that a low packing factor and hence low demand bandwidth is desirable and has a direct impact on program performance. Low packing factor indicates optimal use of the caches and hence better cache design. For a given trace $\mathcal{T}$, packing factor is in turn determined completely by the block size $\mathcal{B}$, and cache size $\mathcal{S}$. Given an application, an automatic design space explorer can generate traces and then optimize the block size, cache size and bandwidth of the cache by calculating the packing factor and hence demand bandwidth for various configurations. Cost for each of these configurations can also be calculated through the cost function F and can be traded off for performance, without the need for time consuming cycle accurate simulation at each point of the design space.

6   Remarks and Conclusions

A central impediment to the continued growth of embedded systems are the twin hurdles of NRE costs and time-to-market pressures. To address these challenges we need fundamental shifts in principles governing hardware-software design that can lead to the major leaps in designer productivity and thence cost reductions to sustain the performance and cost benefits of Moore's Law for future custom embedded systems. In this paper we describe compiler directed design space exploration (DSE) where the design space is spanned by parametric representations of both candidate compiler optimizations and architecture parameters and the navigation of the design space is driven by quantifiable, machine independent metrics. DSE in conjunction with emerging configurable architectures such as coarse grain tiled architectures, microprocessors with customizable ISAs, polymorphic computer architectures, and fine grained FPGAs, offer opportunities for concurrently optimizing the program and the hardware target thereby leveraging optimizing compiler technology with EDA design techniques. We expect such cross-fertilization between EDA and compiler technology will become increasingly important and become a source of significant improvements in designer productivity and system performance.

References

1. Celoxica: http://www.celoxica.com/.
2. Impulse accelerated technologies: http://www.impulsec.com/.
3. LSI logic rapid chip platform ASIC: http://www.lsilogic.com/products/rapidchip_platform_asic/.
4. Quicksilver technology: http://www.qstech.com/.
5. Stretch inc. http://www.stretchinc.com/.
6. Tensilica: http://www.tensilica.com/.
7. M. Baleani, F. Gennari, Y. Jiang, R. B. Y. Patel, and A. Sangiovanni-Vincentelli. HW/SW partitioning and code generation of embedded control applications on a reconfigurable architecture platform. *Proceedings of the tenth international symposium on Hardware/software codesign*, pages 151 – 156, 2002.
8. M. Baron. Stretching performance. *Microprocessor Report*, 18, Apr. 2004.
9. S. Carr, K. S. McKinley, and C.-W. Tseng. Compiler optimizations for improving data locality. *Proceedings of the sixth international conference on Architectural support for programming languages and operating systems*, 1994.
10. T. M. Chilimbi, B. Davidson, and J. R. Larus. Cache-conscious structure definition. *Proceedings of the ACM SIGPLAN 99 Conference on Programming Language Design and Implementation*, 1999.
11. R. Goering. C design goes soft: EETimes: http://www.eetimes.com/story/oeg20010423s003, Apr. 2001.
12. C. R. Hardnett, A. Jayaraj, T. Kumar, K. V. Palem, and S. Yalamanchili. Compiling stream kernels for polymorphous computing architectures. *The Twelfth International Conference on Parallel Architectures and Compilation Techniques*, 2003.
13. H. Jones. How to slow the design cost spiral. *Electronic design Chaing: http://www.designchain.com*, Sept. 2002.
14. J. Liao, W. Wong, and M. Tulika. A model for hardware realization of kernel loops. *Proc. of 13th International Conference on Field Programmable Logic and Application, Lecture Notes of Computer Science*, 2778, Sept. 2003.
15. B. Mei, S. Vernalde, D. Verkest, H. D. Man, and R. Lauwereins. Exploiting loop-level parallelism on coarse-grained reconfigurable architectures using modulo scheduling. *Design, Automation and Test in Europe Conference and Exhibition*, 2003.
16. W. mei W. Hwu, R. E. Hank, D. M. Gallagher, S. A. Mahlke, D. M. Lavery, G. E. Haab, J. C. Gyllenhaal, and D. I. August. Compiler technology for future microprocessors. *Proceedings of the IEEE*, 83(12), Dec. 1995.

17. K. Palem, S. Talla, and W. Wong. Compiler optimizations for adaptive epic processors. *First International Workshop on Embedded Software, Lecture Notes of Computer Science*, 2211, Oct. 2001.
18. K. V. Palem. C-based architecture assembly supports custom design: EETimes: http://www.eet.com/in_focus/embedded_systems/oeg20020208s0058, Feb. 2002.
19. K. V. Palem, D. S. Fussell, and A. J. Welch. High level optimization in a silicon compiler. *Department of Computer Sciences, Technical Report No 215, University of Texas, Austin, Texas*, Nov. 1982.
20. K. V. Palem and R. M. Rabbah. Data remapping for design space optimization of embedded cache systems. *Georgia Institute of Technology, College of Computing Technical Report*, (GIT-CC-02-10), 2002.
21. K. V. Palem, R. M. Rabbah, V. M. III, P. Korkmaz, and K. Puttaswamy. Design space optimization of embedded memory systems via data remapping. *Proceedings of the Languages, Compilers, and Tools for Embedded Systems and Software and Compilers for Embedded Systems (LCTES-SCOPES)*, June 2002.
22. K. V. Palem, S. Talla, and P. W. Devaney. Adaptive explicitly parallel instruction computing. *Proceedings of the 4th Australasian Computer Architecture Conference*, Jan. 1999.
23. R. M. Rabbah and K. V. Palem. Data remapping for design space optimization of embedded memory systems. *In ACM Transactions on Embedded Computing Systems (TECS)*, 2(2), 2003.
24. B. R. Rau and M. Schlansker. Embedded computing: New directions in architecture and automation. *HP Labs technical report: HPL-2000-115*, 2000.
25. K. Sankaralingam, R. Nagarajan, H. Liu, J. Huh, C. Kim, D. Burger, S. Keckler, and C. Moore. Exploiting ilp, tlp, and dlp using polymorphism in the trips architecture. *30th Annual International Symposium on Computer Architecture (ISCA)*, 2003.
26. S.-B. Scholz. Single assignment c – efficient support for high-level array operations in a functional setting. *Journal of Functional Programming*, 13(6), 2003.
27. R. Schreiber, S. Aditya, S. Mahlke, V. Kathail, B. R. Rau, D. Cronquist, and M. Sivaraman. Pico-npa: High-level synthesis of nonprogrammable hardware accelerators. *HP Labs technical report: HPL-2001-249*, 2001.
28. R. Schreiber, S. Aditya, B. R. Rau, V. Kathail, S. Mahlke, S. Abraham, and G. Snider. High-level synthesis of nonprogrammable hardware accelerators. *HP Labs technical report: HPL-2000-31*, 2000.
29. S. P. Seng, K. V. Palem, R. M. Rabbah, W.-F. Wong, W. Luk, and P. Cheung. Pd-xml: Extensible markup language for processor description. *In Proceedings of the IEEE International Conference on Field-Programmable Technology (ICFPT)*, Dec. 2002.
30. B. So, M. W. Hall, and P. C. Diniz. A compiler approach to fast hardware design space exploration in fpga based systems. *ACM SIGPLAN Conference on Programming Language Design and Implementation (PLDI)*, 2002.
31. S. Talla. *Adaptive Explicitly Parallel Instruction Computing*. PhD thesis, New York University, Department of Computer Science, 2000.
32. S. VanderWiel and D. J. Lilja. Data prefetch mechanisms. *ACM Computing Surveys (CSUR)*, 32(2), June 2000.
33. E. Waingold, M. Taylor, D. Srikrishna, V. Sarkar, W. Lee, V. Lee, J. Kim, M. Frank, P. Finch, R. Barua, J. Babb, S. Amarasinghe, and A. Agarwal. Baring it all to software: Raw machines. *IEEE Computer*, Sept. 1997.
34. M. E. Wolf and M. S. Lam. A loop transformation theory and an algorithm to maximize parallelism. *Proceedings of the IEEE Transactions on Parallel and Distributed Systems*, 2(4), Oct. 1991.
35. B. Zahiri. Structured asics: Opportunities and challenges. *21st International Conference on Computer Design*, 2003.

Spectral-Based Document Retrieval

Kotagiri Ramamohanarao and Laurence A. F. Park

Department of Computer Science and Software Engineering,
The University of Melbourne, Australia
{lapark, rao}@csse.unimelb.edu.au

Abstract. The fast vector space and probabilistic methods use the term counts and the slower proximity methods use term positions. We present the spectral-based information retrieval method which is able to use both term count and position information to obtain high precision document rankings. We are able to perform this, in a time comparable to the vector space method, by examining the query term spectra rather than query term positions. This method is a generalisation of the vector space method (VSM). Therefore, our spectral method can use the weighting schemes and enhancements used in the VSM.

1   Introduction

The field of information retrieval has been researched for over half a century. In that time, we have seen many methods of retrieval which have been variants of the bag of words approach (e.g. vector space [1] which take into account term frequencies in document and probabilistic methods [2]) or those that take term positions into account (e.g. term proximity methods [3, 4]). Much research has gone into improving these techniques by adjusting the weighting schemes or adding components such as feedback mechanisms [5] and thesauruses [6]. The most used retrieval method are the vector space and probabilistic methods. This is due to their simplicity and speed.

The vector space method was built from the concept of having a document vector space. In this space each document is a single vector whose topic is determined by the vector direction and content determined by the vector magnitude. Similar documents will point in similar directions, therefore the inner product of the vectors will provide similarity. The vector space dimensionality is determined by the number of terms found in the document set, therefore each element of a document vector represents the appearance of a particular term in the document. The more terms two documents have in common, the greater their inner product will be.

The probabilistic model was derived from a model which uses knowledge about the relevance of documents in the set. In the absence of such knowledge, the model reverts to one which is similar to the vector space model. Each document is seen as a set of probabilities of a term given the document. To obtain the similarity between a document and a query, we calculate the probability of the document being relevant to the given query.

M.J. Maher (Ed.): ASIAN 2004, LNCS 3321, pp. 407–417, 2004.

Both of these popular methods use the term occurrence count to calculate the document scores, but they both ignore term positions in the documents.

In this article, we present a method of spectral-based information retrieval. We will show that we are able to produce the same fast queries that the vector space method produces and we are also able to take the relative query term positions into account when calculating the document scores. In section 2 of the article, we will examine the spectral retrieval process using the Fourier transform. We will then cover the spectral document ranking method using the Discrete Cosine Transform (DCT) in section 3.

2   Spectral Retrieval

The spectral based retrieval method [7–11] calculates document scores based on the query term spectra which are derived from the query term signals. Each term spectra contains two properties; its magnitude, which is related to the term frequencies used in the vector space and probabilistic methods, and its phase, which is related to the term position in the document. By combining these two properties, we are able to calculate a document score which is related to the query term count and relative positions. In this section we will explain how we obtain these term signals and term spectra, and how we calculate the document scores.

2.1   Term Spectra

Each term spectrum $\tilde{F}_{d,t}$ of document d and term t is a sequence of B complex values in the form of:

$$\tilde{F}_{d,t} = [\ F_{d,t,0}\ F_{d,t,1}\ \cdots\ F_{d,t,B-1}\ ] \tag{1}$$

where B is a predefined constant and each spectral component $F_{d,t,b}$ is the bth Fourier coefficient of the corresponding term signal.

Term Signals. A document is a sequence of terms. The order and position of the terms provides the document with its meaning. To take advantage of this flow of terms, we derived a simple structure called a *term signal* to capture a single words approximate position in a document. Each term signal is of the form:

$$\tilde{f}_{d,t} = [\ f_{d,t,0}\ f_{d,t,1}\ \cdots\ f_{d,t,B-1}\ ] \tag{2}$$

where $f_{d,t,b}$ is the count of term t in the bth portion of document d, and $b \in \{0, 1, \ldots, B-1\}$.

If we take the a sample document:

> In the beginning, God created the heavens and the earth. The earth was without form and void, and darkness was over the face of the deep. And the Spirit of God was hovering over the face of the waters. And God said, Let there be light, and there was light. And God saw that the light

was good. And God separated the light from the darkness. God called
the light Day, and the darkness he called Night. And there was evening
and there was morning, the first day.

and we choose eight elements per term signal, we can see that the document is 88
words in length, therefore each element of the term signals from this document
will be the term frequency from each 11 words of the document. If we choose to
build the term signal for the term 'light', we find that in the first 11 words, there
are no occurrences. Therefore, the first element of the term signal is set to zero.
In the second set of 11 words, we also find no occurrences of the term 'light'.
Therefore, the second element of the term signal is zero as well. We find that
'light' first occurs in the fifth set of terms. In this set it occurs twice, therefore
the fifth element of the term signal is 2. There are also two occurrences in the
sixth set of terms, one in the seventh, and zero in the eighth. From this, we have
constructed the following term signal:

$$\tilde{f}_{d,light} = [\, 0 \; 0 \; 0 \; 0 \; 2 \; 2 \; 1 \; 0 \,] \tag{3}$$

We can see that if we were to use only one component ($B = 1$), we would
obtain a single term frequency count ($\tilde{f}_{d,light} = [\, 5 \,]$) found in the vector space
and probabilistic methods of information retrieval.

Once we have our term signals, we can proceed by applying one of the existing
document weighting schemes (e.g. SMART's Lnu.ltu, Okapi's BM25).

The Fourier Transform. The Fourier Transform is an orthogonal transform
that maps a signal from a time or spacial domain to the Fourier spectral domain.
The discrete form of the transform is:

$$F_{d,t,\beta} = \sum_{b=0}^{B-1} f_{d,t,b} \exp\left(-2\pi i \beta b / B\right) \tag{4}$$

where $i = \sqrt{-1}$. The transform components are measures of how much of the
sinusoid of corresponding frequency the original signal is made of. An example
of this is shown in figure 1. The transform is orthogonal, therefore the signal
energy is conserved.

The Fourier transform is used due to its nature of dealing with signal shifts.
If we examine the two signals in the top plot of figure 2, we can see that they
are identical except for a shift. Once we perform the Fourier transform, we can
see in the bottom plot that the magnitude of the spectrum is identical for each
of the signals (the magnitudes are overlapping). The difference appears in the
phase. By shifting a signal, we adjust the phase of the frequency components in
which the signal is made from. Therefore we can use the phase of the spectral
components as a measure of signal proximity which will be term proximity.

The magnitude of each spectral component shows the contribution of the
corresponding sinusoid in the original signal. The phase each spectral component
represents the shift (or delay) of the corresponding sinusoid in the original signal.

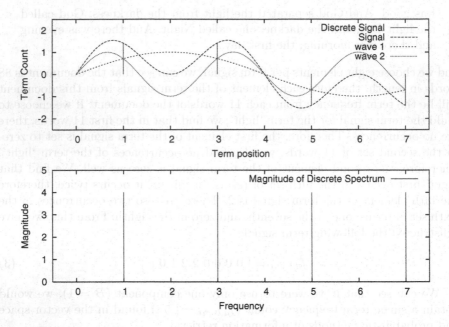

Fig. 1. Fourier transform example. The top plot shows our discrete term signal represented by the solid spikes, the continuous version of the signal and the waves (wave 1 is of frequency 1 and wave 3 is of frequency 2) that when combined, produce our term signal. The bottom plot shows the magnitude of the Fourier transform of the term signal. We see spikes in positions 1 and 3, showing that the term signal is made from waves of frequency 1 and 3. The spikes in frequency position 5 and 7 are the mirrors of positions 1 and 3. This is an artifact of the Fourier transform on real signals

This implies that we can use magnitude to represent the occurrence of the term and the phase to represent the position of the term.

Each spectral component can be represented with a magnitude $H_{d,t,b}$ and phase $\theta_{d,t,b}$ value:

$$F_{d,t,b} = H_{d,t,b} \exp i\theta_{d,t,b} \tag{5}$$

This notation will be used though the rest of the document.

2.2 Spectral Document Scores

To obtain a high precision document retrieval system, we hypothesise that the document scores must not only observe the frequency of the query terms in the document, but also be based on the query term proximity. We are able to obtain this information via calculation of *Magnitude* and *Zero Phase Precision* of the query term spectra. The document score combines these properties over the query terms and is of the form:

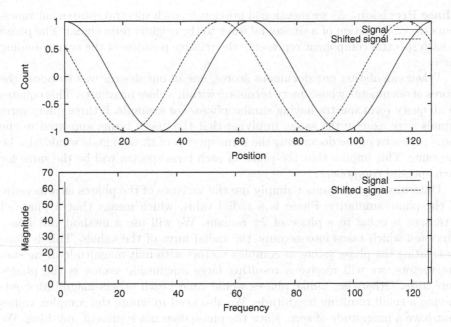

Fig. 2. Signal shift example. The top plot shows a sinusoid and a shifted version of itself. The bottom plot shows the magnitude of the Fourier transform of both signals. Both versions of the signal have exactly the same Fourier spectrum magnitude. By shifting a signal, only the spectral phase changes by an amount relative to the shift

$$S_d = \sum_{b=0}^{B-1} H_{d,b}\bar{\Phi}_{d,b} \tag{6}$$

where $H_{d,b}$ and $\bar{\Phi}_{d,b}$ are the Magnitude and Zero Phase Precision of the bth spectral component of the dth document respectively.

Magnitude. Before we can utilise the complex information of each term spectra component, we must understand its meaning. Each spectral component represents the contribution of a sinusoidal wave to the original term signal. The magnitude of each spectral component represents the contribution of the corresponding wave.

The contribution of each term count to the document score should be independent of all other query terms. Therefore we calculate the Magnitude portion of each spectral component ($H_{d,b}$), we simply add the magnitude of each spectral component of each query term spectrum:

$$H_{d,b} = \sum_{t \in Q} H_{d,t,b} \tag{7}$$

where $H_{d,t,b}$ is the magnitude of the bth spectral component of the tth term in the dth document and Q is the set of unique query terms.

Phase Precision. As we mentioned previously, each spectral component represents the contribution of a sinusoidal wave to the original term signal. The phase of each spectral component represents the relative position of the corresponding wave.

When calculating our document scores, one of our desires was to boost the scores of documents whose query terms are within a close proximity. This equates to all query term spectra having similar phase. For example, if three query term signals were exactly the same, implying that the query terms appeared in the same positions of the document, the term spectra of these signals would also be the same. This implies that the phase of each term spectra will be the same for each spectral component.

Unfortunately, we cannot simply use the variance of the phases as a measure of the phase similarity. Phase is a radial value, which means that a phase of 0 radians is equal to a phase of 2π radians. We will use a method call *Phase Precision* which takes into account the radial form of the values. This is done by treating the phase values as complex vectors with unit magnitude. If we sum the vectors, we will receive a resulting large magnitude vector is the phases were alike. Otherwise, unlike phases would cancel each others magnitudes out, leaving a small resulting magnitude. We also need to ignore the complex values that have a magnitude of zero, since the phase does not represent anything. We will use a variant of Phase Precision called *Zero Phase Precision* ($\Phi_{d,b}$) which is calculated by:

$$\Phi_{d,b} = \frac{\sum_{t \in Q | H_{d,t.b} \neq 0} \exp i\theta_{d,t,b}}{\#(Q)} \tag{8}$$

where $\theta_{d,t,b}$ is the phase of the bth spectral component of the tth term in the dth document and $\#(Q)$ is the cardinality of the set of unique query terms Q.

2.3   Experiments

Experiments were performed to compare out spectral method to various vector space methods and a term proximity retrieval method. We used various weighting schemes on our term signals and generated results from the TREC AP2WSJ2 document set. We chose the titles of queries 51 to 200 as out query key terms. The results are shown in table 1.

The experimental results show that the spectral retrieval method provides a higher precision than the vector space methods, but this comes at the cost of a lengthy query time. For our experiments, we stored the term signals in an index so that they are ready to extract during query time. This means that we had to perform a Fourier transform on each of the query term signals during the query time to obtain the term spectra. The vector space methods takes approximately 0.02 seconds to perform a query, the spectral method takes approximately 1 second for the same query. This means that although the spectral method is producing high precision results, the queries take 100 times longer to complete.

In the next section, we will examine methods of improving the query time.

Table 1. Experiments using the AP2WSJ2 document set from TREC showing the precision at the top 5,10,15 and 20 documents. We can see that the spectral retrieval method obtains better precision than the Vector space methods for several different weighting schemes. This is due to the unique way in which the spectral document retrieval method uses term position information

Method	Weight	Precision 5	Precision 10	Precision 15	Precision 20
Spectral	Lnu.ltu	0.4960	0.4613	0.4391	0.4227
VSM	Lnu.ltu	0.4693	0.4493	0.4356	0.4180
Spectral	BD-ACI-BCA	0.4867	0.4647	0.4440	0.4193
VSM	BD-ACI-BCA	0.4440	0.4247	0.4142	0.3953
Spectral	AB-AFD-BAA	0.4947	0.4673	0.4493	0.4220
VSM	AB-AFD-BAA	0.4880	0.4493	0.4404	0.4217

3   Increasing the Speed

The spectral based retrieval method using the Fourier transform provides high precision results, but that comes at a cost of storage and query speed. If we were to build an index to store the term signals, we would achieve a compact index size, similar to those found in the vector space methods. By doing so, we would also have to perform the Fourier transform on each of the query term signal in every document in the document set during the query time. Therefore, we achieve a compact index by sacrificing the query speed. We could choose to store the term spectra instead. This would lead to fast query times, but the index would have to store eight times as many coefficients which are now complex floating point values. Experiments have shown this to be about 100 times larger than storing term signals.

The storage cost comes from the many components which must be stored in order to attain high precision. To overcome this cost, we look to the Cosine transform as a replacement to the Fourier transform.

3.1   The Cosine Transform

The Cosine transform is an orthogonal transform which maps a signal to a basis set of cosine waves. It is of the form:

$$F_{d,t,\beta} = \sum_{b=0}^{B-1} f_{d,t,b} \cos\left(\frac{\pi\beta(2b+1)}{2B}\right) \tag{9}$$

The benefits of using the Cosine transform are: it maps signals into the real domain rather than the complex domain, therefore we will have less data to store; it is related to the Kahunen Loeve transform (KLT) when the signals are Markov-1 processes with a correlation coefficient between 0.5 and 1, which implies that the whole signal can be approximated by the first few components.

Experimental results have shown that our term signals best fit a Markov-1 process with correlation coefficient of 0.79, therefore we should be able to take advantage of the dimensional reduction properties of the KLT.

3.2   Quantisation and Cropping

The fact that the Cosine transform generates real values and we only need the first few to approximate the whole signal, significantly reduces the data we need to index. We are left with storing the floating point values. Rather than storing the whole four bytes associated to each float, we choose to use quantisation to convert the floats to integer values. Once we obtain the inter values, we can use various compression techniques such as gamma and Bernouli coding.

The quantisation is split into two part; first we deal with the magnitude values, then the phase. The magnitudes are distributed in a logarithmic fashion, where most values are close to zero and only a few achieving high values. To deal with this we want less quantisation error for the values close to zero, since there are many and small differences will count towards the document score. We also do not need accurate high values, since they are large, they will dominate the score even is there is quantisation error involved. Based on these suggestions, we choose to use left geometric quantisation (q_{lg}) for the magnitude values. The phase is a radial value which is evenly distributed, therefore we should use uniform quantisation (q_u) when quantising the phase. The quantisation formulae used were:

$$q_u(x) = \left\lfloor 2^b \frac{x - L}{U - L + \epsilon} \right\rfloor \qquad q_{lg}(x) = \left\lfloor 2^b \frac{\log x - \log L}{\log U - \log L + \epsilon} \right\rfloor \qquad (10)$$

where b is the quantisation bits chosen, L and U are the upper and lower bounds of the data to be quantised, and ϵ is a small positive real value.

Experiments were performed to choose the number of bits required during quantisation. The magnitude values required at least six bits, while the phase required only one (in the case of real numbers, the phase represents the sign of the number).

3.3   Spectral Accumulator

The vector space and probabilistic methods attribute most of their query speed to using an accumulator. The accumulator holds the accumulative document scores, which are generated by adding the available portions of the document score.

The vector space methods use the following equation to calculate document scores:

$$S_d = \sum_{t \in Q} w_{d,t} w_{q,t} \qquad (11)$$

If we obtained a set containing the largest N of $w_{d,t} w_{q,t}$ for our query q and for all d in the document set, we could use these to calculate an approximate set of document score. We will obtain scores for only a few of the documents, but since the weighted values we used were the largest, the document score we obtain will most likely be the largest as well. By doing so, we do not have to calculate all of the document scores.

The accumulator $\mathcal{A} = \{\mathcal{A}_{d_1}, \mathcal{A}_{d_2}, \ldots, \mathcal{A}_{d_N}\}$ is set up to add one $w_{d,t}w_{q,t}$ value at a time to its existing accumulated document scores (initially set to zero) and stops when certain conditions are reached. If the document score for document d does not exist in the accumulator, a new $\mathcal{A}_d$ is created for it ans set to zero.

The stopping conditions can be either:

- Quit : Stop when the N document score accumulators are non-zero.
- Continue : Perform Quit, then continue by ignoring values that come from documents that currently do not exist in the accumulator.

These methods can be used with vector space and probabilistic retrieval methods, but must be modified to be used with spectral methods due to the calculation of phase precision. To calculate Zero phase precision, we must have the phase from a specific spectral component b, for every term t, in a specific document d. The Zero phase precision value is used to weight the magnitude of the spectral component, therefore we must store a magnitude and phase value for every component we wish to use. The document score accumulator becomes:

$$\mathcal{A}_d = \{\mathcal{A}_{d,b}\} \qquad \mathcal{A}_{d,b} = (H_{d,b}, \exp{(i\theta_{d,b})}) \tag{12}$$

where b ranges from 0 to the number of cropped spectral components. As the accumulation process proceeds, we extract the next $H_{d,t,b}$ and $\theta_{d,t,b}$ from the index (where $t \in Q$) and add them to the spectral accumulator. The magnitude portion of the accumulator: $H_{d,b} = H_{d,b} + H_{d,t,b}$ and the phase portion of the accumulator: $\exp{(i\theta_{d,b})} = \exp{(i\theta_{d,b})} + \exp{(i\theta_{d,t,b})}$.

3.4   Experiments

To observe the effect of storing the cosine transformed term spectra and retrieving using a spectral accumulator, we performed experiments using the AP2WSJ2 document set from the TREC collection. In these experiments, the spectral method using the cosine transform used 7 bit quantisation and used only the first two spectral components to calculate the document score. The results can be seen in table 2. Each of the methods used the BD-ACI-BCA weighting scheme.

Further experimental results can be found in [7–11].

We can see in these results that the cosine method produces slightly less precise results when compared to the Fourier spectral method, but this is compensated by the drastic increase in query speed. We can also see that the VSM method query time is slightly faster than the spectral method using the cosine transform, but the spectral method provides much higher precision results.

4   Conclusion

We have presented a new method of information retrieval called spectral-based document retrieval that uses the query term spectra to provide documents with

Table 2. Experiments using the AP2WSJ2 document set from TREC, with BD-ACI-BCA weighting, showing the precision at the top 5,10,15 and 20 documents and average query time. We can see that by using the Cosine transform in the place of the Fourier transform, we are obtaining slightly lower precision. But, by using the Cosine transform, we have also increased our query speed. Queries using the cosine transform, with quantisation, cropping, and a spectral accumulator take only 0.03 seconds. This time is very close to the VSM's 0.02 seconds

Method	Transform	Precision 5	Precision 10	Precision 15	Precision 20	Query time
Spectral	Fourier	0.4867	0.4647	0.4440	0.4193	1 sec
Spectral	Cosine	0.4760	0.4527	0.4236	0.4087	0.03 sec
VSM	n/a	0.4440	0.4247	0.4142	0.3953	0.02 sec

a score. We have shown how the query term spectra can be calculated from the term signals and how term signals are easily extracted from the documents.

Experiments showed that our spectral method using the Fourier transform provided greater precision than the vector space methods using the same weighting scheme, but due to the large amounts of information that needed to be processed, the spectral method was 100 times slower.

To combat this, we showed that the cosine transform would allow us to achieve high precision when processing less data than was needed with the Fourier transform. This allowed us to index the term spectra and use a spectral accumulator. Query times were reduced from 1 second to 0.03 seconds.

By using spectral-based document retrieval, we are able to produce high precision results with a fast query rate.

References

1. Zobel, J., Moffat, A.: Exploring the similarity space. ACM SIGIR Forum **32** (1998) 18–34
2. Jones, K.S., Walker, S., Robertson, S.E.: A probabilistic model of information retrieval: development and comparative experiments. Information Processing and Management **36** (2000) 779–808
3. Clarke, C.L.A., Cormack, G.V.: Shortest-substring retrieval and ranking. ACM Transactions on Information Systems **18** (2000) 44–78
4. Hawking, D., Thistlewaite, P.: Relevance weighting using distance between term occurrences. Technical Report TR-CS-96-08, The Australian National University (1996)
5. Harman, D.: Relevance feedback revisited. In: Proceedings of the 15th annual international ACM SIGIR conference on Research and development in information retrieval, ACM Press (1992) 1–10
6. Jing, Y., Croft, W.B.: An association thesaurus for information retrieval. In: Proc. of Intelligent Multimedia Retrieval Systems and Management Conference (RIAO). (1994) 146–160

7. Park, L.A.F., Palaniswami, M., Kotagiri, R.: Internet document filtering using fourier domain scoring. In de Raedt, L., Siebes, A., eds.: Principles of Data Mining and Knowledge Discovery. Number 2168 in Lecture Notes in Artificial Intelligence, Springer-Verlag (2001) 362–373
8. Park, L.A.F., Ramamohanarao, K., Palaniswami, M.: Fourier domain scoring : A novel document ranking method. IEEE Transactions on Knowledge and Data Engineering **16** (2004) 529–539
9. Park, L.A.F., Palaniswami, M., Ramamohanarao, K.: A novel web text mining method using the discrete cosine transform. In Elomaa, T., Mannila, H., Toivonen, H., eds.: 6th European Conference on Principles of Data Mining and Knowledge Discovery. Number 2431 in Lecture Notes in Artificial Intelligence, Berlin, Springer-Verlag (2002) 385–396
10. Park, L.A.F., Ramamohanarao, K., Palaniswami, M.: A new implementation technique for fast spectral based document retrieval systems. In Kumar, V., Tsumoto, S., eds.: 2002 IEEE International Conference on Data Mining, Los Alamitos, California, USA, IEEE Computer Society (2002) 346–353
11. Park, L.A.F., Palaniswami, M., Ramamohanarao, K.: A novel document ranking method using the discrete cosine transform. IEEE Transactions on Pattern Analysis and Machine Intelligence (2004)

Metadata Inference for Document Retrieval in a Distributed Repository*

P. Rigaux and N. Spyratos

Laboratoire de Recherche en Informatique,
Université Paris-Sud Orsay, France
{rigaux, spyratos}@lri.fr

Abstract. This paper describes a simple data model for the composition and metadata management of documents in a distributed setting. We assume that each document resides at the local repository of its provider, so all providers' repositories, collectively, can be thought of as a single database of documents spread over the network. Providers willing to share their documents with other providers in the network must register them with a coordinator, or *mediator*, and providers that search for documents matching their needs must address their queries to the mediator. The process of registering (or un-registering) a document, formulating a query to the mediator, or answering a query by the mediator, all rely on document content *annotation*.

Content annotation depends on the nature of the document: if the document is atomic then an annotation provided explicitly by the author is sufficient, whereas if the document is composite then the author annotation should be augmented by an *implied annotation*, i.e., an annotation inferred from the annotations of the document's components.

The main contributions of this paper are:

1. Providing appropriate definitions of document annotations;
2. Providing an algorithm for the automatic computation of implied annotations;
3. Defining the main services that the mediator should support.

1 Introduction

In this paper, we propose a simple data model for the composition and metadata management of documents in a distributed setting where a community of *authors* co-operate in the creation of documents to be used also by other authors. Each author is a "provider" of documents to the network but also a "consumer", in the sense that he creates documents based not only on other documents that he himself has created but also on documents that other authors have created and are willing to share. We envisage several possible domain applications for this framework, and in particular *e-Learning systems* where instructors and learners create and share educational material [16, 19, 15]. In a nutshell, our approach can be described as follows.

---

* Research supported by the EU DELOS Network of Excellence in Digital Libraries and the EU IST Project (Self eLearning Networks), IST-2001-39045.

M.J. Maher (Ed.): ASIAN 2004, LNCS 3321, pp. 418–436, 2004.

We distinguish documents into *atomic* and *composite*. Intuitively, an atomic document is any piece of material (text, image, sound, etc.) that can be identified uniquely; its nature and granularity are entirely up to its author. A composite document consists of a set of *parts*, i.e., a set of other documents that can be either atomic or composite. We assume that each document resides at the local repository of its author, so all authors' repositories, collectively, can be thought of as a database of documents spread over the network. Typically, an author wishing to create a new document will use some of the documents in his local database as components and will also search for relevant documents available over the network.

Authors willing to share their documents with other authors in the network must register them with a coordinator, or *mediator*, and authors that search for documents matching their needs must address their queries to the mediator. The process of registering (or un-registering) a document, formulating a query to the mediator, or answering a query by the mediator, all rely on document content *annotations*.

Such annotations are actually sets of terms from a controlled vocabulary, or *taxonomy*, to which all authors adhere. The well known ACM Computing Classification System [1] is an example of such a taxonomy. In this respect, we distinguish three kinds of annotation: the author annotation, the implied annotation and the registration annotation.

During registration of a document at the mediator, its author is required to submit the following items:

1. The document identifier, say d; this can be a URI allowing to access the document.
2. An annotation of the document content, that we call the *author annotation* of d; if d is atomic then the author annotation must be nonempty, whereas if d is composite then the author annotation can be empty.
3. If d is composite, then registration requires, additionally, the submission of all parts of d (i.e., all documents that constitute the document being registered); using the annotations of these parts, the mediator then computes *automatically* an annotation that "summarizes" the annotations of the parts, and that we call the *implied annotation* of d.

To register a document the mediator uses the author annotation augmented by the implied annotation, after removing all redundant terms (i.e., terms that are subsumed by other terms). The final set of terms used for registration is what we call the *registration annotation*.

The mediator is actually a software module that can be seen as one component of a digital library serving a number of subscribers (the "consumers"). A digital library maintains pointers to documents stored at the repositories of their authors. The mediator allows all subscribers to search for and consult documents of interest. Additionally, it allows authors to reuse documents of the library as components of new documents that they create. Of course, apart from the mediator services, a digital library offers a number of other services to its users, such as profiling, recommandations, personalization, and so on. However, in this paper, we focus only on the mediator services.

The mediator, actually, maintains a *catalogue* of registered documents: during registration of a document with identifier d, the mediator inserts in the catalogue a pair (t, d), for each term t in the registration annotation of d. Authors searching for documents that match their needs address their queries to the mediator. In turn, the mediator uses the catalogue to answer such queries.

The main issues addressed in this paper are:

1. Providing appropriate definitions of document annotations;
2. Providing an algorithm for the computation of implied annotations;
3. Defining the main services that the mediator should support.

This paper proposes *generic* solutions to the above issues, i.e., solutions that are valid independently of questions concerning network configuration. In other words, the solutions that we provide are still valid whether the network is configured around a central mediator, or whether it is organized in clusters, each cluster being served by a separate mediator, or even whether there is no mediator but each node plays the role of a mediator for all its connected nodes (as in pure peer-to-peer network [11]).

Work in progress aims at:

1. Validating our model in the context of a prototype, in which the documents are XML documents;
2. Embedding our model in the RDF Suite [2];
3. Integrating our annotation generating algorithms into a change propagation module to be integrated in the mediator.

We stress the fact that, in this paper, we do *not* deal with the management of document content, but only with the management of document annotations based on subject areas. We are aware that, apart from subject area, there are several other dimensions of content description such as the format of the document, its date of creation, its author, the language in which the document content is written (if there is text involved), and so on. However, in this paper, we focus only on the subject area dimension, and when we talk of annotation we actually mean subject area description.

We note that a lot of efforts have been devoted recently to develop languages and tools to generate, store and query metadata. Some of the most noticeable achievements are the RDF language [20], RDF schemas [21], query languages for large RDF databases [14, 2] and tools to produce RDF descriptions from documents [13, 5].

Several metadata standards exist today, such as the Dublin Core [9], or the IEEE Learning Object Metadata [18]. It seems quite difficult to produce them automatically. In this paper, we focus on taxonomy-based annotations to describe the content of documents [4]. Generation of such annotations remains essentially a manual process, possibly aided by acquisition software (see for instance [13, 22, 5], and [10] for a discussion). The fully automatic generation of metadata is hardly addressed in the literature, with few exceptions [17, 25, 8]. A representative work is the Semtag system described in [8] which "tags" web pages with terms from a standard ontology, thanks to text analysis techniques. This is different – and essentially complementary – to our approach, which relies on the structure of composite documents to infer new annotations. Finally the functionalities presented in Section 4 can be seen as an extension of well-known mediation techniques [26, 6, 3, 23], wih specific features pertaining to the management of structured information.

In summary, the novel aspects of our work concern the creation, automatic annotation, management and querying of composite documents distributed over an information network. We are not aware of other approaches in the literature that provide a similar formal framework for handling these functionalities.

In the rest of this paper we first describe the representation of documents (Section 2) and their annotation (Section 3) and then the functionalities supported by the mediator (Section 4).

2   The Representation of a Document

As mentioned earlier, in our model, a document is represented by an identifier together with a set of parts showing how the document is constructed from other, simpler documents. We do not consider the document content itself, but focus only on its representation by an identifier and a set of parts, as this is sufficient for our metadata management and mediation purposes. Therefore, hereafter, when we talk of a document we shall actually mean its representation by an identifier and a set of parts.

In order to define a document formally, we assume the existence of a countably infinite set $\mathcal{D}$ whose elements are used by all authors for identifying the created documents. For example, the set $\mathcal{D}$ could be the set of all URIs. In fact, we assume that the creation of a document is tantamount to choosing a (new) element from $\mathcal{D}$ and associating it with a set of other documents that we call its parts.

Definition 1 (The Representation of a Document). *A document consists of an identifier d together with a (possibly empty) set of documents, called the* parts *of d and denoted as parts(d). If parts(d)* $=$ $\emptyset$ *then d is called* atomic, *else it is called* composite.

For notational convenience, we shall often write $d = d_1 + d_2 \ldots + d_n$ to stand for $parts(d) = \{d_1, d_2, \ldots, d_n\}$. Based on the concept of part, we can now define the concept of component.

Definition 2 (Components of a Document). *Let* $d = d_1 + d_2 \ldots + d_n$. *The set of components of d, denoted as comp(d), is defined recursively as follows:*
if *d is atomic* **then** $comp(d) = \emptyset$
else $comp(d) = parts(d) \cup comp(d_1) \cup comp(d_2) \cup \ldots \cup comp(d_n)$.

In this paper, we assume that a document d and its associated set of components can be represented as a rooted directed acyclic graph (*dag*) with d as the only root. We shall refer to this graph as the *composition graph* of d. Clearly, the choice of parts of a composite document and their arrangement to form a composition graph should be left entirely up to its author. The composition graph of an atomic document consists of just one node, the document identifier itself. We note that the absence of cycles in the composition graph simply reflects the reasonable assumption that a document cannot be a component of itself. Clearly, this does not prevent a document from being a component of two or more distinct documents belonging to the same composition graph, or to different composition graphs.

It is important to note that in our model the ordering of parts in a composite document is ignored because it is not relevant to our purposes. Indeed, as we shall see shortly, deriving the annotation of a composite document from the annotations of its parts does not depend on any ordering of the parts.

3   Annotations of Documents

As we mentioned in the introduction, document content annotations are built based on a controlled vocabulary, or *taxonomy*, to which all authors adhere. A taxonomy consists of a set of terms together with a subsumption relation between terms. An example of a taxonomy is the well known ACM Computing Classification System [1].

Definition 3 (Taxonomy). *A taxonomy is a pair* $(T, \preceq)$ *where* T *is a* terminology, *i.e., a finite and non-empty set of names, or* terms, *and* $\preceq$ *is a reflexive and transitive relation over* T, *called* subsumption.

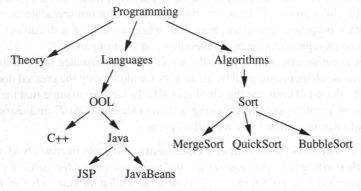

Fig. 1. A taxonomy

If $s \preceq t$ then we say that s is *subsumed* by t, or that t *subsumes* s. A taxonomy is usually represented as a graph, where the nodes are the terms and there is an arrow from term s to term t iff s subsumes t. Figure 1 shows an example of a taxonomy, in which the term Languages subsumes the term OOL, the term Java subsumes the term JavaBeans, and so on. We note that the subsumption relation is *not* antisymmetric, i.e., $(s \preceq t)$ and $(t \preceq s)$ does not necessarily imply $s = t$. Therefore, we define two terms s and t to be *synonyms* iff $s \preceq t$ and $t \preceq s$. However, in this paper, we shall not consider synonyms. From a technical point of view, this means that we work with classes of synonym terms, rather than individual terms. Put it differently, we work with just one representative from each class of synonyms. For example, referring to Figure 1, the term OOL is the representative of a class of synonyms in which one can also find terms such as Object-Oriented Languages, O-O Languages, and so on, that are synonyms of OOL.

However, even if we work only with classes of synonyms, a taxonomy is not necessarily a tree. Nevertheless, most taxonomies used in practice (including the ACM Computing Classification System mentioned earlier) are in fact trees. In this paper, we shall assume that the taxonomy used by all authors to describe the contents of their documents is in fact a tree. We shall simply refer to this tree-taxonomy as "the taxonomy".

Now, in order to make a document sharable, an annotation of its content must be provided, so that users can judge whether the document in question matches their needs.

We define such an annotation to be just a set of terms from the taxonomy. For example, if the document contains the quick sort algorithm written in java then we can choose the terms QuickSort and Java to describe its content. In this case the set of terms {QuickSort, Java} can be used as an annotation of the document.

Definition 4 (Annotation). *Given a taxonomy* $(T, \preceq)$ *we call* annotation *in* T *any set of terms from* T.

However, a problem arises with annotations: an annotation can be redundant if some of the terms it contains are subsumed by other terms. For example, the annotation {QuickSort, Java, Sort} is redundant, as QuickSort is subsumed by Sort. If we remove either Sort or QuickSort then we obtain a non-redundant annotation: either {QuickSort, Java} or {Sort, Java}, respectively. As we shall see later, redundant annotations are undesirable as they can lead to redundant computations during query evaluation. We shall therefore limit our attention to non-redundant, or *reduced annotations*, defined as follows:

Definition 5 (Reduced Annotation). *An annotation* A *in* T *is called* reduced *if for any terms* s *and* t *in* A, $s \npreceq t$ *and* $t \npreceq s$.

Following the above definition one can reduce an annotation in (at least) two ways: removing all but the minimal terms, or removing all but the maximal terms. In this paper we adopt the first approach, i.e., we reduce an annotation by removing all but its minimal terms. The reason for our choice lies in the fact that by removing all but minimal terms we obtain a more accurate annotation. This should be clear from our previous example, where the annotation {QuickSort, Java} is more accurate than {Sort, Java}.

Definition 6 (Reduction). *Given an annotation* A *in* T *we call* reduction *of* A, *denoted* $reduce(A)$, *the set of minimal terms in* A *with respect to the subsumption* $\preceq$.

An annotation can be seen both as a summary of the document content and as a support to find and retrieve the document. In the case of an atomic document the annotation can be provided either by the author or by the system via a semi-automatic analysis of the document content [12]. In the case of a composite document, though, apart from the author annotation, we would like to derive to derive also a second annotation, *automatically*, from the annotations of the document parts. We shall refer to such a derived annotation as the *implied annotation* of the composite document. To get a feeling of the kind of implied annotation that we have in mind, let us see an example.

Example 1. *Let* $d = d_1 + d_2$ *be a composite document with the following annotations of its parts:*

$$A_1 = \{QuickSort, Java\}$$
$$A_2 = \{BubbleSort, C++\}$$

Then the implied annotation of $d = d_1 + d_2$ *will be* {Sort, OOL}, *that summarizes what the two parts have in common.*

We shall come back to this example after the formal definition of implied annotation.

Now, the question is: how can one define the implied annotation of a composite document so as to best reflect the contents of its parts. Roughly speaking, what we propose in this paper is that the implied annotation should satisfy the following criteria:

– it should be reduced, for the reasons explained earlier;
– it should summarize what the parts have in common;
– it should be minimal.

To illustrate points 2 and 3 above, suppose that a composite document has two parts with annotations {QuickSort} and {BubbleSort}. The term Sort is a good candidate for being the implied annotation, as it describes what the two parts have in common. Moreover, as we can see in Figure 1, Sort is the minimal term with these properties. On the other hand, the term Algorithms is not a good candidate because, although it summarizes what the two parts have in common, it is not minimal (as it subsumes the term Sort).

Coming back to Example 1, following the above intuitions, we would like the implied annotation of d to be {Sort, OOL}. Indeed,

– {Sort, OOL} is a reduced annotation;
– the term Sort summarizes what QuickSort and BubbleSort have in common, and OOL summarizes what Java and C++ have in common;
– it is minimal, as any other annotation with the above properties will have terms subsuming either Sort or OOL.

In order to formalize these intuitions, we introduce the following relation on annotations.

Definition 7 (Refinement Relation on Annotations). *Let A and A' be two annotations. We say that A is* finer *than A', denoted $A \sqsubseteq A'$, iff for each $t' \in A'$, there exists $t \in A$ such that $t \preceq t'$.*

In other words, A is finer than A' if every term of A' subsumes some term of A. For example, the annotation $A =$ {QuickSort, Java, BubbleSort} is finer than $A' = $ {Sort, OOL}, whereas A' is not finer than A. To gain some more insight into this ordering, let us see another example. Referring to Figure 1, consider the following reduced annotations:

– $A = $ {JSP, QuickSort, BubbleSort}
– $A' = $ {Java, Sort}

Then $A \sqsubseteq A'$, as each term t' of A' subsumes some term t of A. Indeed, Java subsumes JSP and Sort subsumes QuickSort (of course, Sort also subsumes BubbleSort, but the existence of one term in A subsumed by Sort is sufficient).

Note that, according to this ordering, once we have verified that $A \sqsubseteq A'$ we may add to A as many extra terms as we wish *without* destroying the ordering.

Clearly, $\sqsubseteq$ is a reflexive and transitive relation, thus a pre-ordering over the set of all descripions. However, $\sqsubseteq$ is *not* antisymmetric, as the following example shows. Consider $A_1 = $ {OOL, Java, Sort} and $A_2 = $ {Java, Sort, Algorithms}. It is easy to see that $A_1 \sqsubseteq A_2$ and $A_2 \sqsubseteq A_1$, although $A_1 \neq A_2$. However, as we have explained earlier, for the purposes of this paper, we restrict our attention to reduced annotations only; and, as stated in the following proposition, for reduced annotations, the relation $\sqsubseteq$ becomes also antisymmetric, thus a partial order.

Proposition 1. *The relation $\sqsubseteq$ is a partial order over the set of all reduced annotations.*

Proof. Indeed, assume $A \sqsubseteq A'$ and $A' \sqsubseteq A$, and consider a term t' of A'. Then there is a term t in A such that $t \preceq t'$. We claim that $t' \preceq t$ as well, and therefore that $t = t'$. Otherwise, as $A' \sqsubseteq A$ and t is in A, there is a term t'' (different than t') such that $t'' \preceq t$, and thus $t'' \preceq t'$. Assuming $t'' \neq t'$, we have a contradiction to the fact that A' is a reduced annotation.   □

Now, using this ordering, we can define formally the implied annotation of a composite document so as to satisfy the criteria for a "good" implied desription, given earlier. First, we need the following result:

Theorem 2. *Let $\mathcal{A} = \{A_1, .., A_n\}$ be any set of reduced annotations. Let $\mathcal{U}$ be the set of all reduced annotations S such that $A_i \sqsubseteq S, i = 1, 2, ..., n$, i.e., $\mathcal{U} = \{S | A_i \sqsubseteq S, i = 1, \ldots, n\}$. Then $\mathcal{U}$ has a unique minimal element, that we shall denote as $lub(\mathcal{A}, \sqsubseteq)$.*

Proof. Let $P = A_1 \times A_2 \times \ldots \times A_n$ be the cartesian product of the annotations in $\mathcal{A}$, and suppose that there are k tuples in this product, say $P = \{L_1, L_2, ..., L_k\}$.

Let $A = \{lub_{\preceq}(L_1), lub_{\preceq}(L_2), \ldots, lub_{\preceq}(L_k)\}$, where $lub_{\preceq}(L_i)$ denotes the least upper bound of the terms in L_i, with respect to $\preceq$. As $(T, \preceq)$ is a tree, this least upper bound exists, for all $i = 1, 2, ..., n$. Now, let R be the reduction of A, i.e., $R = reduce(A)$. We shall show that R is the smallest element of $\mathcal{U}$.

Indeed, it follows from the definition of R that $A_i \sqsubseteq R$, for $i = 1, 2, ..., n$. Moreover, let S be any annotation in $\mathcal{U}$, and let t be a term in S. It follows from the definition of $\mathcal{U}$ that there is a term v_i in each annotation A_i such that $v_i \preceq t$. Consider now the tuple $v = < v_1, v_2, ..., v_n >$. By the definition of least upper bound, $lub_{\preceq}(v) \preceq t$, and as $lub_{\preceq}(v)$ is in R, it follows that $R \sqsubseteq S$, and this completes the proof.   □

With this theorem at hand, we can now define the annotation implied by a set of annotations $\mathcal{A} = \{A_1, \ldots, A_n\}$.

Definition 8 (Implied Annotation). *Let $\mathcal{A} = \{A_1, .., A_n\}$ be a set of annotations in T. We call implied annotation of $\mathcal{A}$, denoted $IAnnot(\mathcal{A})$, the least upper bound of $\mathcal{A}$ in $\sqsubseteq$, i.e., $IAnnot(\mathcal{A}) = lub(\mathcal{A}, \sqsubseteq)$*

Theorem 2 suggests the following algorithm for the computation of the implied annotation. Its proof of correctness follows directly from Theorem 2.

Algorithm IANNOT
Input: A set of annotations $A_1, A_2, \ldots, A_n$
Output: The implied annotation
begin
   Compute $P = A_1 \times A_2 \times \ldots \times A_n$

for each tuple $L_k = [t_1^k, t_2^k, \ldots, t_n^k]$ in P,
    compute $T_k = lub_\preceq(t_1^k, t_2^k, \ldots, t_n^k)$
Let $Aux = \{T_1, \ldots, T_l\}$
return $reduce(Aux)$
end

In the algorithm IANNOT, the function $lub_\preceq(t_1^k, \ldots, t_n^k)$ returns the least upper bound of the set of terms $t_1^k, \ldots, t_n^k$ with respect to $\preceq$. In Section 4 we shall use this algorithm to compute automatically the implied annotation of a composite document, based on the annotations of its parts. More precisely, given a composite document $d = d_1, + \ldots + d_n$, the annotations $A_1, \ldots, A_n$ in the above algorithm will be those of the parts $d_1, \ldots, d_n$, respectively, and the implied annotation will then be the implied annotation of d.

We end this section by working out a few examples illustrating how this algorithm works (always referring to the taxonomy of Figure 1).

Example 2. *Consider the document* $d = d_1 + d_2$, *composed of two parts with the following annotations:*

$$A_1 = \{QuickSort, Java\}$$
$$A_2 = \{BubbleSort, C++\}$$

In order to compute the implied annotation, first we compute the cross-product $P = A_1 \times A_2$. *We find the following set of tuples:*

$$P = \begin{cases} L_1 = <QuickSort, \ BubbleSort> \\ L_2 = <QuickSort, \ C++> \\ L_3 = <Java, \ BubbleSort> \\ L_4 = <Java, \ C++> \end{cases}$$

Next, for each tuple L_i, $i = 1, \ldots, 4$, *we compute the least upper bound* T_i *of the set of terms in* L_i:

1. $T_1 = Sort$
2. $T_2 = Programming$
3. $T_3 = Programming$
4. $T_4 = OOL$

We then collect together these least upper bounds to form the set Aux:

$$Aux = \{Sort, Programming, OOL\}$$

Finally we reduce Aux to obtain the implied annotation:

$$Implied \ Annotation = \{Sort, OOL\}$$

In view of our discussions so far, this result can be interpreted as follows: each part of the document concerns both, sorting and object-oriented languages.

Example 3. *Consider now the composite document* $d' = d_1 + d_3$, *with the following annotations of its parts:*

$$A_1 = \{Java, QuickSort\}$$
$$A_3 = \{BubbleSort\}$$

Proceeding similarly, as in Example 2, we find successively:

1. *The cross-product:*

$$P = \begin{cases} L_1 = <QuickSort, \ BubbleSort> \\ L_2 = <Java, \ BubbleSort> \end{cases}$$

2. *The least upper bounds:* $Aux = \{Sort, Programming\}$
3. *The implied annotation:* $reduce(Aux) = \{Sort\}$

The following comments are noteworthy:

1. The term $Java$ is *not* reflected in the implied annotation of Example 3, as it is not something that both parts share.
2. The fact that $Java$ has disappeared in the implied annotation means no loss of information: if a user searches for documents related to java, then d_1 will be in the answer and d' will not, which is consistent.
3. If we had put $Java$ in the implied annotation of d', this would give rise to the following problem: when one searches for documents related to java, the system will return both d_1 and d'. Clearly, this answer is redundant (because d_1 is part of d'), and also somehow irrelevant as only a part of d' concerns java.

Finally we note that the same document will generate different implied annotations, depending on what its "companion" parts are in a composite document. This is illustrated by our last example.

Example 4. *Consider the composite document* $d'' = d_1 + d_4$, *with the following annotations of its parts:*

$$A_1 = \{Java, QuickSort\} \quad A_4 = \{C\text{++}\}$$

Proceeding similarly, as in Example 2, we find successively:

1. *The cross-product:*

$$P = \begin{cases} L_1 = <Java, C\text{++}> \\ L_2 = <QuickSort, C\text{++}> \end{cases}$$

2. $Aux = \{OOL, Programming\}$
3. $reduce(Aux) = \{OOL\}$

Note that, in the two previous examples, d_1 is part of a composite document, but each time with a different "companion part": first with d_3 in d', then with d_4 in d''. It is interesting to note that, depending on the companion part, either the "Sort-aspect" of d_1 or the "OOL-aspect" appears in the implied annotation.

4 The Mediator

As we mentioned in the introduction, we consider that a community of authors co-operate in the creation of documents to be used by other authors. Each author is a "provider" of documents to the community but also a "consumer", in the sense that he creates documents based not only on other documents that he himself has created but also on documents that other authors have created and are willing to share. Each document resides at the local repository of its author, so all authors' repositories, collectively, can be thought of as a database of documents spread over the network. Typically, an author wishing to create a new document will use as components some of the documents from his local database, and will also search for relevant documents that reside at the local databases of other authors – provided that those other authors are willing to share them.

Authors willing to share their documents with other authors in the network must register them with a coordinator, or *mediator*, and authors that search for documents matching their needs must address their queries to the mediator. The mediator is actually a software module supporting the sharing of documents. It provides a set of services, among which the following basic services:

- query evaluation
- registration of a document
- un-registration of a document
- annotation modification

In this section, we discuss these basic services and outline their interconnections.

The implementation of all the above services relies on document annotations that are provided to the mediator during document registration. Indeed, during registration of a document, its author is required to submit the document identifier, say d, and an annotation of d that we call the *author annotation* of d, denoted as $AAnnot(d)$. If d is atomic then the author annotation must be nonempty, whereas if d is composite the author annotation can be empty. However, if d is composite the author is also required to submit all parts of d. Based on the annotations of the parts, the mediator then computes (automatically) the implied annotation of d. Finally, to register d, the mediator uses the author annotation augmented by the implied annotation, after removing all redundant terms. The final set of terms used for registration is what we call the *registration annotation* of d.

Definition 9 (Registration Annotation). *The Registration Annotation of a document* $d = d_1 + \ldots + d_n$, *denoted* $RAnnot(d)$, *is defined recursively as follows:*

- *if d is atomic, then* $RAnnot(d) = reduce(AAnnot(d))$
- *else* $RAnnot(d) = reduce(AAnnot(d) \cup IAnnot(RAnnot(d_1), \ldots, RAnnot(d_n)))$

One may wonder why the author annotation is not sufficient for document registration, and why we need to augment it by the implied annotation. The answer is that the author of a composite document d may not describe the parts of d in the same way as the authors of these parts have done. Let us see an example. Suppose that two documents, d_1 and d_2, have been created by two different authors, with the following author annotations:

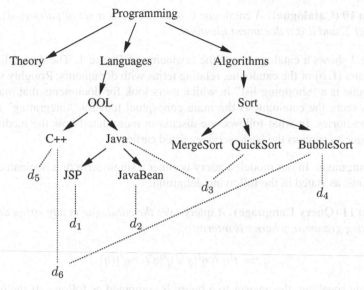

Fig. 2. A catalogue

$$AAnnot(d_1) = \{\texttt{QuickSort},\texttt{Java}\}$$
$$AAnnot(d_2) = \{\texttt{BubbleSort},\texttt{C++}\}$$

Assume now that, after browsing their content, a third author considers d_1 and d_2 as examples of good programming style, and decides to use them as parts of a new, composite document $d = d_1 + d_2$. Consequently, the author of d provides the following author annotation:

$$AAnnot(d) = \{\texttt{GoodProgrammingStyle}\}$$

Although this author annotation might be accurate for the author's own purposes, the document d still can serve to teach (or learn) java and sorting algorithms. This information will certainly be of interest to users searching for documents containing material on java and sorting algorithms. Therefore, before registration, the author annotation should be completed, or augmented by the implied annotation of d, i.e., $\{\texttt{OOL},\texttt{Sort}\}$, to obtain the following registration annotation:

$$\{\texttt{GoodProgrammingStyle},\texttt{OOL},\texttt{Sort}\}$$

This annotation expresses *all* annotations, i.e., the one given by the author of d *and* those given by the authors of the parts of d.

During document registration, the registration annotation of d is what is actually stored by the mediator in a repository. Conceptually, the repository can be thought of as a set of pairs constructed as follows: during registration of a document d, the mediator stores a pair (t, d) for each term t appearing in the registration annotation of d. The set of all such pairs (t, d), for all documents that are (currently) registered is what we call the *Catalogue*.

Definition 10 (Catalogue). *A catalogue C over $(T, \preceq)$ is a set of pairs (t, d), where t is a term of T and d is a document identifier.*

Figure 2 shows a catalogue over the taxonomy of Figure 1. The dotted lines indicate the pairs (t, o) of the catalogue, relating terms with documents. Roughly speaking, the catalogue is a "shopping list" in which users look for documents that match their needs. As such, the catalogue is the main conceptual tool for "integrating" all documents repositories. In what follows, we discuss in more detail how the mediator uses the catalogue to support the basic services listed earlier.

Query Language. In our model, a query is either a single term or a boolean combination of terms, as stated in the following definition.

Definition 11 (Query Language). *A query over the catalogue is any string derived by the following grammar, where t is a term:*

$$q ::= t|q \wedge q'|q \vee q'|q \wedge \neg q'|(q)$$

Roughly speaking, the answer to a query is computed as follows. If the query is a single term, then the answer is the set of all documents related either to t or to a term subsumed by t. If the query is not a single term then we proceed as follows. First, for each term appearing in the query, replace the term by the set of all documents computed as explained above; then replace each boolean combinator appearing in the query by the corresponding set-theoretic operator; finally, perform the set-theoretic operations to find the answer. These intuitions are reflected in the following definition of answer, where the symbol $tail(t)$ stands for the set of all terms in the taxonomy strictly subsumed by t, i.e., $tail(t) = \{s\backslash s \preceq t\}$.

Definition 12 (Query Answer). *The answer to a query q over a catalogue C, denoted by $ans(q)$, is a set of documents defined as follows, depending on the form of q (refer to Definition 11):*

> **Case 1:** *q is a single term t from T, i.e., $q = t$*
> $ans(t) =$ **if** $tail(t) = \emptyset$
>   **then** $\{d\backslash(t, d) \in C\}$
>   **else** $\bigcup\{ans(s)|s \in tail(t)\}$
> **Case 2:** *q is a general query*
> $ans(q) =$
>   **if** $q = t$ **then** $ans(t)$
>   **else**
>   **begin**
>   **if** $q = q_1 \wedge q_2$, $ans(q) = ans(q_1) \cap ans(q_2)$
>   **if** $q = q_1 \vee q_2$, $ans(q) = ans(q_1) \cup ans(q_2)$
>   **if** $q = q_1 \wedge \neg q_2$, $ans(q) = ans(q_1)\backslash ans(q_2)$
>   **end**

Example 5. *Consider the query* $q = C++ \vee \texttt{Sort}$. *Referring to Figure 2 and apply-ing the above definition, we find* $ans(q) = \{d_5, d_6\} \cup \{d_3, d_4, d_6\} = \{d_3, d_4, d_5, d_6\}$. *Similarly, for the query* $q = C++ \wedge\neg \texttt{BubbleSort}$ *we find* $ans(q) = \{d_5\}$.

Registration

An author wishes to make a document available to other users in the network.

To make a document available to other users in the network, its author must submit the following three items to the mediator:

1. The document identifier, say d;
2. The author annotation of d, which must be nonempty if d is atomic;
3. The identifiers of the parts of d, if d is composite.

The mediator then computes the registration annotation of d on which the actual registration will be performed. To do this, the mediator uses the following algorithm, whose correctness is an immediate consequence of Definition 9. The algorithm takes as input the above items, and updates the catalogue (i.e., the old catalogue is augmented by a set of pairs (t, d), one for each term t in the registration annotation of d).

Algorithm REGISTRATION
Input: A document d, the author annotation $AAnnot(d)$,
        the parts $\{d_1, d_2, \ldots, d_n\}$ of d
begin
  $\mathcal{A} = \emptyset$
  for each $d_i \in parts(d)$ **do**
    if d_i is registered **then**
      $R_i = \{t | (t, d_i) \in \mathcal{C}\}$
    else
      Input the author annotation $AAnnot(d_i)$
      $R_i = Registration\ (d_i, AAnnot(d_i), parts(d_i))$
    endif
    $\mathcal{A} = \mathcal{A} \cup \{R_i\}$
  end for
  Let $R = reduce(IAnnot(\mathcal{A}) \cup AAnnot(d))$
  for each t in R, insert the pair (t, d) in the catalogue
end

Note that, if d is atomic then its registration annotation reduces to the reduction of its author annotation (because $parts(d) = \emptyset$, and thus $IAnnot(\emptyset) = \emptyset$). From a practical point of view, the following scenarios can be envisaged for providing the inputs to the algorithm REGISTRATION; they depend on the nature of the parts of d, as well as on whether these parts have been registered beforehand or not:

 – if a part d_i of d has already been registered then its registration annotation is taken from the catalogue, independently of whether d_i is atomic or composite.

- else if d_i is composite and not yet registered, then its registration annotation is recursively computed from the registration annotations of the parts of d_i;
- else if d_i is atomic then its author annotation is required as input, and its registration annotation is the reduction of its author annotation.

We assume that a document d, whether atomic or composite, can be registered only if its registration annotation is nonempty. This assumption is justified by the fact that search for documents of interest is based on annotations. As a consequence, if we allow registration of a document with an empty annotation, then such a document would be inaccessible. Therefore, the mediator needs at least one term t, in order to insert the pair (t, o) in the catalogue, and make d accessible. This is ensured by the following constraint.

Constraint 1 (Registration). *A document can be registered only if its registration annotation is nonempty*

For atomic documents this is tantamount to requiring that the author decription be nonempty.

Constraint 2 (Atomic Documents Registration). *An atomic document can be registered only if its author annotation is nonempty*

If the document is composite, then Constraint 1 implies that either the author annotation must be nonempty or the implied annotation must be nonempty. A sufficient condition for the implied annotation to be nonempty (and thus for Constraint 1 to be satisfied) is that *all* parts of the document be registered, or (reasoning recursively) that all atomic components of the document be registered. Indeed, if all atomic components have already been registered, then the mediator will be able to compute a nonempty implied annotation, and thus a nonempty registration annotation, independently on whether the author annotations of one or more components are missing. Therefore the following sufficient condition for the registration of a composite document:

Constraint 3 (Composite Document Registration). *If every atomic component of a composite document is registered then the document can be registered*

Figure 3 shows an example of composite document registration. As shown in the figure, two atomic documents, d_3 and d_4 have already been registered in the catalogue $\mathcal{C}$, and so has the composite document d_2, whose parts are d_3 and d_4. The author annotations of all three documents are shown in the figure. Although the author annotation of d_2 is empty, its registration was possible as both its parts have nonempty author annotations. Note that the registration annotations of d_3 and d_4 coincide with their author annotations (since both documents are atomic and their author annotations happen to be reduced). The registration annotation of d_2 is easily seen to be $\{OOL\}$.

Now, suppose that an author wishes to reuse d_2 (and its parts) in order to create a new document d, composed of two parts: d_1 and d_2, where d_1 is an atomic document from the author's local database. Suppose now that in order to register d, the author provides to the mediator author annotations for d and d_1, as shown in the figure. Based

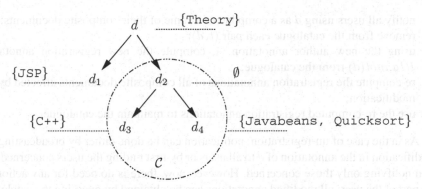

Fig. 3. Registration annotation of a composite document

on the author annotations of d and d_1, and the registration annotation of d_2 (computed above), the mediator will then compute the registration annotation of d – which is easily seen to be {OOL, Theory}. Finally, the mediator will enter in the catalogue the two pairs (OOL, d) and (Theory, d).

Unregistration

An author wishes to remove from the catalogue one of his registered documents

To unregister a document, its author must submit to the mediator the document identifier, say d. The mediator then performs the following tasks:

- notify all users using d as a component in their composite documents;
- remove from the catalogue each pair (t, d);
- re-compute the registration annotations of all composite documents affected by the removal of d;
- use the re-computed registration annotations to maintain the catalogue.

We note that notification can be done either by broadcasting the removal of d to all users, or by first finding the users concerned and then notifying only those concerned. The first solution is probably cheaper but may create inconvenience to those users not concerned, whereas the second avoids inconvenience but requires searching the catalogue for finding the users concerned (assuming that the mediator keeps track of the "foreign documents" used by each user). In any case, once notified of the (pending) unregistration of d, the users concerned have the option of first creating (in their local database) a copy of d and then proceeding to re-register all composite documents in which d appears as a component. Otherwise, the registration annotation of such documents might become empty.

Annotation Modification

An author wishes to modify the annotation of one of his registered documents

To modify the annotation of a document, its author must submit to the mediator the document identifier, say d, and the new author annotation, say A. The mediator then performs the following tasks:

- notify all users using d as a component in some of their composite documents;
- remove from the catalogue each pair (t, d);
- using the new author annotation A, compute the new registration annotation $RIannot(d)$ from the catalogue;
- re-compute the registration annotations of all composite documents affected by the modification;
- use the re-computed registration annotations to maintain the catalogue.

As in the case of un-registration, notification can be done either by broadcasting the modification in the annotation of d to all users or by first finding the users concerned and then notifying only those concerned. However, now, there is no need for any action on the part of the user: all modified annotations can be obtained by querying the catalogue.

5   Concluding Remarks

We have presented a model for composing documents from other simpler documents and we have seen an algorithm for computing implied annotations of composite documents based on the annotations of their parts.

In our model, a document is represented by an identifier together with a composition graph which shows how the document is composed from other, simpler documents. The annotation of each document is a set of terms from the taxonomy. We have distinguished three kinds of annotation:

1. The author annotation, i.e., the annotation provided to the mediator explicitly by the author;
2. The implied annotation, i.e., the annotation implied by the annotations of the parts (and computed by the mediator automatically during registration);
3. The registration annotation, i.e., the annotation computed from the previous two annotations and used by the mediator to register the document.

We have also outlined the main functionalities of the mediator, a software module that acts as a central server, registering or unregistering sharable documents, notifying users of changes, maintaining the catalogue and answering queries by authors.

Work in progress aims at:

1. Validating our model in the context of a prototype, in which the documents are XML documents;
2. Embedding our model in the RDF Suite [2];
3. Integrating our annotation generating algorithms into a change propagation module to be integrated in the mediator.

The basic assumption underlying our work is the existence of a network-wide taxonomy according to which documents are described and queries are formulated. As a result, the local repositories can be seen as peers served by a central catalogue (a "super-peer").

Future work aims at relaxing this assumption, in order to arrive at a pure peer-to-peer network. This will be done in two steps, as follows.

First, we will assume each author, or group of authors to have their own (possibly non-standard) taxonomy, for describing their documents locally and for formulating their queries to the mediator. This will require the establishment of *articulations*, i.e., semantic mappings between each local taxonomy and the central taxonomy. Work in that direction will be based on previous work on mediation [23, 24].

Second, we will assume that the role of the mediator can be played by any local taxonomy. Indeed, in principle, any local taxonomy can play the role of a mediator for all other local taxonomies that are articulated to it.

Another line of future research concerns a personalized interaction with the network. Indeed, from a conceptual point of view, all one has to do is to let the network user express his needs in terms of a set of named queries, or *views* of the form:

```
<term-name> = <query-to-the mediator>
```

The set of terms thus declared (plus, eventually, a user-defined subsumption relation) will then constitute the user-defined taxonomy, that will serve as the *personalized* interface to the network. Queries to this personalized taxonomy can be answered by simple substitution, based on the user declarations defining the terms of the personalized taxonomy. Work on the personalization aspects is ongoing and will be reported later.

References

1. The ACM Computing Classification System. ACM, 1999. http://www.acm.org/class/.
2. S. Alexaki, V. Christophides, G. Karvounarakis, D. Plexousakis, and K. Tolle. The ICS-FORTH RDFSuite: Managing Voluminous RDF Description Bases. In *Proc. Intl. Conf. on Semantic Web*, 2001.
3. J. Ambite, N. Ashish, G. Barish, C. Knoblock, S. Minton, P. Modi, I. Muslea, A. Philpot, and S. Tejada. ARIADNE: a System for Constructing Mediators for Internet Sources. In *Proc. ACM SIGMOD Symp. on the Management of Data*, pages 561–563, 1998.
4. R. Baeza-Yates and B. Ribeiro-Neto, editors. *Modern Information Retrieval*. Addison-Wesley, 1999.
5. F. Ciravegna, A. Dingli, D. Petrelli, and Y. Wilks. User-System Cooperation in Document Annotation based on Information Extraction. In V. R. B. A. Gomez-Perez, editor, *Proc. of the Intl. Conf. on Knowledge Engineering and Knowledge Management (EKAW02)*, Lecture Notes in Artificial Intelligence 2473, Springer Verlag, 2002.
6. S. Cluet, C. Delobel, J. Simeon, and K. Smaga. Your Mediators need Data Conversion. In *Proc. ACM SIGMOD Symp. on the Management of Data*, 1998.
7. S. Decker, S. Melnik, F. van Harmelen, D. Fensel, M. Klein, J. Broekstra, M. Erdmann, and I. Horrocks. The Semantic Web: The Roles of XML and RDF. *IEEE Expert*, 15(3), 2000.
8. S. Dill, N. Eiron, D. Gibson, D. Gruhl, R. Guha, A. Jhingran, T. Kanungo, S. Rajagopalan, and A. Tomkins. SemTag and seeker: bootstrapping the semantic web via automated semantic annotation. In *Proc. Intl. World Wide Web Conference (WWW)*, pages 178–186, 2003.
9. Dublin Core Metadata Element Set. Technical Report, 1999. http://dublincore.org/.
10. M. Erdmann, A. Maedche, H. Schnurr, and S. Staab. From Manual to Semi-automatic Semantic Annotation: About Ontology-based Semantic Annotation Tools. In *Proc. COLING Intl. Workshop on Semantic Annotation and Intelligent Context*, 2000.

11. H. Garcia-Molia. Peer-to-peer Data Management. In *Proc. IEEE Intl. Conf. on Data Engineering (ICDE)*, 2002.
12. S. Handschuh, S. Staab, and R. Volz. On deep annotation. In *Proc. Intl. World Wide Web Conference (WWW)*, pages 431–438, 2003.
13. J. Kahan and M. Koivunen. Annotea: an Open RDF Infrastructure for Shared Web Annotations. In *Proc. Intl. World Wide Web Conference (WWW)*, pages 623–632, 2001.
14. G. Karvounarakis, S. Alexaki, V. Christophides, D. Plexousakis, and M. Scholl. RQL: A Declarative Query Language for RDF. In *Proc. Intl. World Wide Web Conference (WWW)*, pages 623–632, 2002.
15. K. Keenoy, G. Papamarkos, A. Poulovassilis, D. Peterson, and G. Loizou. Self e-Learning Networks – Functionality, User Requirements and Exploitation Scenarios. Technical report, SeLeNe Consortium, 2003. www.dcs.bbk.ac.uk/selene/.
16. B. Kieslinger, B. Simon, G. Vrabic, G. Neumann, J. Quemada, N. Henze, S. Gunnersdottir, S. Brantner, T. Kuechler, W. Siberski, and W. Nejdl. ELENA Creating a Smart Space for Learning. In *Proc. Intl. Semantic Web Conference*, volume 2342 of *LNCS*. Springer Verlag, 2002.
17. E. D. Liddy, E. Allen, S. Harwell, S. Corieri, O. Yilmazel, N. E. Ozgencil, A. Diekema, N. McCracken, J. Silverstein, and S. Sutton. Automatic Metadata Generation and Evaluation. In *Proc. ACM Symp. on Information Retrieval*, Tempere, Finland, 2002. Poster session.
18. Draft Standard for Learning Objects Metadata. IEEE, 2002.
19. W. Nejdl, B. Worlf, C. Qu, S. Decker, M. Sintek, A. Naeve, M. Nilsson, M. Palmer, and T. Risch. EDUTELLA: a P2P networking Infrastruture Based on RDF. In *Proc. Intl. World Wide Web Conference (WWW)*, page 604:615, 2002.
20. Resource Description Framework Model and Syntax Specification. World Wide Web Consortium, 1999.
21. Resource Description Framework Schema (RDF/S). World Wide Web Consortium, 2000.
22. S. Staab, A. Maedche, and S. Handschuh. An Annotation Framework for the Semantic Web. In *Proc. Intl. Workshop on Multimedia annotation*, 2001.
23. Y. Tzitzikas, N. Spyratos, and P. Constantopoulos. Mediators over Ontology-based Information Sources. In *Proc. Intl. Conf. on Web Information Systems Engineering (WISE'01)*, 2001.
24. Y. Tzitzikas, N. Spyratos, and P. Constantopoulos. Query Evaluation for Mediators over Web Catalogs. In *Proc. Intl. Conf. on Information and Communication Technologies and Programming*, Primorsko, Bulgaria, 2002.
25. J. Wang and F. Lochovsky. Data extraction and label assignment for web databases. In *Proc. Intl. World Wide Web Conference (WWW)*, pages 187–196, 2003.
26. G. Wiederhold. Mediators in the Architecture or Future Information Systems. *IEEE Computer*, 25, 1992.

A Simple Theory of Expressions, Judgments and Derivations

Masahiko Sato

Graduate School of Informatics, Kyoto University

Abstract. We propose a simple theory of expressions which is intended to be used as a foundational syntactic structure for the Natural Framework (NF). We define expression formally and give a simple proof of the decidability of α-equivalence. We use this new theory of expressions to define judgments and derivations formally, and we give concrete examples of derivation games to show a flavor of NF.

1 Introduction

Abstraction and *instantiation* are two basic operations on expressions. Although the intuitive meanings of an abstract like $B \equiv (x)A(x)$ and its instance $A(t)$ (which is obtained from $A(x)$ by substituting t for x) are clear, the actual process of computing the expression $A(t)$ from B is very subtle as it sometimes requires the renaming of bound variables in B to avoid the capturing of free variables in t during the process of instantiation.

In this paper, we will present a simple theory of expressions in which expressions form a structure equipped with the operations of abstraction and instantiation. The novelty of the structure is that the instantiation operation can be carried out without renaming bound variables. Because of this property, we can define α-equivalence of expressions without relying on renaming of variables, and the proof of the decidability of α-equivalence becomes very simple. We remark that, as pointed out by Vestergaard [7], to give a rigorous proof of it is a delicate problem. We will give an informal account of the theory of expressions in 2 and give a formal presentation of the theory in 3.

We will then use the theory of expressions to reformulate the theory of judgments and derivations introduced in [5]. This theory forms the basis of the Natural Framework (NF), and in NF we can define various mathematical systems in a uniform and convenient way using *derivation games*. In 4, we will define the concepts of judgments, derivations and derivations, and will show that these concepts can be rigorously defined by using the higher-order abstract syntax provided by the simple theory of expressions.

2 Informal Theory of Expressions

In this section we present our theory of expression in an informal way. The presentation is informal only because we will take the notion of α-equivalence

M.J. Maher (Ed.): ASIAN 2004, LNCS 3321, pp. 437–451, 2004.

for granted. As we usually take this notion for granted, the reader should be able
to grasp the essence of the theory by reading this section.

2.1 Expressions

For each n ($n = 0, 1, 2, \cdots$), we assume a countably infinite set V_n of *variables*
(x, y, z). For each n ($n = 0, 1, 2, \cdots$), we assume a countably infinite set C_n of
constants (c, d). We assume that all these sets are mutually disjoint, so that given
any variable x (or constant c) we can uniquely determine a natural number such
that $x \in V_n$ ($c \in C_n$, resp.).

We will say that a variable (constant) is of *arity* n if it is in V_n (C_n, resp.).
A variable is *higher-order* if its arity is positive and it is *first-order* if its arity is
0, and similarly for a constant.

We define expressions as follows, where e:exp will mean that e is an *expression*.
We identify α-equivalent expressions.

$$\frac{x \in V_n \quad a_1 : \mathsf{exp} \quad \cdots \quad a_n : \mathsf{exp}}{x[a_1, \ldots, a_n] : \mathsf{exp}} \; \mathsf{var}$$

$$\frac{c \in C_n \quad a_1 : \mathsf{exp} \quad \cdots \quad a_n : \mathsf{exp}}{c[a_1, \ldots, a_n] : \mathsf{exp}} \; \mathsf{const}$$

$$\frac{x \in V_n \quad a : \mathsf{exp}}{(x)a : \mathsf{exp}} \; \mathsf{abs}$$

We will understand that $x[a_1, \ldots, a_n]$ ($c[a_1, \ldots, a_n]$) stands for x (c, resp.)
when $n = 0$. We will write $(x_1, \cdots, x_n)a$ for $(x_1) \cdots (x_n)a$, and when $n = 0$, this
stands for a. We will also write $(x)[a]$ for $(x)a$ when we wish to emphasize that
x is the binding variable and its scope is a.

Note that a variable standing by itself is not an expression if its arity is
positive. For each variable $x \in V_n$, we associate an expression $x^* :$ exp as follows.

1. $x^* :\equiv x$ if $n = 0$.
2. $x^* :\equiv (x_1, \ldots, x_n)x[x_1, \ldots, x_n]$ if $n > 0$, where $x_1, \ldots, x_n$ are all of arity 0.

We will sometimes simply write x as a short hand for x^*.

2.2 Environments

We define environments which are used to instantiate abstract expressions and
also to define substitution. Let x be an n-ary variable. We say that an expression
e is *substitutable for* x if e is of the form $(x_1, \ldots, x_n)a$ where $x_1, \ldots, x_n$ are all
0-ary variables. So, any expression is substitutable for a 0-ary variable, but,
only expressions of the form $(x, y)e$ (x, y are 0-ary) are substitutable for 2-ary
variables.

If x is a variable of arity n and e is substitutable for x, then $x = e$ is a
definition, and a set of definitions $\rho = \{x_1 = e_1, \ldots, x_k = e_k\}$ is an *environment*
if $x_1, \ldots, x_k$ are distinct variables, and its *domain* $|\rho|$ is $\{x_1, \ldots, x_k\}$.

2.3   Instantiation

Given an expression e and an environment ρ, we define an expression $[e]_\rho$ as follows. We choose fresh local variables as necessary.

1. $[x]_\rho :\equiv e$ if x is of arity 0 and $x = e \in \rho$.
2. $[x[a_1, \ldots, a_n]]_\rho :\equiv [e]_{\{x_1=[a_1]_\rho, \ldots, x_n=[a_n]_\rho\}}$ if $n > 0$ and $x = (x_1, \ldots, x_n)e \in \rho$.
3. $[x[a_1, \ldots, a_n]]_\rho :\equiv x[[a_1]_\rho, \ldots, [a_n]_\rho]$ if $x \notin |\rho|$.
4. $[c[a_1, \ldots, a_n]]_\rho :\equiv c[[a_1]_\rho, \ldots, [a_n]_\rho]$.
5. $[(x)[a]]_\rho :\equiv (x)[[a]_\rho]$.

We can check the well-definedness of the above definition as follows.

An environment ρ is *first-order* if all the variables in $|\rho|$ are first-order, and it is *higher-order* if $|\rho|$ contains at least one higher-order variable. If the given environment is first-order, then the above definition is an ordinary inductive definition. Now, since we know that the above definition is well-defined for first-order environments, we can carry out the above definition for higher-order environments.

It is essential to distinguish first-order variables and higher-order variables. Without the distinction, evaluation of expressions may fail to terminate as can be seen by the following example.

$$[x[x]]_{\{x=(y)y[y]\}} \equiv [y[y]]_{\{y=[x]_{\{x=(y)y[y]\}}\}} \equiv [y[y]]_{\{y=(y)y[y]\}} \equiv \cdots$$

However, since we do have the distinction of first-order and higher-order variables, the above computation is not possible. Namely, by our definition of environment, y must be of arity 0, since y occurs as a binder in $(y)y[y]$. But y must be also of arity 1 because of the first occurrence of y in $y[y]$. This is a contradiction.

3   Formal Theory of Expressions

We now present our theory of expressions formally. To do this, we first extend the notion of variables as follows.

3.1   Variable References

If $x \in V_n$ and $k \in \mathbb{N}$, where $\mathbb{N}$ is the set of natural numbers, then $\sharp^k x$ is a *variable reference* and k is called the *level* of the variable reference. In particular, if $k = 0$, then $\sharp^k x$ is the variable x. We use the letter r as a meta variable ranging over variable references.

For each arity n we choose and fix an n-ary variable and write it as v_n.

3.2   Expressions

We define expressions inductively as follows. If the judgment $e{:}\mathsf{exp}$ can be derived by the following rules, e is said to be a *expression*.

$$\frac{k \in \mathsf{N} \quad x \in \mathsf{V}_n \quad a_1 : \mathsf{exp} \quad \cdots \quad a_n : \mathsf{exp}}{\sharp^k x[a_1, \ldots, a_n] : \mathsf{exp}} \text{ varref}$$

$$\frac{c \in \mathsf{C}_n \quad a_1 : \mathsf{exp} \quad \cdots \quad a_n : \mathsf{exp}}{c[a_1, \ldots, a_n] : \mathsf{exp}} \text{ const}$$

$$\frac{x \in \mathsf{V}_n \quad a : \mathsf{exp}}{(x)a : \mathsf{exp}} \text{ abs}$$

For a variable reference $r \equiv \sharp^k x$ ($x \in \mathsf{V}_n$) we define an expression r^* as follows.

$$(\sharp^k x)^* :\equiv \begin{cases} \sharp^k x & \text{if } n = 0, \\ (x_1, \ldots, x_n)\sharp^k x[x_1, \ldots, x_n] & \text{if } n > 0, \text{ where } x_i \equiv \mathsf{v}_0 \ (1 \le i \le n). \end{cases}$$

For each expression a we assign its *size*, $|a|$, as follows.

1. $|\sharp^k x[a_1, \ldots, a_n]| :\equiv |a_1| + \cdots + |a_n| + 1$.
2. $|c[a_1, \ldots, a_n]| :\equiv |a_1| + \cdots + |a_n| + 1$.
3. $|(x)a| :\equiv |a| + 1$.

For each variable reference r and expression a we associate a set $\mathrm{occ}(r, a)$ of *free occurrences of r in a* as follows. An occurrence is represented by a string in N^*. If $S \subseteq \mathsf{N}^*$ and $n \in \mathsf{N}$ we put $n \cdot S :\equiv \{n \cdot \sigma \mid \sigma \in S\}$, where $n \cdot \sigma$ denotes the concatenation of the natural number n (regarded as a character) with the string σ.

1. $\mathrm{occ}(\sharp^k x, \sharp^\ell y[a_1, \ldots, a_n])$
$:\equiv \begin{cases} \{0\} \cup 1 \cdot \mathrm{occ}(\sharp^k x, a_1) \cup \cdots \cup n \cdot \mathrm{occ}(\sharp^k x, a_n) & \text{if } k = \ell \text{ and } x \equiv y, \\ 1 \cdot \mathrm{occ}(\sharp^k x, a_1) \cup \cdots \cup n \cdot \mathrm{occ}(\sharp^k x, a_n) & \text{otherwise.} \end{cases}$
2. $\mathrm{occ}(\sharp^k x, c[a_1, \ldots, a_n]) :\equiv 1 \cdot \mathrm{occ}(\sharp^k x, a_1) \cup \cdots \cup n \cdot \mathrm{occ}(\sharp^k x, a_n)$.
3. $\mathrm{occ}(\sharp^k x, (y)a) :\equiv \begin{cases} \mathrm{occ}(\sharp^{k+1} x, a) & \text{if } x \equiv y, \\ \mathrm{occ}(\sharp^k x, a) & \text{if } x \not\equiv y. \end{cases}$

For example, assuming that c is a binary constant and d is a 4-ary constant, we have

$$\mathrm{occ}(x, c[x, (x)d[x, \sharp^1 x, \sharp^2 x, y]]) = \{1, 2 \cdot 2\},$$
$$\mathrm{occ}(\sharp^1 x, c[x, (x)d[x, \sharp^1 x, \sharp^2 x, y]]) = \{2 \cdot 3\}.$$
$$\mathrm{occ}(\sharp^2 x, c[x, (x)d[x, \sharp^1 x, \sharp^2 x, y]]) = \emptyset \text{ and}$$
$$\mathrm{occ}(y, c[x, (x)d[x, \sharp^1 x, \sharp^2 x, y]]) = \{2 \cdot 4\}.$$

We put $\mathrm{FV}(a) :\equiv \{r \mid \mathrm{occ}(r, a) \neq \emptyset\}$ and call it the set of *free variable occurrences* in a.

An expression a is *closed* if $\mathrm{FV}(a) = \emptyset$. For example, $(x)\sharp^2 x$ is an expression but is not closed since $\mathrm{FV}((x)\sharp^2 x) = \{\sharp^1 x\}$, and $(x)x$ is a closed expression since $\mathrm{FV}((x)x) = \emptyset$.

3.3 Environments

We modify the notion of environments and define it as follows. Firstly, a *definition* is defined in the same way as before. Namely, a definition is an expression of the form $x = (x_1, \ldots, x_n)a$ where $x \in V_n$ and $x_1, \ldots, x_n$ are all arity 0 variables. Secondly, we will call any variable x a *declaration*. Then an *environment* is a list of definitions and declarations of the form:

$$\rho = (x_1 = a_1, \ldots, x_k = a_k, y_1, \ldots, y_\ell) \quad (k, \ell \in \mathbb{N}).$$

If $k = l = 0$, then $\rho = ()$ is called the *empty environment*. An environment is said to be *first-order* if x is of arity 0 for any definition $x = a$ in the environment, and it is said to be *higher-order* otherwise.

We will also modify the process of instantiation, and for that we first define the process of *pushing* an expression *through* a variable reference. Namely, for any expression a and any variable reference $\sharp^k z$, we define an expression $a \uparrow \sharp^k z$ as follows and call it the result of *pushing a through $\sharp^k z$*.

1. $\sharp^m x[a_1, \ldots, a_n] \uparrow \sharp^k z := \begin{cases} \sharp^m x[a_1 \uparrow \sharp^k z, \ldots, a_n \uparrow \sharp^k z] & \text{if } m < k \text{ and } x \equiv z, \\ \sharp^{m+1} x[a_1 \uparrow \sharp^k z, \ldots, a_n \uparrow \sharp^k z] & \text{if } m \geq k \text{ and } x \equiv z, \\ \sharp^m x[a_1 \uparrow \sharp^k z, \ldots, a_n \uparrow \sharp^k z] & \text{if } x \not\equiv z. \end{cases}$

2. $c[a_1, \ldots, a_n] \uparrow \sharp^k z :\equiv c[a_1 \uparrow \sharp^k z, \ldots, a_n \uparrow \sharp^k z]$.

3. $(x)[a] \uparrow \sharp^k z := \begin{cases} (x)[a \uparrow \sharp^{k+1} z] & \text{if } x \equiv z, \\ (x)[a \uparrow \sharp^k z] & \text{if } x \not\equiv z. \end{cases}$

Next, given an environment ρ and a variable reference $\sharp^k x$ we define $\rho(\sharp^k x)$ as follows and call it the *value* of $\sharp^k x$ in ρ.

1. $\rho(\sharp^k x) :\equiv \sharp^k x$, if $\rho = ()$.

2. $(\rho, z)(\sharp^k x) := \begin{cases} \sharp^k x & \text{if } k = 0 \text{ and } x \equiv z, \\ \rho(\sharp^{k-1} x) \uparrow z & \text{if } k > 0 \text{ and } x \equiv z, \\ \rho(\sharp^k x) \uparrow z & \text{if } x \not\equiv z. \end{cases}$

3. $(\rho, z = a)(\sharp^k x) :\equiv \begin{cases} a & \text{if } k = 0 \text{ and } x \equiv z, \\ \rho(\sharp^{k-1} x) & \text{if } k > 0 \text{ and } x \equiv z, \\ \rho(\sharp^k x) & \text{if } x \not\equiv z. \end{cases}$

We note that the actual value is computed by (1) the first item of the above definition, or by (2) the first subcase of the second item, or by (3) the first subcase of the third item. If it is computed by (1), then $\sharp^k x$ is said to be *free* in ρ, and if computed by (2), then it is said to be *declared* in ρ, and if computed by (3), then it is said to be *defined* in ρ,

3.4 Instantiation and Substitution

Now, given an expression a and an environment ρ, we can define the *instantiation* $[a]_\rho$ of a in the environment ρ as follows. $[a]_\rho$ is also called the *ρ-instance* of a.

1. $[\sharp^k x[a_1, \ldots, a_n]]_\rho$

$$:\equiv \begin{cases} [b]_{(x_1=[a_1]_\rho, \ldots, x_n=[a_n]_\rho)} & \text{if } \sharp^k x \text{ is defined in } \rho \text{ and } \rho(\sharp^k x) \equiv (x_1, \ldots, x_n)b, \\ \sharp^m x[[a_1]_\rho, \ldots, [a_n]_\rho] & \text{if } \sharp^k x \text{ is free or declared in } \rho \text{ and } \rho(\sharp^k x) \equiv \sharp^m x. \end{cases}$$

2. $[c[a_1, \ldots, a_n]]_\rho :\equiv c[[a_1]_\rho, \ldots, [a_n]_\rho]$.

3. $[(x)\,[b]]_\rho :\equiv (x)\,[[b]_{(\rho,x)}]$.

It should be noted that because of item 3, we have to extend the notion of environment by allowing declarations to be its part, and also that the definition in item 3 does not rely on the notion of α-equivalence. The above definition can be seen to be well-defined, by first considering the case where the environment ρ is first-order, and then considering the case where ρ is higher-order. It can also be seen that $[a]_\rho$ is an expression for any environment ρ.

If x and y are variables of the same arity, then we can easily see that for any expression a, $|[a]_{(x=y^*)}| = |a|$. If b is an expression and a is an expression subsitutable for x, then $[b]_{(x=a)}$ is called the result of *substituting a for x in b*.

3.5 α-Equivalence

In this subsection, we define the notion of α-*equivalence* and show that it is a decidable relation with the expected property that α-equivalence is preserved by substitution.

The judgment $a \equiv_\alpha b$ which is characterized by the following rules will be read 'a is α-equivalent to b'. Two expressions a and b are α-*equivalent* if and only if the judgment $a \equiv_\alpha b$ can be derived by the following rules.

$$\frac{k \in \mathsf{N} \quad x \in \mathsf{V}_n \quad a_1 \equiv_\alpha b_1 \quad \cdots \quad a_n \equiv_\alpha b_n}{\sharp^k x[a_1, \ldots, a_n] \equiv_\alpha \sharp^k x[b_1, \ldots, b_n]} \text{ varref}$$

$$\frac{c \in \mathsf{C}_n \quad a_1 \equiv_\alpha b_1 \quad \cdots \quad a_n \equiv_\alpha b_n}{c[a_1, \ldots, a_n] \equiv_\alpha c[b_1, \ldots, b_n]} \text{ const}$$

$$\frac{x, y \in \mathsf{V}_n \quad \mathrm{occ}(x, a) = \mathrm{occ}(y, b) \quad [a]_{(x=\mathsf{v}_n^*)} \equiv_\alpha [b]_{(y=\mathsf{v}_n^*)}}{(x)a \equiv_\alpha (y)b} \text{ abs}$$

We can show that the α-equivalence is indeed an equivalence relation in a straight forward way by induction on the size of expressions. The inductive argument works for the abs-rule case since $|[a]_{(x=\mathsf{v}_n^*)}| = |a|$. We can similarly show that the α-equivalence relation is a decidable relation.

We give two simple examples of derivations assuming that x and y are distinct variables of arity 0. In the second example below, we note that $[(x)\sharp^1 x]_{(x=\mathsf{v}_0^*)} \equiv x$.

$$\frac{x, y \in \mathsf{V}_0 \quad \mathrm{occ}(x, x) = \mathrm{occ}(y, y) = \{0\} \quad \dfrac{0 \in \mathsf{N} \quad \mathsf{v}_0 \in \mathsf{V}_0}{\mathsf{v}_0 \equiv_\alpha \mathsf{v}_0} \text{ var}}{(x)x \equiv_\alpha (y)y} \text{ abs}$$

$$\frac{}{(x)\sharp^1 x \equiv_\alpha (y)x} \quad x,y \in V_0 \quad occ(x,\sharp^1 x) = occ(y,x) = \emptyset$$

$$\frac{0 \in \mathsf{N} \quad x \in V_0}{x \equiv_\alpha x} \text{ var}$$

$$\text{abs}$$

We can also prove the basic theorem which establishes that α-equivalence is preserved by substitution of α-equivalent expressions.

Theorem 1. *If $a \equiv_\alpha b$ and $a' \equiv_\alpha b'$, then $[a]_{(z=a')} \equiv_\alpha [b]_{(z=b')}$.*

Proof. The theorem is obtained as a corollary to the following lemma.

Lemma 1. *If $(x_1,\ldots,x_m)s \equiv_\alpha (y_1,\ldots,y_m)t$, $\rho = (x_1 = a_1,\ldots,x_m = a_m)$, $\sigma = (y_1 = b_1,\ldots,y_m = b_m)$ and $a_i \equiv_\alpha b_i$ $(1 \le i \le m)$, then $[s]_\rho \equiv_\alpha [t]_\sigma$.*

Proof. We first prove the lemma for the case where ρ is an essentially first-order environment by induction on $|s|$. Here, by an essentially first-order environment we mean an environment such that for each definition $x = a$ in a, either x is of arity 0 or a is of the form r^*.

We do the case analysis on the shape of s.

1. $s \equiv \sharp^k x[s_1,\ldots,s_n]$. In this case, t is of the form $\sharp^\ell y[t_1,\ldots,t_n]$ and $[s_i]_\rho \equiv_\alpha [t_i]_\sigma$ $(1 \le i \le n)$.
   (a) $\sharp^k x$ is defined in ρ. In this case we have $\rho(\sharp^k x) \equiv a_i$ and $\sigma(\sharp^\ell y) \equiv b_i$ for some i $(1 \le i \le m)$. Since ρ is essentially first-order, we see that $[s]_\rho \equiv a_i$ and. $[t]_\sigma \equiv b_i$; or else $\rho(\sharp^k x) \equiv \sigma(\sharp^\ell y) \equiv r^*$ for some r since $\rho(\sharp^k x) \equiv_\alpha \sigma(\sharp^\ell y)$. So, we have $[s]_\rho \equiv_\alpha [t]_\sigma$.
   (b) $\sharp^k x$ is free or declared in ρ and $\rho(\sharp^k x) \equiv \sharp^{k'} x$. By induction hypothesis, we have

$$[s]_\rho \equiv \sharp^{k'} x[[s_1]_\rho,\ldots,[s_n]_\rho] \equiv_\alpha \sharp^{k'} x[[t_1]_\sigma,\ldots,[t_n]_\sigma] \equiv [t]_\sigma.$$

2. $s \equiv c[s_1,\ldots,s_n]$. In this case, t is of the form $c[t_1,\ldots,t_n]$ and, by induction hypothesis, we see that

$$[s]_\rho \equiv c[[s_1]_\rho,\ldots,[s_n]_\rho] \equiv_\alpha c[[t_1]_\sigma,\ldots,[t_n]_\sigma] \equiv [t]_\sigma.$$

3. $s \equiv (x)[s']$. In this case t is of the form $(y)[t']$ and $[s']_{(x=v_n*)} \equiv_\alpha [t']_{(y=v_n*)}$. We also have $[s]_\rho \equiv [(x)[s']]_\rho \equiv (x)[[s']_{(\rho,x)}]$ and $[t]_\sigma \equiv [(y)[t']]_\sigma \equiv (y)[[t']_{(\sigma,x)}]$. So, if we can show that $[[s']_{(\rho,x)}]_{(x=v_n*)} \equiv_\alpha [[t']_{(\sigma,y)}]_{(y=v_n*)}$, we will be done. We can show this as follows. By Lemma 2 which we prove next, we have $[[s']_{(\rho,x)}]_{(x=v_n*)} \equiv [s']_{(\rho',x=v_n*)}$ and $[[t']_{(\sigma,y)}]_{(x=v_n*)} \equiv [t']_{(\sigma',y=v_n*)}$ where

$$\rho' := (x_1 = [a_1]_{(x=v_n*)},\ldots,x_m = [a_m]_{(x=v_n*)})$$

and

$$\sigma' := (y_1 = [b_1]_{(y=v_n*)},\ldots,y_m = [b_m]_{(y=v_n*)})$$

Since $|s'| < |s|$, by induction hypothesis, we have $[s']_{(\rho'.x=v_n*)} \equiv_\alpha [t']_{(\sigma',y=v_n*)}$. Therefore, we have $[[s']_{(\rho,x)}]_{(x=v_n*)} \equiv_\alpha [[t']_{(\sigma,y)}]_{(y=v_n*)}$.

Next, we prove the lemma for the case where ρ is a higher-order environment by induction on $|s|$. We do the case analysis on the shape of s.

1. $s \equiv \sharp^k x[s_1, \ldots, s_n]$. In this case, t is of the form $\sharp^k x[t_1, \ldots, t_n]$ and $s_i \equiv_\alpha t_i$ $(1 \leq i \leq n)$.

   (a) $\sharp^k x$ is defined in ρ and $\rho(\sharp^k x) \equiv (x_1, \ldots, x_n)a$. In this case, we have $[s]_\rho \equiv [a]_{(x_1=[s_1]_\rho, \ldots, x_n=[s_n]_\rho)}$. We also see that $\sharp^k x$ is defined in σ and $\sigma(\sharp^k x)$ is of the form $(y_1, \ldots, y_n)b$ and hence we have $[t]_\sigma \equiv [b]_{(y_1=[t_1]_\rho, \ldots, y_n=[t_n]_\rho)}$. Here, both $(x_1 = [s_1]_\rho, \ldots, x_n = [s_n]_\rho)$ and $(y_1 = [t_1]_\rho, \ldots, y_n = [t_n]_\rho)$ are first-order, and by induction hypothesis, $[s_i]_\rho \equiv_\alpha [t_i]_\sigma$ for $(1 \leq i \leq m)$. So, we have

$$[s]_\rho \equiv [a]_{(x_1=[s_1]_\rho, \ldots, x_n=[s_n]_\rho)} \equiv_\alpha [b]_{(y_1=[t_1]_\rho, \ldots, y_n=[t_n]_\rho)} \equiv [t]_\sigma.$$

   (b) $\sharp^k x$ is free or declared in ρ and $\rho(\sharp^k x) \equiv \sharp^m x$. Same as 1(b) in the first-order case.

2. $s \equiv c[s_1, \ldots, s_n]$. Same as 2 in the first-order case.

3. $s \equiv (x)s'$. Same as 3 in the first-order case.

Lemma 2 (Substitution Lemma). *If* $\rho = (x_1 = a_1, \ldots, x_m = a_m)$ *and* $\rho' = (x_1 = [a_1]_\rho, \ldots, x_m = [a_m]_\rho)$, *then* $[[b]_{(\rho,x)}]_{(x=a)} \equiv [b]_{(\rho',x=a)}$.

Proof. By induction on $|b|$. We actually show $[[b]_{(\rho,x,\bar{z})}]_{(x=a)} \equiv [b]_{(\rho',x=a,\bar{z})}$ where $\bar{z}$ is a sequence of variables.

4　Natural Framework

In this section we introduce the Natural Framework (NF) which was originally given in Sato [5]. In [5], NF was developed based on a restricted theory of expressions. In this section we revise and extend NF by using the simple theory of expressions we have just defined.

NF is a computational and logical framework which supports the formal development of mathematical theories in the computer environment, and it has been implemented by the author's group at Kyoto University and has been successfully used as a computer aided education tool for students [4].

Based on the theory of expressions we just presented we now define judgements and derivations. In doing so, we first introduce the fundamental concept of *derivation context*. This concept is fundamental since, in general, an expression a containing free variables does not have a fixed meaning since its meaning depends on the meaning of its free variables while free variables do not have fixed meaning. However, in mathematical reasoning we often treat expressions containing free variables. In order to cope with this situation, we will make use of derivation contexts. Namely we will treat expressions containing free variables always *under* a derivation context Γ such that all the free variables in these expressions are declared in Γ.

Although it is possible and actually it is more natural and simpler to use the formal theory from a formal point of view, we will present our theory of judgments and derivations using the informal theory for the sake of readability.

In the following, we will use the following specific constants. Nil, zero, succ (arity 0), s (arity 1), Pair, ::, ⇒, HD, : (arity 2), and CD (arity 3).

We use the following notational convention.

$$<> :\equiv \texttt{Nil},$$
$$<e \mid f> :\equiv \texttt{Pair}[e, f],$$
$$<e_1, e_2, \ldots, e_n> :\equiv <e_1 \mid <e_2 \mid \cdots <e_n \mid \texttt{Nil}> \cdots >>,$$
$$X :: J :\equiv ::[X, J],$$
$$H \Rightarrow J :\equiv \Rightarrow[H, J].$$

An expression of the form $<e_1, e_2, \ldots, e_n>$ is called a *list* and we will define concatenation of two lists by:

$$<e_1, \ldots, e_m> \oplus <f_1, \ldots, f_n> :\equiv <e_1, \ldots, e_m, f_1, \ldots, f_n>.$$

If $V = <x_1, \ldots, x_n>$ is a list of distinct variables, then an expression e is a *V-expression* if $(V)e$ is a closed expression.

4.1   Judgments and Derivations

We first define the notion of *judgment*.

Definition 1 (Judgment). *We will call any expression a* judgment. *A judgment of the form* $H \Rightarrow J$ *is called a* hypothetical judgment *and a judgment of the form* $(x)[J]$ *is called a* universal judgment. *A judgment J is called a V-judgment if V is a list of variables and $(V)[J]$ is closed.*

Thus, formally speaking, any expression is a judgment. However, in order to make a judgment, or, in order to *assert* a judgment, we must *prove* it. Namely, we have to construct a derivation whose conclusion is the judgment. Below, we will make the notion of derivation precise. To this end, we first define *derivation context*.

Definition 2 (Derivation Context). *We define a* derivation context Γ *together with its* general variable part $GV(\Gamma)$ *and* variable part $V(\Gamma)$.

1. Empty context. *The empty list $<>$ is a derivation context. Its general variable part is $<>$ and variable part is $<>$.*
2. General variable declaration. *If Γ is a derivation context, and x is a variable not declared in Γ, then $\Gamma \oplus <x>$ is a derivation context. Its general variable part is $GV(\Gamma) \oplus <x>$ and variable part is $V(\Gamma) \oplus <x>$.*
3. Derivation variable declaration. *If Γ is a derivation context, H is a $V(\Gamma)$-expression, and X is a 0-ary variable not declared in Γ, then $\Gamma \oplus <X :: J>$ is a derivation context. Its general variable part is $GV(\Gamma)$ and variable part is $V(\Gamma) \oplus <X>$.*

We now define derivation games.

Definition 3 (Derivation Game). *A list of the form* $<c_1 :: R_1, \ldots, c_n :: R_n>$ *is called a* derivation game *if each c_i is a 0-ary constant and R_i is a closed judgment $(1 \le i \le n)$. Each R_i is called a* rule *of the game and c_i is called the* name *of the rule R_i. c_i's must be all distinct.*

Derivation games are used to define mathematical or logical theories and also to define computation systems. We will give some examples of derivation games later, but see [5] for more examples of derivation games.

Any closed expression R can be uniquely written in the form

$$(x_1, \ldots, x_m) [H_1 \Rightarrow \cdots \Rightarrow H_n \Rightarrow J]$$

where J is *not* a hypothetical judgment. This expression can be instantiated as follows. We fix a list of variables V. If $\rho = (x_1 = e_1, \ldots, x_m = e_m)$ where each e_j $(1 \le j \le m)$ is a V-expression, then $[R]_\rho$ is called a V-instance of R and we write $R(e_1, \ldots, e_m)$ for it. Since $R(e_1, \ldots, e_m)$ is of the form $H_1' \Rightarrow \cdots \Rightarrow H_n' \Rightarrow J'$, we will write

$$\frac{H_1' \quad \cdots \quad H_n'}{J'} \; R(e_1, \ldots, e_m)$$

for $R(e_1, \ldots, e_m)$. We will sometimes write R itself as

$$\frac{H_1 \quad \cdots \quad H_n}{J} \; R(x_1, \ldots, x_m)$$

in order to make the role of R as a rule clear. Actually, R is a rule-schema, and as we will see in the definition of derivations below, instances of R are used as inference rules when we construct derivations.

We can now proceed to the definition of derivations. Derivations are defined with the following informal meanings of judgments in mind. A hypothetical judgment $H \Rightarrow J$ means that we can derive the judgment J whenever H is derivable. A universal judgment of the form $(x) [J]$ means that we can derive the judgment $[J]_{(x=e)}$ for any expression e which is substitutable for x.

Definition 4 (Derivation). *Let G be a derivation game. We define a G-derivation* relative to a derivation context Γ *as follows. We define its conclusion at the same time. In the following definition, Γ stands for an arbitrary derivation context. We can see from the definition below, that if D is a G-derivation under Γ, then its conclusion is a $\mathrm{GV}(\Gamma)$ expression.*

1. Derivation variable. *If X is a derivation variable and $X :: H$ is in Γ, then*

$$X$$

is a G-derivation under Γ and its conclusion is H.

2. Composition. *Suppose that R is a rule in G and c is the name of the rule R. If $D_1, \ldots, D_n$ are G-derivations under Γ such that their conclusions are $H_1, \ldots, H_n$, respectively, and*

$$\frac{H_1 \quad \cdots \quad H_n}{J} \ R(e_1, \ldots, e_m)$$

is a $GV(\Gamma)$-instance of R, then

$$\frac{D_1 \quad \cdots \quad D_n}{J} \ c(e_1, \ldots, e_m),$$

which is an abbreviation of the expression $CD[J, <c, e_1, \ldots, e_m>, <D_1, \ldots, D_n>]$, *is a G-derivation and its conclusion is J.*

3. Hypothetical derivation. *If D is a G-derivation under $\Gamma \oplus <X :: H>$ and its conclusion is J, then*

$$(X :: H) [D],$$

which is an abbreviation of the expression $HD[H, (X) [D]]$, *is a G-derivation under Γ and its conclusion is $H \Rightarrow J$.*

4. Universal derivation. *If D is a G-derivation under $\Gamma \oplus <x>$ and its conclusion is J, then*

$$(x) [D]$$

is a G-derivation under Γ and its conclusion is $(x) [J]$.

We will write

$$\Gamma \vdash_G D :: J$$

if D is a derivation in G under Γ whose conclusion is J.

A very simple example of a derivation game is the game Nat:

$$\text{Nat} :\equiv \ < \text{zero} :: 0 : \text{Nat}, \ \text{succ} :: (n) [n : \text{Nat} \Rightarrow s(n) : \text{Nat}] >,$$

and, by using obvious notational convention, we can display the two rules of this game as follows. We write $s(x)$ for $s[x]$.

$$\frac{}{0 : \text{Nat}} \ \text{zero}() \qquad \frac{n : \text{Nat}}{s(n) : \text{Nat}} \ \text{succ}(n)$$

In Nat, we can have the following derivation

$$\vdash_{\text{Nat}} D :: s(s(0)) : \text{Nat}.$$

NF provides another notation which is conveniently used to input and display derivations on a computer terminal. In this notation, instead of writing $\Gamma \vdash_G D :: J$ we write:

$$\Gamma \vdash J \ \text{in} \ G \ \text{since} \ D.$$

Also, when writing derivations in this notation, a derivation of the form

$$\frac{D_1 \quad \cdots \quad D_n}{J} \; R(e_1, \ldots, e_m)$$

will be written as:

$$J \text{ by } R(e_1, \ldots, e_m) \; \{D_1; \ldots; D_n\}$$

Here is a complete derivation in Nat in this notation.

```
⊢ (x)[x:Nat ⇒ s(s(x)):Nat] in Nat since
(x)[(X::x:Nat)[
    s(s(x)):Nat by succ(s(x)) {
        s(x):Nat by succ(x) {X}
    }
]]
```

The conclusion of the above derivation asserts that for any expression x, if x is a natural number, then so is $s(s(x))$, and the derivation shows us how to actually construct a derivation of $s(s(x))$:Nat given a derivation X of x:Nat.

We can prove the following basic properties of derivations in the same way as in [5].

Theorem 2 (Decidability). *If G is a derivation game, Γ is a derivation context, D is a $V(\Gamma)$-expression, J is a $GV(\Gamma)$-expression, then it is primitive recursively decidable whether $\Gamma \vdash_G D :: J$ or not.*

For a derivation game G, we let $D(G)$ be the set of G-derivations under the empty context. Then, we have the following corollary which fulfills Kreisel's dictum.

Corollary 1. *For any derivation game G, $D(G)$ is a primitive recursive subset of E.*

We can also check the following properties of derivations.

Proposition 1. *If $\Gamma \vdash D :: J$, then any free variable in J is declared in $GV(\Gamma)$, and any free variable in D is declared in $V(\Gamma)$.*

Derivation games enjoy the following fundamental properties.

Theorem 3. *The following properties hold for any derivation game G.*

1. *Weakening. If $\Gamma \vdash_G D :: J$ and $\Gamma \oplus \Gamma'$ is a context, then $\Gamma \oplus \Gamma' \vdash_G D :: J$.*
2. *Strengthening for General Variable. If $\Gamma \oplus \langle x \rangle \oplus \Gamma' \vdash_G D :: J$, and $x \notin FV(\Gamma') \cup FV(D) \cup FV(J)$, then $\Gamma \oplus \Gamma' \vdash_G D :: J$.*

3. *Strengthening for Derivation Variable. If $\Gamma \oplus <X :: H> \oplus \Gamma' \vdash_G D :: J$, and $X \notin FV(D)$, then $\Gamma \oplus \Gamma' \vdash_G D :: J$.*

4. *Substitution for Derivation Variable. If $\Gamma \oplus <X :: H> \oplus \Gamma' \vdash_G D :: J$ and $\Gamma \vdash_G D' :: H$, then $\Gamma \oplus \Gamma' \vdash_G [D]_{(X=D')} :: J$.*

5. *Substitution for General Variable. If $\Gamma \oplus <x> \oplus \Gamma' \vdash_G D :: J$, and e is a $GV(\Gamma)$-expression substitutable for x, then $\Gamma \oplus [\Gamma']_{(x=e)} \vdash_G [D]_{(x=e)} :: [J]_{(x=e)}$.*

6. *Exchange. If $\Gamma \oplus <e, f> \oplus \Gamma' \vdash_G D :: J$, and $\Gamma \oplus <f, e> \oplus \Gamma'$ is a derivation context, then $\Gamma \oplus <f, e> \oplus \Gamma' \vdash_G D :: J$.*

These basic properties of derivations imply that it is possible to implement a system on a computer that can manipulate these symbolic expressions and decide the correctness of derivations. At Kyoto University we have been developing a computer environment called CAL (for Computation And Logic) [4] which realizes this idea.

There are already several powerful computer systems for developing mathematics with formal verification, including Isabelle [3], Coq [1] and Theorema [2]. NF/CAL is being developed with a similar aim, but at the same time it is used as an education system for teaching logic and computation.

4.2   Lambda Calculus in NF

As an example of a derivation that requires higher-order variables in the defining rules, we define the untyped $\lambda\beta$-calculus LambdaBeta as follows. We introduce the following new constants. λF, appF, β, refl, sym, trans, appL, appR, ξ (arity 0), λ (arity 1), $=$, app (arity 2).

Term. The λ-terms are defined by the game Term which consists of the following two rules.

$$\frac{(x)\,[x : \mathtt{Term} \Rightarrow M[x] : \mathtt{Term}]}{\lambda[(x)\,[M[x]]] : \mathtt{Term}} \; \lambda\mathrm{F}(M) \qquad \frac{M : \mathtt{Term} \quad N : \mathtt{Term}}{\mathtt{app}[M, N] : \mathtt{Term}} \; \mathrm{appF}(M, N)$$

We can see from the form of the rule that the variable M in the λF-rule is a unary variable, and we can instantiate M by an expression of the form $(x)e$.

An example of a λ-term is given by the following derivation, where we will write $\lambda(x)\,[M]$ for $\lambda[(x)\,[M]]$ and $\mathrm{app}(M, \ N)$ for $\mathrm{app}[M, N]$.

```
Y::y:Term ⊢ λ(x)[app(x,y)]:Term in Term since
λ(x)[app(x,y)]:Term by λF {
    (x)[(X::x:Term)[
        app(x,y):Term by appF {X; Y}
    ]]
}
```

EqTerm. We can define the β-equality relation on λ-terms by the game EqTerm which is defined by the following rules, where we write $M = N$ for $= [M, N]$.

$$\frac{M : \text{Term}}{M = M} \; \text{refl}(M) \qquad \frac{M = N}{N = M} \; \text{sym}(M, N) \qquad \frac{M = N \quad N = L}{M = L} \; \text{trans}(L, M, N)$$

$$\frac{M = N \quad Z : \text{Term}}{\text{app}[M, Z] = \text{app}[N, Z]} \; \text{appL}(M, N, Z) \qquad \frac{Z : \text{Term} \quad M = N}{\text{app}[Z, M] = \text{app}[Z, N]} \; \text{appR}(M, N, Z)$$

$$\frac{(x) \, [x : \text{Term} \Rightarrow M[x] = N[x]]}{\lambda[(x) \, [M[x]]] = \lambda[(x) \, [N[x]]]} \; \xi(M, N)$$

We can see from the form of the rule that the variable M and N in the ξ-rule is a unary variable.

We can now define the untyped $\lambda\beta$-calculus LambdaBeta by putting:

$$\text{LambdaBeta} :\equiv \text{Term} \oplus \text{EqTerm}.$$

We give below an example of a formal derivation of a reduction in the $\lambda\beta$-calculus.

```
Y::y:Term ⊢ app(λ(x)[app(x,x)],y) = app(y,y) in LambdaBeta
since app(λ(x)[app(x,x)],y) = app(y,y) by β((x)[app(x,x)],y) {
   λ(x)[app(x,x)]:Term by λF {
      (x)[(X::x:Term)[
            app(x,y):Term by appF {X; Y}
      ]]
   };
   Y
}
```

In the above derivation, the β-rule is instantiated by the environment $\rho = (M = (x) \, [\text{app}[x, x]], N = y)$. Hence $M[N]$ is instantiated as follows:

$$[M[N]]_{\rho} \equiv [\text{app}[x, x]]_{(x = [N]_{\rho})} \; \text{since} \; (M = (x) \, [\text{app}[x, x]]) \in \rho$$
$$\equiv [\text{app}[x, x]]_{(x = y)}$$
$$\equiv \text{app}[y, y]$$

5 Conclusion

We have introduced a simple theory of expressions equipped with the operations of abstraction and instantiation. Abstraction is realized by a syntactic constructor but instantiation is realized by an external operation. In the usual systems of expression with named variables for binders, it is necessary to rename local binding variables to avoid unsolicited capture of free variables. In our system, we have introduced variable references which can refer to any surrounding variable. Variable references are already introduced in [6], but the definition of substitution in it is very complicated. We could simplify the definition by introducing the extended notion of environment.

The theory of expressions introduced here is a modification of our previous theory of expressions given in [5]. The previous theory did not have the notion of arity, and simpler than the current theory. However the previous theory could not define derivation games as objects in the theory. In the current theory it is possible to define rules of derivation games by closed expressions.

We have also shown that using this new theory of expressions, we can reformulate the theory of judgments and derivations introduced in [6].

Acknowledgements

The author wishes to thank Bruno Buchberger, Murdoch Gabbay, Atsushi Igarashi, Yukiyoshi Kameyama, Per Martin-Löf, Koji Nakazawa, Takafumi Sakurai, and René Vestergaard, for having fruitful discussions on expressions with the author.

References

1. Y. Bertot and P. Castéran, *Interactive Theorem Proving and Program Development, Coq'Art: The Calculus of Inductive Constructions*, Texts in Theoretical Computer Science, Springer, 2004.
2. B. Buchberger, C. Dupre, T. Jebelean, F. Kriftner, K. Nakagawa, D. Vasaru, W. Windsteiger, The Theorema Project: A Progress Report, in *Symbolic Computation and Automated Reasoning (Proceedings of CALCULEMUS 2000, Symposium on the Integration of Symbolic Computation and Mechanized Reasoning, August 6-7, 2000, St. Andrews, Scotland)*, M. Kerber and M. Kohlhase (eds.), A.K. Peters, Natick, Massachusetts, pp. 98-113.
3. T. Nipkow, L.C. Paulson and M. Wenzel, *Isabell/HOL — A Proof Assistant for Higher-Order Logic*, Lecture Notes in Computer Science, **2283**, Springer 2002.
4. M. Sato, Y. Kameyama and I. Takeuti, CAL: A computer assisted learning system for computation and logic, in Moreno-Diaz, R., Buchberger, B. and Freire, J-L. eds., *Computer Aided Systems Theory – EUROCAST 2001*, Lecture Notes in Computer Science, **2718**, pp. 509 – 524, Springer 2001.
5. M. Sato, Theory of Judgments and Derivations, in Arikawa, S. and Shinohara, A. eds., *Progress in Discovery Science*, Lecture Notes in Artificial Intelligence **2281**, pp. 78 – 122, Springer, 2002.
6. M. Sato, T. Sakurai, Y. Kameyama, *A Simply Typed Context Calculus with First-Class Environments*, Journal of Functional and Logic Programming, Vol. 2002, No. 4, March 2002.
7. R. Vestergaard, *The primitive proof theory of the λ-calculus*, PhD thesis, School of Maths and Computer Sciences, Heriot-Watt University, 2003.

Reactive Framework for Resource Aware Distributed Computing*

Rajesh Gupta** and R.K. Shyamasundar

Tata Institute of Fundamental Research,
Mumbai 400 005, India

In Honour of Jean-Louis Lassez

Abstract. Rapid strides in technology have lead to pervasive computing in a spectrum of applications such as crisis management systems,
distributed critical systems, medical therapy systems, home entertainment etc. One of the common features in the spectrum of applications
has been the reactivity of the systems and context of the environments.
The environmental context has become highly sophisticated due to rapid
advances in sensor technology and deployment. In this paper, we propose
a reactive framework to enable the development of pervasive computing
applications for different context environments. One of the novelties has
been to use contexts as observables between components. Some of the
context features that are observable are: the classical communications,
termination, clock-time, suspension of actions based on presence or absence of signals, location, resource parameters, etc. The new observables
provide a framework for the development of appropriate flexible middleware for a spectrum of applications. Further, it leads to the development
of an implementational model for a spectrum of applications that can be
effectively hooked onto available component implementations with appropriate interfaces. The new observables are suspensive in nature with
respect to communications and locations and allow to model varieties of
distributed applications that include sensor technology, wireless environments etc.

1 Introduction

Rapid strides in technology have lead to a confluence of various technologies such
as wide-area network architectures, grid technology & services, sensor architectures, pervasive computing environments, etc. The confluence of technologies
has lead to a wide variety of applications such as distributed crisis management systems, network security, logistics & supply chain management, home

* The work was done under grant from NSF (CCR-0098335) and SRC (2003-HJ-
1117). Support from Qualcomm and California Institute of Telecommunications and
Information Technology at UCSD is gratefully acknowledged.
** Department of Computer Science and Engineering, University of California at San
Diego, La Jolla, CA 92093-0114, USA.

M.J. Maher (Ed.): ASIAN 2004, LNCS 3321, pp. 452–467, 2004.

entertainment, cellular services etc. Evidently, the underlying software and system abstractions for such hardware platforms range from geographically distributed systems to synchronous hardware. Each of these platforms have a well developed range of methodologies, and programming environments for development, reasoning and assurance. Further, due to trends in hardware technology, wide variety of applications are being developed exploiting notions such as location or space awareness of the objects, resource awareness in the above hardware platforms. While such a trend of integrated hardware platform presents a very promising development, success of such developments will need the development of methodologies and frameworks that would allow wide ranging abstractions of the hardware/sensors/actuators etc and allow the development of software/middleware for a spectrum of applications with the convenience available for each of the specific cases; for instance, in critical applications, it is necessary to keep the formal verifiability and the use of off-the-shelf tools as an important criterion along with simulatability.

In this paper, we propose a reactive framework that provides a platform that enables the integrated development of processors, services, embedded sensors/actuators taking into account resource constraints, physical parameters such as locations and clock time etc. The framework uses contextual awareness primitives as observables. Due to technological innovations and software architectures, it is indeed possible to use such primitives *a la* classical observables such as communication, clocks and termination property. The framework provides a flexible formalism for a spectrum of applications for deriving an implementational model from available component implementations with appropriate hooks for interaction among the different components. This also enables the validation of the implementation relative to the underlying quantitative and functional constraints. We illustrate how the new observables make it possible to model hidden/exposed terminal problem of wireless networks. Another interesting aspect of such a framework is that it leads to a formal model of the system that can be effectively simulated and verified. After describing the framework referred to as Reactive Mobile Communicating Processes (RMCoP), we shall illustrate it with typical applications to pervasive computing.

2   Resource Aware Distributed Computing: Need

Before we go into our framework, let us analyze informally the requirements that would be demanded by such systems. We choose to look at distributed crisis management systems as that represents a dynamic evolutionary system that encompasses various technological, computational, and non-computational processes . Distributed crisis management system, can be considered as a typical distributed control application that abstracts out various aspects of dynamism, and asynchronous nature of interaction over participating agents (including human interaction– an example of a non-computable/predictable component). As articulated in [8], the task forces for crisis management is very dynamic that could even possibly be set up after the crisis and could evolve as the crisis unfolds. In other words, a functional representation is infeasible as feedback is essential for overcoming the crisis. A broad structure of such a system is shown in Figure 1.

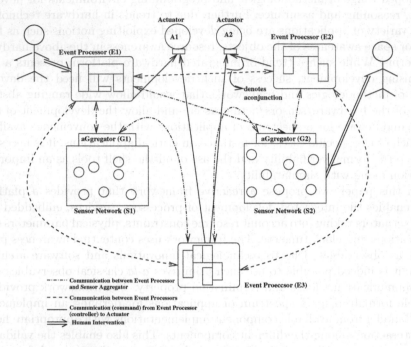

Fig. 1. Distributed Critical System Management

One of the important aspects in the design of such systems is that the problem is not necessarily intended to solve an algorithmic computational problem but to provide a service by performing tasks via computational frameworks. Important characteristics of such systems are:

1. Input and output are modeled essentially as nondeterministic dynamic streams wherein the values could depend on the earlier values (in other words, feedback needs to be integrated in the model),
2. The environment dynamically supplies the input streams to the computational system, and consumes the output values from the system.
3. Naturally, the computational engine works concurrently with the environment (that consists of sensors, actuators and perhaps non-computable elements such as human beings whose behaviour is not possible to compute or predict (treatable as oracle)).

Before we look at models for such systems, let us see what are the observable features of such systems wherein the heterogeneous components treated as black boxes. They encompass the classical distributed models [9] as well as global computing models [6]. In the classical systems, naturally communications, locations (for parallel actions) and termination of the system are observable. In the global computing models failure and unpredictability of the system are features of the systems. Thus, naturally, failure, bandwidth, congestion, as well as geographical locations are observables. Thus, it is imperative that our model should be capable of observations of these observables.

An important aspect that needs to be kept in mind while looking for models to fit the above requirements is that while the process is dynamic, it should have *every means to avoid adding crisis to crisis*. This demands the feature to withdraw/preempt/suspend commands. The natural characteristics that become evident are that the system must be event-driven[1] and reactive[2]. Further, the abstract model should relate to the implementation as closely as possible in a flexible way. Of course, the most important aspect of any such model is that it should be possible to establish the trustworthiness of the system in the environment.

A general description of such systems captured in Figure 1 is given below:

1. A set of event processors, that would take input from sensors, and command actuators.
   (a) Event Processors: These need to be reactive relative to its inputs from sensors and control actuators in an unambiguous way.
   (b) Sensors collect the required parameters (time, temperature, pressure, etc) and pass it to the event processors possible after defined aggregation.
   (c) actuators: act as per the command of event processors.
   (d) All these components behave in a predictable manner.
2. Communication and interfaces between event processors and sensors, event processors and event processors, event processors and actuators may vary widely and use a wide range of communication mechanisms due to various resource constraints and requirements.
3. Human Intervention

Any abstract computing structure for such a framework needs to specify

1. Processes through which the goal is to be achieved.
2. The behavior of each component.
3. The type of
   - *communication or interface between components*: range from synchronous to asynchronous, and
   - *observables between components*: The basic specification naturally would establish the observable events between components. In the context of web, naturally we have observables like the rate limit of transmission, timeouts etc. Sensor technologies have made it possible to have geographical location information and thus, location itself would be a possible observable between components; the latter finds extensive applications in pervasive computing environments.

---

[1] At this point, we don't articulate whether an event represents an exact value of a phenomenon or the change from its previous quiescence value etc.

[2] Reactive systems were introduced by David Harel and Amir Pnueli.

- In summary, we can observe actions between different components as *suspensive* actions based on the presence/absence of certain observable signals/actions. For instance, based on the bandwidth or the rate the communication action can be suspended that can be resumed later (ftp is a simple example). We refer to these as *suspensive observables*.

4. Human intervention: treated as an oracle as it is not predictable.

Now, let us look at the above requirements and frameworks that exist for the standalone components described above. The core of the system is the set of event processors from the perspective of control. Event processors being reactive must exhibit predictability. Response of the event processors needs to be predictable and hence, demand orthogonal features like logical concurrency (to master complexity), broadcast communication, preemption, priority, verifiability etc. Thus, suspension/preemption (similar to withdrawing a command and overriding another command) and priority are very important features that are not affordable to be missed in such frameworks; these features are lacking in framework of the when-then rule abstraction envisaged in [8]. There are several candidate models like the synchronous language models [4], Statecharts of David Harel for components such as event processors. Let us look at models that have been proposed for general communication/mobile systems. Some of the candidate models are based on π calculus [13] such as ambient calculus, spatial logic [7] etc. These models are not reactive, use synchronous channels (and hence demands atomicity for realization of synchronous communication across geographically distant nodes), spatial features are used to make distinction between concurrency and nondeterminism. For these reasons, it is not possible to use reactive suspensive/preemptive actions across distributed locations. Another candidate model is the model referred to as CRP [3] that has been proposed for a network of synchronous nodes communicating via synchronous channels. Though it is reactive, it suffers from synchronous communication requirements and also does not support reactive suspensive/preemptive communication actions across distributed locations.

In this paper, we shall describe a new framework referred to as Reactive Mobile Communicating Processes (RMCoP) that consists of

1. A set of interacting modules (hardware/software)
2. Each module could be composed of various components/modules.
3. The observer can observe the interaction of the modules through *observables* such as communication, location, QoS, rate-limit, clock time, etc.
4. Note that the interaction mechanism across the modules and within a module need not be uniform. The layers can be shrunk (the hiding operator, denoted [ ] could be interpreted in this way).

Such a distinction, enables us to arrive at obligations of the layers in satisfying various constraints such as real-time, communication bandwidth and possibly motion-planning constraints. In other words, the framework provides a rich plethora of abstractions in an integrated manner. Further, the framework is based on firm semantic foundations enabling the assessment of efficient techniques of implementation that can lead to development of middleware for the appropriate classes of applications. The semantic framework also makes it possible to have formal reasoning of the system using off-the-shelf systems.

3   Reactive Mobile Communicating Processes (RMCoP)

Our model is reflective of asynchronous network of distributed reactive processes whose distinct characteristics are briefed below:

 - The network consists of a network of reactive processes that can communicate with each other through asynchronous communication mechanisms.
 - Remote communication at different locations is suspensive and is observable.
 - Communication is observable along with location information.
 - Nodes may carry unique names/identies.

In the following, we shall describe the basic model informally without the primitive for dynamic creation of processes.

Notation

 - P, Q, R (and subscripts) denotes programs
 - M (and subscripts) denotes a program module.
 - $\mathcal{A}$ (and subscripts) denotes identity information - could be location (this again be in several forms), ip-addresses, just a label etc. T (and its subscripts) denotes tasks.
 - x,y, z (and subscripts) denote program variables and e1,e2, ... denote expressions

RMCoP

```
P ::=   𝒜₁ : M₁   //     𝒜ₙ : Mₙ
M ::= S, T | [M]  (* [M] denotes that one can observe only the external
interface *)
                        (* This is the classical hiding operator *)
T ::= task Task_name (x₁, ..., xₙ)
        | task Task_name (x₁, ..., xₙ) return (y₁,..yₙ) | T, T
S ::=    Hooks | Reactive

Hooks::= Comm | Location | Load | Select

Comm::=  send (m) to  Mᵢ | receive (x) from  Mᵢ

Location::= match_loc(a1,b1)      (* coordinate matching of a1 and b1 *)
          |match_range(a1,b1) (*Is b1 within the range of a1? *)
          |match_containment(a1,b1)  (* Is b1 contained within a1? *)
          |match_sourcewithin(sourcename, a1) (* Is  sourcename  in a1? *)
          | check_comm(location A1): (* check communication status at A1 *)
       (* The match operations are grouped under the class Match_ontology *)
```

```
|find_acoustic(self,a1): (*Distance to a1 with acoustic estimate *)
|find_loc(a1)          (* find coordinates of a1 *)
|find_range(a1,b1)     (* find the distance of b1 from a1 *)
|findall_range(a1, source_type, L) (* find all the resources of
                            source_type within a range of L from a1 *)
Load ::= bandwidth(self, M2) (* rate of communication from self to M1 *)

Select::= if [match_ontology] → S
          ‖ else  skip
          if


          |if
             b1; g1  → S1
             ‖b2;g2  → S2
                   . . .
             ‖bn; gn → Sn
          fi
g::=     receive from M | Location | load | Time
Time:= delay e (* e denotes time expressions on a global reference *)
Reactive::= | skip (nothing)  | halt   | X:=E
            | emit S | S1; S2 | loop S endloop
            | stat1 ‖ stat2
            | present e then S1 else s2 end
            | abort S when e end
            | suspend P when immediate s end
            | trap T in S end | exit T
            | execute task_name (...)
            | signal S in stat end
```

For lack of space, we shall only provide an informal semantics of RMCoP. The reader is referred to [10] formal semantics and details.

Informal Interpretation

Program denoted, P, is a set of modules, M_i that is prefixed with label $\mathcal{A}_i$; in other words, $\mathcal{A}_i$, denotes a way of identifying a module through means such as geographical location, hardware identity, ip-address, or just a label etc; for such purposes, it is understood that there are devices/techniques that would make such information extraction possible. *It must be noted that our associating a name to a location does not mean that we will use only static naming; it only means that wherever needed we shall use the static names to distinguish computations; otherwise, we shall use dynamic naming to distinguish computations.* Some of the operations for getting the location, etc., are described above. For simplicity, we often refer to M_i without the address marker like $\mathcal{A}_i$.

Each module, M, consists of a reactive program S and a set of tasks that are either explicitly available or assumed to be available through the usual notion of import. The tasks may return either signals or messages. Through the notion of a wrapper, one can assume that tasks return signals uniformly. Note that tasks may return signals or need not return signals (events). We use the structure [M] to denote that M is an encapsulated black box and only the external interfaces are observable.

The reactive program consists of commands for interaction between modules as well as reactive commands. The commands that capture interaction between modules are referred to as *Hook Commands* keeping the operational intuition that these commands provide hooks between modules. Some of the relevant hook commands for pervasive or ubiquitous computing requirements are captured through the syntactic category, H, that is categorized[3] into

1. Communication Commands (*Comm*):
   (a) send (m) to M_i: send message m to module M_i. This is an asynchronous send operation and is always enabled. I.e., the operation is non-blocking.
   (b) receive (x) from M_i: receive a message and assign it to variable x from module M_i. Note that this operation needs to wait for a message from another module and hence, the operation is blocking.
2. Location Commands (*Location*): Some of the location finding operations that are relevant for keeping track of various location identity requirements for ubiquitous computing are given below: Location matching operators can be treated as ontology related location finding operators. We refer to these operators as `match_ontology` operators.
   (a) *match_loc*($a1, b1$): returns **true** if a1 and b1 match exactly (geographical coordinates) and **false** otherwise.
   (b) *match_range*($a1, b1$): returns true if b1 is *within* the range of a1.
   (c) *match_containment*($a1, b1$): returns **true** if $a1$ is inside $b1$; other wise, the operation returns **false**. Such an operation allows for building hierarchical (topological) spaces.
   (d) match source within(sourcename, a1): returns **true** if resource named *sourcename* is contained within $a1$; other wise, it returns **false**.
   (e) check_comm(location A1): checks communication status at location A1.
   (f) findacoustic(self,a1): estimates distance to, a1, from itself (*self*), through acoustic signals.
   (g) find_local(a1): yields the coordinates of a1.
   (h) find_range(a1,b1): finds the distance of b1 from a1.
   (i) findall_range(a1, source_type, L): the operator finds all the resources of source_type within a range of L from a1.
3. Resource Aware Commands (*R_load*):
   (a) *bandwidth*($self, M2$): yields the bandwidth between *self* and $M2$.

Selection commands (*Select*):
Basic Location Matching:

> if [match_ontology] → S fi

If command of the class *match_ontology* returns **true** then the command *behaves* as S; other wise, the behavior is the same as **skip**. This is essentially the same as the if-then-else statement.

---

[3] Based on the application, sensors, and embedded devices one can get into more specific commands as needed in the context.

Guarded Selection: if $\|_a^n$ bi; gi $\rightarrow$ Si fi

This is the classical guarded command (as introduced in CSP-R [12]) where the guards could be any one of the following: (i) receive from M [4], (ii) boolean location commands, (iii) boolean Rload operations and (iv) Time delay denoted delay d; note that b_i refers to boolean expressions of local variables. Without the time delay, the command corresponds to nondeterministic selection from the open guards. If all are false, the command aborts. With the delay operator, the command is interpreted as follows: compute all the open delay expressions. Let the minimum delay be *min* units of time; the reference is to global clock time. Out of the open guards, select any of the enabled guard within *min* units of time. If none of the other other guards are enabled, choose any of the guards with the delay equal to *min*. Note that there may be several guards with the same delay value. In such a case, any one branch is chosen nondeterministically.

3.1   Interpretation of Reactive Commands

In the following, we provide an informal interpretation of reactive commands. For the formal semantics of these commands, the reader is referred to [4].

nothing: it does nothing and terminates instantaneously.

halt: it does nothing and does not terminate; this is the only statement that takes time.

emit s: causes an instantaneous broadcast emission of signal s and terminates instantaneously.

p;q: the sequencing operator ";" takes no time by itself. Thus, the q is started in the last instant of p.

loop p end: The loop statement never terminates but could be exited due to preemptions. the body is assumed to have an execution of at least one instant.

tt present s then p else q end: The statement tests for the presence of signal s and executes either statement p or statement q depending on whether signal s is present or absent. Both the test and the transfer of control to either statement p or statement q are done synchronously with the clock.

p || q: Both the component statements are initiated synchronously with the initiation of the construct and the construct terminates synchronously with the termination of the component that terminates last.

signal s in p end: The statement p is started immediately with a fresh signal s local to the block p overriding other bindings. It ensures that any reference to signal s within the statement block p is bound to local signal s and this signal is not exported outside the block.

abort stat when s end: Statement stat is started immediately and terminates either when statement stat terminates normally or whenever the signal s occurs in the future, in which case the whole statement gets preempted instantaneously.

suspend statp when immediate S: stat is initiated when the control arrives at the statement, provided signal S is absent. If stat terminates in the same instant

---

[4] The *send* is excluded as it is non-blocking (or always enabled unless the the number of messages are unbounded.

the statement terminates. If stat does not terminate in the same instant, the execution of stat continues until termination or the presence of S is detected. On the occurrence of S, the execution is suspended and remains as long as S is present. The execution resumes on the detection of the absence of S. The process continues till termination of stat.

trap T in stat end: the body stat is run normally until it executes an "exit T" statement. Then execution of *stat* is preempted and the whole trap construct terminates. The body of a trap statement can contain parallel components; the trap is exited as soon as one of the components executes an "exit T" statement, the other components being preempted.

exit T: The statement exits the trap T instantaneously terminating the corresponding trap statement unless an enclosing trap is exited synchronously.

3.2   Are These Observables Realizable?

We shall take a very brief look at the sensor technology developments that indeed make it possible to realize the above mentioned suspensive observables. Sensors can provide varieties of information including location information. There are a variety of low cost sensors that can provide GPS information (Radio Frequency Identifier (RFID)) being one of the simplest and cheapest; these are available both in active and passive mode) to various degrees of accuracy. In general, sensors are used by controllers and expect at least the following services such as:

1. Location Awareness: This could be a range of location information including applicable coordinates.
   - Could be the value of the location in the applicable coordinates.
2. State Awareness:
   - Make aware of the state of component so that other components can use the knowledge for appropriate action; it could be used for finding the status for establishing communication, QOS of the component etc.
   - Could also indicate the state of the component with which it is associated; e.g., ready-com(location) could be used to find out whether location is ready to communicate or not. Note that with sensors (even RFIDs), one can associate a password and thus, have a secure communication.
3. Synchronization service: This can be used for synchronization for participating components (real-time could be a typical parameter).
4. Context-awareness: The required context could be aggregated through a network of devices.

A typical API structure of a sensor with controllers is shown in Figure 3 and the services can be interpreted on the lines of classical sockets.

The above discussion should throw light on the way protocols for our suspensive communication observables can be discovered.

4   Illustrative Examples

Here, we shall illustrate examples the power of RMCoP.

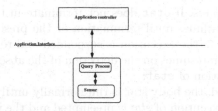

Fig. 2. API Structure for a Sensor

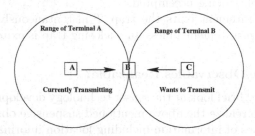

Fig. 3. Hidden Terminal Problem

4.1   Modelling Hidden/Exposed Terminal Problem

Consider the *hidden terminal* problem depicted in Figure 2. In the instance shown in the figure, the wireless terminal A is talking to terminal B in its range. Terminal C wants to talk to terminal B but its range is such that it is unable to know that A is already communicating with B. Thus, C starts talking to B resulting in collisions, wasting bandwidth and energy. A model of the hidden terminal problem in RMCoP is given below:

```
A: loop                            B:loop
   compute;                           compute;
   suspend                            if receive(x, A)→ skip
      send(m,B)                       [
      on not ready-com(location(B))      receive(y,C)→ skip
      end                             fi
   end                             end

   C:loop
      compute;
      suspend
         send m to B
         on not ready-com(location(B))
         end
      end
```

In this setup, terminal A when it wants to communicate with B remains suspended when terminal B is busy. Here, location (B) corresponds to the geographical location. The command `ready-com(location(B))` denotes the observability of communication with location B. Similarly, the behaviour of terminal C follows. In this model, absence of a carrier implies an idle medium. Assuming such an observable is possible, the interference can be avoided dynamically.

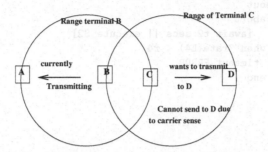

Fig. 4. Exposed Terminal Problem

Exposed Terminal Problem. In the instance shown in Figure 4, B is communication with A and C wants to communicate with D. Classically, it finds out the carrier as there is an intersection of the range of C and B,and hence decide not to communicate under the assumption that D is communicating.

Now, a similar solution can be obtained to overcome this problem using the location awareness primitive as shown below:

```
B: loop                          A:loop
     compute;                         compute;
     suspend                          if
       send(m,B)                      ‖ receive(x, A) → skip
     on not ready-com(location(B))    fi
     end
                                      end
   C:loop                        D: loop
     compute;                         compute;
     suspend                          if
       send(m,B)                      ‖ receive(y,C)→ skip
     on not ready-com(location(B))    fi
     end                              end
   end
```

Note that, it is no longer necessary to confuse the presence of a carrier to imply busy medium.

4.2   Rate Limit Based Scheduler

Consider a simple adaptive scheduler that has to switch from location L1 to L2 whenever the rate value falls below r1 Mbps at L1 to L2. If the rate at L4 falls below r4 Mbps from L1 to L4 then it is to be treated as an error. A model in RMCoP for this problem is given below:

```
loop
   abort
      [await t1 secs || execute S1]
   when ?rate(L1) < r1
   timeout
      abort
         [await t2 secs || execute S2]
      when ?rate(L4) < r4
       timeout ERROR
      end
   end
end
```

Verification

1. Task S1 is being executed and whenever the rate of transmission between L1 to L2 falls below $r1$, it switches to task $S2$. If you are given that the rate will not fall below $r4$ ($< r1$), it can be proved that there will be a continuous service. S1 could be a video service and $S2$ could be a text based service. Considering the program and the model, this property can be formally established.

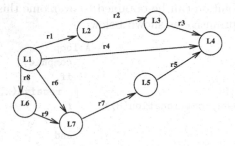

Fig. 5. Scheduling Constraints

Thus, for maintaining the above property, the required obligation is: *It is always the case that either the rate between L1 and L2 will be at least r1 or the rate between L1 and L4 will be at least r2.* That is $\square service \rightarrow \square[[rate(L1 - L2) > r1] \vee [rate(L1 - L4) > r2]]$

2. Now, let general graph shown in Figure 5 referring to locations L1, L2, and L4 among others describe the layer of interaction among the components. In such a scenario, we need to establish that under the given scheduling strategy the above property is always satisfied (or prove otherwise).

3. With the commitments (1) and (2), we can establish that the property is indeed satisfied.

4. In fact, it is also possible to consider it as precedence graph. This also can be verified using the classical techniques or the synchronous methodology proposed in [14].

The above example also illustrates how the quantitative and functional constraints can be validated against specification among components.

4.3   A Perimeter Security Problem

The availability of standardized low cost wireless connectivity such as 802.11 miniaturization computing devices, and availability of Global Positioning System (GPS) has opened the possibility of using the same for a plethora of applications. How, this can be effectively used is demonstrated through the following example. Let us consider the perimeter security problem in a high information security zone as shown in Figure 6. Any authorized person carrying a mobile computing device is able to communicate as long as he is within the perimeter and he cannot access once the device moves outside the region. A model in in RMCoP is shown below (tracking the devices is modeled as a task):

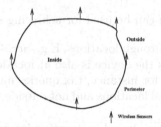

Fig. 6. Perimeter Security

```
Controller_for_the_perimeter@L::
        begin
           exec Checkdevice;
        loop
        await_request(mobile_device);
          present mobile_device_inside then
                   suspend
                   sustain mobile_device_inside
                   when mobile_device_outside
                   end suspend;
               end;
           end
       end
   [Mobile_device]:   newtick  mobile_device_inside in
                       loop
                         compute_what_you_want
                       endloop
                     end newtick;
```

```
[tracking_Sensor]: task checkdevice(L,device) in
        begin
         loop
           await_device_movement (device);
           if match_containment(L, device) & authorized  then
              emit mobile_device_inside(device);
           else emit mobile_device_ouside(device)
        end
```

Note: To indicate the interaction of the device with the environment through signal mobile_device_inside, we have used the keyword newtick as in Multi-clock Esterel; this is translatable to the classical monochronous setup..

The request of the mobile_device that is authorized is enabled for the usage as long as it is inside the perimeter recognized through the sensor. If the device moves outside the perimeter, the emitting of the clock to the mobile device is cut off and hence, it cannot transact with the network. To allow for keeping the environment in tact for transient movements the emission of the clock is kept suspended.

In general, the location can be used for achieving security and privacy to:

- Define access control through locations. E.g., access to data base at Location L can be denied unless the device is also at location L. This way, if one has restrictions about say, for instance, tax queries must be handled exclusively at the designated official locations and not outside, the policy can be followed in letter and spirit.
- Associate database access based on location information.

5   Discussions

We have described a framework referred to as, RMCoP, for pervasive computing and illustrated how the new observables enable the modelling of distributed pervasive applications effectively. The new observables have been arrived through amalgamation of reactivity, communication and sensor technology. We have shown, how RMCoP provides a modelling of various applications including hidden/exposed terminal problem. In fact, this example serves as a concrete real-life scenario to illustrate the need of localities to distinguish concurrent activities. An important distinguishing feature of the framework from a practical perspective has been the integration of heterogeneous frameworks for modelling complex distributed reactive systems without loosing the verifiability of the individual frameworks. Furthermore, the layered approach also shows how it is possible to enforce functional and quantitative constraints across layers (cf. rate limit based scheduler example). In fact, our initial study shows that the proposed formalism nicely formalizes and generalizes the informal implementations proposed in [1], provides a paradigm for spatial programming [5], provides a model for studying distributed cooperative control [11], and provide a formalism for specifying wireless control test-beds as envisaged in [2]. Basic prototype implementation is being planned in the context of spatial programming. Further analysis of the system and the need of additional features such as dynamic creation and migration is underway.

References

1. A. Acharya, M. Ranganathan, J. Saltz (1997), *Sumatra; A Language for Resource-aware Mobile Programs*, LNCS, 1222, 111-130
2. G. Baliga, S Graham, L Sha, PR Kumar (2004), *Etherware: Domainware for Wireless Control Networks*,
3. G.Berry, S. Ramesh, and R.K. Shyamasundar (1993), *Communicating Reactive Processes*, 20th ACM Symposium on principles of Programming Languages, 85-99.
4. G. Berry and G. Gonthier (1992), *The Esterel synchronous programming language: Design, semantics, Implementation*, Science of Programming.
5. C. Borcea et al. (2004), *Spatial Programming using Smart Messages: Design and Implementation*, 24th ICDCS, March.
6. L. Cardelli (1999), *Wide area computation*, ICALP99, LNCS, 1644, 10-24
7. L. Cardelli, A. Gordon (1998), *Mobile Ambients*, FoSSaCS98, LNCS, 1378, 140-155.
8. K. Mani Chandy, Brian Emre Aydemir, Elliot Michael Karpilosky, S.M. Zimmerman (2003), *Event Webs for Crisis Management*, IASTED Conference.
9. C.A.R. Hoare (1985), *Communicating Sequential Processes*, Prentice-Hall.
10. Rajesh Gupta, R.K. Shyamasundar (2004), *Reactive Mobile Communicating Processes*, T.R., Computer Science Department, UCSD, September.
11. E. Klavins and R.M. Murray (2004), *Distributed Algorithms for Cooperative Control*, Pervasive Computing.
12. R. Koymans, R.K. Shyamasundar, W.P de Roever et al. (1998), Compositional Semantics for Real-Time Distributed Programming Languages, Information and Computation,79, 210-256.
13. R. Milner, *Communicating and Mobile Systems: the π-Calculus*, Cambridge University Press, 1999.
14. R.K. Shyamasundar and J.V. Aghav, *Validating Real-Time Constraints in Embedded Systems*, IEEE PRDC, pp. 347-355, 2001.

The Feature Selection and Intrusion Detection Problems

Andrew H. Sung and Srinivas Mukkamala

Department of Computer Science, New Mexico Tech, Socorro, NM 87801, U.S.A.
{sung, srinivas}@cs.nmt.edu

Abstract. Cyber security is a serious global concern. The potential of cyber terrorism has posed a threat to national security; meanwhile the increasing prevalence of malware and incidents of cyber attacks hinder the utilization of the Internet to its greatest benefit and incur significant economic losses to individuals, enterprises, and public organizations. This paper presents some recent advances in intrusion detection, feature selection, and malware detection.

In intrusion detection, stealthy and low profile attacks that include only few carefully crafted packets over an extended period of time to delude firewalls and the intrusion detection system (IDS) have been difficult to detect. In protection against malware (trojans, worms, viruses, etc.), how to detect polymorphic and metamorphic versions of recognized malware using static scanners is a great challenge.

We present in this paper an agent based IDS architecture that is capable of detecting probe attacks at the originating host and denial of service (DoS) attacks at the boundary controllers. We investigate and compare the performance of different classifiers implemented for intrusion detection purposes. Further, we study the performance of the classifiers in real-time detection of probes and DoS attacks, with respect to intrusion data collected on a real operating network that includes a variety of simulated attacks.

Feature selection is as important for IDS as it is for many other modeling problems. We present several techniques for feature selection and compare their performance in the IDS application. It is demonstrated that, with appropriately chosen features, both probes and DoS attacks can be detected in real time or near real time at the originating host or at the boundary controllers.

We also briefly present some encouraging recent results in detecting polymorphic and metamorphic malware with advanced static, signature-based scanning techniques.

1 Introduction

Intrusion detection is a problem of great importance to protecting information systems security, especially in view of the worldwide increasing incidents of cyber attacks. Since the ability of an IDS to identify a large variety of intrusions in real time with accuracy is of primary concern, we will in this paper consider performance measures of learning machine based IDSs in the critical aspects of classification accuracy, training time, testing times, and scalability.

One of the main problems with IDSs is the overhead, which can become prohibitively high. To analyze system logs, the operating system must keep information regarding all the actions performed, which invariably results in huge

M. J. Maher (Ed.): ASIAN 2004, LNCS 3321, pp. 468–482, 2004

amounts of data, requiring disk space and CPU resource. Next, the logs must be processed and converted into a manageable format and then compared with the set of recognized misuse and attack patterns to identify possible intrusions. Further, the stored patterns need be continually updated, which would normally involve human expertise. Detecting intrusions in real time, therefore, is a difficult task.

Several artificial intelligence techniques have been utilized to automate the intrusion detection process to reduce human intervention, they include neural networks, fuzzy inference systems, evolutionary computation machine learning, etc. Several data mining techniques have been introduced to identify key features or parameters that define intrusions [1-3]. A summary of intrusion detection techniques is given in [4,5]. Some previous work applied neural networks as classifiers to detect low level probes; and a summary of different port-scan detection techniques is also available [6,7].

In this paper, we implement and evaluate the performance of an intelligent agent based IDS to detect, among other attacks, probes at the originating host and DoS attacks at the boundary controllers. Intelligent agents that encapsulate different AI paradigms including Support Vector Machines (SVM) [8], Multivariate Adaptive Regression Splines (MARS) [9] and Linear Genetic Programming (LGP) [10] are utilized. The data we use in our experiments is collected on a real network at New Mexico Tech that includes normal activity and several classes of probing attacks and DoS attacks [11] generated using scripts available on the Internet. We perform experiments to classify the network traffic in real-time into "Normal", "Probe" and "DoS".

It is demonstrated that with appropriately chosen population size, program size, crossover rate and mutation rate, linear genetic programs outperform other artificial intelligent techniques in terms of detection accuracy.

An introduction to computation techniques used for experiments is given in section 2. A brief introduction to our intelligent agents based architecture is given in section 3. In section 4 we present using different classifiers for IDSs and compare their performance. Section 5 presents three feature ranking algorithms. Real-time data collection and feature extraction are described in section 6, as well as performance evaluation of probes and DoS attacks detection. In section 7 we present a robust technique for signature-based malware (viruses, worms, trojans, etc.) detection, with emphasis on obfuscated (polymorphic) malware and mutated (metamorphic) malware. The conclusions are given in section 8.

2   Computational Paradigms for Intrusion Detection

Various techniques have been used for buidling IDSs. In this paper we study the performance of three paradigms: Multivariate Adaptive Regression Splines (MARS), Support Vector Machines (SVMs), and Linear Genetic Programs (LGPs). These are general techniques that can be utilized to perform classification, as well as feature ranking, for the intrusion detection problem.

2.1   Multivariate Adaptive Regression Splines (MARS)

Splines can be viewed as a mathematical process for complicated curve drawings and function approximation. To develop a 2-dimensional spline the X-axis is broken into a convenient number of regions. The boundary between regions is also known as a knot. With a sufficiently large number of knots virtually any shape can be well approximated. While it is easy to draw a spline in 2-dimensions by keying on knot locations (approximating using linear, quadratic or cubic polynomial etc.), manipulating the mathematics in higher dimensions is best accomplished using basis functions. The MARS model is a regression model using basis functions as predictors in place of the original data. The basis function transform makes it possible to selectively blank out certain regions of a variable by making them zero, and allows MARS to focus on specific sub-regions of the data. It excels at finding optimal variable transformations and interactions, and the complex data structure that often hides in high-dimensional data [12].

Given the number of records in most data sets, it is infeasible to approximate the function $y = f(x)$ by summarizing y in each distinct region of x. For some variables, two regions may not be enough to track the specifics of the function. If the relationship of y to some $x's$ is different in 3 or 4 regions, for example, the number of regions requiring examination becomes larger than 34 billion with only 35 variables. Given that the number of regions cannot be specified *a priori*, specifying too few regions in advance can have serious implications for the final model. A solution is needed that accomplishes the following two criteria:

- judicious selection of which regions to look at and their boundaries
- judicious determination of how many intervals are needed for each variable

Given these two criteria, a successful method will essentially need to be adaptive to the characteristics of the data. Such a solution will probably ignore quite a few variables (affecting variable selection) and will take into account only a few variables at a time (also reducing the number of regions). Even if the method selects 30 variables for the model, it will not look at all 30 simultaneously. Such simplification is accomplished by a decision tree at a single node, only ancestor splits are being considered; thus, at a depth of six levels in the tree, only six variables are being used to define the node [12].

2.2 Support Vector Machines (SVMs)

The SVM approach transforms data into a feature space F that usually has a huge dimension. It is interesting to note that SVM generalization depends on the geometrical characteristics of the training data, not on the dimensions of the input space [13,14]. Training a support vector machine (SVM) leads to a quadratic optimization problem with bound constraints and one linear equality constraint. Vapnik shows how training a SVM for the pattern recognition problem leads to the following quadratic optimization problem [15].

$$\text{Minimize: } W(\alpha) = -\sum_{i=1}^{l} \alpha_i + \frac{1}{2} \sum_{i=1}^{l} \sum_{j=1}^{l} y_i y_j \alpha_i \alpha_j k(x_i, x_j) \tag{1}$$

$$\text{Subject to } \sum_{i=1}^{l} y_i \alpha_i \qquad \forall i : 0 \le \alpha i \le C \tag{2}$$

Where l is the number of training examples, α is a vector of l variables and each component α_i corresponds to a training example (x_i, y_i). The solution of (1) is the vector α^* for which (1) is minimized and (2) is fulfilled.

2.3 Linear Genetic Programs (LGPs)

LGP is a variant of the Genetic Programming (GP) technique that acts on linear genomes [10]. The linear genetic programming technique used for our experiments is based on machine code level manipulation and evaluation of programs. Its main characteristics in comparison to tree-based GP lies in that the evolvable units are not the expressions of a functional programming language (like LISP), but the programs of an imperative language (like C).

In the Automatic Induction of Machine Code by Genetic Programming, individuals are manipulated directly as binary code in memory and executed directly without passing an interpreter during fitness calculation. The LGP tournament selection procedure puts the lowest selection pressure on the individuals by allowing only two individuals to participate in a tournament. A copy of the winner replaces the loser of each tournament. The crossover points only occur between instructions. Inside instructions the mutation operation randomly replaces the instruction identifier, a variable or the constant from valid ranges. In LGP the maximum size of the program is usually restricted to prevent programs without bounds. As LGP could be implemented at machine code level, it can possibly be fast enough to be able to detect intrusions in near real time.

3 Computationally Intelligent Agents Based IDS Architecture

This section presents an intelligent agent-based IDS [16]. The system consists of modules that will be implemented by agents in a distributed manner. Communication among the agents is done utilizing the TCP/IP sockets. Agent modules running on the host computers consist of data collection agents, data analysis agents, and response agents. Agents running on the secure devices consist of the agent control modules that include agent regeneration, agent dispatch, maintaining intrusion signatures and information regarding the features to be used in attack detection.

Host Agents: Reside on the hosts of the internal network and perform the tasks specified by the server or master agent. These agents are implemented to be read/execute only and fragile. In the event of tampering or modification the agent reports to the server agent and automatically destroys itself.

Server Agents: Reside on the secure server of the network. Controls the individual host agents for monitoring the network and manages communication between the agents. These agents manage the life cycle and also update the host agents with new detection, feature extraction, response and trace mechanisms.

Host agents and server agents contain a variety of modules, as described below,

- Agent controller: Manages the agents' functionality, life cycle, communication with other agents and the response mechanisms.
- Attack signatures module: Maintains all the attack signatures and updates agents in the event of a new attack signature.
- Data collection: Extracts features required by the detection algorithm to decide whether the activity is malicious or normal.
- Intrusion detection module: Performs intrusion detection and classification.

- Response module: Decides weather a proactive (honey pots, decoys, traps) or a reactive (block the activity at the boundary) response should be initiated upon detection of intrusion.
- Trace back initiation: Initiates trace back.

The advantages of our proposed model include:

- With knowledge of the device and user profiles of the network, specific agents can be designed and implemented in a distributed fashion.
- Feature collection, intrusion determination, and attack response are performed in distributed fashion, speeding up the IDS.
- Agents can be implemented and dispatched to respond to specific new threats.
- Efficient detection algorithms can be implemented for early intrusion detection.
- Rapid intrusion response and trace back can be performed more easily with the agents communicating with each other.
- Adjustable detection thresholds can be implemented.

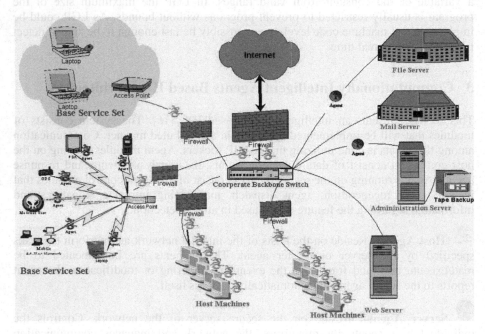

Fig. 1. Computational intelligent agents IDS architecture [16,17]

In section 6, we implement a portion of the proposed architecture by demonstrating that classes of attacks, including probes and denial of service attacks, can be detected at, respectively, the originating host and the network boundary.

4 IDS Using Classifiers

This section presents our work in implementing and evaluating IDSs using different classifiers, including SVMs, MARS, and LGPs. (Our previous comparative study of using neural networks for IDS indicated that neural networks are relatively uncompetitive, especially in their scalability.) [18]

4.1 Attack Taxonomy

A subset of the DARPA intrusion detection data set is used for offline analysis. In the DARPA intrusion detection evaluation program, an environment was set up to acquire raw TCP/IP dump data for a network by simulating a typical U.S. Air Force LAN. The LAN was operated like a real environment, but being blasted with multiple attacks [19,20]. For each TCP/IP connection, 41 various quantitative and qualitative features were extracted [1] for intrusion analysis. Attacks are classified into the following types.

4.1.1 Probes
Probing is a class of attacks where an attacker scans a network to gather information for the purpose of exploiting known vulnerabilities. An attacker with a map of machines and services that are available on a network can use the information to look for exploits. There are different types of probes: some of them abuse the computer's legitimate features; some of them use social engineering techniques. This class of attacks is the most common and requires little technical expertise.

4.1.2 Denial of Service Attacks
Denial of Service (DoS) is a class of attacks where an attacker makes some computing or memory resource too busy or too full to handle legitimate requests, thus denying legitimate users access to a system. There are different ways to launch DoS attacks: by abusing the computers' legitimate features; by targeting the implementation bugs; or by exploiting the system's misconfigurations. DoS attacks are usually classified based on the service(s) that an attacker renders unavailable to legitimate users.

4.1.3 User to Root Attacks
User to root or user to super-user (U2Su) exploits are a class of attacks where an attacker starts out with access to a normal user account on the system and then exploits vulnerability to gain root access. Most common exploits in this class of attacks are regular buffer overflows, which are caused by regular programming mistakes and incorrect environment assumptions.

4.1.4 Remote to User Attacks
Remote to local (R2L) is a class of attacks where an attacker sends packets to a machine over a network, then exploits the system's vulnerability to illegally gain local access as a user. There are different types of R2L attacks; the most common attacks in this class are done using social engineering.

4.2 Performance Results

Using the DARPA dataset and including all 41 features, the performance of LGP, MARS, and SVM based IDS is summarized in the following table.

Table 1. Performance of classifiers on all features (41)

Class	Classifier Accuracy (%)		
	LGPs	MARS	SVMs
Normal	99.89	96.08	99.55
Probe	99.85	92.32	99.70
DoS	99.91	94.73	99.25
U2Su	99.80	99.71	99.87
R2L	99.84	99.48	99.78

5 Feature Ranking and Selection

The 41 features extracted fall into three categorties: "intrinsic" features that describe the individual TCP/IP connections can be obtained form network audit trails; "content-based" features that describe payload of the network packet can be obtained from the data portion of the network packet; and "traffic-based" features that are computed using a specific window (connection time or no of connections).

Feature selection is an important issue in intrusion detection. Of the large number of features that can be monitored for intrusion detection purpose, which are truly useful, which are less significant, and which may be useless? The question is relevant because the elimination of useless features (the so-called audit trail reduction) enhances the accuracy of detection while speeding up the computation, thus improving the overall performance of IDS. In cases where there are no useless features, by concentrating on the most important ones we may well improve the time performance of an IDS without affecting the accuracy of detection in statistically significant ways.

The feature selection problem for intrusion detection is similar in nature to various engineering problems that are characterized by:

▪ Having a large number of input variables $\mathbf{x} = (x_1, x_2, ..., x_n)$ of varying degrees of importance to the output $\mathbf{y}$; i.e., some elements of $\mathbf{x}$ are essential, some are less important, some of them may not be mutually independent, and some may be useless or noise (in determining the value of $\mathbf{y}$).

▪ Lacking an analytical model that provides the basis for a mathematical formula that precisely describes the input-output relationship, $y = F(x)$.

▪ Having available a finite set of experimental data, based on which a model (e.g. intelligent systems) can be built for simulation and prediction purposes.

Due to the lack of an analytical model, one can only seek to determine the relative importance of the input variables through empirical methods. A complete analysis would require examination of all possibilities, e.g., taking two variables at a time to analyze their dependence or correlation, then taking three at a time, etc. This, however, is both infeasible (requiring 2^n experiments!) and not infallible (since the available data may be of poor quality in sampling the whole input space).

We describe in the following feature ranking algorithms based on SVM, LGP, and MARS.

5.1 SVM-Specific Feature Ranking Method

Information about the features and their contribution towards classification is hidden in the support vector decision function. Using this information one can rank their significance, i.e., in the equation

$$F(X) = \Sigma W_i X_i + b \qquad (1)$$

The point X belongs to the positive class if F(X) is a positive value. The point X belongs to the negative class if F(X) is negative. The value of F(X) depends on the contribution of each value of X and W_i. The absolute value of W_i measures the strength of the classification. If W_i is a large positive value then the ith feature is a key factor for positive class. If W_i is a large negative value then the ith feature is a key factor for negative class. If W_i is a value close to zero on either the positive or the negative side, then the ith feature does not contribute significantly to the classification. Thus, a ranking can be done by considering the support vector decision function.

Support Vector Decision Function Ranking:
The input ranking is done as follows: First the original data set is used for the training of the classifier. Then the classifier's decision function is used to rank the importance of the features. The procedure is:

1. Calculate the weights from the support vector decision function.
2. Rank the importance of the features by the absolute values of the weights.

5.2 LGP Specific Ranking Algorithm

The performance of each of the selected input feature subsets is measured by invoking a fitness function with the correspondingly reduced feature space and training set and evaluating the intrusion detection accuracy. Once the required number of iterations are completed, the evolved high ranked programs are analyzed for how many times each input appears in a way that contributes to the fitness of the programs that contain them. The best feature subset found is then output as the recommended set of features to be used in the actual input for the intrusion detection model.

In the feature selection problem the main interest is in the representation of the space of all possible subsets of the given input feature set. Each feature in the candidate feature set is considered as a binary gene and each individual consists of fixed-length binary string representing some subset of the given feature set. An individual of length d corresponds to a d-dimensional binary feature vector Y, where each bit represents the elimination or inclusion of the associated feature. Then, $y_i = 0$ represents elimination and $y_i = 1$ indicates inclusion of the ith feature. Fitness F of an individual program p is calculated as the mean square error (*MSE*) between the predicted output (O_{ij}^{pred}) and the desired output (O_{ij}^{des}) for all n training samples and m outputs [21].

$$F(P) = \frac{1}{n \cdot m} \sum_{i=1}^{n} \sum_{j=1}^{m} (O_{ij}^{pred} - O_{ij}^{des})^2 + \frac{w}{n} CE = MSE + w \cdot MCE \qquad (2)$$

Classification Error (*CE*) is computed as the number of misclassifications. Mean Classification Error (*MCE*) is added to the fitness function while its contribution is proscribed by an absolute value of weight (W).

5.3 MARS Specific Ranking Algorithm

Generalized cross-validation (GCV) is an estimate of the actual cross-validation which involves more computationally intensive goodness of fit measures. Along with the MARS procedure, a generalized cross-validation procedure is used to determine the significant input features. Non-contributing input variables are thereby eliminated.

$$GCV = \frac{1}{N}\sum_{i=1}^{N}[\frac{y_i - f(x_i)^2}{1 - k/N}] \tag{3}$$

where N is the number of records and x and y are independent and dependent variables respectively. k is the effective number of degrees of freedom whereby the GCV adds penalty for adding more input variables to the model. The contribution of the input variables may be ranked using the GCV with/without an input feature [12].

5.4 Feature Ranking Performance Results

Description of the most important features as ranked by three feature-ranking algorithms (SVDF, LGP, MARS) is given in Table 2. The (training and testing) data sets contain 11982 randomly generated points from the five classes, with the number of data from each class made proportional to its size, except that the two smallest classes (R2L and U2Su) are completely included. The normal data belongs to class 1, probe is class 2, DoS belongs to class 3, U2Su is class 4, and R2L is class 5. Where attack data comprises a collection of 22 different types of attack instances that belong to the four classes of attacks. A different, randomly selected, set of 6890 points from the whole data set (11982) is used for testing. Classifier performance using all the 41 features were given in table 1; and their performance using the most important 6 features as inputs to the classifier is given in table 3 below.

Table 2. Most important features as ranked by (SVDF, LGP, MARS)

Ranking Algorithm	Feature Description
SVDF	▪ source bytes: number of bytes sent from the host system to the destination system ▪ dst_host_srv_count: : number of connections from the same host with same service to the destination host during a specified time window ▪ count: number of connections made to the same host system in a given interval of time ▪ protocol type: type of protocol used to connect (e.g. tcp, udp, icmp, etc.) ▪ srv_count: number of connections to the same service as the current connection during a specified time window ▪ flag: normal or error status of the connection

Table 2. (*continued*)

LGP	<ul><li>dst_host_diff_srv_rate: % of connections to different services from a destination host</li><li>rerror_rate: % of connections that have REJ errors</li><li>srv_diff_host_rate: % of connections that have same service to different hosts</li><li>logged in: binary decision</li><li>service: type of service used to connect (e.g. fingure, ftp, telnet, ssh, etc.)</li><li>source bytes: number of bytes sent from the host system to the destination system</li></ul>
MARS	<ul><li>dst_host_diff_srv_rate: % of connections to different services from a destination host</li><li>dst_host_srv_count: : number of connections from the same host with same service to the destination host during a specified time window</li><li>source bytes: number of bytes sent from the host system to the destination system</li><li>dst_host_same_srv_rate: % of connections to same service ports from a destination host</li><li>srv_count: : number of connections to the same service as the current connection during a specified time window</li><li>rerror_rate: % of connections that have REJ errors</li></ul>

Table 3. Performance of classifiers on most important features (6)

Class	Classifier Accuracy (%)		
	LGPs	MARS	SVMs
Normal	99.77	94.34	99.23
Probe	99.87	90.79	99.16
DoS	99.14	95.47	99.16
U2Su	99.83	99.71	99.87
R2L	99.84	99.48	99.78

6   Detection of Probes and DoS Attacks

As DoS and probe attacks involve several connections in a short time frame, whereas R2L and U2Su are often embedded in the data portions of a single connection, "traffic-based" features play an important role in deciding whether a particular network activity is engaged in probing or DoS.

In this section, we describe an implementation of part of the CIA-based IDS proposed in section 3. It is shown that probe attacks can be detected at the originating hosts, and the class of DoS attacks can be detected at the network boundary.

6.1   Real-Time Data Collection and Feature Extraction

Experiments were performed on a real network using two clients and the server that serves the New Mexico Tech Computer Science Department network. The network packet parser uses the WINPCAP library to capture network packets and extracts the relevant features required for classification. The output of the parser for probe classification includes seven features: 1. duration of the connection to the target machine, 2. protocol used to connect, 3. service type, 4. number of source bytes, 5. number of destination bytes, 6. number of packets sent, and 7. number of packets received. The output summary of the parser includes eleven features for DoS classification: 1. duration of the connection to the target machine, 2. protocol used to connect, 3. service type, 4. status of the connection (normal or error), 5. number of source bytes, 6. number of destination bytes, 7. number of connections to the same host as the current one during a specified time window (in our case .01seconds), 8. number of connections to the same host as the current one using same service during the past 0.01 seconds, 9. percentage of connections that have SYN errors during the past .01 seconds, 10. percentage of connections that have SYN errors while using the same service during the past .01 seconds, and 11. percentage of connections to the same service during the past .01 seconds.

The output form the intrusion classifier is either (normal or probe) or (normal or DoS) for each connection. A variety of probes including SYN stealth, FIN stealth, ping sweep, UDP scan, null scan, xmas tree, IP scan, idle scan, ACK scan, window scan, RCP scan, and list scan with several options are targeted at the server. Normal data included multiple sessions of ftp, telnet, SSH, http, SMTP, pop3 and imap. Network data originating form a host to the server that included both normal and probes is collected for analysis.

In the experiments performed to analyze DoS detection more than 24 types of DoS attacks, including all 7 types of DoS from the DARPA data set that still exist, and 17 additional types. Network data originating from a host to the server is collected for analysis. The set of features selected for stealthy probe detection and DoS detection are based on our own feature ranking algorithms and obtained using the DARPA intrusion data set. The classifiers used in our experiments are SVMs (Support Vector Machines), MARS (Multivariate Adaptive Regressive Spline) and LGP (Linear Genetic Programming).

6.2   Performance Evaluation

Network packets contain information of protocol and service used to establish the connection between a client and the server. Network services have an expected number of bytes of data to be passed between a client and the server. If data flow is too little or too much it raises a suspicion about the connection as to whether it is a misuse. Using this information normal, probing and DoS activities can be separated.

In our evaluation, we perform binary classification (Normal/Probe) and (Normal/DoS). The (training and testing) data set for detecting probes contains 10369 data points generated from normal traffic and probes. The (training and testing) data set for detecting DoS contains 5385 data points generated from normal traffic and

DoS attack traffic. Table 4 and table 5 summarizes the overall classification accuracy (Normal/Probe) and (Normal/DoS) using MARS, SVM and LGP respectively.

Table 4. Performance comparison of testing for detecting probes

Class \ Machine	SVMs	LGPs	MARS
Normal	99.75%	100%	99.12%
Probe	99.99%	100%	100%

Table 5. Performance comparison of testing for 5-class classifications

Class\ Learning Machine	Normal SVM /LGP/MARS	DoS SVM/LGP/MARS	Overall Accuracy SVM/LGP/MARS
Normal	2692 / 2578 / 1730	14 / 128 / 976	99.48/ 95.26/ 63.9
DoS	538 / 153 / 0	2141 / 2526 / 2679	79.91/ 94.28/ 100
Accuracy (%) SVM/ LGP/ MARS	83.77/ 99.08/ 63.9	80.44/ 99.06/ 73.2	

The Table 5 below (containing three "confusion matrices" for the different classifiers used in experiments) gives the performance in terms of DoS detection accuracy:

The top-left entry of Table 5 shows that 2692, 2578, and 1730 of the actual "normal" test set were detected to be normal by SVM, LGP and MARS respectively; the last column indicates that 99.46, 95.26 and 63.9 % of the actual "normal" data points were detected correctly by SVM, LGP and MARS, respectively.

7 Malware Detection

Malware detection is becoming a very important task for assuring information system security. Software obfuscation, a general technique that can be used to protect the software from reverse engineering, can be easily used to produce polymorphic versions of malware to circumvent the current detection mechanisms (anti virus tools). Further, malware may be modified or evolved to produce metamorphic versions that are potentially even more harmful. Current static scanning techniques for malware detection have serious limitations; on the other hand, sandbox testing does not provide a complete solution either due to time constraints (e.g., time bombs cannot be detected before its preset time expires).

In our recent work we developed robust and unique signature-based malware (viruses, worms, trojans, etc.) detection, with emphasis on detecting obfuscated (or polymorphic) malware and mutated (or metamorphic) malware. The hypothesis is that all versions of the same malware share a common or core signature–possibly a second-order signature that is a combination of several features of the code. After a particular malware has been first identified (through sandbox testing or other means); it can be analyzed to extract the signature which provides a basis for detecting variants and mutants of the same malware in the future. The detection algorithm is based on calculating the 'similarity' of the code under scanning and the signatures; for more details of the detection algorithm, refer to [22,23].

The following table shows the preliminary results of our recent investigation of the MyDoom worm and several other recent worms and viruses, using eight different (commercial) scanners and proxy services. (✓ indicates detection, ✗ indicates failure to detect, and ? indicates only an "alert"; all scanners used are most current and updated version).

The obfuscation techniques used to produce the polymorphic versions of different malware tested in the experiments include control flow modification (e.g. Mydoom V2, Beagle V2), data segment modification (e.g., Mydoom V1, Beagle V1), and insertion of dead code (e.g., Bika V1). Our ongoing experiments also include investigation of metamorphic versions.

As can be seen from the last column, our scanner named *SAVE*, a signature based detection algorithm, performs the most accurate detection. The signature used in our algorithm is the sequence of API calls invoked, combined with a suitably defined "similarity measure" [23].

Table 6. Polymorphic malware detection using different scanners

	N	M^1	M^2	D	P	K	F	A	SAVE
W32.Mydoom.A	✓	✓	✓	✓	✓	✓	✓	✓	✓
W32.Mydoom.A V1	✗	✓	✓	✗	✗	✓	✓	✗	✓
W32.Mydoom.A V2	✓	✗	✗	✗	✗	✗	✗	✗	✓
W32.Mydoom.A V3	✗	✗	✗	✗	✗	✗	✗	✗	✓
W32.Mydoom.A V4	✗	✗	✗	✗	✗	✗	✗	✗	✓
W32.Mydoom.A V5	✗	?	✗	✗	✗	✗	✗	✗	✓
W32.Mydoom.A V6	✗	✗	✗	✗	✗	✗	✗	✗	✓
W32.Mydoom.A V7	✗	✗	✗	✗	✗	✗	✗	✗	✓
W32.Bika	✓	✓	✓	✓	✓	✓	✓	✓	✓
W32.Bika V1	✗	✗	✗	✓	✗	✓	✓	✓	✓
W32.Bika V2	✗	✗	✗	✓	✗	✓	✓	✓	✓
W32.Bika V3	✗	✗	✗	✓	✗	✓	✓	✓	✓
W32.Beagle.B	✓	✓	✓	✓	✓	✓	✓	✓	✓
W32.Beagle.B V1	✓	✓	✓	✗	✗	✓	✓	✗	✓
W32.Beagle.B V2	✓	✗	✗	✗	✗	✗	✗	✗	✓
W32. Blaster.Worm	✓	✓	✓	✓	✓	✓	✓	✓	✓
W32. Blaster.Worm V1	✗	✓	✓	✓	✓	✓	✓	✗	✓
W32. Blaster.Worm V2	✓	✓	✓	✗	✗	✓	✓	✗	✓
W32. Blaster.Worm V3	✓	✓	✓	✓	✓	✗	✗	✗	✓
W32. Blaster.Worm V4	✗	✗	✗	✗	✗	✓	✓	✗	✓

N – Norton, M^1 – McAfee UNIX Scanner, M^2 – McAfee, **D** – Dr. Web, **P** – Panda, **K** – Kaspersky, **F** – F-Secure,
A – Anti Ghostbusters, **SAVE** – *Static Analyzer for Vicious Executables*

7 Conclusions

We presented in this paper a summary of our recent work on intrusion detection, feature ranking, IDS architecture design, and malware detection. Details of our work can be found in the references.

Cyber security is a current and global issue of tremendous importance, and there are a great many challenging research problems that will likely require multi-disciplinary approaches for their satisfactory solution. In our work we have, for example, investigated using learning machines and classifiers for intrusion detection and compared their performance. Based on the experience, our current research is focused on malware detection.

Acknowledgements

Support for this research received from ICASA (Institute for Complex Additive Systems Analysis, a division of New Mexico Tech), a DoD IASP, and an NSF SFS Capacity Building grants are gratefully acknowledged. We would also like to acknowledge many insightful discussions with Dr. Jean-Louis Lassez that helped clarify our ideas. From the learning machines to the proposed IDS architecture, Dr. Lassez's enthusiastic teaching and invaluable contributions are reflected in our work.

References

1. Stolfo, J., Wei, F., Lee, W., Prodromidis, A., Chan, P.K.: Cost-based Modeling and Evaluation for Data Mining with Application to Fraud and Intrusion Detection. Results from the JAM Project by Salvatore (1999)
2. Mukkamala, S., Sung, A.H.: Feature Selection for Intrusion Detection Using Neural Networks and Support Vector Machines. Journal of the Transportation Research Board of the National Academics, Transportation Research Record No 1822. (2003) 33-39
3. Mukkamala, S., Sung, A.H.: Identifying Significant Features for Network Forensic Analysis Using Artificial Intelligence Techniques. In International Journal on Digital Evidence, IJDE. Vol.3 (2003)
4. Denning, D.: An Intrusion-Detection Model. IEEE Transactions on Software Engineering, Vol.13(2). (1987) 222-232
5. Kumar, S., Spafford, E.H.: An Application of Pattern Matching in Intrusion Detection. Technical Report CSD-TR-94-013. Purdue University (1994)
6. Staniford, S., Hoagland, J., McAlerney, J.: Practical Automated Detection of Stealthy Port scans. Journal of Computer Security; Vol.10 (1/2). (2002) 105-136
7. Basu, R., Cunningham, K.R., Webster, S.E., Lippmann, P.R.: Detecting Low-Profile Probes and Novel Denial of Service Attacks. Proceedings of the 2001 IEEE Workshop on Information Assurance (2001)
8. Cristianini, N., Taylor, S.J.: An Introduction to Support Vector Machines. Cambridge University Press (2000)
9. Friedman, J.H.: Multivariate Adaptive Regression Splines. Annals of Statistics; Vol.19. (1991) 1-141

10. Banzhaf, W., Nordin, P., Keller, E.R., Francone, F.D.: Genetic Programming: An Introduction on the Automatic Evolution of Computer Programs and its Applications. Morgan Kaufmann Publishers, Inc (1998)
11. Computer Science Department website, New Mexico Tech, USA, http://www.cs.nmt.edu
12. Steinberg, D., Colla, P.L., Kerry.: MARS User Guide. Salford Systems, San Diego (1999)
13. Joachims, T.: Making Large-Scale SVM Learning Practical. LS8-Report, University of Dortmund (2000)
14. Joachims, T.: SVMlight is an Implementation of Support Vector Machines (SVMs) in C. Collaborative Research Center on Complexity Reduction in Multivariate Data (SFB475), University of Dortmund (2000)
15. Vladimir, V.N.: The Nature of Statistical Learning Theory. Springer (1995)
16. Sung, A.H., Mukkamala, S., Lassez, J-L., and Dawson, T.: Computationally Intelligent Agents for Distributed Intrusion Detection System and Method of Practicing. United States Patent Application No: 10/413,462 (Pending) (2003)
17. Mukkamala, S., Sung, A.H., Abraham, A.: Distributed Multi-Intelligent Agent Framework for Detection of Stealthy Probes, Third International Conference on Hybrid Intelligent Systems, Design and Application of Hybrid Intelligent Systems, IOS Press. (2003) 116-125
18. Mukkamala, S., Sung, A.H.: A Comparative Study of Techniques for Intrusion Detection. Proceedings of 15th IEEE International Conference on Tools with Artificial Intelligence, IEEE Computer Society Press; (2003) 570-579
19. Kendall, K.: A Database of Computer Attacks for the Evaluation of Intrusion Detection Systems. Master's Thesis, Massachusetts Institute of Technology (1998)
20. Webster, S.E.: The Development and Analysis of Intrusion Detection Algorithms. Master's Thesis, Massachusetts Institute of Technology (1998)
21. Brameier, M, Banzhaf, W.: A Comparison of Linear Genetic Programming and Neural Networks in Medical Data Mining. IEEE Transactions on Evolutionary Computation; Vol. 5(1). (2001) 17-26
22. Sung, A.H., Xu, J., Ramamurthy, K., Chavez, P., Mukkamala, S., Sulaiman, T., Xie, T.: Static Analyzer for Vicious Executables (SAVE). Presented in Work-in-progress Section of IEEE Symposium on Security and Privacy (2004)
23. Sung, A.H., Xu, J., Chavez, P., Mukkamala, S.: Static Analyzer for Vicious Executables (SAVE). To appear in the Proceedings of 20th Annual Computer Security Applications Conference, ACSAC (2004)

On the BDD of a Random Boolean Function

Jean Vuillemin and Frédéric Béal

Ecole Normale Supérieure, 45 rue d'Ulm, 75005 Paris France

Abstract. The *Binary Decision Diagram* BDD, the *Binary Moment Diagram* BMD and the *Minimal Deterministic Automaton* MDA are three canonical representations for Boolean functions. Exact expression $a(i)$ and $w(i)$ are provided for the average and worst size BDD over $\mathbf{B}^i \mapsto \mathbf{B}$, and they are proved equal to those for the average and worst size BMD. The expressions $\overline{a}(i)$ and $\overline{w}(i)$ for MDA are just slightly bigger since

$$1 = \lim_{i \mapsto \infty} \frac{\overline{a}(i)}{a(i)} = \lim_{i \mapsto \infty} \frac{\overline{w}(i)}{w(i)}.$$

The significant differences between worst and average sizes are shown located on levels c and $c+1$: the *critical depth* $c = \lfloor r \rfloor$ is the integer part of the root $r \in \mathbf{R}$ of $i = r + 2^r$. The analysis shows that the average to worst size ratios $\frac{a(i)}{w(i)}$ and $\frac{\overline{a}(i)}{\overline{w}(i)}$ oscillate between

$$1 - \frac{1}{2e} = 0.81606\cdots \text{ and } 1 \text{ as } i \mapsto \infty.$$

Successive minima are found for $i = n + 2^n$ and $n \in \mathbf{N}$, where the gain between average and worst sizes is about 18%. Yet, such numbers are far in between: the gain far less than 1%, with probability one over the integers. The BDD/BMD/MDA sizes of a random function with a random number i of inputs are all equivalent to those of the worst size structure $a(i) \simeq \overline{a}(i) \simeq w(i) \simeq \overline{w}(i)$.

The results presented confirm, sharpen and extend those of [5, 6].

1 Introduction

The MDA, BDD and BMD are classical [1–3] *strong normal forms* for representing Boolean functions, and they are key to modern circuit verification.

This paper provides an exact analysis for the worst and average size of all these data structures. The worst case analysis relates to classical counting arguments for circuits [10]. Our contribution lies in the exact and asymptotic analysis of the worst and average sizes. The techniques used are somewhat sharper than those used in [5, 6]. They reveal the first order average phenomenon, which is the same for all structures, and they provide a better information on the structure specific second order term which is quasi-periodic in all cases.

M.J. Maher (Ed.): ASIAN 2004, LNCS 3321, pp. 483–493, 2004.
© Springer-Verlag Berlin Heidelberg 2004

The example function is defined by the expression $E = x_1' x_2 + x_1 x_3$, with $x_k' = 1 - x_k$.

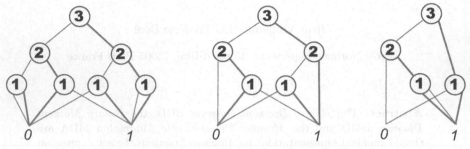

Graphs of $MDA_3(E)$, $BDD_3(E)$ and $BMD_3(E)$.

$$BDD_3(E) = \{0, 1, x_1, x_1', x_2 x_1', x_1 \cup x_2, E\},$$
$$BMD_3(E) = \{0, 1, x_1, x_1', x_2 x_1', E\}.$$

Fig. 1. Example MDA, BDD and BMD

2   Boolean Functions and Canonical Forms

Let $\mathbf{B}_i = \mathbf{B}^i \mapsto \mathbf{B}$ denote the set formed by the $\beta_i = |\mathbf{B}_i| = 2^{2^i}$ Boolean functions having $i \in \mathbf{N}$ inputs and a single output. An *abstract* Boolean function $f^e = [\![e]\!] \in \mathbf{B}_i$ is represented by some *concrete* expression (a.k.a. circuit net-list)

$$e \in \mathbf{E}_i = \langle 0, 1, x_1, \cdots, x_i, \neg, \cap, \cup, \oplus \rangle$$

composed from $0, 1$, inputs $\overline{x} = x_1 \cdots x_i$ and say, the logical *not, and, or, xor*.

In turn, each expression $e \in \mathbf{E}_i$ denotes a unique function $f^e(\overline{x}) = [\![e]\!] \in \mathbf{B}_i$. The Boolean value $f^e(\overline{b}) = [\![e_{\overline{x} = \overline{b}}]\!] \in \mathbf{B} = \{0, 1\}$ is computed by substituting $\overline{b} = b_1 \cdots b_i \in \mathbf{B}^i$ for $\overline{x} = x_1 \cdots x_i$ in e and by simplifying, to either 0 or 1.

Let $e \in \mathbf{E}_i$ be some Boolean expression with i inputs, and $f^e = [\![e]\!] \in \mathbf{B}_i$ be the denoted Boolean function. For $i > 0$ and $b \in \mathbf{B}$, the partial substitutions $e_{x_i = b}$ project f^e on two *prefix functions* with $i - 1$ inputs:

$$f^e_{x_i = b} = [\![e_{x_i = b}]\!] \in \mathbf{B}_{i-1}.$$

A function $f \in \mathbf{B}_i$ is thus uniquely expressed by Shannon's prefix decomposition:

$$f = x_i f_{x_i = 1} + (1 - x_i) f_{x_i = 0}.$$

The MDA for $f \in \mathbf{B}_i$ can be constructed as follows:

1. Recursively apply Shannon's prefix decomposition down to the constant functions $0, 1 \in \mathbf{B}_0$; the result is a complete binary decision tree of depth i, whose 2^i leaves are labelled by the 2^i bits: $f(\overline{b}) \in \mathbf{B}$ for $\overline{b} \in \mathbf{B}^i$.
2. Systematically share the nodes which represent equal Boolean functions, at all levels p for $1 \leq p \leq i$ in the decision tree.

The resulting MDA data structure is a *directed acyclic graph* in which: all paths from the root to a leaf have the same length i; distinct nodes $e \neq e' \in \mathbf{E}_p$ at depth p represent distinct Boolean functions: $f^e(\overline{b}) = 1 - f^{e'}(\overline{b})$ for some $\overline{b} \in \mathbf{B}^p$.

The *prefix closure* of $f \in \mathbf{B}_i$ is the set $MDA_i(f)$ of Boolean functions which label nodes in the MDA for f. It is recursively defined by:

$$MDA_0(f) = \{f\} \text{ for } f = 0 \text{ and } f = 1,$$
$$i > 0: MDA_i(f) = \{f\} \cup MDA_{i-1}(f_{x_i=0}) \cup MDA_{i-1}(f_{x_i=1}).$$

The partial derivative of $f \in \mathbf{B}_i$ with respect to variable x_i is defined by

$$\frac{\partial f}{\partial x_i} = (f_{x_i=0} \oplus f_{x_i=1}), \text{ so that } (\frac{\partial f}{\partial x_i} = 0) \Leftrightarrow (f = f_{x_i=0} = f_{x_i=1}).$$

Condition $\frac{\partial g}{\partial x_p} = 0$ detects when g is independent of input bit x_p. We let $\mathbf{B}'_p = \{g \in \mathbf{B}_p : \frac{\partial g}{\partial x_p} \neq 0\}$ denote the $\beta'_p = |\mathbf{B}'_p| = \beta_p - \beta_{p-1}$ functions in $\mathbf{B}_p$ which effectively depend upon input bit x_p.

The BDD for $f \in \mathbf{B}_i$ may be constructed from the MDA, by simplifying away all nodes $g \notin \mathbf{B}'_p$ with are independent of x_p. It is recursively defined by:

$$BDD_0(f) = \{f\},$$
$$\frac{\partial f}{\partial x_i} = 0: BDD_i(f) = BDD_{i-1}(f_{x_i=0}),$$
$$\frac{\partial f}{\partial x_i} \neq 0: BDD_i(f) = \{f\} \cup BDD_{i-1}(f_{x_i=0}) \cup BDD_{i-1}(f_{x_i=1}).$$

A dual of Shannon's decomposition is the Reed-Muller decomposition:

$$f = f_{x_i=0} \oplus x_i \frac{\partial f}{\partial x_i}.$$

The BMD for $f \in \mathbf{B}_i$ is constructed by recursively applying Reed-Muller's decomposition and by systematically sharing all common sub-expressions:

$$BMD_0(f) = \{f\},$$
$$\frac{\partial f}{\partial x_i} = 0: BMD_i(f) = BMD_{i-1}(f_{x_i=0}),$$
$$\frac{\partial f}{\partial x_i} \neq 0: BMD_i(f) = \{f\} \cup BMD_{i-1}(f_{x_i=0}) \cup BMD_{i-1}(\frac{\partial f}{\partial x_i}).$$

3   Worst Case Analysis

In defining the size of our structures, it is convenient to only count internal nodes and to exclude the two leaf nodes $0, 1 \in \mathbf{B}_0$, as in the tables in Figure 2.

Definition 1. *Let the worst size MDA, BDD, BMD over $f \in \mathbf{B}_i$ be:*

$$W^{mda}(i) = \max\{|MDA_i(f)| : f \in \mathbf{B}_i\},$$
$$W^{bdd}(i) = \max\{|BDD_i(f)| : f \in \mathbf{B}_i\},$$
$$W^{bmd}(i) = \max\{|BMD_i(f)| : f \in \mathbf{B}_i\}.$$

$i\backslash p$	1	2	3	4	$w(i)$
1	1				1
2	2	1			3
3	2	2	1		5
4	2	4	2	1	9
5	2	8	4	2	17
6	2	12	8	4	29
7	2	12	16	8	45
8	2	12	32	16	77

$i\backslash p$	1	2	3	4	$\overline{w}(i)$
1	1				1
2	2	1			3
3	4	2	1		7
4	4	4	2	1	11
5	4	8	4	2	19
6	4	16	8	4	35
7	4	16	16	8	51
8	4	16	32	16	83

Table for $w_p(i)$ and $w(i)$.   Table for $\overline{w}_p(i)$ and $\overline{w}(i)$.

Fig. 2. Worst Size tables

The analysis for $i \in \mathbf{N}$ relates to the unique root $r = r(i) \in \mathbf{R} \geq 0$ of[1]

$$i = r + 2^r. \tag{1}$$

The *critical depth* $c = c(i) \in \mathbf{N}$ of i is the integer part $c = \lfloor r \rfloor$ of r.

3.1   Exact Analysis

In the worst case structures of depth $i \in \mathbf{N}$, the nodes above $c+1$ form a complete binary tree. The nodes beneath c enumerate all the Boolean functions in $\mathbf{B}_c$ - in a redundant way within the MDA, non-redundant within the BDD/BMD.

Proposition 1. *The worst BDD and BMD over $f \in \mathbf{B}_i$ have equal sizes*

$$w(i) = W^{bdd}(i) = W^{bmd}(i) = 2^{i-c} + 2^{2^c} - 3 \tag{2}$$

and $c = c(i)$ is the critical depth of i. The worst MDA has the related size

$$\overline{w}(i) = W^{mda}(i) = w(i) + \beta''_{c-1} \tag{3}$$

for $\beta''_j = \sum_{0 < p \leq j} \beta_p$.

Proof: Let $\overline{w}_p(i)$ count the nodes at depth p in the worst size MDA_i, so that $\overline{w}(i) = \sum_p \overline{w}_p(i)$. There are 2^{i-p} nodes at depth p in a complete binary decision tree of depth i. In the worst case, each node represents a different Boolean function and $\overline{w}_p(i) = 2^{i-p}$. This is true for as long as there are enough functions to choose from, namely $2^{i-p} \leq \beta_p$. Otherwise, the worst case $\overline{w}_p(i) = \beta_p$ takes place when each Boolean function in $\mathbf{B}_p$ is represented by some node at depth p in MDA_i. In summary

$$\overline{w}_p(i) = \min(2^{i-p}, \beta_p).$$

---

[1] The root of (1) is related to that $x = L(y)$ of Lambert's transcendental equation $xe^x = y$ by $r = i - \frac{2^i}{e^{L\left(\ln(2)2^i\right)}}$ - function L is called *LambertW* by [9].

The number of nodes at depth p in the worst BDD_i or BMD_i is then

$$w_p(i) = \min(2^{i-p}, \beta'_p),$$

since $\beta'_p = \beta_p - \beta_{p-1}$ counts there the Boolean functions in $\mathbf{B}'_p$.

The sign of $2^{i-p} - \beta_p$ is equal to the sign of $d_p = i - p - 2^p$ and to the sign of $p - r$ since $d_p = p - r + 2^{p-r}$. The following equivalence

$$(2^{i-p} < \beta_p) \Leftrightarrow (d_p > 0) \Leftrightarrow (p > r) \Leftrightarrow (2^{i-p} < \beta'_p)$$

is simply derived for $p > 0$. Substituting in the above expressions gives us:

$$\overline{w}_p(i) = \begin{cases} \beta_p & \text{if } p \le r, \\ 2^{i-p} & \text{if } p > r, \end{cases}$$

$$\text{and } w_p(i) = \begin{cases} \beta'_p & \text{if } p \le r, \\ 2^{i-p} & \text{if } p > r. \end{cases}$$

Summing up $w(i) = \sum w_p(i)$ and $\overline{w}(i) = \sum \overline{w}_p(i)$ yield (2,3).   Q.E.D.

3.2   Asymptotic Analysis

One classical [10] asymptotic equivalent to $w(i)$ is $\frac{2^i}{i}$; yet Figure 3 indicates that the ratio $\frac{iw(i)}{2^i}$ has no limit for $i \mapsto \infty$.

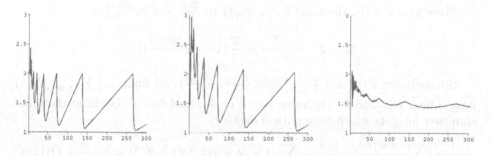

Fig. 3. Graphs of $\frac{w(i)}{\beta_{r(i)}}$, $\frac{iw(i)}{2^i}$ and $\frac{1}{i}\sum_{1 \le n \le i} \frac{nw(n)}{2^n}$

Proposition 2. *For i large enough, the size $\overline{w}(i)$ of worst MDA is equivalent to that $w(i)$ of worst BDD/BMD:* $1 = \lim_{i \mapsto \infty} \frac{\overline{w}(i)}{w(i)}$.

The ratios between worst-size and $\frac{2^i}{i}$ (or $\beta_r = 2^{2^r}$ for $r + 2^r = i$) oscillate:

$$1 = \liminf_{i \mapsto \infty} \frac{w(i)}{\beta_r} \text{ and likewise for } \frac{iw(i)}{2^i}, \frac{\overline{w}(i)}{\beta_r} \text{ and } \frac{i\overline{w}(i)}{2^i};$$

$$2 = \limsup_{i \mapsto \infty} \frac{w(i)}{\beta_r} \text{ and likewise for the above ratios.}$$

The average ratio is half way between the limiting values:

$$\frac{3}{2} = \lim_{i \mapsto \infty} \frac{1}{i} \sum_{1 \le n \le i} \frac{w(n)}{\beta_r(n)} \text{ and likewise for the above ratios.}$$

Proof: Expression (3) gives the equivalence

$$\frac{\overline{w}(i)}{w(i)} = 1 + \frac{\beta''_{c-1}}{w(i)} = 1 + O(\frac{1}{\beta_{c-1}}) = 1 + O(\frac{1}{\sqrt{w(i)}})$$

whose limit is sharply 1 for $i, c \mapsto \infty$.

Let $f = r - c$ be the fractional part of $r = r(i)$. From (2), we derive that
$w(i) + 3 = 2^{i-c} + \beta_c = \beta_r(2^f + \beta_c^{1-2^f}) = \beta_r g(2^f - 1, 1/\beta_c)$ where

$$g(t, x) = 1 + t + x^t.$$

Function g decreases from $g(0, x) = 2$ to $g_{min} = g(t_x, x)$ as t increases from 0
to t_x; g then increases from g_{min} to its maximum $g(1, x) = 2 + x$ as t goes from
t_m to 1. The minimum is reached at $x^{t_x} \log \frac{1}{x} = 1$ and $g_{min} = 1 + O\left(\frac{\log \log \frac{1}{x}}{\log \frac{1}{x}}\right)$.
Since $\lim_{x \mapsto 0} g(1, x) = 2$ and $\lim_{x \mapsto 0} g(t_x, x) = 1$, we conclude that

$$1 = \liminf_{x_n \mapsto 0} g(t_n, x_n) \text{ and } 2 = \limsup_{x_n \mapsto 0} g(t_n, x_n)$$

over all real sequences $t_n \in [0, 1]$, hence the claimed limits for $\frac{w(i)}{\beta_r} = g(t, x)$.

Since $\overline{w}(i) \simeq w(i)$, the same limits apply to $\frac{\overline{w}(i)}{\beta_r}$ and to $\frac{iw(i)}{2^i}$ as

$$\beta_r = 2^{2^r} = \frac{2^i}{i - r} = \frac{2^i}{i}(1 + O(\frac{\log(i)}{i})).$$

Similarly for $E(n) \in \{\frac{w(n)}{\beta_r(n)}, \frac{nw(n)}{2^n}, \frac{\overline{w}(n)}{\beta_r(n)}, \frac{n\overline{w}(n)}{2^n}\}$, all $\lim_{i \mapsto \infty} \frac{1}{i} \sum_{1 \le n \le i} E(n)$
have a limit equal to L. The value $L = \frac{3}{2}$ is obtained for $c \mapsto \infty$, from the finite
sum over integers which have critical depth c:

$$\frac{1}{1 + 2^c} \sum_{c(n)=c} \frac{nw(n)}{2^n} = \frac{1}{1 + 2^c} \sum_{d \le 2^c} (1 + (c + d)2^{-c})(1 + 2^{-d}) = 1 + \frac{1}{2} + O(1/c).$$

The analysis confirms the intuition from Figure 3: in the limit, the ratio $\frac{iw(i)}{2^i}$
tends to the piece-wise-linear pseudo-periodic function $\rho_\infty(i) = \frac{i}{c + 2^c}$, where
$c \in \mathbf{N}$ is the critical integer such that $C(c) = c + 2^c \le i < C(c + 1)$.   *Q.E.D.*

4   Average Case Analysis

Definition 2. *Let the average size MDA, BDD, BMD over $f \in \mathbf{B}_i$ be:*

$$A^{mda}(i) = \frac{1}{\beta_i} \sum_{f \in \mathbf{B}_i} |MDA_i(f)|,$$

$q\backslash p$	1	2	3	4 $\cdots$	$oa(q)$
1	1				1
2	1.7	1			2.75
3	2.7	1.9	1		5.7
4	3.6	3.6	2	1	10.2
5	4.0	6.5	4.0	2.0	17.4
6	4.0	10.3	7.9	4.0	29.2
7	4.0	14.0	15.5	8.0	48.5
8	4.0	15.7	30.1	16.0	80.9

Table for $\bar{a}_p(q)$ and $\bar{a}(q)$.

$q\backslash p$	1	2	3	4 $\cdots$	$a(q)$
1	0.5				0.5
2	0.9	0.8			1.6
3	1.4	1.5	0.9		3.8
4	1.8	2.7	1.9	1.0	7.4
5	2.0	4.8	3.7	2.0	13.5
6	2.0	7.7	7.4	4.0	24.1
7	2.0	10.5	14.6	8.0	42.
8	2.0	11.8	28.3	16.0	73.

Table for $a_p(q)$ and $a(q)$.

Fig. 4. Average size tables

$$A^{bdd}(i) = \frac{1}{\beta_i} \sum_{f \in \mathbf{B}_i} |BDD_i(f)|,$$

$$A^{bmd}(i) = \frac{1}{\beta_i} \sum_{f \in \mathbf{B}_i} |BMD_i(f)|.$$

Let $A_p^{mda}(i)$ count the nodes at depth p in $A^{mda}(i) = \sum_p A_p^{mda}(i)$, and similarly for BDD and BMD.

The average analysis is related to that of *hashing* [8]. Let $\mathcal{N} = [n_1 \cdots n_k]$ be some sequence of k integers chosen at random in $\{0 \cdots m - 1\}$. The probability that an integer j such that $0 \le j < m$ belongs to $\mathcal{N}$ is

$$\Pr(j \in \mathcal{N}) = 1 - (1 - \frac{1}{m})^k = h(\frac{1}{m}, k). \qquad (4)$$

The *hash* function to analyze here is $h(x, y) = 1 - (1 - x)^y$.

4.1   Exact Analysis

Proposition 3. *The average BDD and BMD have the same size at depth p:*

$$a_p(i) = A_p^{bdd}(i) = A_p^{bmd}(i) = \beta'_p h(x_p, y_p) \qquad (5)$$

for all $1 \le p \le i$; here, $x_p = \frac{1}{\beta_p}$ and $y_p = 2^{i-p}$. The size of the average MDA is

$$\bar{a}_p(i) = A_p^{mda}(i) = \beta_p h(x_p, y_p). \qquad (6)$$

Proof: The probability that some Boolean function $g \in \mathbf{B}_p \cap MDA_i(f)$ is among the 2^{i-p} prefixes of a random $f \in \mathbf{B}_i$ amounts to $h(x_p, y_p)$ by (4). The average number of nodes at depth p in $MDA_i(f)$ is the sum $\bar{a}_p(i) = \beta_p h(x_p, y_p)$ of these probabilities over $\mathbf{B}_p$. Summing over $\mathbf{B}'_p$ yields $a_p(i) = \beta'_p h(x_p, y_p)$ for $BDD_i(f)$ and as well for $BMD_i(f)$. $\qquad$ Q.E.D.

4.2   Asymptotic Analysis

The limit (if any) of the hash function $h(x, y) = 1 - (1 - x)^y$ for $x \mapsto 0$ and $y \mapsto \infty$ depends on that (if any) of the product $p = xy$: $h(x, y) \mapsto 0$ if $p \mapsto 0$; $h(x, y) \mapsto 1$ if $p \mapsto \infty$; finally $h(x, y) \mapsto 1 - \frac{1}{e^p}$ if $xy \mapsto p \in \mathbf{R} > 0$.

Throughout this paper, the reader should see θ_n whenever *she/he/it* reads the letter θ. The *invisible index* n is the number of occurrences of θ before this point in the text. Each variable θ_n represents a real number such that $0 < \theta_n < 1$, and no relation is assumed beyond that. The usage of θ is restricted to a single occurrence per real-valued expression.

Lemma 1. *Let* $i \in \mathbf{N}$ *have critical depth* $c = \lfloor r(i) \rfloor$; *let* $x_p = \frac{1}{\beta_p} < 1$ *and* $y_p = 2^{i-p} \geq 1$. *The number* $h(x_p, y_p) = 1 - (1 - x_p)^{y_p} = 1 - e^{y_p \log(1 - x_p)}$ *equals*

$$
h(x_p, y_p) = \begin{cases}
1 - \theta x_{c+1} & \text{if } 1 \leq p < c; \\
1 - e^{-x_p y_p} - \theta x_p & \text{if } p = c; \\
x_p y_p (1 - \frac{x_p y_p}{2}(1 - \frac{\theta}{6})) & \text{if } p = c+1; \\
x_p (y_p - \theta) & \text{if } c+1 < p \leq i.
\end{cases}
$$

Proof: By (1), the sign of $d_p = i - p - 2^p$ is equal to the sign of $r - p$ since

$$
d_p = \log_2(x_p y_p) = (r - p) + (2^r - 2^p).
$$

– The condition $p < c \Leftrightarrow p \leq r - 1 \Leftrightarrow d_p > 2^p$ implies that

$$
1 - e^{y_p \log(1 - x_p)} = 1 - e^{-x_p y_p/(1+\theta)} = 1 - \theta e^{-x_p y_p/2} = 1 - \theta x_{c+1};
$$

indeed, $x_p y_p / 2 = 2^{d_p - 1} > 2^{c+1}$ follows from $d_p - 1 \geq 2^{c-1} > c + 1$ which is true for $c > 2$. A computer verification confirms the expression for $c \leq 2$.

– The condition $p = c \Leftrightarrow r - 1 < p \leq r \Leftrightarrow 0 \leq d_p \leq 2^c$ implies that

$$
e^{y_c \log(1 - x_c)} = e^{-x_c y_c + \frac{\theta}{2} x_c^2 y_c} = e^{-x_c y_c}(1 + \theta x_c^2 y_c) = e^{-x_c y_c} + \theta x_c
$$

since $\frac{\theta}{2} x_c^2 y_c = \frac{\theta}{2} 2^{d_c - 2^c} < 1/2$ and $x_c y_c < e^{x_c y_c}$.

– The condition $p > c \Leftrightarrow p > r \Leftrightarrow d_p < 0 \Leftrightarrow x_p y_p < 1$ implies that

$$
1 - e^{-x_p y_p + \theta x_p} = 1 - e^{-x_p y_p} - \theta x_p = x_p y_p (1 - \frac{x_p y_p}{2}(1 - \frac{\theta}{6})).
$$

Condition $p > c + 1 \Leftrightarrow p > r + 1 \Leftrightarrow d_p + 1 < -2^{p-1}$ finally entails

$$
h(x_p, y_p) = x_p y_p - \theta(x_p y_p)^2 = x_p y_p - \theta x_p
$$

since $x_p y_p < x_{p-1}$ implies that $(x_p y_p)^2 < x_{p-1}^2 = x_p$.   *Q.E.D.*

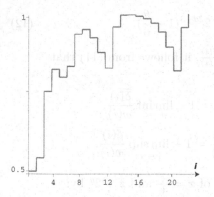 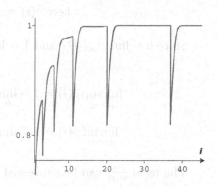

Fig. 5. Graphs of $\rho(i) = \frac{a(i)}{w(i)}$ and $\overline{\rho}(i) = \frac{\overline{a}(i)}{\overline{w}(i)}$

4.3   Average Versus Worst Sizes

The tables in Figures 2 and 4 indicate that the average and worst cases have almost equal sizes, except near the critical depth $c = c(i)$. Indeed, Lemma 1 implies that the differences $w_p(i) - a_p(i)$ and $\overline{w}_p(i) - \overline{a}_p(i)$ are infinitesimals unless $p = c$ or $p = c + 1$. In other words, the average size structure is the same as the worst size structure at all levels, except sometimes at depths c or $c + 1$.

The difference between worst and average size is maximized when the number i of inputs takes the critical form $i = n + 2^n$ for some $n \in \mathbf{N}$. In the critical case, the ratio between worst and average size quickly approaches $1 - \frac{1}{2e} = 0.8160\cdots$ for a maximal gain of 18%. Yet, it follows from (10) that this ratio is greater than 0.999 for *almost all* $i \in \mathbf{N}$: on average, the gain is less than 0.1%.

Proposition 4. *For i large enough, the average BDD, BMD and MDA all have the same relative size:*

$$1 = \lim_{i \to \infty} \frac{\overline{a}(i)}{a(i)}, \tag{7}$$

The average to worst size ratios $\rho(i) = \frac{a(i)}{w(i)}$ and $\overline{\rho}(i) = \frac{\overline{a}(i)}{\overline{w}(i)}$ are such that:

$$1 = \limsup_{i \to \infty} \rho(i) = \limsup_{i \to \infty} \overline{\rho}(i), \tag{8}$$

$$1 - \frac{1}{2e} = \liminf_{i \to \infty} \rho(i) = \liminf_{i \to \infty} \overline{\rho}(i), \tag{9}$$

$$1 = \lim_{i \to \infty} \frac{1}{i} \sum_{1 \le n \le i} \rho(n) = \lim_{i \to \infty} \frac{1}{i} \sum_{1 \le n \le i} \overline{\rho}(n). \tag{10}$$

Proof: Let $c = c(i)$ be the crital depth, so that $i = c + d + 2^c$ and $0 \le d \le 2^c$. We split the sum $\overline{a}(i) = \sum_{p<c} \overline{a}_p(i) + \overline{a}_c(i) + \overline{a}_{c+1}(i) + \sum_{p>c+1} \overline{a}_p(i)$ and replace each term $\overline{a}_p(i) = \beta_p h(x_p, y_p)$ by its equivalent from Lemma 1 to find that

$$\overline{a}(i) = \overline{w}(i) - \overline{\delta}(i) - \theta i \tag{11}$$

$$\text{where } \overline{\delta}(i) = \beta_c e^{-2^d} + 2^{2d-3}(1 - \frac{\theta}{6}). \tag{12}$$

Since $0 = \lim_{i \mapsto \infty} \frac{\theta i}{\overline{w}(i)}$ and $1 = \lim_{i \mapsto \infty} \frac{\overline{w}(i)}{w(i)}$, it follows from (11) that

$$\limsup_{i \mapsto \infty} \overline{\rho}(i) = 1 - \liminf_{i \mapsto \infty} \frac{\overline{\delta}(i)}{\overline{w}(i)} = 1 - \liminf_{i \mapsto \infty} \frac{\overline{\delta}(i)}{w(i)},$$

$$\liminf_{i \mapsto \infty} \overline{\rho}(i) = 1 - \limsup_{i \mapsto \infty} \frac{\overline{\delta}(i)}{\overline{w}(i)} = 1 - \limsup_{i \mapsto \infty} \frac{\overline{\delta}(i)}{w(i)}.$$

The ratio $\frac{\overline{\delta}(i)}{w(i)}$ can be expressed in terms of $x = \frac{1}{\beta_c}$ and $z = 2^d$ by:

$$\frac{\overline{\delta}(i)}{w(i)} = k(x, z) = \frac{e^{-z}}{1+z} + \frac{xz}{8}(1 - \frac{\theta xz}{3}). \tag{13}$$

For $x < 1$ fixed, function $k(x, z)$ is uni-modal in the interval $1 \le z \le \frac{1}{x}$: it is exponentially decreasing from $k(x, 1) = \frac{1}{2e} + \theta x$ to some infinitesimal $k_{min} = k(x, z_x) = O(x \log(x))$ in the interval $1 \le z \le z_x = O(\log(x))$. It is (almost linearly) increasing from k_{min} to $k(x, \frac{1}{x}) < \frac{1}{8} < k(x, 1)$ in $z_x \le z \le \frac{1}{x}$, hence

$$0 = \liminf_{i \mapsto \infty} \frac{\overline{\delta}(i)}{w(i)} \text{ and } \frac{1}{e} = \limsup_{i \mapsto \infty} \frac{\overline{\delta}(i)}{w(i)}.$$

By a similar evaluation through Lemma 1, the average BDD/BMD has size

$$a(i) = w(i) - \delta(i) - \theta i$$

$$\text{where } \delta(i) = \beta'_c e^{-2^d} + 2^{2d-3}(1 - \frac{\theta}{6}) = \overline{\delta}(i) - \beta_{c-1} e^{-2^d}.$$

Equality $\frac{\delta(i)}{w(i)} = \frac{\overline{\delta}(i)}{w(i)} - \frac{\beta_{c-1} e^{-2^d}}{w(i)} = \frac{\overline{\delta}(i)}{w(i)} - \theta/\beta_{c-1}$ implies that the limits as $i \mapsto \infty$ are equal to the above for $\frac{\overline{\delta}(i)}{w(i)}$, and (8,9) are proved. Limit (7) follows from:

$$\frac{\overline{a}(i)}{a(i)} = \frac{\overline{w}(i) - \overline{\delta}(i) - \theta i}{w(i) - \delta(i)} = 1 + \frac{\beta_{c-1}(\theta + e^{-2^d})}{w(i) - \delta(i)} = 1 + O(\frac{1}{\beta_{c-1}}).$$

The limit (10) is established by proving that $0 = \lim_{c \mapsto \infty} 2^{-c} S(c)$ for the sum

$$S(c) = \sum_{1 \le d \le 2^c} \frac{\overline{\delta}(c + d + 2^c)}{w(c + d + 2^c)} = \sum_{1 \le d \le 2^c} k(x_c, 2^d).$$

We evaluate $S(c)$ by (13) to find that

$$0 < S(c) < \sum_d \frac{e^{-2^d}}{1 + 2^d} + \frac{x_c}{8} \sum_{1 \le d \le 2^c} 2^d = 0.154 \cdots + \frac{x_c}{8}(2\beta_c - 3) < 1$$

and the convergence of $2^{-c} S(c)$ to 0 is exponentially fast.   *Q.E.D.*

5 Conclusion

We leave as a first open question to derive an explicit quasi-periodic term à la Delange [4] for the second order terms in the above analysis. Succeeding may be plausible since the above derivations reduce the analysis to the levels above and beneath the critical one, and all neglected terms have an exponentially smaller contribution.

A second open question is to generalize the analysis to function distributions having known entropy. For example, consider the average BDD size over functions whose truth-table has bit 0 distributed with probability $0 < p < 1$. It is shown in [7] that the BDD representation can be optimally encoded be a sequence of size proportional to Shannon's entropy

$$h(p) = -p \log_2(p) - (1 - p) \log_2(1 - p).$$

We further conjecture that the average BDD size over biased distributions is also proportional to the entropy. This hypothesis is backed by a number of experimental simulations.

References

1. R. E. Bryant. Graph-based algorithms for boolean function manipulation. *IEEE Trans. on Computers*, 35:8:677–691, 1986.
2. R. E. Bryant. Symbolic boolean manipulations with ordered binary decision diagrams. *ACM Comp. Surveys*, 24:293–318, 1992.
3. R. E. Bryant and Y.-A. Chen. Verification of arithmetic functions with binary moment diagrams. *Design Automation Conf.*, pages 535–541, 1995.
4. H. Delange. Sur la fonction sommatoire de la fonction somme des chiffres. *Enseignement Math'ematique*, 21:31–47, 1975.
5. C. Gröpl. *Binary Decision Diagrams for Random Boolean Functions*. 126 pages, Dissertation, Humboldt University, Berlin, 1999.
6. C. Gröpl, H. J. Prömel, and A. Srivastav. Size and structure of random ordered binary decision diagrams. *15th Annual Symposium on Theoretical Aspects of Computer Science*, LLNCS 1373, Springer:238–248, 1998.
7. J. C. Kiefer, E. Yang, G. J. Nelson, and P. Cosman. Universal lossless compression via multilevel pattern matching. *IEEE Transactions on Information Theory*, 46:1227–1245, 2000.
8. D. E. Knuth. *The Art of Computer Programming, vol. 3, Sorting and Searching*. Addison Wesley, 1971.
9. MapleSoft. *Maple 9 Guide*. Waterloo Maple Inc., 2003.
10. I. Wegener. *The Complexity of Boolean Functions*. John Wiley and Sons Ltd, 1987.

Concurrent Constraint-Based Memory Machines: A Framework for JAVA Memory Models (Summary)

Vijay Saraswat

IBM T.J. Watson Research Lab,
PO Box 704,
Yorktown Heights,
NY 10598

Abstract. A central problem in extending the von Neumann architecture to petaflop computers with a shared memory and millions of hardware threads is defining the *memory model* [1, 2, 3]. Such a model must specify the behavior of concurrent (conditional) reads and writes to the same memory locations. We present a simple, general framework for the specification of memory models based on an abstract machine that uses sets of *order* and *value* constraints to communicate between threads and main memory. Memory is permitted to specify exactly those linkings (mappings from read events to write events) which result in a unique solution for the constraints, and hence forces a unique patterns of bits to flow from writes to reads.

We show that this single principle accounts for almost all the "causality test cases" proposed for the JAVA memory model. It may be used as the basis for developing a formal memory model for the JAVA programming language.

An implementation of CCMs in a Prolog-based constraint language has been developed by Tom Schrijvers and Bart Demoen [4]. This paper is a summary of [5].

1 Introduction

The evolution of programming languages has been marked by increasing sophistication in application models, in machine architecture and in *programming models* that bridge the gap between the two.

For many years the von Neumann *Sequential Imperative Programming Machine* has defined an elegant and simple framework for computation. Abstractly, a programmer could think of the state of the computation as a *store* (a mapping from variables to values) together with instructions to extend the store (create new variables) and modify the store. Programs could be understood in terms of the effect they had on the store.

The situation is substantially more complicated for parallel computation. Lamport [1] first identified a simple set of rules (*Sequential Consistency*, SC) for

M.J. Maher (Ed.): ASIAN 2004, LNCS 3321, pp. 494–508, 2004.

parallel imperative programming: Intuitively, an ensemble of parallel programs may be thought of as executing one step at a time. In each step, only one program accesses memory, completing its operation before the system takes the next step. Thus, basic memory operations are *interleaved*.

However, as computer designers scale their systems to millions of hardware threads directed at achieving petaflops of computational power, the cost of satisfying sequential consistency are becoming apparent [2, 3]. Parallelism in computation creates a demand for parallelism in the access to memory. Fundamentally, at any given instant such threads may have millions of interacting memory operations outstanding, in various stages of completion across intervening buses, bypass buffers, caches, memory banks etc. One needs a *concurrent* model of interaction.

The *memory model* problem is thus the problem of defining the semantics of concurrent reads and writes to shared memory locations. To support standard compiler optimization techniques, the model must account for *conditional writes*, i.e. allow conditions to be specified on write operations. The condition may arise from *speculative execution*: statements under a conditional are being executed speculatively without their condition being discharged. Or it may arise from *out of order* execution: a statement after a conditional may be submitted ahead of time under the assumption that the conditional will not succeed. The thread may submit *data-dependent* writes, specifying the value to be written as a function of other values to be read. The thread may also submit ordering constraints on events specifying, e.g., that a write must (appear to) execute before a read (i.e. in "program order").

The memory model must specify which of these reads can be answered by which of these writes, while respecting the logical properties of conditional execution and the ordering constraints between events.

The issue of formalizing memory models has recently been raised in connection with JAVA concurrency [6]. The following high-level requirements seem reasonable [7]:

M1: Programs that do not exhibit data races[1] in SC executions must behave as if they are SC. ("Correctly synchronized programs cannot go wrong.")

M2: Each read of a variable must see a value written by a write to that variable. ("No thin air reads.")

M3: The model must support efficient implementation on current multiprocessor architectures based on processors such as the PowerPC, IA64, IA32, and SPARC.

M4: Removing useless synchronization should be semantically valid.

Additionally, we propose that the model should be simple to understand for programmers. It should be framed as an *abstract machine*. The model should be generative. That is, given a program it should be possible to use the model to generate all possible execution sequences of the program.

---

[1] A program has a data race if there is an execution sequence in which one thread writes to a location and another reads/writes to it without mutual synchronization.

1.1   Some Motivating Examples

The test cases discussed below are taken from [8]. Below we use a hopefully self-evident syntax for programs. (Details may be found in [5].)

Example 1 (Out of Order Writes). Consider the program:

```
init A=0;
init B=0;
thread { r2=A; B=1; } | thread { r1=B; A=2; }
```

It is easy to see that an SC execution cannot yield r2==2 and r1==1.

Suppose we extend the model so that it may execute several instructions at the same time, restricted only by the requirement that the value read from a local variable is the last value written into that variable. Now the machine for Thread 1 may issue a read for variable A from memory, and without waiting for it to complete, may go ahead and issue a write for variable B. Similarly the machine for Thread 2 may issue the read for B and simultaneously the write for A. That is, the system behaves as if the program being executed is:

```
init A=0;
init B=0;
thread { B=1; r2=A; } | thread { A=2; r1=B; }
```

Now memory may respond by matching the read on B with 1 and the read on A with 2.

Example 2 (Promoting Conditional Writes, Test 6, [9]). Consider:

```
init x=0;
init y=0;
thread { r1=x; if (r1==1) y=1; }  |
thread { r2=y; if (r2==1) x=1; if (r2==0) x=1; }
```

An SC execution must yield 0 for either r1 (Thread 1 goes first) or r2 (Thread 2 goes first). However, there is a plausible execution which can lead to r1==r2==1. Thread 1 communicates to main memory an unconditional write r1=x, and a conditional write y=1 provided that r1==1. At the same time, Thread 2 communicates an unconditional write of y to r2, and a conditional write of 1 on x if r2 is 0 or 1. Now main memory may choose to conditionally answer the read of x with the writes in the body of the (r2 == 1) conditional and the r2 == 0) conditional. The read of y may be answered by the write under r1==1 (if that condition succeeds) and by the initial value otherwise. This linking produces a unique solution, namely r1==r2==1.

Example 3 (No Thin-Air Reads.). Consider Test 4.

```
init x=0;
init y=0;
thread { r1=x; y=r1;} | thread { r2=y; x=r2;}
```

Arguably, the behavior `r1==r2==1` should *not* be exhibited. Two threads are copying values from one variable to the other in parallel. The variables are initialized to 0. Consider now that memory has simultaneously received all four read/write events. It may satisfy the read for x from Thread 1 by answering it with the write for x by Thread 2, and the read for y in Thread 2 with the write in Thread 1. This imposes no conditions on what values are transferred. So, theoretically, it might be possible to say that any value could be transferred, e.g. 1. However, the semantics should not allow this.

1.2   Concurrent Constraint-Based Memory Machines

We desire to formulate a simple shared memory abstract machines that can be used to generate all possible behaviors of a given program. We call such an abstract machine a *concurrent constraint-based memory machine* (CCM Machine).

There are two major approaches to modeling concurrent memory accesses.

The first choice is to permit a single step by a thread to initiate a memory action (read/write) but not complete it. For instance, an assignment step may be modeled by adding events (corresponding to the reads and writes in the step) to the store, without requiring that the effects of these events be reflected in the new store. Over time, such a store would autonomously pick up "incomplete" events, and add additional information (e.g. ordering information) to "link" it into the rest of the store. Thus a step in the computation may be initiated by a thread in the program or by memory, and a memory operation may take several steps to complete.

While such an approach is feasible [6–Chapter 17] (and, arguably, directly reflects implementation considerations) it can lead to intricate models with a lot of irrelevant detail. Instead we propose an *invariant-oriented* approach. We model the store as a collection of events and associated pieces of information that represents a set of *completed* memory operations. Such a store must be *valid*: it must satisfy a set of invariants arising from the memory model we are trying to formalize. A transition takes a valid store to a valid store. To correctly model concurrent operations on memory, a transition must represent a *set* of memory operations (e.g. read, write, synchronize), possibly arising from different threads, that are to be thought of as executing "at the same time". The new store obtained on completion of the transition must reflect the successful completion of *all* of these operations. In contrast with the incomplete step semantics, such an approach does not require autonomous memory transitions.

We shall call the set of events and associated pieces of information communicated by one or more threads to memory an *action set*. An action set may specify an ordering on events, conditional introduction of events, and dependent writes.

Given such an action set, memory first adds the action set to the store. Second, it adds ordering information between events in the action set, arising from different threads, so that the invariants associated with a valid store may be satisfied. For instance, if the action set contains two `lock` events from different threads targeted at the same location, it must ensure that these events are ordered. It adds ordering information between events in the old store and events in the new store (to record that the events in the action set occur after the events in the store, which represents a record of past events). Third, it finds and adds a *linking*, which matches each read event in the input action set to a write event on the same location. The write event may already have occurred in the past (i.e. may be in the store) or may be supplied in the input action set. Crucially, the write event must be *visible* to the read event according to the *ordering rule* between different events specified by the memory model. Further, the linking must unambiguously specify the *value* (i.e. pattern of bits) that is read from a write event. Thus in the new store every read will be satisfied by a write on the same location that is visible to the read, and all necessary ordering relations between synchronization events be established.

This completes one basic step of execution. Execution begins in an "empty store" (reflecting no past interactions), and proceeds by accepting an action set from threads, computing the new store and repeating until done.

Let us make explicit certain operations that memory must *not* do. It must not add ordering information between events in the action set supplied by the same thread (either now or in the past); such information can only be supplied by the thread. It must not add ordering information between events that have already occurred in the past; for that would be tantamount to changing the past. Note also that most operations are targeted at a single memory location (e.g. reads and writes, but not *fences*) and to implement them memory should only add ordering information between events targeted to that location. Finally, memory should not introduce "magic writes" – i.e. answer reads by pulling values (patterns of bits) out of thin air. It must merely *link* reads to reads, not create values.

Constraints. The central question that remains is how these action sets are to be specified. Since stores are just an "integral" of action sets over time, this will also determine how stores are specified.

We shall use *constraints* to specify the linkages between reads and writes. The use of constraints for communication and control in concurrent programming was discussed at length in [10] (see specifically Section 1.4.2; and [5] for more details appropriate to the current context). An action set (and hence the store) will be modeled as a constraint. We shall require the invariant that the store uniquely specify the value of all read and write operations; that is, it has a *unique solution* for program variables. Additional invariants are associated with the particular choice of *Order Model* (see below). Typically, all Order Models require that the ordering on events be acyclic, and that all events targeted at a particular location by a particular thread are totally ordered.

1.3   Examples Revisited

Example 4 (Out of Order Writes). The first thread sends two write events and a read event, with the associated constraints w(R2)=r(A1r), w(B1w)=1. These constraints may be read as saying that the value written into the register R2 is the value read by Thread 1 from location A, and the value written into location B by Thread 1 is 1.

The second thread similarly sends $w(R1)=r(B2r)$, $w(A2w)=2$.

Now main memory can choose to answer each read from the corresponding write received from the other thread, through the linking

$r(A1r)=w(A2w),r(B2r)=w(B1w)$

The conjunction of these constraints establishes $r(R1)=2,r(R2)=1$.

The next example shows that memory can respond to conditional writes by making conditional linkages.

Example 5 (Test 6). Thread 1 can be seen as communicating

$w(R1w)=r(X1r), (r(R1r)=1 \rightarrow w(Y1w)=1)$

and Thread 2 can be seen as communicating

$w(R2w)=r(Y2r), (r(R2r)=1 \rightarrow w(X2w1)=1), (r(R2r)=0 \rightarrow w(X2w2)=1)$

Here we use $\rightarrow$ to stand for the implication operation. Given these action sets the store can establish the linking

$r(X1r)=(r(R2r)=1)?w(X2w1):((r(R2r)=0)?w(X2w2):w(Xi)),$
$r(Y2r)=(r(R1r)=1)?w(Y1w):w(Yi),$
$r(R1r)=w(R1w),$
$r(R2r)=w(R2w),$

(Here we use JAVA syntax for conditional expressions: the term $c?s:t$ evaluates to s if c is true and to t otherwise.) This linking states that X1r should receive the value from X2w1 if that write action executes successfully, and if not, from X2w2 (if that action succeeds), and if not, from Xi. Similarly for Y2r. This linking together with the other constraints establishes $r(R2r)=1$ and $r(R1r)=1$. No other solutions are possible.

Example 6 (Test 4). The two threads may be seen to communicate

$w(R1w)=r(X1r), w(Y1w)=r(R1r)$

and

$w(R2w)=r(Y2r), w(X2w)=r(R2r)$

respectively. However, no linkage can be used to establish the desired result. Consider for example the linkage

$r(X1r)=w(X2w),r(Y2r)=w(Y1w), r(R1r)=w(R1w), r(R2r)=w(R2w)$

This succeeds in establishing $r(R1r)=r(R2r)$ (all the variables are equated to each other). But this is vacuous – every valuation is a solution. Hence this linkage must be rejected.

1.4   Comparison with Related Work

We believe the major advantage of our approach is that detailed considerations of specific architectural features and compiler optimizations may be completely omitted. Instead this approach focuses on establishing and using simple logical invariants associated with the relevant operations. Indeed, our approach is adequate to give an account of all the test cases presented in [7].

The use of constraints distinguishes our approach from the models of Manson/Pugh [11] and Sarita Adve [12]. We believe that ideas in this paper can be used to considerably simplify their models. For instance, the Manson/Pugh model requires that there be a total order over actions called a *justification order*. The value returned by a read on a location must be the value written into that location by a prior write event (in the justification order). The conditions on justification orders require another complex notion, i.e. the set of forbidden prefixes. Adve's proposal (the SC- model) requires that the out of order writes be justified by another execution of the system which must be "similar" to the current execution in many ways but which should have enough information content to justify the out of order writes.

We believe that the details of these two models are very brittle: seemingly simple variations in the definitions can lead to different results. These ideas appear to be in need of some underlying systematic theory.

In contrast, we believe that constraint-based communication, with the unique solutions criterion, offers a simple, operationally-motivated approach that will appeal to programmers. A more detailed technical comparison between our approach and these models must await completion of the technical description of these models (which are still being worked on by their authors).

Examples of the incomplete step approach are the CRF approach [13] and the Uniform Memory model of [14]. CRF provides a small rewriting-based target language for compilers which exposes both data replication and instruction reordering. UMM uses an abstract machine coupled with a transition table for executing instructions. We believe that our formalization in terms of memory as a store of constraints permits a simpler and more elegant treatment of similar intuitions. Our approach permits multiple memory instructions to be executed in "one step", thus obtaining the effect of instruction reordering.

2   Concurrent Constraint-Based Memory Machines

2.1   Constraint System

A constraint system suitable for formalizing the structure needed for [8] may be found in [5]. It formalizes that part of the semantics of the Java Virtual Machine (JVM) that corresponds to interactions between threads and main memory.

We review the basic idea. We assume some underlying undefined (infinite) sets of *threads* **T**, *locations* **L** and *events* **E**. **T** has no special structure other than a special constant $\mathbf{0} \in \mathbf{T}$ that designates the "main" thread.

All locations are thought of as existing in shared memory, even those accessed only by a single thread. Locations may be structured. A location may correspond to an object and may have fields (which are other locations). An object value is represented by a *literal* $\{f_1 = k_1, \ldots, f_n = k_n\}$ specifying the values of fields. A location may correspond to an array, whose elements (corresponding to other locations) are accessed through an index. **L** is classified into *normal* or *volatile* locations.

E has a strict partial order (i.e. a binary relation that is transitive, asymmetric and irreflexive) which is the interpretation of the predicate $\_ \ll \_$. Each event is associated with exactly one thread T and may be associated with zero or more locations. **E** supports read, write, lock, unlock, array creation, object creation, thread creation and other events. For instance, a read event is represented by the atomic formula r(X,T,L), a write event by w(X,T,L), a lock event by l(X,T,L) etc., where X is an *event identifier*, T is a *thread identifier* and L is a location. Some events have associated functions, e.g. r(E) (w(E)) is the value read (written) by a read (write) event E, Some events (e.g. reads and writes on volatile variables, lock and unlock events) are considered as *synchronization events*.

Equality, arithmetic operators and logical connectives are interpreted in the standard way.

The set of terms $r(E)$ for E a read event, closed up under the given primitive operations, is called the set of *read terms*. We shall often say that $r(E)$ is a *read (program) variable* and $w(E)$ is a *write (program) variable* and call such variables *program variables*.

Definition 1 (Valuation, Satisfaction). *A valuation is a mapping of variables to values in the underlying domain of interpretation (i.e. E for events, L for locations, the integers for integer program variables etc). A valuation realizes a constraint if the constraint is true under this mapping. A set of constraints s entails a constraint d, written $s \vdash d$, if every valuation that realizes s realizes d. The completion of a set of constraints s, written $[s]$ is the set of all constraints entailed by s.*

Note that $s \subseteq [s]$. An *inconsistent* constraint is one which cannot be realized by any valuation. As is standard in logic, an inconsistent constraint entails every constraint.

2.2 Configuration

The state of a CM machine is defined by $k \geq 0$ *threads*, and a *store*.

Thread. A thread contains all the private state (e.g. local variables, stack) necessary to execute a JVM-like thread of control. It communicates with the shared store through a collection of reads and (conditional) write operations, via an *action set* as described below. As a result, the store is changed (because writes are committed), and values are obtained (through read operations) into private variables local to the thread. Based on these values, the thread may perform certain expression evaluations, decide the value of predicates, take some branches, throw an exception, invoke a method etc. Subsequently it may again interact with memory by communicating another action set, wait for (some or all of) the results before continuing etc.

Store. The set of events defined by a particular CM Machine depends on the underlying Order Model (Section 2.4).

Definition 2 (Action Set). *An* action set *is a set a of constraints satisfying the Order Invariance for action sets (Definition 15).*

Store. A store is an action set with additional *linking* constraints. Before defining these constraints, we need to introduce the notion of a *write term*.

Given that an action set may contain conditional writes, it must be possible for the memory to specify a linking between reads and writes that is also conditional. For instance, suppose the store contains two write events for a location x, with associated write variables Y and Z. Suppose the first is a conditional write, with constraint c. Then it should be possible for the store to specify that a read should be answered from location Y if the condition c is true, and from location Z otherwise. This motivates the following definition.

Definition 3 (Write Term). *A* write term *for location L is the term $w(E)$ for any event E on location L or the term $d\,?\,w_1 : w_2$, where d is a condition and w_1, w_2 are write terms for location L.*

Definition 4 (Store). *A* store *is a set of constraints containing an action set and* linking *constraints, which are of the form $r(E) = w$ for a read event E for location L and a write term w for L.*

Since we allow conditional events, we must have a way of determining when an event is *active*. Intuitively, an event is active if its associated condition (if any) is satisfied.

Definition 5 (Active Event). *We say that a read event E is* active *in a store s if $s \vdash r(E, T, L)$ for some T, L. Similarly for write events.*

By the remark above, every event is active in an inconsistent store.

Definition 6 (Extension of a Store). *An* extension *of a store s is any set augmenting s with equations of the form $r(E) = k$, for k a literal. An extension is* total *if it forces every active read event to have a unique value.*

For any store s, let $o(s)$ be the set of all event constraints and order constraints entailed by s. We can use the underlying Order Model to answer questions about visibility of a write event at a read event in $o(s)$:

Definition 7 (Unconditional Write Visibility). *In a store s, each read event r on a line $m(l, t)$ can (unconditionally) see a write event w on a line $m(l, t')$ if w is a maximal element in the set of all write events that occur $\ll$-before r in s or w is not $\ll$-related to r.*

However, we must deal with conditional events, and hence need to extend the notion of visibility to write terms.

Definition 8 (Conditional Write Visibility). *Given a store s, a read event E for location L,*

- *$w(E')$ is visible to E if for every consistent extension s' of s, E' is visible to E in $o(s')$.*
- *$d\,?\,w_1 : w_2$ is visible to E if w_1 is visible to E in every consistent extension of s that entails d and w_2 is visible to E in every consistent extension of s that entails $!(d)$.*

We say that a set of linkings l is *valid* in a store s if for every link $r(E) = w \in l$, w is visible to E in s.

Proposition 1. *Let s be a store and l a set of linkings valid in s. Then l is valid in $s \cup l$.*

The definition above has been chosen carefully. As we see below (Definition 11), memory is required to add linkings that are valid in the union of the store and action set. This *independence* condition ensures that each constraint $r(E)=w$ will link E to a write event which is visible to it, *regardless* of other linkings introduced for other events.

Definition 9 (Events Associated with a Store). *Let s be a store. The set $e(s)$ of events associated with s is defined as the set of all events E such that some consistent extension of s entails an event constraint for E.*

Definition 10 (Valid Store). *A store s is said to be valid if*

1. *the store satisfies Order Invariance,*
2. *every linking in s is valid in s, and*
3. *s forces $r(E)$ to have a unique value for every active read event E in s (Unique Solution Criterion).*
4. *if $s \vdash w(E) = r$ and some read term $r(E)$ occurs in r, then $s \vdash r(E,T,L)$ for some T, L (Read Closure).*
5. *for every condition d and event n, if $d \to n \in s$ and $s \vdash d \to n$ then for every term $r(E)$ occurring in d and every event E' in n, $s \vdash E \ll E'$.*

Proposition 2 (Consistency). *Every valid store is consistent.*

Proposition 3 (Empty Store). *The empty set is a valid store.*

Execution of a CCM Machine begins in the empty store.

2.3 Transition

The CM Machine moves from one state to another as the result of two kinds of moves.

Silent Transitions. Based on its control state, and data received into private (logical) variables from the shared memory, a thread may make a private transition (affecting only its local state) to another state. For instance it may perform arithmetic operations, take a branch, thrown an exception, allocate a new object on the heap etc. This change of state is entirely unobservable by the shared store or by any other thread.

Shared Transitions. The machine may also make a step based on a simultaneous interaction between $k \geq 1$ threads and the store s.

In such a step, each thread T_i participating in that step communicates an action set S_i to the store obtained from the next set of statements to be executed by the thread. The size of the action set is entirely up to the thread. The action set may contain a single read event or a single write event or dozens of such events. Roughly speaking, the size of this set is correlated to the amount of prescient computation permitted by the thread. An execution in which each shared transition corresponds to one thread executing one event is an SC execution.

We now define how to obtain in one step a new store from an existing store and an input action set. This step will determine values for all the reads in the action set and communicate them to the relevant threads.

Definition 11 (New Store). *Let s be a valid store and a an action set such that $e(s)$ and $e(a)$ are disjoint. We say that s' is a new store for s on input a and write $s, a \triangleright_O s'$ provided that $s' = s \cup a \cup o_h \cup o_i \cup l$ is a valid store where:*

1. *o_h is the (unique) set of constraints $E_1 \ll E_2$ where $E_1 \in e(s)$ and $E_2 \in e(a)$ are defined on the same thread (sequencing constraints),*
2. *o_i is a set of constraints $E_1 \ll E_2$ for E_1, E_2 synchronization events in $e(a)$ on different threads, (synchronization constraints), and,*
3. *l is a valid linking in $s \cup a \cup o_h \cup o_i$ (linking constraints).*

Above, o_i should be chosen to be minimal, *that is, it should add only those order relations that need to be added in order to obtain a valid store. (That is, we require that there be no subset o_i' of o_i such that $s \cup a \cup o_h \cup o_i' \cup l$ satisfies all the conditions above.)*

We say $s \longrightarrow s'$ if there is some action set a such that $s, a \triangleright_O s'$.

The synchronization constraints capture ordering constraints introduced to resolve competing synchronized accesses to shared variables; as such there can be zero, one or more choices of o_i for a given s, a and o_h.[2]

With s, a, o_h, o_i fixed, there may still be many choices of linkings l – but only if there are data races. Indeed, there will be at least one linking corresponding

---

[2] There may be zero choices if a offers events that cannot be performed in the current state, e.g. a thread T may offer a *lock* operation on a variable x, but in s the lock is held by some other thread T' and a does not contain an *unlock* event on $m(T, x)$. It is up to the underlying Order Model to specify conditions on input action sets such that if these conditions are satisfied then it is always possible to obtain a new store from the current store.

to the values that would be returned by an SC execution. There is exactly one linking if there are no data races.

Propositions. The following propositions are true for CCMs.

Proposition 4 (Accumulativity of Store). *Suppose s, s' are valid stores such that $s \longrightarrow s'$. Then $s' \vdash s$.*

Proposition 5 (Conservation of Order Relations). *Suppose s, s' are valid stores such that $s \longrightarrow s'$. Then $s' \vdash E \ll E'$ for $E, E' \in e(s)$ iff $s \vdash E \ll E'$.*

Proposition 6 (Conservation of Linkings). *Suppose l is a valid linking in s and $s \longrightarrow s'$. Then l is a valid linking in s'.*

This follows from Proposition 5. The new information cannot add any new order relation on old variables, hence cannot affect the visibility of write terms to read events.

At each step, the store resolves enough of the non-determinism in the events offered by participating threads to determine the order of synchronization operations, and specify the precise value returned by each read operation.

2.4   Order Models

We now define the *Happens Before* (HB) Order Model, [8–P.17].[3]
The following definitions will be useful below.

Definition 12 (*Event(L)*, etc.). *For L a location and T a thread we define the following sets:*

- Event(L) *is the set of all event constraints on L.*
- Synch(L) *is the set of all synchronization event constraints on L.*
- Thread(T) *is the set of all event constraints whose thread is T.*
- $m(T, L)$ *is the set of all event constraints whose thread is T and location is L. (Note that $m(T, L) = \text{Thread}(T) \cap \text{Event}(L)$.)*

Let s be a set of constraints, and O a set of event constraints. Then the restriction of s to O, written $s \downarrow O$ is defined to be the set

$$(s \cap O) \cup \{E \ll E' \mid E, E' \in s \cap O, s \vdash E \ll E'\}$$

Definition 13 (Totally Ordered). *Let s be a store. s is said to be* totally ordered *if for every consistent total extension s' of s, and every pair of events $E, E' \in e(s')$ either $s' \vdash E \ll E'$ or $s' \vdash E' \ll E$.*

(See Definition 6.)

---

[3] [5] defines a more general framework where CCMs are parameterized by Order Models, and HB is just one Order Model.

Definition 14 (Lock Condition for Action Sets). *An action set s satisfies the lock condition for a location L if for all threads T, and extensions s' of s, $o(s') \downarrow (\mathrm{Synch}(L) \cap \mathrm{Thread}(T))$ is a total order in which every lock (unlock) event is followed (if at all) by an unlock (lock) event E', and for every lock event E followed by an unlock event E', $t(E) = t(E')$.*

Definition 15 (Order Invariance for Action Sets). *An action set s must satisfy the conditions:*

HB-A1: *for all threads T and total extensions s' of s, $s' \downarrow \mathrm{Thread}(T)$ is totally ordered, and,*

HB-A2: *the lock condition for action sets for all normal locations L.*

Definition 16 (Proper Initialization). *A valid store s is said to be properly initialized if for every consistent extension s' of s and every location L such that s' entails some event constraint on L, s entails a minimal event constraint n on L, and n is a write event.*

Definition 17 (Lock Condition for Stores). *A valid store s satisfies the lock condition for a location L if $o(s) \downarrow \mathrm{Synch}(L)$ is a total order, whose minimal event (if any) is a lock event, and in which every lock (unlock) event is followed (if at all) by an unlock (lock) event E', and for every lock event E followed by an unlock event E', $t(E) = t(E')$.*

Definition 18 (Ordering Criterion for Stores). *An store s must satisfy the conditions:*

HB-S1:   *$s \downarrow \mathrm{Thread}(T)$ is totally ordered, for every thread T,*

HB-S2:   *the locking condition for all normal locations,*

HB-S2v: *for all volatile locations V, if $s \vdash r(E') = w(E)$ for two events E, E' for V, then $s \vdash E \ll E'$*

HB-S3:   *the store is properly initialized.*

We now consider the work that needs to be done to move to a new store from a given store s and an action set z. First, o_h must be chosen as for sequential consistency. Second, o_i must be chosen to satisfy the lock condition. Note that the lock condition may not always be satisfiable. Third, an order relation must be introduced between every write on a volatile variable and a read which is answered by that write.

[5] lists all the test cases in [9] and [8] and discusses their results for CCM.

2.5   Properties

No Thin Air Reads. This property requires that all reads of a variable must return a value that was communicated by some thread to memory in a write event on that variable, and this write event must be visible to the read event, per the ordering rules.

This property is structurally guaranteed by CCMs because they can only establish "variable-variable" value constraints, linking a read event to a write event to the same location. Such linkings may be thought of as introducing "flows" in a graph, not supplying sources or sinks. Because a linking must satisfy the Unique Solutions Criterion, each read returns a concrete value.

Adding Threads Does Not Invalidate Executions. More threads can only lead to more writes and hence more ways of answering reads. Previous ways of answering reads continue to be valid.

Specifically, CCM Machines satisfy *linking monotonicity*. Suppose a CCM machine produces a set of linking constraints l in a valid store s in response to valid input action set $a = a_0 \cup \ldots \cup a_{k-1}$ with each a_i received from thread z_i. Let a_k be a new action set received from a new thread z_k such that $a \cup a_k$ is a valid input action set for s. Then there must be a new store s' such that $s, a \cup a_k \rhd s'$ and $l \subseteq s'$.

Instrumentation Reads are Benign. Any linking established by memory cannot be invalidated because of the presence of more reads. Therefore if the program is modified only by adding more reads all previous executions continue to be valid. This is true even if an originally properly-synchronized program now contains data-races.

Properly Synchronized Programs Have SC Executions. We outline the proof. Let s be a valid store reached at some stage of the execution, a an input action set, and o_h and o_i sequencing and synchronization constraints respectively. Let l be a valid linking such that for $s' = s \cup a \cup o_h \cup o_i \cup l$, we have $s, a \rhd s'$. We observe that if there are no data races, then all valid linkings are equivalent, that is, for any other store $s'' \supseteq s \cup a \cup o_h \cup o_i$ such that $s, a \rhd s''$, we have $s' \vdash r(E) = k$ iff $s'' \vdash r(E) = k$. Further, we observe that if s forces a read event E to be answered by a write event E' then $s \vdash E' \ll E$. (Why? Because proper initialization ensures that there is always such a dominated write that can be used to answer a read in every valid store. If there are no data races then there are no other writes that can be visible at the read in s', which has resolved all the conditionals.) Next we observe that any scheduling (= total order) of the events in the program that respects the $\ll$ relation in s' will return the values for the reads determined by s' and corresponds to a sequentially consistent execution of the program.

3   Conclusion

We have presented a simple framework for concurrent access to shared memory, the Concurrent Constraint-based Memory Model. We believe the semantic framework is naturally motivated and easy to reason with informally. We have worked out test cases in [9] and [8]. Of the 30-odd cases, our proposed theory agrees with all but one of the test cases. For the one test case (Test 18), we believe there are sound semantic reasons for the desired result of the test case to be changed; however, if necessary the case can be taken care of with a simple extension of this theory.

We believe that the structure of CCM Machines may be usefully pulled back into programming language design, in the form of appropriate type systems and *assertional* systems. For instance, it may make sense for the programmer to directly assert ordering constraints on events and value constraints on locations in the program syntax. Conceptually, these constraints would be transmitted to Main Memory at run-time. For instance, a programmer may specify that a location can only take on the value 0 and 42. The compiler may use such assertions to introduce early writes that could permit certain behaviors that may not have been permissible without these assertions. In effect, certain linkings that would have been ruled out by Memory because they do not uniquely specify values for reads may now be permitted because the additional constraints force a unique valuation.

Acknowledgements. My thanks to Doug Lea, Vivek Sarkar, Guang Gao, Zhang Yuan, Robert O'Callahan, Perry Cheng, Julian Dolby, Kemal Ebcioglu, Tom Schrijvers and Bart Demoen for discussions. Also to Bill Pugh, Hans Boehm, Jeremy Manson, Sarita Adve and other participants of the Java Memory Model mailing list.

References

1. Lamport, L.: How to make a multiprocessor computer that correctly executes multiprocess programs. IEEE Transactions on Computers **28** (1979)
2. Adve, S., Gharachorloo, K.: Shared Memory Consistency Models: A tutorial. Technical report, Digital Western Research Laboratory (1995)
3. Adve, S., Pai, V.S., Ranganthan, P.: Recent Advances in Memory Consistency Models for Hardware Shared-Memory Systems. Proceedings of the IEEE **87** (1999) 445–455
4. Schrijvers, T., Demoen, B.: JMMSOLVE: a generative reference implementation of CCM Machines. Technical Report Report CW 379, Katholieke Universiteit Leuven (2004)
5. Saraswat, V.: Concurrent Constraint-based Memory Machines: A framework for JAVA Memory Models. Technical report, IBM T.J.Watson Research Center, Hawthorne NY (2004) Available at www.saraswat.org.
6. Gosling, J., Joy, W., Steele, G., Bracha, G.: The Java Language Specification. Addison Wesley (2000)
7. Pugh, W.: Proposal for Java Memory Model and Thread Specification Revision (2001) JSR 133, http://www.jcp.org/en/jsr/detail?id=133.
8. Pugh, W.: Java Memory Model and Thread Specification Revision (2004) JSR 133, http://www.jcp.org/en/jsr/detail?id=133.
9. Pugh, W.: Java Memory Model Causality Test Cases. Technical report, U Maryland (2004) http://www.cs.umd.edu/ pugh/java/memoryModel/.
10. Saraswat, V.: Concurrent Constraint Programming. Doctoral Dissertation Award and Logic Programming. MIT Press (1993)
11. Manson, J., Pugh, W.: The Manson/Pugh model. Technical report, U Maryland (2004) http://www.cs.umd.edu/ pugh/java/memoryModel/.
12. Adve, S.: Sc-. Technical report, University of Illinois Urbana-Champaign (2004) http://www.cs.umd.edu/ pugh/java/memoryModel/.
13. Maessen, J.W., Arvind, Shen, X.: Improving the Java Memory Model Using CRF. In: OOPSLA. (2000)
14. Yang, Y., Gopalakrishna, G., Lindstrom, G.: A Generic Operatonal Memory Model Specification Framework for Multithreaded Program Verification. Technical report, School of Computing, U. of Utah (2004)

Author Index

Lecture Notes in Computer Science

For information about Vols. 1–3225

please contact your bookseller or Springer